31st Annual Meeting of the Division of Particles and Fields of the Mexican Physical Society (AMDPF 2017)

Journal of Physics: Conference Series Volume 912

Mexico City, Mexico
24 - 26 May 2017

Editors:

I. Bautista
E. de la Cruz Burelo
A. Fernandez-Tellez
G. Lopez-Castro
M. Rodriguez-Cahuantzi
P. Roig

ISBN: 978-1-5108-5064-4
ISSN: 1742-6588

Printed from e-media with permission by:

Curran Associates, Inc.
57 Morehouse Lane
Red Hook, NY 12571

CURRAN ASSOCIATES INC. proceedings.com

Some format issues inherent in the e-media version may also appear in this print version.

Printed by Curran Associates, Inc. (2017)

Institute of Physics
Dirac House, Temple Back
Bristol BS1 6BE UK

Phone: 44 1 17 929 7481
Fax: 44 1 17 920 0979

techtracking@iop.org

Additional copies of this publication are available from:

Curran Associates, Inc.
57 Morehouse Lane
Red Hook, NY 12571 USA
Phone: 845-758-0400
Fax: 845-758-2633
Email: curran@proceedings.com
Web: www.proceedings.com

Table of contents

Volume 912

31st Annual Meeting of the Division of Particles and Fields (DPyC) of the Mexican Physical Society

24–26 May 2017, Mexico City, Mexico

Accepted papers received: 12 October 2017
Published online: 28 October 2017

Preface

Invited and contributed talks

Posters

IOP Conf. Series: Journal of Physics: Conf. Series **912** (2017) 011001 doi:10.1088/1742-6596/912/1/011001

Preface

I. Bautista[1], E. de la Cruz Burelo[2], A. Fernández-Téllez[1], G. López-Castro[2], M. Rodríguez-Cahuantzi[1], P. Roig[2]

[1] Facultad de Ciencias Físico-Matemáticas, Benemérita Universidad Autónoma de Puebla, Av. San Claudio y 18 Sur, Edif. EMA3-231, Ciudad Universitaria 72570, Puebla, Mexico

[2] Departamento de Física, Centro de Investigación y de Estudios Avanzados del IPN, Apdo. Postal 14-740, 07000 Ciudad de México, México

The 31st Annual Meeting of the Division of Particles and Fields (DPyC) of the Mexican Physical Society was held at the José Ádem Auditorium in Cinvestav-IPN, Mexico City, Mexico, on 24-26 May 2017. We are extremely glad to have had a record of 180 participants attending the meeting, in which 82 presentations (39 posters and 53 talks including 40 contributed, 12 invited and a closing talk) were given. These numbers and the unquestionable scientific quality of the research work presented show the success of this meeting. In the days immediately before the meeting (22-23 May 2017) we organized a Tau Physics mini-Workshop, which can be considered the third edition of a series started in Krakow [1, 2], and where we had 12 invited seminars fostering discussions between the local Community working on Tau Physics and the invited foreign speakers, who participated in our national meeting as well. All presenters of both meetings were invited to contribute and roughly half of them accepted. We thank them for this effort, which made these proceedings possible.

References

[1] The first Tau Physics Workshop was held in Krakow (Poland) in May, 2012, https://indico.cern.ch/event/188018/timetable/?view=standard.

[2] The second Tau Physics Workshop was held in Krakow (Poland) in September, 2013, https://indico.cern.ch/event/261349/.

Committees

Local organizing committee of the III Tau Physics mini-Workshop:

- Eduard de la Cruz Burelo (Cinvestav-DF).
- Iván Heredia de la Cruz (Cinvestav-DF).
- Gabriel López Castro (Cinvestav-DF).
- Pedro Podesta Lerma (FCFM-UAS).
- Pablo Roig Garcés (Cinvestav-DF, Chairman).

Local organizing committee of the 31st Annual Meeting of the Division of Particles and Fields of the Mexican Physical Society:

- Eduard de la Cruz Burelo (Cinvestav-DF).
- Arturo Fernndez Téllez (FCFM-BUAP).
- Iván Heredia de la Cruz (Cinvestav-DF, Co-chairman).
- Gabriel López Castro (Cinvestav-DF).
- Pablo Roig Garcés (Cinvestav-DF, Chairman).

National organizing committee of both events:

- Arturo Fern´andez Téllez (FCFM-BUAP, Chairman).
- Mauro Napsuciale Mendívil (DF-UG).
- Pedro Podesta Lerma (FCFM-UAS).
- Ildefonso León Monzón (FCFM-UAS).
- David Delepine (DF-UG).
- J. Lorenzo Díaz Cruz (BUAP).
- Eduard de la Cruz Burelo (Cinvestav-DF).
- Mariana Kirchbach (IF-UASLP).
- Alfredo Aranda (UCOL).
- Saúl Ramos (IF-UNAM).
- Adnan Bashir (IFM-UMSNH).
- Alexis Aguilar Arévalo (ICN-UNAM).

IOP Conf. Series: Journal of Physics: Conf. Series **912** (2017) 011002 doi:10.1088/1742-6596/912/1/011002

- Salvador Carrillo Moreno (UIA).
- Karen Salomé Caballero Mora (UNACH).
- María Elena Tejeda Yeomans (UNISON).

IOP Conf. Series: Journal of Physics: Conf. Series **912** (2017) 011003 doi:10.1088/1742-6596/912/1/011003

Acknowledgements

The editors would like to thank Iván Heredia de la Cruz (Conacyt and Cinvestav-IPN) for his work in the organization of the Annual Meeting of the Division of Particles and Fields of the Mexican Physical Society and the Tau Physics mini-Workshop. We are also thankful to Pedro Podesta (UAS) for helping us organizing the mini-Workshop. We are grateful as well to the members of the national organizing committee of both events.

We are indebted to CONACYT for financial support through RED-FAE, 'Proyecto Fronteras de la Ciencia' (296) and SMF. We acknowledge the assistance and help of the Secretariats and staff of both events, particularly of Flor Ibáñez, Elizabeth Mote and Guadalupe Saucedo. We are very glad to thank Christian Ignacio Balderas, whose extremely professional assistance saved us a number of times.

Participants

III Tau Physics mini-Workshop:

(i) Dr. ARROYO UREA, Marco Antonio BUAP
(ii) Mr. AVENDAÑO, Eduardo Cinvestav-IPN
(iii) Dr. BANERJEE, Swagato University of Louisville
(iv) Mr. BECERRIL, Hugo Cinvestav-IPN
(v) Dr. BOLAÑOS, Azucena Facultad de Física UV
(vi) Ms. CHAMORRO SOLANO, SINDY CINVESTAV
(vii) Dr. DE LA CRUZ BURELO, Eduard Cinvestav-IPN
(viii) Mr. DE YTA, Alejandro Cinvestav-IPN
(ix) Dr. DELEPINE, David IF-UG
(x) Dr. DOMÍNGUEZ JIMÉNEZ, Isabel Universidad Autónoma de Sinaloa
(xi) Dr. EPIFANOV, Denis BINP
(xii) Dr. ERLER, Jens IF-UNAM
(xiii) Dr. FERNÁNDEZ-RAMÍREZ, César ICN-UNAM
(xiv) Dr. FLORES BAEZ, Francisco V. FCFM-UANL
(xv) Dr. FLORES TLALPA, Alain IF-UNAM
(xvi) Dr. GARCÉS, Estela Cinvestav-IPN
(xvii) Dr. GOLTERMAN, Maarten San Francisco State University
(xviii) Dr. GONZÁLEZ SPRINBERG, Gabriel República Univ.
(xix) Dr. GUEVARA ESCALANTE, Adolfo Cinvestav-IPN
(xx) Dr. GUTIERREZ GUERRERO, Laura Xiomara CENTRO MESOAMERICANO DE FÍSICA TEORICA (MCTP)
(xxi) Mr. GUTIÉRREZ SANTIAGO, Jorge Luis Cinvestav-IPN
(xxii) Dr. GÓMEZ ÁVILA, Selim Universidad Autónoma del Estado de Hidalgo
(xxiii) Dr. GÓMEZ-BOCK, Melina UDLAP
(xxiv) Dr. HENTSCHINSKI, Martin UDLAP
(xxv) Dr. HEREDIA DE LA CRUZ, Iván Cinvestav-IPN
(xxvi) Mr. HERNÁNDEZ JUÁREZ, Alan Ignacio BUAP
(xxvii) Dr. HERNÁNDEZ PINTO, Roger Universidad Autónoma de Sinaloa
(xxviii) Dr. HERNÁNDEZ TOMÉ, Gerardo Cinvestav-IPN

(xxix) Mr. HERNÁNDEZ VILLANUEVA, Michel Cinvestav-IPN
(xxx) Mr. JUNYA, Sasaki Tokyo Univ.
(xxxi) Dr. LÓPEZ CASTRO, Gabriel Cinvestav
(xxxii) Dr. LÓPEZ LOZANO, Lao-Tse Universidad Autónoma del Estado de Hidalgo
(xxxiii) Ms. MARES DAMAS, Palmira Universidad Autónoma de Sinaloa
(xxxiv) Dr. MARTÍNEZ ORTEGA, Jorge Cinvestav-IPN
(xxxv) Dr. MOYOTL, Agustín Cinvestav-IPN
(xxxvi) Mr. PACHECO, Jorge Cinvestav-IPN
(xxxvii) Dr. PASSEMAR, Emilie Indiana Univ.
(xxxviii) Dr. PODESTA LERMA, Pedro Universidad Autónoma de Sinaloa
(xxxix) Mr. RABADAN TREJO, Raul Iraq CINVESTAV
(xl) Dr. RICHTER-WAS, Elzbieta IFJ PAN & CERN
(xli) Mr. RODRÍGUEZ PÉREZ, David Universidad Autónoma de Sinaloa
(xlii) Dr. ROIG, Pablo Cinvestav
(xliii) Mr. ROJAS RAMOS , Antonio Instituto de Física UNAM
(xliv) Ms. SALINAS IBAÑEZ, Mónica Benemérita Universidad Autónoma de Puebla
(xlv) Dr. SHEKHOVTSOVA, Olga IFJ-PAN
(xlvi) Mr. SÁNCHEZ GARCÍA, Marxil IF-UNAM
(xlvii) Dr. TOLEDO SÁNCHEZ, Genaro IF-UNAM
(xlviii) Dr. WAS, Zbigniew IFJ PAN & CERN
(xlix) Mr. ZAREMBA, Jakub IFJ PAN

31st Annual Meeting of the Division of Particles and Fields of the Mexican Physical Society:

(i) Ms. Sc. ABADIA SARMIENTO, JOSE IVAN FACULTAD DE CIENCIAS EN FISICA Y MATEMATICAS UNACH
(ii) Dr. AGUILAR-AREVALO, Alexis Instituto de Ciencias Nucleares, UNAM
(iii) Dr. AHMAD, Aftab Gomal university (Pakistan) and IFM-UMSNH, Morelia
(iv) Ms. Sc. ALVARADO, Ricardo BUAP
(v) Ms. Sc. AMOR QUIROZ, David Arturo ICN-UNAM
(vi) Dr. ARANDA-SÁNCHEZ, Jorge Isidro Facultad de Ciencias Físico-MatemÁticas de la Universidad Michoacana de San NicolÁs de Hidalgo
(vii) Dr. ARANDA, Alfredo Universidad de Colima
(viii) Ms. Sc. ARCEO, Luis División de Ciencias e Ingenierías campus León - Universidad de Guanajuato
(ix) Ms. Sc. ARELLANO, Concepción BUAP
(x) Mrs. ARQUEZ, Sair Instituto Politecnico Nacional
(xi) Ms. Sc. AVENDAÑO OCAMPO, Jesús Eduardo Cinvestav-IPN
(xii) Ms. Sc. AYALA SÁNCHEZ, Gabriel Cinvestav
(xiii) Mrs. BARBOZA MENDOZA, Carolina Cinvestav

IOP Conf. Series: Journal of Physics: Conf. Series **912** (2017) 011004 doi:10.1088/1742-6596/912/1/011004

(xiv) Dr. BAUTISTA, irais BUAP
(xv) Ms. Sc. BECERRIL GONZALEZ, Hugo Alberto Cinvestav
(xvi) Dr. BELLO MARTÍNEZ, HÉctor BUAP
(xvii) Ms. Sc. BERMUDEZ, Rocio Universidad de Sonora
(xviii) Dr. BESPROSVANY, Jaime IF-UNAM
(xix) Dr. BIETENHOLZ, Wolfgang ICN, UNAM
(xx) Dr. BIJKER, Roelof Instituto de Ciencias Nucleares, UNAM
(xxi) Dr. BOLAÑOS, Azucena Facultad de Física UV
(xxii) CABALLERO-MORA, Karen SalomÉ Universidad Autónoma de Chiapas
(xxiii) Ms. Sc. CABANILLAS-NORIS, Juan Carlos BUAP-UAS
(xxiv) Ms. Sc. CARLOS, Cervantes Universidad de Guanajuato
(xxv) Dr. CARRILLO, Salvador Universidad Iberoamericana
(xxvi) Dr. CAÑAS ORDUZ, Blanca Cecilia CINVESTAV
(xxvii) Ms. Sc. CHAMORRO SOLANO, SINDY CINVESTAV
(xxviii) Ms. Sc. CHAVEZ, Daniel Universidad de Guanajuato / Texas A&M University
(xxix) Ms. Sc. CHÁVEZ ESTRADA, Marisol ICN-UNAM
(xxx) Dr. COBOS-MARTINEZ, Javier Universidad Cruzeiro do Sul
(xxxi) Ms. Sc. CRUZ VENEGAS, Noel Alberto Instituto de Física UNAM
(xxxii) Ms. Sc. CRUZ, Eligio U.M.S.N.H.
(xxxiii) Ms. Sc. DE HARO SANTOS, Juan Carlos Universidad de Guanajuato
(xxxiv) Prof. DE LA CRUZ BURELO, Eduard CINVESTAV-IPN Mexico
(xxxv) Dr. DE LA FUENTE ACOSTA, Eduardo Universidad de Guadalajara
(xxxvi) Ms. Sc. DE YTA HERNÁNDEZ, Alejandro CINVESTAV
(xxxvii) Dr. DEGOLLADO, Juan Carlos ICN-UNAM
(xxxviii) Prof. DIAZ-CRUZ, Lorenzo FCFM-BUAP
(xxxix) Dr. DOMÍNGUEZ JIMÉNEZ, Isabel Universidad Autónoma de Sinaloa
(xl) Ms. Sc. DOMÍNGUEZ, Santiago IF-UNAM
(xli) Dr. DUARTE GALVÁN, Carlos UAS
(xlii) Ms. Sc. DURAN OSUNA, M. Cecilia CINVESTAV
(xliii) Dr. EPIFANOV, Denis BINP
(xliv) Ms. Sc. ESCAMILLA TORRES, Luis AdriÁn Cinvestav
(xlv) Ms. Sc. ESCOBAR, Sofia Universidad Nacional Autónoma de Honduras
(xlvi) Ms. Sc. ESPINOSA, DAVID FCFM-UMSNH
(xlvii) Dr. ESPINOZA-HERNANDEZ, Maria Catalina Conacyt / IFUNAM
(xlviii) Ms. Sc. ESTRADA JIMÉNEZ, Fidel Facultad de Ciencias Fisica Matematicas de Unach
(xlix) Ms. Sc. EYSERMANS, Jan BUAP
(l) Ms. Sc. FABELA ENRIQUEZ, Brenda Autonomous University of Puebla
(li) Ms. Sc. FERMIN, PEREZ UNIVERSITY OF GUANAJUATO
(lii) Prof. FERNANDEZ TELLEZ , Arturo Facultad de Ciencias Físico MatemÁticas BUAP, Mexico
(liii) FERNÁNDEZ RANGEL, Luis Albino UMSNH

(liv) Dr. FERNÁNDEZ-RAMÍREZ, CÉsar ICN-UNAM
(lv) Ms. Sc. FERRO, Rodolfo Universidad de Guanajuato
(lvi) Ms. Sc. FIERRO ROJAS, Pablo BUAP FCFM
(lvii) Dr. FLORES-MENDIETA, RubÉn UASLP
(lviii) Ms. Sc. FLORES, Luis CINVESTAV
(lix) FLORES, Yesica Sonia Instituto PolitÉcnico Nacional
(lx) Mrs. FORTUNA MONTECILLO, Fabiola Elena Cinvestav
(lxi) Dr. GARCES, Estela A. CINVESTAV
(lxii) Ms. Sc. GARCIA DE LA VEGA, Leon Manuel Instituto de Física UNAM
(lxiii) Ms. Sc. GARCIA, Roberto BUAP
(lxiv) Dr. GOMEZ BOCK, Melina Universidad de las AmÉricas Puebla
(lxv) Ms. Sc. GONZALEZ HERNANDEZ, Emma FCFM-BUAP
(lxvi) Ms. Sc. GONZÁLEZ HERNÁNDEZ, Emma BUAP
(lxvii) Dr. GONZÁLEZ SPRINBERG, Gabriel República Univ.
(lxviii) Ms. Sc. GONZÁLEZ, Darío Facultad de Ciencias Físico MatemÁticas BUAP
(lxix) Ms. Sc. GUERRERO IBARRA, JosÉ Irvin Universidad Autónoma de Zacatecas
(lxx) Ms. Sc. GUERRERO NAVARRO, Gustavo Hazel Univ Nacl Autonoma de Mexico
(lxxi) Ms. Sc. GUEVARA, Adolfo Cinvestav
(lxxii) Ms. Sc. GUIOT, Marianne Universidad Veracruzana
(lxxiii) Dr. GUTIERREZ GUERRERO, Laura Xiomara Centro Mesoamericano de Física Teorica (MCTP)
(lxxiv) Ms. Sc. GUTIÉRREZ CANO, Gustavo CINVESTAV
(lxxv) Ms. Sc. GUTIÉRREZ SANTIAGO, Jorge Luis Cinvestav-IPN
(lxxvi) Ms. Sc. GUTIÉRREZ SÁNCHEZ, Raúl Alejandro Universidad de Guanajuato
(lxxvii) Dr. GÓMEZ, Selim UAEH
(lxxviii) Dr. HENTSCHINSKI, Martin Universidad de las AmÉricas Puebla
(lxxix) Dr. HEREDIA DE LA CRUZ, Ivan CINVESTAV / CONACyT
(lxxx) Ms. Sc. HERNÁNDEZ JUÁREZ, Alan Ignacio BUAP
(lxxxi) Ms. Sc. HERNÁNDEZ SEGURA, Miguel Angel Instituto de Física, UNAM
(lxxxii) Ms. Sc. HERNÁNDEZ VILLANUEVA, Michel CINVESTAV
(lxxxiii) HERNÁNDEZ-PINTO, Roger JosÉ FCFM-UAS
(lxxxiv) Dr. HERNÁNDEZ-TOMÉ, Gerardo Departamento de Física, CINVESTAV
(lxxxv) Ms. Sc. HERNÁNDEZ, JosÉ Luis Instituto de Física
(lxxxvi) Dr. HERRERA CORRAL, Gerardo Cinvestav
(lxxxvii) Ms. Sc. HERRERA, Karla Universidad de Guanajuato
(lxxxviii) Prof. HESS, Peter Instituto de Ciencias Nucleares, Universidad Nacional Autonoma de Mexico
(lxxxix) Dr. LAO TSÉ, López Lozano Universidad Autónoma del Estado de Hidalgo
(xc) Ms. Sc. LARIOS, Bryan BUAP
(xci) Ms. Sc. LOPEZ GUERRERO, VICTOR MANUEL FCFM-BUAP

IOP Conf. Series: Journal of Physics: Conf. Series **912** (2017) 011004 doi:10.1088/1742-6596/912/1/011004

(xcii) Ms. Sc. LUCINA, Espinoza UAS
(xciii) Ms. Sc. LUVIANO-VALENZUELA, Uriel AdriÁn IF UNAM
(xciv) Dr. LÓPEZ CASTRO, Gabriel Cinvestav
(xcv) Dr. LÓPEZ-CRUZ, Omar INAOE-Astrofísica
(xcvi) Ms. Sc. MARES DAMAS, Palmira Sujey Universidad Autónoma de Sinaloa
(xcvii) MARTINEZ CORTES, Tania BUAP
(xcviii) Dr. MARTINEZ, oscar fcfm-buap
(xcix) Dr. MARTÍNEZ HERNÁNDEZ, Mario IvÁn BUAP
(c) Dr. MARTÍNEZ ORTEGA, Jorge Cinvestav
(ci) Ms. Sc. MARTÍNEZ VEGA, Alfredo División de Ciencias e Ingenierías, Campus León, Universidad de Guanajuato
(cii) Ms. Sc. MARTÍNEZ, Angelo Instituto de Física y MatemÁticas, UMSNH
(ciii) MARTÍNEZ, Humberto Physics Department, CINVESTAV-IPN
(civ) Dr. MAURY CUNA, Humberto Universidad de Guanajuato
(cv) MEJIA, Jhovanny CINVESTAV-IPN, Physics department
(cvi) Ms. Sc. MIGUEL COLIN, Osvaldo Univ. Iberoamericana
(cvii) Ms. Sc. MILLAN, Fernando r Universidad de Guanajuato
(cviii) Ms. Sc. MIRANDA HERNÁNDEZ, Jesús Alejandro Cinvestav
(cix) Dr. MODAK, Sujoy Universidad de Colima
(cx) Dr. MONTES DE OCA YEMHA, Jose Halim FESC-UNAM
(cxi) Ms. Sc. MONTIEL HERNÁNDEZ, NÉstor AndrÉs Cinvestav
(cxii) Ms. Sc. MONTOYA SOTO, Gaspar Ricardo Universidad Autónoma de Sinaloa
(cxiii) Dr. MORALES MACÍAS, María Facultad de Estudios Superiores CuautitlÁn, UNAM
(cxiv) Ms. Sc. MORALES ÁLVAREZ, JosÉ Antonio Facultad de Ciencias, UNAM
(cxv) Ms. Sc. MORENO PALACIOS, Oscar Eduardo Universidad de Guanajuato
(cxvi) Dr. MOYOTL-ACUAHUITL, Agustin Cinvestav
(cxvii) Dr. MURILLO QUIJADA, Javier Alberto CMS / Cinvestav
(cxviii) Ms. Sc. NAVA BLANCO, Miguel Ángel BenemÉrita Universidad Autónoma de Puebla
(cxix) Ms. Sc. NAVA, Alejandro IFUAP-BUAP
(cxx) Dr. NÚÑEZ, Darío ICN-UNAM
(cxxi) Ms. Sc. OLGUÍN TREJO, Yessenia IF UNAM
(cxxii) Prof. ORDUZ-DUCUARA, Javier-Andres FESAc-UNAM
(cxxiii) Dr. OROPEZA BARRERA, Cristina Universidad Iberoamericana
(cxxiv) Ms. Sc. PACHECO ZAMUDIO, Jorge Cinvestav
(cxxv) Ms. Sc. PADRON STEVENS, Roberto Universidad Autónoma de San Luis Potosí
(cxxvi) Ms. Sc. PAREDES MALDONADO, Ricardo BUAP
(cxxvii) Ms. Sc. PARRA FLORES, Elsa Alejandra Facultad de Ciencias Físico MatemÁticas, BUAP
(cxxviii) Dr. PASSEMAR, Emilie Indiana Univ.
(cxxix) Dr. PELAYO, Rodrigo UPIITA-IPN
(cxxx) Dr. PEREZ ANGON, Miguel Angel Depto de Fisica, Cinvestav

IOP Conf. Series: Journal of Physics: Conf. Series **912** (2017) 011004 doi:10.1088/1742-6596/912/1/011004

(cxxxi) Dr. PEREZ-LORENZANA, Abdel Physics Department, CINVESTAV-IPN
(cxxxii) Dr. PODESTA LERMA, Pedro Luis Manuel FCFM-UAS (Sabatical at Cinvestav)
(cxxxiii) Ms. Sc. PSIHAS, Fernanda FNAL & Indiana Univ.
(cxxxiv) Ms. Sc. PÉREZ DE LEÓN, Mario Aldair FCFM-BUAP
(cxxxv) Ms. Sc. PÉREZ FIGUEROA, Omar IF UNAM
(cxxxvi) Ms. Sc. PÉREZ NAVARRO, Daniel Alejandro cinvestav
(cxxxvii) Ms. Sc. RABADAN TREJO, Raul Iraq CINVESTAV
(cxxxviii) Ms. Sc. RAMIREZ GARCIA, Mateo Cinvestav-IPN
(cxxxix) Ms. Sc. RAMIREZ GUADARRAMA , Dalia Lucero Facultad de Ciencias
(cxl) Ms. Sc. RAMIREZ-SANCHEZ, FRANCISCO Universidad Autónoma de Zacatecas
(cxli) Dr. RAMOS-SANCHEZ, Saul IF-UNAM
(cxlii) Ms. Sc. RAMÍREZ URIBE, Norma Universidad Autónoma de Sinaloa
(cxliii) RAYA, KhÉpani IFM-UMSNH
(cxliv) Ms. Sc. REGULES MEDEL, HÉctor David BUAP
(cxlv) Mrs. REGULES MEDEL, Karen BUAP
(cxlvi) Ms. Sc. REYES ALMANZA, Rogelio CINVESTAV
(cxlvii) Ms. Sc. REYES, Humberto Instituto de Fisica, UNAM
(cxlviii) Dr. RICARDO, Gaitan FES Cuautitlan Universidad Nacional Autónoma de MÉxico
(cxlix) RIVERA AGUDELO, Diana Carolina CINVESTAV-IPN
(cl) Dr. RODRÍGUEZ CAHUANTZI, Mario BUAP
(cli) Ms. Sc. RODRÍGUEZ PÉREZ, David FCFM-UAS
(clii) Dr. ROIG, Pablo Cinvestav
(cliii) Ms. Sc. ROJAS RAMOS, Antonio Instituto de Física UNAM
(cliv) Dr. ROMERO, Ricardo UAM - I
(clv) Ms. Sc. ROSAS, Francisco Javier Universidad de Guanajuato
(clvi) Ms. Sc. SALAZAR ARIAS, JosÉ GermÁn Cinvestav
(clvii) Dr. SALDAÑA SALAZAR, Ulises FCFM-BUAP
(clviii) Ms. Sc. SALINAS IBAÑEZ, Mónica BenemÉrita Universidad Autónoma de Puebla
(clix) Ms. Sc. SANCHEZ, Edgar IFUNAM
(clx) Ms. Sc. SANDOVAL CÁRDENAS, Ilya Orson ICN-UNAM
(clxi) Ms. Sc. SARKIS MOBARAK, youssef ICN
(clxii) Ms. Sc. SAULO, HernÁndez López Cinvestav
(clxiii) Dr. STUART, Alexander James Colima Univ.
(clxiv) Ms. Sc. SÁNCHEZ GARCÍA, Marxil Instituto de Física UNAM
(clxv) SÁNCHEZ GONZÁLEZ, Mayra Jacaranda ESFM-IPN
(clxvi) Dr. SÁNCHEZ, Alberto Cinvestav
(clxvii) Ms. Sc. TAME-NARVAEZ, Karla FCFM-BUAP
(clxviii) Dr. TEJEDA MUÑOZ, Guillermo BUAP
(clxix) Ms. Sc. TOSTADO ROBLEDO, Sergio CINVESTAV
(clxx) Ms. Sc. TÉLLEZ TOVAR, Luis Osvaldo Cinvestav

IOP Conf. Series: Journal of Physics: Conf. Series **912** (2017) 011004 doi:10.1088/1742-6596/912/1/011004

(clxxi) Ms. Sc. VALDIVIA GARCÍA, Marco Alan University of Guanajuato
(clxxii) Dr. VALENCIA PALOMO, Lizardo UNACH
(clxxiii) Ms. Sc. VALERIO LIZARRAGA, Cristhian Alfonso UAS
(clxxiv) Ms. Sc. VARELA, Enrique FCFM-BUAP
(clxxv) Dr. VAZQUEZ-JAUREGUI, Eric IF-UNAM
(clxxvi) Mrs. VAZQUEZ, Fabiola Universidad Iberoamericana
(clxxvii) Ms. Sc. VENZOR VALDIVIA, Jorge Alberto CINVESTAV
(clxxviii) VÁZQUEZ RUEDA, Omar ICN-UNAM
(clxxix) Ms. Sc. YEBRA, Azarael Universidad de Guanajuato
(clxxx) ZEPEDA FERNÁNDEZ, Cristian Heber Centro de Investigación y de Estudios Avanzados del I.P.N.

All data have been extracted from the indico databases of both events and correspond to the registrants input.

Peer review statement

All papers published in this volume of *Journal of Physics: Conference Series* have been peer reviewed through processes administered by the proceedings Editors. Reviews were conducted by expert referees to the professional and scientific standards expected of a proceedings journal published by IOP Publishing.

Tau magnetic moment

G. A. González-Sprinberg [1], J.Vidal [2]

[1] Instituto de Física, Facultad de Ciencias, Universidad de la República, Iguá 4225, 11400 Montevideo, Uruguay

[2] Departament de Física Teòrica Universitat de València, E-46100 Burjassot,València, Spain and IFIC, Centre Mixt Universitat de València-CSIC, València, Spain

E-mail: gabrielg@fisica.edu.uy

Abstract. The τ lepton magnetic moment theoretical predictions and measurements are reviewed. While it is believed that such a high mass particle is a good candidate to show up new physics, this is not the case up to now. The magnetic moment of elementary fermions, and in particular the anomalous magnetic moment of the electron, had an historical impact both in relativistic quantum mechanics and in quantum field theories. Besides, many new physics models were discarded when confronted with these magnitudes. More recently, the discrepancy of the experiments and the theoretical predictions for the muon anomalous magnetic moment is still an open issue. For the τ lepton, instead, while the theoretical prediction is well known for the standard model and some new physics models, the data are very far of determining even its sign or the first figure. We will discuss the most important theoretical aspects of the τ magnetic moment, and also the current accepted measurements and future perspectives, in particular related to B-factories.

1. Introduction

The magnetic moment of elementary fermions allows to write a first order interaction of a fermion with a magnetic field. This electromagnetic interaction, for stable fermions in the presence of a magnetic field, provides a way to measure some of these magnetic moments with a unusual precision. For the electron this precision is more than ten figures and the relative precision is 0.7 ppb [1]. This method and a similar precision are not possible in the case of an unstable particle, such as the τ lepton, with a mean life $(290.3 \pm 0.3) \times 10^{-15} s$. On the other side, theoretical predictions for the magnetic moments, and in particular for the quantum corrections that give origin to the anomalous magnetic moment (AMM), can be computed with extraordinary precision in QED, and also in electroweak theory. More recently higher order hadronic corrections where also computed.

The magnetic moment anomaly for a spin one-half fermion f was first computed to leading order in QED by J.Schwinger in 1948 [3], and this is the famous result:

$$a_f \equiv \frac{g_f - 2}{2} = \frac{\alpha}{2\pi} \simeq 0.00116 \tag{1}$$

This expression is flavour independent and in the case of the electron only higher QED corrections are necessary in order to approach the experimental result. For other fermions, and in particular for the muon, corrections proportional to mass enter in the computation of the electroweak and hadronic contributions.

The AMM allows a stringent test for new physics and it was deeply investigated by many authors (see, for example, [4] and cites therein). The discrepancy of several sigma between the theoretical and experimental value of the muon AMM gave origin to a series of impressive experiments of the Muon $(g-2)$ Collaboration that resulted in a 0.7 ppm measurement [5, 6, 7]; besides a large number of theoretical work was done in order to understand this discrepancy [8, 9, 10].

The case of the tau is very different. While the theoretical prediction in the standard model is well known with many figures, the experimental precision is far from the first figure and even the sign of the AMM is unknown. The short mean life of the tau produces promptly decays and it is through these secondary particles that the AMM should be traced back. The PDG value for the tau AMM is [1]:

$$-0.052 < a_\tau < 0.013 \;\; (95\% C.L.) \tag{2}$$

while the theoretical prediction in the standard model is [11]

$$a_\tau = 117721(5) \times 10^{-8} \tag{3}$$

Note that the PDG value is taken from the DELPHI Collaboration[12] in 2003 and has not changed since then. There, and also in other experiments, the bounds are obtained from total cross section data, like $e^+e^- \to e^+e^-\tau^+\tau^-$ for DELPHI or $e^+e^- \to Z \to e^+e^-\tau$, also in LEP. Note that these total cross sections observables provide a rather indirect bound on the AMM and not a direct measurement of this quantity. Besides, in the same papers, bounds on the CP-violating electric dipole moment of the τ are also computed, using the total cross section which is a CP-even observable.

On the other side, the theoretical prediction in [11] updated the QED and electroweak corrections and took into account the leading order hadronic ones (based on e^+e^- data from BaBar, CMD-2, KLOE and SND), and the light by light contributions. But the comparison between experiment and theory is disappointing: the measurements provide bounds that are one order of magnitude above the theory, and not even the sign of the AMM is known. As expressed some years ago by M.Perl[2] in the section *Dreams and odd ideas in tau research:* "... It would be very nice to measure μ_τ with enough precision to check this (the Schwinger term $\alpha/2\pi$), as it was checked for the e and the μ years ago. At present such precision is a dream."

In the next section we reviewed some of the results in the theoretical side, next we addressed the ideas already investigated in order to measure the AMM, and finally we presented our conclusions.

2. Theory

The magnetic moment of a particle f can be defined by the gyromagnetic ratio as:

$$\vec{\mu}_f = g_f \frac{e}{2m} \vec{s} \tag{4}$$

For a spin 1/2 particle f the Dirac's equation predicts $g_f = 2$, and quantum corrections to this quantity are usually expressed by means of the AMM a_f defined in Eq.(1), that quantifies the difference with the prediction of the Dirac equation. The quantum corrections, dominated by QED, slightly increases the value of a_f. Then, the magnetic moment of a spin 1/2 particle of negative charge $-e$ is just the "Bohr magneton" (but replacing m_e with the particle mass) times $(1 + a_f)$. For the electron, where the mass is well below the electroweak and new physics scales, the AMM provides a strong test of QED and also the best measurement [13] of the fine structure constant α. The $\vec{\mu} \cdot \vec{B}$ term naturally arises in the non-relativistic limit of the Dirac's equation. This allows to define the AMM in a precise way in this domain, that is useful for stable particles that can be put in interaction with an external magnetic field. As already stated, the AMM for the tau is dominated by the Schwinger term, shown in Eq.(1), and all the QED, electroweak and hadronic contributions add up to less than 2%. Due to the high precision experiments for the electron and the muon those contributions played a crucial important role. In general, the other particles in the theory (with mass M) contribute with terms proportional to m_f^2/M^2, and this is true for the standard model and many new physics models. Other dependencies on the new masses and/or scales can also be found in some models. Then, the muon AMM is an scenario where the contributions of all the standard model spectrum can show up. The enhancement factor for the muon with respect to the electron is $m_\mu^2/m_e^2 \simeq 4 \times 10^4$ while for the tau these factors are $m_\tau^2/m_e^2 \simeq 1,1 \times 10^7$ and $m_\tau^2/m_\mu^2 \simeq 282$. Then, it is expected that the new electroweak contributions may appear in the AMM of heavier leptons, but this also depends on the precision of the measurements.

The theoretical prediction of the tau AMM was first addressed taking into account higher order QED corrections and electroweak and hadronic corrections as well in ref.[14]. Latter, the computation was updated using more data and new orders and terms in ref.[11], so that they obtained the result quoted in Eq.(3).

In the quantum field theory framework the definition of the magnetic moment and the AMM can be written in several ways. For example, a lagrangian term can be written that reproduces the correct magnetic tensor contributions to the matrix elements. These are originated from loop corrections to the tree level lagrangian of the theory. However, in this approach the particles are implicitly assumed to be on their mass-shell and the coefficients of the lagrangian (i.e., the AMM times some constants) should not depend on the scale of the process. This can be understood as an effective lagrangian approach [15]. Then, necessarily this approach can not have absorptive parts. The magnetic moments and AMM are gauge invariant quantities and for all the particles on-shell are physical quantities. As long as this on-shell condition is relaxed the situation may change, as was investigated for the electroweak theory in ref.[16]. A more precise and general approach in order to define the AMM is just to resort to the on-shell matrix elements, using all the symmetries that are appropriate for the case. Then, one ends with a matrix element with a few terms that are gauge invariant and represent physical properties of the particles. We then define the AMM as the coefficient of a particular term in the Lorentz tensor structure of the matrix element.

The most general Lorentz invariant structure describing the interaction of an on-shell vector boson V with two on-shell fermions $f\bar{f}$ can be written in terms of six form factors:

$$\begin{aligned}\langle f(p_-)\bar{f}(p_+)|\, J^\mu(0)\,|0\rangle = e\,\bar{u}(p_-)\Big[&(F_1 + F_4\gamma_5)\gamma^\mu \\ &+\frac{1}{2m_f}(i\,F_2 + F_3\gamma_5)\sigma^{\mu\nu}q_\nu + \frac{1}{2m_f}(i\,F_5 + F_6\gamma_5)q^\mu\,\Big]v(p_+)\end{aligned} \tag{5}$$

where $q = p_+ + p_-$. Since the two fermions are on-shell the form factors F_i appearing in Eq. (5) are functions of q^2 and m_f^2 only.

In addition, if the current J^μ is conserved, we must have

$$i\frac{q^2}{2m_f}\,F_5 + \left(\frac{q^2}{2m_f}F_6 - 2m_f\,F_4\right)\gamma_5 = 0 \quad \Rightarrow \left\{\begin{array}{l} F_5 = 0 \\ F_6 = \dfrac{4m_f^2}{q^2}F_4 \end{array}\right. \tag{6}$$

so that the final expression for the gauge invariant $f\bar{f}\gamma$ vertex reduces to:

$$\begin{aligned}&\langle f(p_-)\bar{f}(p_+)|\, J^\mu(0)\,|0\rangle = \\ &e\,\bar{u}(p_-)\left[\gamma^\mu\,F_1 + \frac{1}{2m_f}(i\,F_2 + F_3\gamma_5)\sigma^{\mu\nu}q_\nu + \left(q^2\gamma^\mu - q^\mu\,\not{q}\right)\gamma_5 F_A\right]v(p_+)\end{aligned} \tag{7}$$

In this expression, F_1 parametrizes the vector part of the electromagnetic current ($F_1(0) = 1$), and it is identified with the charge, $F_A = -F_4/q^2$ is the so-called anapole moment, F_3 parametrize the electric dipole moment

$$d_f = \frac{e}{2m_f} F_3(0), \tag{8}$$

while $F_2(0)$ defines defines the AMM:

$$F_2(0) = a_f; \quad \mu_f = (1 + a_f)\frac{e}{2m_f} \tag{9}$$

Note that this is true for particles on-shell, while for off-shell particles usually the matrix elements (or the effective vertex) contains many form factors, depending on q^2 and all other possible scalar invariants, and besides can be not gauge-invariant, as it is well known [17]. Similar expression can be written for other vector bosons, such as the Z, where the weak magnetic moments are now well defined at the Z-peak, i.e. at $q^2 = M_Z^2$. In this case the magnetic moment develops also an absorptive part that can also be measured using suitable observables.

The AMM that is measured in the experiments contains all the contributions of the well established standard model and possible new physics. This last contributions are in general suppressed by the scale of the new physics, typically $m^2{}_f/\Lambda^2$. Then, these contributions are enhanced for high mass particles and promotes the AMM as a probe for new physics.

The tensor structure of the magnetic terms flips chirality, and in the standard model fermion masses are the only source of chirality flips, so one should expect that the AMM appears proportional to mass. Besides, observables that are exactly zero when chirality is conserved are the best candidates in order to measure magnetic moments. These observables only be sensitive to fermion masses and magnetic moments. Furthermore, they will depend linearly on magnetic moments. A measurement using observables of this kind is what one should call a direct observable related to the AMM.

3. Experiments

The main new ingredient for the tau AMM is its short mean life, that prevents to perform similar experiments and ideas that were successful for other leptons. Then, the AMM should be searched in the tau decay products, using the several channels that are open for this lepton. In particular, this can be done by means of the energy distribution of some of the these particles [26]. The properties of the weak interaction decays allow to study the AMM using both linear and correlations polarizations of the taus as spin analyzers. Then, the AMM propagates in the linear polarizations and correlations modifying the angular distribution of these decaying particles. This can be useful for both the electromagnetic AMM and also for the weak AMM [27, 28, 29, 30]. In this way some direct observables can be identified and investigated, that are closely related to the anomalous magnetic moments. Besides, other indirect observables, such as the ones coming from total cross sections can also give partial information and bounds on the AMM.

The current PDG limit was obtained by DELPHI [12] using data from the total cross section $e^+e^- \to e^+e^-\tau^+\tau^-$ at LEP2, where they selected 2390 events at energies in the range $183\,GeV < \sqrt{s} < 208\,GeV$. The resulting bound Eq.(2) can also be expressed as:

$$a_\tau = 0.018 \pm 0.017 \tag{10}$$

to be compared with the theoretical prediction in the standard model Eq.(3). Clearly the experiment is far from determining the anomaly for the tau in the standard model. The leading Feynman diagram for this process is shown in Fig.1. The interference of this diagram with the

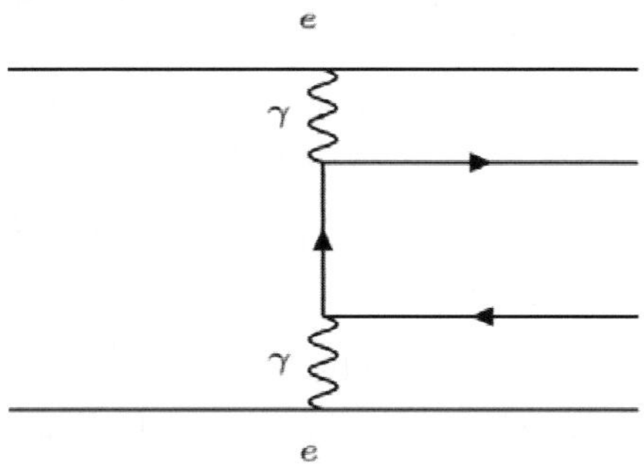

Figure 1. Feynman diagram for the process in DELPHI at LEP2.

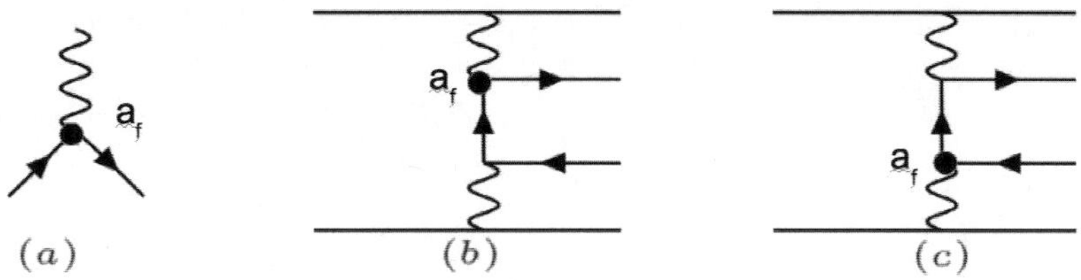

Figure 2. (a) AMM effective vertex insertion; (b) and (c) Feynman diagram with and AMM insertion that interferes with the diagram in Fig.1 for the process in DELPHI at LEP2.

two similar diagrams but with an insertion of the AMM in the $\tau\tau\gamma$ vertex, shown in Fig.2, produces contributions in the total cross section that are proportional to two chirality flipping terms, such as AMM times a mass or the squared value of the AMM. These ideas were studied by the authors of ref.[18]. Note that two particles in the effective vertex in diagrams (b) and (c) in Fig.2 are off-shell: the photon and also one tau. This means that the question of gauge invariance should be addressed and, besides, this effective vertex depends on the two scalar quantities that the vertex depends on. The bound was obtained in the range of energies quoted above, and in each of the events the scalar variables in the vertex are different. Furthermore, the effective AMM vertex is not the only one that can be written for this vertex, in which two of the three lines are off-shell, and several form factor can be associated with the vertex (see for example ref.[17]). Related to these topics, it is appropriate to ask oneself if just a bound or a measurement of the (on-shell) AMM predicted in the standard model would ever be found with this kind of observables. Even in the hypothesis of a very high statistics, will the observables allow to determine any of the figures in Eq.(3)? The answer is no, no matter how high the statistics would be. In any case, this result is just a bound, under the hypothesis that all the other form factors do not contribute. This bound is still sensible as long as it is not close to the one loop standard model contributions to the form factors of the vertex; in particular, for the AMM. Then, this bound may be interpreted as a bound on new physics contributions to the AMM, under the hypothesis mentioned before.

Other ideas related to the tau AMM were also investigated, but they were not taken into account by the PDG. Some of them share the same kind of problems in order to interpret their meaning. The authors of ref.[20] explored the total cross section $e^+e^- \to \tau^+\tau^-$ at PETRA using data with $q^2 < (37\, GeV)^2$ looking for bounds on the AMM and also for a scale of compositeness. They found a limit $\mid a_\tau \mid < 0.02$; in ref.[34, 35] bounds were investigated from the radiative decay

$Z \to \tau^+ \tau^- \gamma$ at LEP1, where only the anomalous coupling a_γ was taken into account and the contributions coming from the tau Z-magnetic coupling a_Z was neglected. Using this approach, with the inclusion of only linear terms in a_γ in the cross section (were the τ which emits the photon is off-shell) the analysis of the L3 Collaboration [23] obtained

$$-0.056 < a_\tau < 0.044 \quad (95\% \, C.L.) \tag{11}$$

while the OPAL Collaboration [24] obtained the bound:

$$-0.068 < a_\tau < 0.065 \quad (95\% \, C.L.) \tag{12}$$

In ref.[21, 22] they explored the bounds one can obtain from the process $e^+e^- \to \tau^+\tau^-\gamma$ that was latter measured at L3 [23]; in ref.[18, 19] they studied in $\gamma\gamma \to \tau^+\tau^-$ the limit that can be found in $\gamma\gamma$ colliders and in hadron colliders such as the LHC; in ref.[32, 33] they investigated the bounds at the Z-peak in the channel $Z \to \tau^+\tau^-$ using an effective lagrangian approach and without taking into account the weak magnetic moments. In [25] they recently explored the radiative decay of the Higgs boson $H \to \tau^+\tau^-\gamma$. Some of these authors and experiments also investigated the bounds one could get from the same (CP-even) observables on the tau electric dipole moment. Recently, the possibility of obtaining bounds for the AMM in B-factories was also addressed in ref.[37]. Here, they studied the possibility of obtaining bounds for the tau AMM and also for the electric dipole moment using a leptonic radiative decay of the tau lepton, with a photon in the final state. The vertex in which this photon participates may have information associated to the AMM. They concluded that a sensitivity of the order of 0.012 could be obtained with the planned full set of Belle II data. This bound is below (but close) to the DELPHI result, 0.017, already mentioned.

We have also investigated an approach to the problem that took the effective lagrangian approach fully into account in ref.[31]. There, a direct bound on the electromagnetic and weak AMM were obtained on general grounds and in a model independent way. These bounds came from the observation that in general extensions of the standard model it is very difficult to generate a magnetic moment for a lepton without originating a coupling of the Z boson to the lepton, of the same order of magnitude. Then, as the last is strongly bounded by LEP, one may obtain a rather strong bound on the γ-magnetic moment. A complete analysis of the magnetic moment couplings of the tau to the photon, the Z and the W bosons, using the complete amount of data coming from tau-lepton production at LEP1, SLD and LEP2, and data on W decays into tau leptons from LEP2 and pp collider, was done. LEP1 and SLD are more sensitive the weak magnetic coupling, while LEP2 is sensitive to both the electromagnetic and weak AMM, and data from W decays into tau in colliders provide some bounds on the magnetic $W\tau\nu_\tau$ coupling.

The effective Lagrangian considered is,

$$\mathcal{L}_{eff} = \alpha_B \mathcal{O}_B + \alpha_W \mathcal{O}_W + \text{h.c.} \,, \tag{13}$$

where the couplings α_B and α_W real (note that complex couplings will break CP conservation and would lead to the electric and weak-electric dipole moments). The two operators in Eq.(13) are:

$$\mathcal{O}_B = \frac{g'}{2\Lambda^2} \overline{L_L} \varphi \sigma_{\mu\nu} \tau_R B^{\mu\nu} \,, \tag{14}$$

and

$$\mathcal{O}_W = \frac{g}{2\Lambda^2} \overline{L_L} \vec{\tau} \varphi \sigma_{\mu\nu} \tau_R \vec{W}^{\mu\nu} \,. \tag{15}$$

Here $L_L = (\nu_L, \tau_L)$ is the tau leptonic doublet, φ is the Higgs doublet, $B^{\nu\nu}$ and $\vec{W}^{\mu\nu}$ are the $\text{U}(1)_Y$ and $\text{SU}(2)_L$ field strength tensors, and g' and g are the gauge couplings. After spontaneous symmetry breaking, the Higgs gets a vacuum expectation value and the interactions in Eq.(13) can be written in terms of the gauge boson mass eigenstates, A^μ and Z^μ, as

$$\begin{aligned}\mathcal{L}_{eff} &= \epsilon_\gamma \frac{e}{2m_Z}\overline{\tau}\sigma_{\mu\nu}\tau F^{\mu\nu} + \epsilon_Z \frac{e}{2m_Z s_W c_W}\overline{\tau}\sigma_{\mu\nu}\tau Z^{\mu\nu} \\ &+ \left(\epsilon_W \frac{e}{2m_Z s_W}\overline{\nu_{\tau L}}\sigma_{\mu\nu}\tau_R W_+^{\mu\nu} + \text{h.c.}\right) , \end{aligned} \tag{16}$$

where $F_{\mu\nu}$ is the electromagnetic field strength tensor, $Z_{\mu\nu} = \partial_\mu Z_\nu - \partial_\nu Z_\mu$ and $W_+^{\mu\nu} = \partial_\mu W_\nu^+ - \partial_\nu W_\mu^+$ are the corresponding strength tensors for the Z and W gauge bosons. We have not written the non-abelian couplings involving more than one gauge boson because they are not relevant to our purposes.

The Lagrangian (16) gives additional contributions to the magnetic moments of the tau, which are usually expressed in terms of a_τ and similar parameters for the corresponding weak magnetic moments for the Z-boson, a_Z [27, 28, 29] and the W-boson, κ^W [26]. They can be expressed as follows,

$$a_\tau = \frac{2\,m_\tau}{m_Z}\epsilon_\gamma \, , \tag{17}$$

$$a_Z = \frac{2\,m_\tau}{m_Z}\frac{1}{s_W\,c_W}\epsilon_Z \, , \tag{18}$$

$$\kappa^W = \sqrt{2}\frac{2\,m_\tau}{m_Z}\epsilon_W \, . \tag{19}$$

where the dimensionless couplings are

$$\epsilon_\gamma = (\alpha_B - \alpha_W)\frac{u m_Z}{\sqrt{2}\Lambda^2} \, , \tag{20}$$

$$\epsilon_Z = -(\alpha_W c_W^2 + \alpha_B s_W^2)\frac{u m_Z}{\sqrt{2}\Lambda^2} \, , \tag{21}$$

$$\epsilon_W = \alpha_W \frac{u m_Z}{\Lambda^2} = -\sqrt{2}\left(\epsilon_Z + s_W^2 \epsilon_\gamma\right) \, . \tag{22}$$

One remarkable result from this effective lagrangian approach is that the three magnetic moments, namely the electromagnetic one a_τ, the weak magnetic a_Z and the non-diagonal week W-boson coupling, κ^W, appear as a result of just the two effective operators in Eqs.(14,15), and only depend on the two coefficients α_B and α_W. Note that the three magnetic moments are not independent of each other. Besides, in the effective Lagrangian approach, exactly the same couplings that contribute to processes at high energies also contribute to the magnetic moment form factors, $F^{new}(q^2)$, at $q^2 = 0$. The difference $F^{new}(q^2) - F^{new}(0)$ only comes from higher dimension operators whose effect is suppressed by powers of q^2/Λ^2, as long as $q^2 \ll \Lambda^2$, as needed for the consistence of the effective Lagrangian approach.

Combining the data cited above (see details in [31]) we obtained the contour plots of Fig.3 and Fig.4.

These two figures represent the same fit, but just expressed in terms of different parameters. Projecting into the axes, one can find bounds to the non-standard contributions to the anomalous electromagnetic and weak magnetic moments a_τ, a_Z:

$$(1\sigma) \rightarrow \begin{cases} -0.005 < a_\tau < 0.002 \, , \\ -0.0007 < a_Z < 0.0019 \, , \end{cases} \tag{23}$$

$$(2\sigma) \rightarrow \begin{cases} -0.007 < a_\tau < 0.005 \, , \\ -0.0024 < a_Z < 0.0025 \, . \end{cases} \tag{24}$$

IOP Conf. Series: Journal of Physics: Conf. Series **912** (2017) 012001 doi:10.1088/1742-6596/912/1/012001

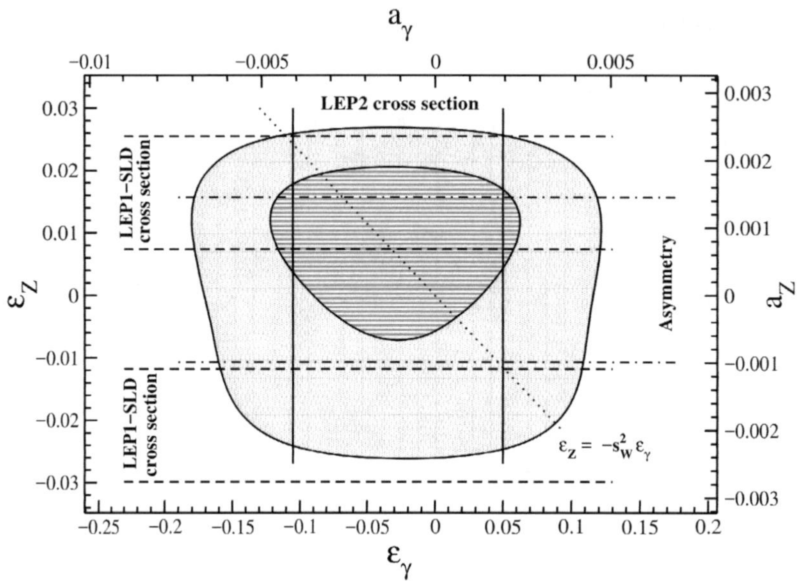

Figure 3. Global fit including all constraints discussed in ref.[31]. 95% CL and 68% CL contours are shown. The bands between straight lines show the allowed regions (1σ) coming from the different experiments: solid (LEP2-189 GeV), dashed (LEP1-SLD cross section), dot-dashed (asymmetry). We also have plotted the line $\epsilon_Z = -s_W^2 \epsilon_\gamma$ (dotted line) that corresponds to the results of refs.[32, 33]. This relationship appears when only the operator $\mathcal{O}_B$ contributes.

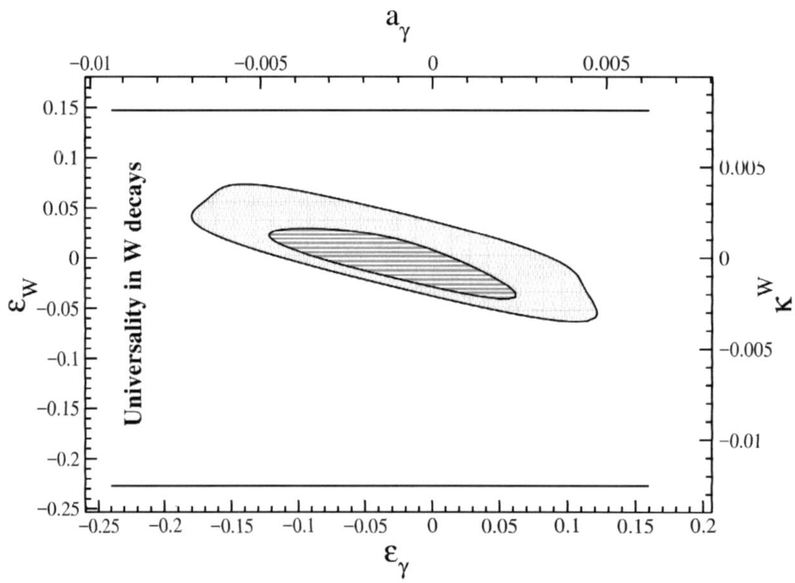

Figure 4. The global fit of Fig. 3 is now plotted in the plane ϵ_γ and ϵ_W to show the combined bounds on ϵ_W. As in Fig. 3 95% CL and 68% CL contours are shown. For comparison we also draw as straight lines the direct 1σ bounds obtained from universality tests in W decays.

For a_γ these limits are only about one order of magnitude larger than the standard model contribution in Eq.(3) and well below the PDG bound.

Using the relationship among ϵ_γ, ϵ_Z, α_B and α_W at a given value of the scale of new physics, one can easily obtain bounds on α_B and α_W. Alternatively, by assuming that $\alpha_B/4\pi$ or $\alpha_W/4\pi$ are order unity one can obtain bounds on the scale of new physics Λ ($\Lambda > 9$ TeV).

The above bounds are completely model independent and no assumption has been made on the relative size of couplings α_B and α_W in the effective Lagrangian (13). For the sake of comparison with published data in ref.[32, 33] we present now the limits that can be found by considering separately only operator $\mathcal{O}_B$ or only operator $\mathcal{O}_W$ in the Lagrangian (13). Consider that only $\mathcal{O}_B$ is present, as in Ref. [32, 33], it is equivalent to impose the relation $\epsilon_Z = -s_W^2\epsilon_\gamma$. Thus, from Fig. 3, it is straightforward to obtain that the bounds on the anomalous magnetic moment (at 2σ) are reduced to $-0.004 < a_\tau < 0.003$, while little change is found on the weak-magnetic moment $-0.0019 < a_Z < 0.0024$.

In most of the above mentioned experiments some of the particles in the vertex are off-shell. The interpretation of off-shell form factors is then problematic since they can hardly be isolated from other contributions and gauge invariance can be a problem. In the effective Lagrangian approach all those problems are solved because form factors are directly related to couplings, which are gauge invariant. Besides, as already mentioned, the difference $F^{new}(q^2) - F^{new}(0)$ only comes from higher dimension operators whose effect is suppressed by q^2/Λ^2. In this way a stringent and clear bound on the AMM was obtained.

More recently, some ideas were investigated having in mind the potential of the high statistics (in tau pairs) B-factories, close to 10^{12}. They now have by far the highest statistics of tau pairs, as it is the case for BaBar and Belle, while BELLE-II soon will also produce a huge number of tau pairs. Some time ago, the SuperB [36] project considered the possibility of vertex detectors and polarized electron beams, but the project was finally canceled.

We investigated the potential of the B-factories in order to get bounds on the tau AMM in ref.[29, 30]. There, the taus are produced $q^2 \simeq (10\,GeV)^2$, and, then, the problem of gauge invariance also appears. Tau pairs are obtained by both resonant and direct production. Cross section and asymmetries related to the linear and spin-spin correlations were studied on top of the Υ resonances. In particular, total cross section and the normal τ polarization -for the case of unpolarized electron beams- and the transverse and longitudinal ones -for the electron polarized beams- were investigated.

When attempting to extract the value of $F_2(0)$ from scattering experiments there are additional contributions of other Feynman diagrams, not related to the magnetic form factor. In the case of $e^+e^- \longrightarrow \tau^+\tau^-$ there are contributions not only from the usual s-channel one-loop vertex corrections but also from box diagrams. The contributions of the latter may interfere in the experimental determination of what we call $F_2(q^2)$, i.e. the magnetic part coming *only* from the vertex, and should be somehow "subtracted out". This may be done either by computing the box contributions and subtracting them from the cross-section, or by performing the measurement in a kinematic region where the boxes happen to be numerically subleading, i.e. on top of the Υ resonances. In this kinematic regime the non-resonant box diagrams are numerically negligible, and only one loop corrections to the $\gamma f\bar{f}$ vertex are relevant.

The direct computation of the magnetic part of the standard one-loop QED vertex yields

$$F_2(s) = \left(\frac{\alpha}{2\pi}\right)\frac{2m_\tau^2}{s}\frac{1}{\beta}\left(\log\frac{1+\beta}{1-\beta} - i\,\pi\right), \quad \text{for} \quad q^2 = s > 4m_\tau^2, \tag{25}$$

where α is the fine structure constant and $\beta = \left(1 - 4m_\tau^2/s\right)^{1/2}$ is the velocity of the τ. For $M_\Upsilon \sim 10\,\text{GeV}$, we obtained:

$$F_2(M_\Upsilon^2) = (2.65 - 2.45\,i) \times 10^{-4}. \tag{26}$$

Note that at this energy the real and imaginary parts are of the same order of magnitude. The above expression for $F_2(s)$ is a QED gauge-independent quantity, despite being an off-shell amplitude. This fact may be verified through an explicit calculation of the vertex diagram. The gauge-independence of $F_2(s)$ may also be understood in terms of the way the gauge-cancellations organize themselves in the QED S-matrix elements: first, the vacuum polarization is gauge-independent by itself; the gauge-dependence of the direct box cancels exactly against that of the crossed box and finally the gauge-dependence of the vertex correction can therefore cancel only against the fermion self-energy graphs renormalizing the external (on-shell) fermions. The latter however are proportional to γ_μ. Therefore the contribution of the vertex proportional to $\sigma_{\mu\nu}q^\nu$ must be individually gauge invariant. With these ideas in mind we investigated the QED gauge invariant magnetic tau form factor that can be accessed at B-factories.

The direct production mechanism is shown in Fig.3 while the resonant one is depicted in Fig.3. The interference of the diagrams shown in each figure gives the leading contribution to terms in $F_2(q^2)$. In this way, an expression for the differential and total cross section can be

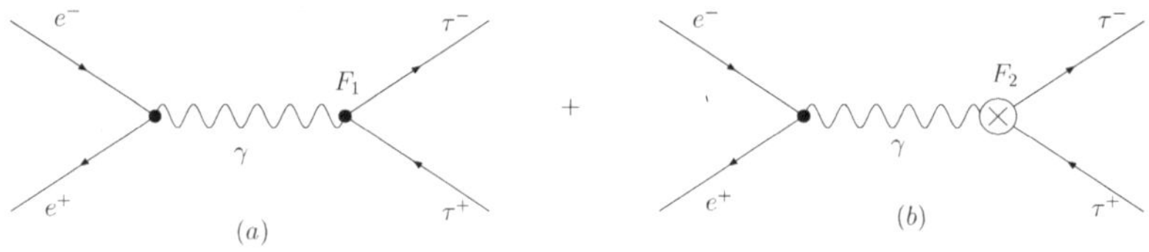

Figure 5. Diagrams: (a) direct γ exchange, (b) F_2 in γ exchange.

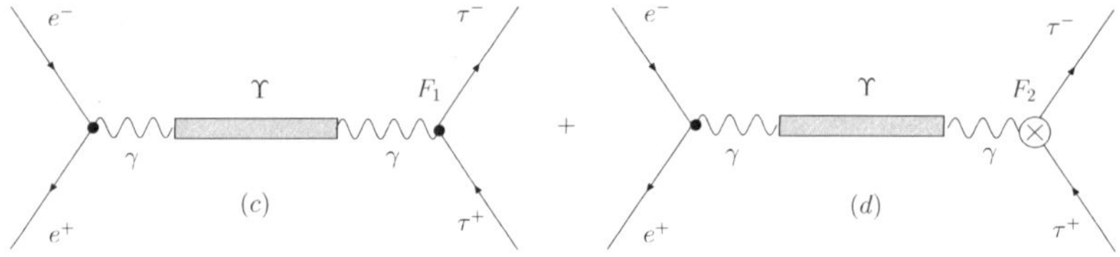

Figure 6. Diagrams: (c) Υ production, (d) F_2 in Υ production.

computed, where only the real part of F_2 appears. The normal polarization instead, can be obtained through an asymmetry constructed with the angle distribution of the decay products of the tau (see details in ref.[29, 30]), and results proportional to the imaginary part of F_2. Using polarized electron beams more possibilities are allowed that will not be discussed here.

Also, Z contamination through the interference with the resonant and direct diagrams shown in Fig.3 can be properly addressed: it is either very suppressed or it does not contributes (for example to the normal polarization asymmetry). Table 1 shows the upper limit to the sensitivity to the real and imaginary parts of F_2 assuming different integrated luminosities for BaBar, SuperB and Belle. No detector efficiencies are taken into account and only statistical errors are considered.

Other observables related to spin-spin correlations give sensitivities similar to the ones shown here, and can be found in ref.[30]. Note that some of these observables depend on the possibility of having polarized electron beams. In ref.[37] the authors discussed the energy resolution of the beams that may limit the possibility of disentangling the direct and resonant productions, and also the fact that the beam energy at Belle II will be far from some of the Υ resonances. All these facts may have some impact in these ideas.

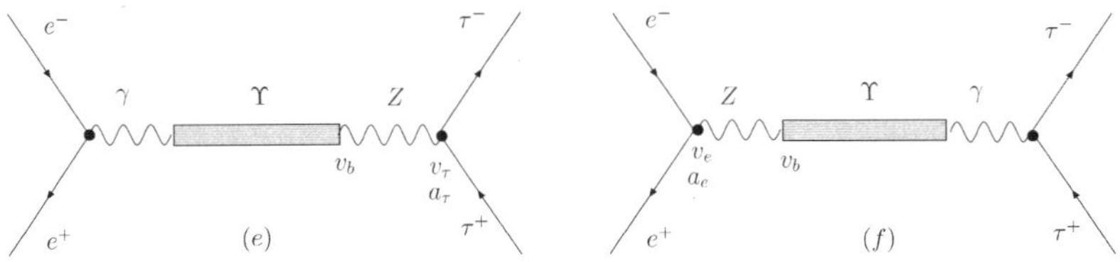

Figure 7. γ and Z interchange on Υ production.

Table 1. Sensitivity of the F_2 measurement at the Υ energy (ab = attobarn = $10^{-18}b$). * means polarized electron beam required.

EXPERIMENT ⇓	OBSERVABLE		
	Cross Section	Normal Asymmetry	Transverse and Longitudinal Asymmetry combined*
	$Re(F_2)$	$Im(F_2)$	$Re(F_2)$
Babar+Belle $2ab^{-1}$	4.6×10^{-6}	2.1×10^{-5}	1.0×10^{-5}
Super B/Flavor Factory (1 yr. running) $15ab^{-1}$	1.7×10^{-6}	7.8×10^{-6}	3.7×10^{-6}
Super B/Flavor Factory (5 yrs. running) $15ab^{-1}$	7.5×10^{-7}	3.5×10^{-6}	1.7×10^{-6}

This approach shows that a high sensitivity and several figures in the magnetic properties of the tau lepton can be obtained using these ideas.

4. Conclusions

We have review the main experiments and theoretical studies related to the tau AMM. The interpretation of these results, as well of the appropriate way to express the physics in the different formalisms, were discussed. The tau AMM is still a largely unknown quantity, despite the fact that many of the tau electroweak properties are well known [38]. It appears to be that we are far from the time where the tau AMM could play a role similar to the one that the AMM for the electron and for the muon have played. However, tau physics, being the tau such a high mass lepton, is a wonderful scenario where new physics may show up. More efforts should be addressed to this physics and there is a great opportunity for the new Belle II experiment for discoveries.

4.1. Acknowledgements

This work has been supported, in part, by the Ministerio de Economía y Competitividad (MINECO), Spain, under grants FPA2014-54459-P and SEV-2014-0398; by Generalitat Valenciana, Spain, under grant PROMETEOII2014-087. G.A.G-S. acknowledges the support of CSIC and Pedeciba, Uruguay.

References

[1] Patrignani C *et al.* [Particle Data Group] 2016 *Chin. Phys.* C **40** no.10, 100001

[2] M. L. Perl 1998 *Preprint* hep-ph/9812400
[3] Schwinger J S 1948 *Phys. Rev.* **73** 416
[4] Roberts B L, Marciano W J 2009 *Adv. Ser. Direct. High Energy Phys.* **20** 1
[5] Brown H N *et al.* [Muon g-2 Collaboration] 2001 *Phys. Rev. Lett.* **86** 2227 *Preprint* hep-ex/0102017
[6] Bennett G W *et al.* [Muon g-2 Collaboration] 2002 *Phys. Rev. Lett.* **89** 101804; Erratum 2002: *Phys. Rev. Lett.* **89** 129903 *Preprint* hep-ex/0208001
[7] Bennett G W *et al.* [Muon g-2 Collaboration] 2004 *Phys. Rev. Lett.* **92** 161802 *Preprint* hep-ex/0401008
[8] Czarnecki A, Marciano W J 2001 *Phys. Rev. D* **64** 013014 *Preprint* hep-ph/0102122
[9] Passera M 2005 *J. Phys. G* **31** R75 *Preprint* hep-ph/0411168
[10] Calmet J, Narison S, Perrottet M and de Rafael E 1977 *Rev. Mod. Phys.* **49** 21
[11] Eidelman S and Passera M 2007 *Mod. Phys. Lett.* A **22** 159 *Preprint* hep-ph/0701260
[12] Abdallah J*et al.* [DELPHI Collaboration] 2004 *Eur. Phys. J.* C **35** 159 *Preprint* hep-ex/0406010
[13] Hanneke D, Fogwell S and Gabrielse G 2008 *Phys. Rev. Lett.* **100** 120801 *Preprint* arXiv:0801.1134 [physics.atom-ph]
[14] Samuel M A, Li G w and Mendel R 1991 *Phys. Rev. Lett.* **67** 668; Erratum 1992: *Phys. Rev. Lett.* **69** 995
[15] Buchmuller W and Wyler D 1986 *Nucl. Phys.* B **268** 621
[16] Fujikawa K, Lee B W and Sanda A I 1972 *Phys. Rev.* D **6**, 2923
[17] Weinberg S 1995 *The Quantum theory of fields.* Vol. 1: Foundations (New York: Cambridge Univ. Press)
[18] Cornet F and Illana J I 1996 *Phys. Rev.* D **53** 1181 *Preprint* hep-ph/9503466
[19] Peressutti J and Sampayo O A 2012 *Phys. Rev.* D **86** 035016
[20] Silverman D J and Shaw G L 1983 *Phys. Rev.* D **27** 1196
[21] Gau S S, Paul T C, Swain J and Taylor L 1998 *Nucl. Phys.* B **523**, 439 *Preprint* hep-ph/9712360
[22] Swain J 2001 *Nucl. Phys. Proc. Suppl.* **98** 351
[23] Acciarri M *et al.* [L3 Collaboration] 1998 *Phys. Lett.* B **434**, 169
[24] Ackerstaff K *et al,* 1998 *Physics Letters* B **431** 188
[25] Galon I, Rajaraman A and Tait T M P 2016 *JHEP* **1612** 111 *Preprint* [arXiv:1610.01601 [hep-ph].
[26] Rizzo T G, 1997 *Phys. Rev.* D **56** 3074 *Preprint* hep-ph/9704337
[27] Bernabeu J, González-Sprinberg G A and Vidal J 1994 *Phys. Lett.* B **326** 168
[28] Bernabeu J, González-Sprinberg G A, Tung, M and Vidal J 1995 *Nucl. Phys.* B **436** 474 it Preprint hep-ph/9411289
[29] Bernabeu J, González-Sprinberg G A, Papavassiliou J and Vidal J 2008 *Nucl. Phys.* B **790** 160 *Preprint* arXiv:0707.2496 [hep-ph]
[30] Bernabeu J, González-Sprinberg G A and Vidal J 2009 *JHEP* **0901** 062 *Preprint* arXiv:0807.2366 [hep-ph]
[31] González-Sprinberg G A, Santamaria A and Vidal J 2000 *Nucl. Phys.* B **582** 3 *Preprint* hep-ph/0002203
[32] Escribano R and E. Massó E 1993 *Physics letters* B **301** 419
[33] Escribano R and E. Massó E 1997 *Physics letters* B **395** 369
[34] Grifols J A and Méndez A 1991 *Physics Letters* B **255** 611
[35] Biebel J and T. Riemann T 1997 *Zeitschrift für Physik* C **76** 53
[36] O'Leary B *et al.* SuperB Collaboration 2010 *Preprint* arXiv:1008.1541 [hep-ex]
[37] Eidelman S, Epifanov D, Fael M, Mercolli L and Passera M 2016 *JHEP* **1603** 140 *Preprint* arXiv:1601.07987 [hep-ph]
[38] Pich A 2014 *Prog. Part. Nucl. Phys.* **75** 41 *Preprint* arXiv:1310.7922 [hep-ph]

Study of five-body leptonic decays of tau at Belle experiment

Junya Sasaki, and Belle Collaboration

Department of Physics, University of Tokyo, 7-3-1 Hongo, Bunkyo-ku, Tokyo 113-0033, Japan

E-mail: sasaki@hep.phys.s.u-tokyo.ac.jp

Abstract.
To perform an extended test of the Lorentz structure of the charged weak interaction, a study of five-body leptonic decays $\tau^- \to l^- l'^+ l'^- \nu_\tau \bar{\nu}_l$ $(l, l' = e,\, \mu)$ is ongoing with data sample which contains about 0.91×10^9 $\tau^+\tau^-$ pairs, collected at Belle. The Standard Model predicts that the Lorentz structure has a V-A structure and this can be tested through the measurement of Michel parameters. In our study we try to give tighter constraints to Michel parameters through the measurement of branching fraction of five-body leptonic decays of tau. With an embedded formalism of total differential decay width which has recently been published, Monte Carlo event generator has been developed within TAUOLA program. In this paper, we finally report our preliminary result of systematic uncertainties of the branching fractions.

1. Introduction

Standard Model (SM) predicts a maximum asymmetry between left-handed and right-handed fundamental fermions, which indicates that the Lorentz structure of the charged weak current has a V-A structure. The most general, Lorentz invariant, derivative-free and lepton-number-conserving four-lepton point interaction matrix element of the $\tau^- \to l^- \bar{\nu}_l \nu_\tau$[1] decay (its SM Feynman diagram is shown in Fig. 1) is given by:

$$\mathcal{M} = \frac{4G_F}{\sqrt{2}} \sum_{\substack{i=S,V,T \\ j,k=L,R}} g^i_{jk} \left[\bar{u}_j(l^-)\Gamma^i v_\xi(\bar{\nu}_l)\right]\left[\bar{u}_\kappa(\nu_\tau)\Gamma_i u_k(\tau^-)\right],$$

$$\Gamma^S = 1,\ \Gamma^V = \gamma^\mu,\ \Gamma^T = \frac{i}{2\sqrt{2}}(\gamma^\mu\gamma^\nu - \gamma^\nu\gamma^\mu). \tag{1}$$

Here, ξ and κ are the chiralities of neutrinos and g^i_{jk} is dimensionless coupling constant. In the SM, $g^V_{LL} = 1$ is the only non-zero constant. The differential decay width of the $\tau^- \to l^- l'^+ l'^- \nu_\tau \bar{\nu}_l$ (its SM diagrams are shown in Fig. 2) is written as [1]:

$$\frac{d\Gamma_5}{d^3p_1 d^3p_2 d^3p_3} \propto \frac{1}{E_1E_2E_3}[(Q_{LL}T^Q_{LL} + Q_{RL}T^Q_{RL} + B_{RL}T^B_{RL} + L \leftrightarrow R) + I_\alpha \Re(T^I_\alpha) + I_\beta \Re(T^I_\beta)], \tag{2}$$

[1] Unless specified otherwise, charge-conjugated decays are implied throughout the paper.

IOP Conf. Series: Journal of Physics: Conf. Series **912** (2017) 012002 doi:10.1088/1742-6596/912/1/012002

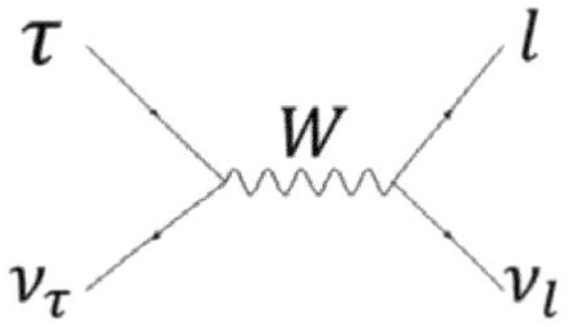

Figure 1: Feynman diagram of the $\tau^- \to l^-\bar{\nu}_l\nu_\tau$ decay in the SM.

Figure 2: Two diagrams for the $\tau^- \to l^-l'^+l'^-\nu_\tau\bar{\nu}_l$ in the SM.

where (E_1, p_1), (E_2, p_2) and (E_3, p_3) are the energies and momenta of l^-, l'^+ and l'^-, respectively; T^i_{jk} $(i = Q,\ B,\ I;\ j, k = L,\ R)$ is a known function of the kinematical variables, Q_{ij} $(i, j = L,\ R)$, B_{LR}, B_{RL}, I_α and I_β are eight Michel parameters [2]. The SM predicts that only $Q_{LL} = 1$ has a non-zero value. If some of the other Michel parameters have non-zero values, we can say there are contributions from New Physics (NP) models. Precision measurement of the branching fractions of $\tau^- \to l^-l'^+l'^-\nu_\tau\bar{\nu}_l$ decays allows one to constrain Michel parameters. The theoretical formula for the branching fraction (BR) can be obtained integrating Eq. (2):

$$BR_{\rm exp} = BR_{\rm SM}[Q_{LL} + bQ_{LR} + cB_{LR} + Q_{RR} + dQ_{RL} + eB_{RL} + \Re(f)I_\alpha + \Re(g)I_\beta] + BR_{\rm NLO}. \quad (3)$$

Here, $BR_{\rm exp}$ is experimentally measured branching fraction, $BR_{\rm SM}$ is the branching fraction predicted in the SM (see Table 1), $BR_{\rm NLO}$ is radiative correction [3, 4]. The coefficients b, c, d, e, f, and g are the integrated terms of the T^Q_{RL}, T^B_{RL}, T^Q_{LR}, T^B_{LR}, T^I_α, and T^I_β functions, which is renormalized by the $BR_{\rm SM}$, respectively. $BR_{\rm SM}$ is the value of integrated terms related to T^Q_{LL} functions. Since the T^i_{jk} functions have symmetry under the transformation $L \leftrightarrow R$, the coefficients b, c, d, and e have a relation $b = d$, $c = e$ [1]. Here the $BR_{\rm NLO}$ is negligibly small in our study: $BR_{\rm NLO}$ becomes about $O(0.1)\%$ of $BR_{\rm SM}$ in our study [5]. We therefore do not need to consider the effect of $BR_{\rm NLO}$ for the constraints of Michel parameters.

Table 1: Branching fractions of the $\tau^- \to l^-l'^+l'^-\nu_\tau\bar{\nu}_l$ decays predicted by the SM.

Mode	BR from Ref. [1]
$\tau^- \to e^-e^+e^-\bar{\nu}_e\nu_\tau$	$(4.21 \pm 0.01) \times 10^{-5}$
$\tau^- \to \mu^-e^+e^-\bar{\nu}_\mu\nu_\tau$	$(1.984 \pm 0.004) \times 10^{-5}$
$\tau^- \to e^-\mu^+\mu^-\bar{\nu}_e\nu_\tau$	$(1.247 \pm 0.001) \times 10^{-7}$
$\tau^- \to \mu^-\mu^+\mu^-\bar{\nu}_\mu\nu_\tau$	$(1.183 \pm 0.001) \times 10^{-7}$

2. Formula of Branching Fraction Depending on Michel Parameters

To obtain the value of coefficients b, c, d, e, f, and g as introduced in Eq. (3), we perform a numerical integral of Eq. (2).

$$b \sim g = \frac{BR_{\rm NP}^{b\sim g}}{BR_{\rm SM}} = \frac{1}{\Gamma_{\rm SM}} \int d(PS) d\Gamma_{\rm NP}^{b\sim g}. \tag{4}$$

Here the $d\Gamma_{\rm NP}^{b\sim g}$ is the differential decay width related to T^i_{jk} function. Eq. (4) is written by:

$$\begin{aligned} b \sim g &= \frac{1}{BR_{\rm SM}} BR_{\rm NP}^{b\sim g} = \frac{1}{\Gamma_{\rm SM}} \int d\Gamma_{\rm NP}^{b\sim g} d(PS) = \frac{1}{\Gamma_{\rm SM}} \int \frac{d\Gamma_{\rm NP}^{b\sim g}}{d\Gamma_{\rm SM}/\Gamma_{\rm SM}} [(d\Gamma_{\rm SM}/\Gamma_{\rm SM}) d(PS)] \\ &= \frac{1}{\Gamma_{\rm SM}} \int \frac{d\Gamma_{\rm NP}^{b\sim g}}{d\widetilde{\Gamma}_{\rm SM}} [(d\widetilde{\Gamma}_{\rm SM}) d(PS)] \approx \frac{1}{\Gamma_{\rm SM}} \frac{1}{N_{\rm gen}} \sum_{\mathbf{x}\in\Omega} \frac{d\Gamma_{\rm NP}^{b\sim g}(\mathbf{x})}{d\widetilde{\Gamma}_{\rm SM}(\mathbf{x})} = \frac{1}{N_{\rm gen}} \sum_{\mathbf{x}\in\Omega} \frac{d\Gamma_{\rm NP}^{b\sim g}(\mathbf{x})}{d\Gamma_{\rm SM}(\mathbf{x})}, \end{aligned} \tag{5}$$

where $d\widetilde{\Gamma}_{\rm SM} = d\Gamma_{\rm SM}/\Gamma_{\rm SM}$ is a normalized differential decay width of the SM, Ω is the allowed phase space (PS), $\mathbf{x}$ follows the distribution of $d\Gamma_{\rm SM}$, and $N_{\rm gen}$ is the number of generated events. The detail of method is described in the Appendix A. The calculated result for the $\tau^- \to l^- l'^+ l'^- \nu_\tau \bar{\nu}_l$ decays are given as follows.

$$\begin{aligned} BR_{\rm exp}^{\tau^-\to e^- e^+ e^- \bar{\nu}_e \nu_\tau} = BR_{\rm SM}^{\tau^-\to e^- e^+ e^- \bar{\nu}_e \nu_\tau} \{ [Q_{LL} + (1.051 \pm 0.036) Q_{LR} + (-0.2053 \pm 0.1431) B_{LR} \\ + L \leftrightarrow R] + (0.2416 \pm 0.0002) I_\alpha + (0.8606 \pm 0.0001) I_\beta \}. \end{aligned} \tag{6}$$

$$\begin{aligned} BR_{\rm exp}^{\tau^-\to \mu^- e^+ e^- \bar{\nu}_\mu \nu_\tau} = BR_{\rm SM}^{\tau^-\to \mu^- e^+ e^- \bar{\nu}_\mu \nu_\tau} \{ [Q_{LL} + (1.220 \pm 0.049) Q_{LR} + (-0.8717 \pm 0.1957) B_{LR} \\ + L \leftrightarrow R] + (181.3 \pm 0.1) I_\alpha + (104.4 \pm 0.1) I_\beta \}. \end{aligned} \tag{7}$$

$$\begin{aligned} BR_{\rm exp}^{\tau^-\to e^- \mu^+ \mu^- \bar{\nu}_e \nu_\tau} = BR_{\rm SM}^{\tau^-\to e^- \mu^+ \mu^- \bar{\nu}_e \nu_\tau} \{ [Q_{LL} + (1.226 \pm 0.001) Q_{LR} + (-0.8456 \pm 0.0001) B_{LR} \\ + L \leftrightarrow R] + (0.2253 \pm 0.0001) I_\alpha + (0.5231 \pm 0.0001) I_\beta \}. \end{aligned} \tag{8}$$

$$\begin{aligned} BR_{\rm exp}^{\tau^-\to \mu^- \mu^+ \mu^- \bar{\nu}_\mu \nu_\tau} = BR_{\rm SM}^{\tau^-\to \mu^- \mu^+ \mu^- \bar{\nu}_\mu \nu_\tau} \{ [Q_{LL} + (1.216 \pm 0.005) Q_{LR} + (-0.8459 \pm 0.0005) B_{LR} \\ + L \leftrightarrow R] - (18.00 \pm 0.01) I_\alpha + (197.3 \pm 0.1) I_\beta \}. \end{aligned} \tag{9}$$

The large error in some coefficients in the first two modes derives from a peculiarity in the matrix elements related to factor $1/q^2_{e^+e^-}$ (q, is a 4-momentum of off shell γ $(q = p_{e^+} + p_{e^-})$). The terms related to $1/q^2_{e^+e^-}$ become very large when the q has small value ($O(m_e) \sim O(\text{MeV})$) and this causes large error. The explicit expressions through the coupling constants g^i_{jk} can be found in Appendix B.

3. Selections, background

The selection process is organized in two stages. The first-stage selection suppresses beam background and rejects most of the background from the non-$\tau^+\tau^-$ processes, and select the candidate of signal-events. In the second stage, we select the samples enriched with the signal events. Each event is divided into two hemispheres using the plane perpendicular to the direction of the thrust axis in the center-of-mass system (c.m.s.). One charged track is reconstructed in one hemisphere (tag side) and three charged tracks identified as electrons or muons - in the other hemisphere (signal side).

Full MC simulation of the signal modes and the $\tau^+\tau^-$ generic MC sample were used to study detection efficiencies of the signal events and background contamination from the other τ decays, see Table 2. Recently published analytical formalism for the $\tau^- \to l^- l'^+ l'^- \bar{\nu}_l \nu_\tau$ differential decay width [1] was used to develop Monte Carlo event generator in the framework of the KKMC+TAUOLA generator [6, 7]. A Monte Carlo (MC) sample of 4 million signal decays was used for evaluating the background and calculate efficiencies. The detector response is simulated

by a GEANT3-based program [8].
Expected numbers of the signal events with the whole Belle statistics are calculated taking into account MC detection efficiencies, branching fractions, and the number of the $\tau^+\tau^-$ pairs at Belle ($N_{\tau\tau} = 0.91 \times 10^9$). Figure 4 shows the distribution of the invariant mass of three charged outgoing leptons for the selected events.
For the selection, we use a parameter $\sum_{i<j} \cos\theta_{ij}$ which was introduced in Ref. [9], is a sum of $\cos\theta_{ij}$ where θ_{ij} is an angle between two leptons in the signal-hemisphere (Fig. 3).

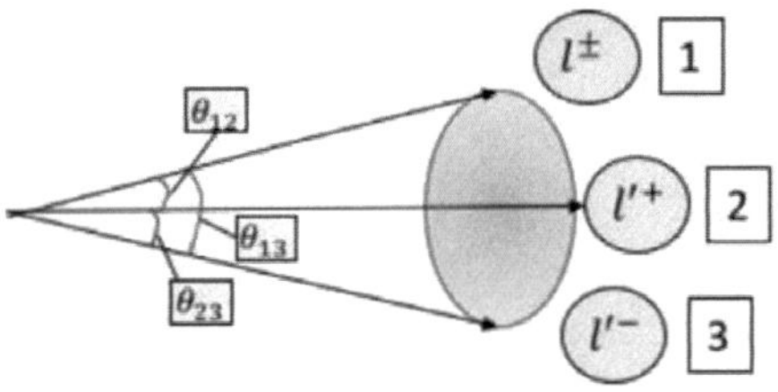

Figure 3: Explanation of θ_{ij}.

We apply the cut $\sum_{i<j} \cos\theta_{ij}$ should be larger than 2.90, 2.93, 2.70, and 2.85 for $e^-e^+e^-\bar{\nu}_e\nu_\tau$, $\mu^-e^+e^-\bar{\nu}_\mu\nu_\tau$, $e^-\mu^+\mu^-\bar{\nu}_e\nu_\tau$, and $\mu^-\mu^+\mu^-\bar{\nu}_\mu\nu_\tau$, respectively. For the modes $\tau^- \to e^-e^+e^-\bar{\nu}_e\nu_\tau$ and $\tau^- \to \mu^-e^+e^-\bar{\nu}_\mu\nu_\tau$, this selection suppresses the backgrounds such as $\tau^- \to \pi^-\pi^0\nu_\tau \to \pi^-(\gamma\gamma)\nu_\tau \to \pi^-((e^+e^-)\gamma)\nu_\tau$ and $\tau^- \to \pi^-\pi^0\nu_\tau \to \pi^-(e^+e^-\gamma)\nu_\tau$. For the modes $\tau^- \to e^-\mu^+\mu^-\bar{\nu}_e\nu_\tau$ and $\tau^- \to \mu^-\mu^+\mu^-\bar{\nu}_\mu\nu_\tau$, this selection suppresses the backgrounds such as $\tau^- \to \pi^-\pi^+\pi^-\nu_\tau$ and $\tau^- \to \pi^-\pi^+\pi^-\pi^0\nu_\tau$. In these backgrounds, the pion is misidentified as electron or muon. The particle identification (PID) we used is described in Appendix C.
Because of the gamma conversion in the material of the detector, a radiative leptonic decay $\tau^- \to l^-\bar{\nu}_l\nu_\tau\gamma$ becomes a dominant background for the $\tau^- \to e^-e^+e^-\bar{\nu}_e\nu_\tau$ and $\tau^- \to \mu^-e^+e^-\bar{\nu}_\mu\nu_\tau$ modes. To suppress this background, we apply the cut that the reconstructed gamma conversion point r_{xy} (is a distance to the beam axis (z-axis)) should be smaller than 1.6 cm and 1.5 cm, respectively.
Since two muons in the modes $\tau^- \to e^-\mu^+\mu^-\bar{\nu}_e\nu_\tau$ and $\tau^- \to \mu^-\mu^+\mu^-\bar{\nu}_\mu\nu_\tau$ are generated by off-shell photon conversion, the invariant mass of $\mu^+\mu^-$ tends to be smaller. To suppress $\tau^- \to \pi^-\pi^0\nu_\tau \to \pi^-(\gamma\gamma)\nu_\tau \to \pi^-((e^+e^-)\gamma)\nu_\tau$ and $\tau^- \to \pi^-\pi^+\pi^-\nu_\tau$ processes whose branching fraction is large, we require the invariant mass of $\mu^+\mu^-$ should be smaller than 0.4 GeV only for $\tau^- \to e^-\mu^+\mu^-\bar{\nu}_e\nu_\tau$.
Also the number of photons are used. For the $\tau^- \to e^-e^+e^-\bar{\nu}_e\nu_\tau$ and the $\tau^- \to e^-\mu^+\mu^-\bar{\nu}_e\nu_\tau$ modes, we require the number of photons in the signal-hemisphere should be less than one, and when the case of one we require the energy of photon should be less than 0.5 GeV. For the $\tau^- \to \mu^-e^+e^-\bar{\nu}_\mu\nu_\tau$, we require the number of photons in the (signal+tag)-hemisphere should be less than five, and the sum of energy of photon in the signal-hemisphere should be less than 0.3 GeV. For the $\tau^- \to \mu^-\mu^+\mu^-\bar{\nu}_\mu\nu_\tau$, we require there should be no photon in the signal-hemisphere. Since the number of selections we use is many, we introduced some important selections to suppress the backgrounds. Table 2 shows the result after applying all selections.

Table 2: Summary of the signal detection efficiencies and background contaminations.

τ^- decay mode	$e^-e^+e^-\bar{\nu}_e\nu_\tau$	$\mu^-e^+e^-\bar{\nu}_\mu\nu_\tau$	$e^-\mu^+\mu^-\bar{\nu}_e\nu_\tau$	$\mu^-\mu^+\mu^-\bar{\nu}_\mu\nu_\tau$
Detection efficiency	(1.769±0.004)%	(1.204±0.003)%	(3.561±0.006)%	(1.674±0.004)%
Main backgrounds	$e^-\bar{\nu}_e\nu_\tau\gamma$ $\to e^-\bar{\nu}_e\nu_\tau(e^+e^-)$ $\pi^-\pi^0\nu_\tau$ $\to \pi^-(\gamma\gamma)\nu_\tau$ $\to \pi^-((e^+e^-)\gamma)\nu_\tau$ (mis-ID π as e) $e^-\bar{\nu}_e\nu_\tau$	$\mu^-\bar{\nu}_\mu\nu_\tau\gamma$ $\to \mu^-\bar{\nu}_\mu\nu_\tau(e^+e^-)$ $\pi^-\pi^0\nu_\tau$ $\to \pi^-(e^+e^-\gamma)\nu_\tau$ $\pi^-\pi^0\pi^0\nu_\tau$ $\to \pi^-(\gamma\gamma)(\gamma\gamma)\nu_\tau$ $\to \pi^-((e^+e^-)\gamma)(\gamma\gamma)\nu_\tau$ (mis-ID π as μ)	$\pi^-\pi^0\nu_\tau$ $\to \pi^-(\gamma\gamma)\nu_\tau$ $\to \pi^-((e^+e^-)\gamma)\nu_\tau$ $\pi^-\pi^+\pi^-\nu_\tau$ (mis-ID π as μ, e)	$\pi^-\pi^0\nu_\tau$ $\to \pi^-(\gamma\gamma)\nu_\tau$ $\to \pi^-((e^+e^-)\gamma)\nu_\tau$ $\pi^-\pi^+\pi^-\nu_\tau$ (mis-ID π as μ)
Expected number of signal events	1300	430	8	4
Fraction of the signal	47%	50%	37%	16%

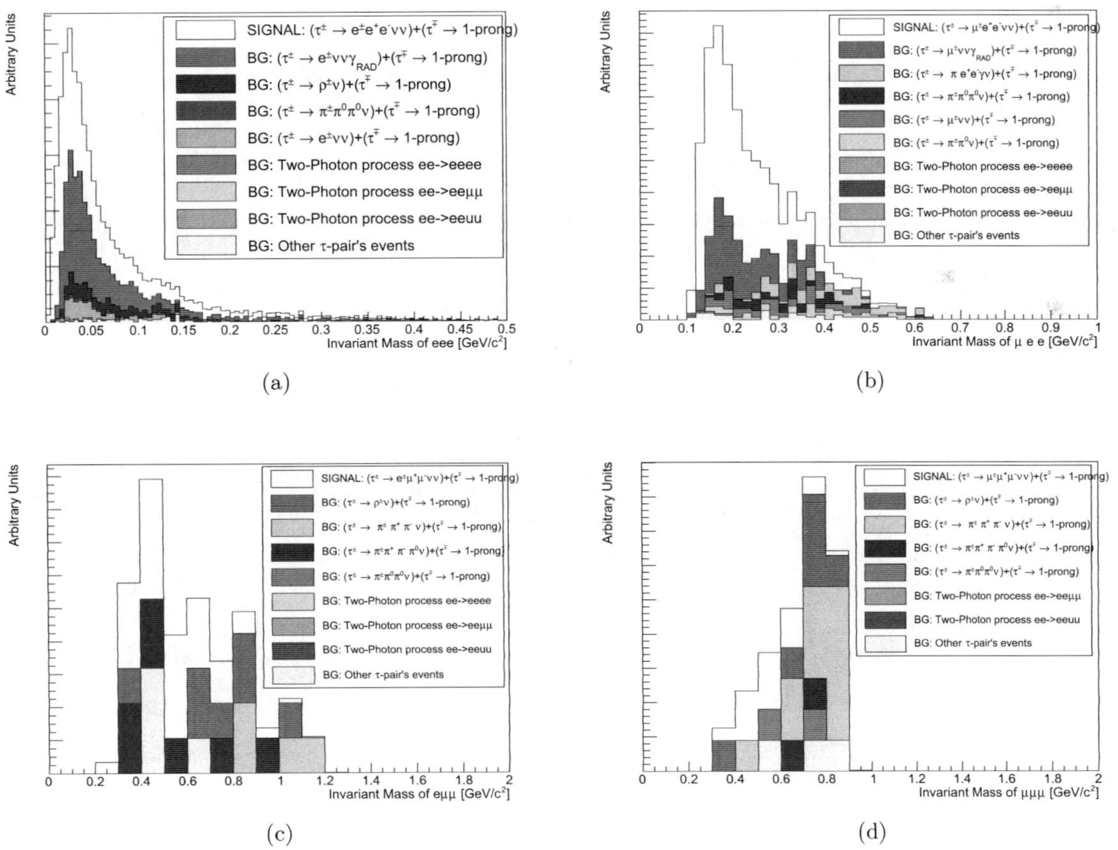

Figure 4: Distribution of the invariant mass of three charged outgoing leptons, $M_{ll'l'}$, for the selected events: (a) $\tau^- \to e^-e^+e^-\bar{\nu}_e\nu_\tau$, (b) $\tau^- \to \mu^-e^+e^-\bar{\nu}_\mu\nu_\tau$, (c) $\tau^- \to e^-\mu^+\mu^-\bar{\nu}_e\nu_\tau$, and (d) $\tau^- \to \mu^-\mu^+\mu^-\bar{\nu}_\mu\nu_\tau$. Open histograms show signal events, hatched histograms - background contributions.

4. Preliminary Result of Systematic Uncertainties of Branching Fractions

Following systematic uncertainties are taken into account:

- **Particle Identification (PID) correction**
  This error is considered because the efficiency of PID is different between MC samples and experimental data.
- **Tracking efficiency**
  This error considers the effect of difference of charged track's efficiency between MC samples and experimental data.
- **Trigger Correction**
  We also consider the difference of trigger efficiency's correction between MC samples and data.
- **Tag-side (1-prong side)**
  This error is for tag-side.
- **Luminosity**
  This error is for an uncertainties of luminosity.
- **Background (BG)**
  The systematic error is estimated as the statistical error of backgrounds.
- **Selection Cut**
  The systematic error is estimated from the fluctuation of the number of events between experimental data and MC by shifting the cut point of selection.

Table 3, 4, 5, and 6 show the preliminary result of systematic uncertainties of branching fraction of $\tau^- \to e^-e^+e^-\bar{\nu}_e\nu_\tau$, $\tau^- \to \mu^-e^+e^-\bar{\nu}_\mu\nu_\tau$, $\tau^- \to e^-\mu^+\mu^-\bar{\nu}_e\nu_\tau$, and $\tau^- \to \mu^-\mu^+\mu^-\bar{\nu}_\mu\nu_\tau$, respectively. Since we have mainly two kinds of data set called SVD1 and SVD2, we show the systematic uncertainties for each data sets. The "Selection Cut" has not been estimated yet. For this estimation, we used the data of SVD1's integrated luminosity about 140 fb^{-1} ($\Upsilon(4S)$ on-resonance) and the data of SVD2's integrated luminosity 560 fb^{-1} ($\Upsilon(4S)$ on-resonance) [2].

5. Conclusion

With the purpose of constraining the Michel parameters, we study the five-body leptonic decays of tau $\tau^- \to l^-l'^+l'^-\bar{\nu}_l\nu_\tau$ at Belle. As our results, we obtained the formula of branching fraction of $\tau^- \to l^-l'^+l'^-\bar{\nu}_l\nu_\tau$ expressed by the Michel parameters. Also the selections and background's contamination are determined through the full Monte Carlo simulation. The estimation of systematic uncertainties of branching fractions is ongoing. After finishing the estimation, we measure the branching fractions and constrain the Michel parameters.

6. Acknowledgements

We strongly appreciate Pablo Roig and Denis Epifanov for the helpful discussion.

[2] In Belle, there are mainly two data sets called SVD1 and SVD2. In this study, we used only $\Upsilon(4S)$ on-resonance data in SVD1 and SVD2 whose total integrated luminosity is about 700 fb^{-1}. Including the other data not $\Upsilon(4S)$ on-resonance, the total integrated luminosity becomes about 1000 fb^{-1}.

IOP Conf. Series: Journal of Physics: Conf. Series **912** (2017) 012002 doi:10.1088/1742-6596/912/1/012002

Table 3: Systematic uncertainties of the $\tau^- \to e^- e^+ e^- \bar{\nu}_e \nu_\tau$ (preliminary)

contents	syst. error (SVD1)	syst. error (SVD2)
PID correction	7.3%	5.0%
Tracking efficiency	1.1%	1.1%
Trigger correction	0.1%	0.1%
Tag-side	0.35%	0.35%
Luminosity	1.4%	1.4%
Background	5.5%	2.8%
Selection Cut	–	–
Total	12.1%	6.0%

Table 4: Systematic uncertainties of the $\tau^- \to \mu^- e^+ e^- \bar{\nu}_\mu \nu_\tau$ (preliminary)

contents	syst. error (SVD1)	syst. error (SVD2)
PID correction	6.9%	6.4%
Tracking efficiency	1.1%	1.1%
Trigger correction	0.1%	0.1%
Tag-side	0.35%	0.35%
Luminosity	1.4%	1.4%
Background	9.6%	4.8%
Selection Cut	–	–
Total	13.0%	8.2%

Table 5: Systematic uncertainties of the $\tau^- \to e^- \mu^+ \mu^- \bar{\nu}_e \nu_\tau$ (preliminary)

contents	syst. error (SVD1)	syst. error (SVD2)
PID correction	8.7%	7.4%
Tracking efficiency	1.1%	1.1%
Trigger correction	0.1%	0.1%
Tag-side	0.35%	0.35%
Luminosity	1.4%	1.4%
Background	71.0%	35%
Selection Cut	–	–
Total	72.0%	36.0%

Table 6: Systematic uncertainties of the $\tau^- \to \mu^- \mu^+ \mu^- \bar{\nu}_\mu \nu_\tau$ (preliminary)

contents	syst. error (SVD1)	syst. error (SVD2)
PID correction	6.2%	8.4%
Tracking efficiency	1.1%	1.1%
Trigger correction	0.1%	0.1%
Tag-side	0.35%	0.35%
Luminosity	1.4%	1.4%
Background	71.0%	35%
Selection Cut	–	–
Total	72.0%	36.0%

Appendix A. Formula of Differential Decay Width and Coefficients $b \sim g$

To perform the numerical integration, we generate the MC events which follow a distribution of SM-case. The differential decay width of $\tau^- \to l^- l'^+ l'^- \nu_\tau \bar{\nu}_l$ decays given by [1]:

$$\frac{d\Gamma_5}{dp_1 d\Omega_1 dp_2 d\Omega_2 dp_3 d\Omega_3} = \frac{p_1^2 p_2^2 p_3^2}{3 \cdot 2^{18} \cdot \pi^{10} M E_1 E_2 E_3} |\mathcal{M}_{\text{total}}|^2, \tag{A.1}$$

Including the NP parts, $|\mathcal{M}_{\text{total}}|^2$ is given by [1]:

$$|\mathcal{M}_{\text{total}}|^2 = e^4 |G_{ll'}|^2 [(Q_{LL} T_{LL}^Q + Q_{RL} T_{RL}^Q + B_{RL} T_{RL}^B + L \leftrightarrow R) + I_\alpha \Re(T_\alpha^I) + I_\beta \Re(T_\beta^I)], \tag{A.2}$$

where $|G_{ll'}|^2$ is a scale factor, and T_{jk}^i are the known function of kinematical variables as described in Sec. 1. In the SM, only $Q_{LL} T_{LL}^Q$ remains ($Q_{LL} = 1$), i.e., $T_{LL}^Q = T_{\text{SM}}$. In Ref. [1], $T_{RL}^Q, T_{RL}^B, T_\alpha^I$, and T_β^I are given by,

$$T_{RL}^Q = \frac{1}{3}(16 T_{RL}^S - T_{RL}^V), T_{RL}^B = \frac{4}{3}(T_{RL}^V - 4 T_{RL}^S), T_\alpha^I = 4 T_{LRRL}^{SV}, T_\beta^I = 2 T_{LLRR}^{SV}. \tag{A.3}$$

where the $T_{RL}^S, T_{RL}^V, T_{LRRL}^{SV}$, and T_{LLRR}^{SV} are known functions which are given in Ref. [1]. We calculate the integral of $T_{RL}^Q, T_{RL}^B, T_\alpha^I$, and T_β^I from the integral of $T_{RL}^S, T_{RL}^V, T_{LRRL}^{SV}$, and T_{LLRR}^{SV}. For the convenience of calculation, we perform a transformation:

$$dp_1 d\Omega_1 dp_2 d\Omega_2 dp_3 d\Omega_3 \to dp_{3\text{body}} d\Omega_{3\text{body}} d\widetilde{M}_{3\text{body}}^2 d\widetilde{\Omega}_3 d\widetilde{M}_{2\text{body}}^2 d\widetilde{\Omega}_2, \tag{A.4}$$

where, $p_{3\text{body}}$ is a momentum of three leptons frame $(l^- l'^+ l'^-)$ in a tau-rest frame, $\Omega_{3\text{body}}$ is a solid angle of three leptons frame $(l^- l'^+ l'^-)$ in a tau-rest frame, $\widetilde{M}_{3\text{body}}$ is a invariant mass of three leptons frame $(l^- l'^+ l'^-)$, $\widetilde{\Omega}_3$ is a solid angle of two leptons frame $(l'^+ l'^-)$ in a 3-body $(l^- l'^+ l'^-)$ rest frame, $\widetilde{M}_{2\text{body}}$ is a invariant mass of two leptons frame $(l'^+ l'^-)$, and $\widetilde{\Omega}_2$ is a solid angle of one lepton (l'^+ or l'^-) in a 2-body $(l'^+ l'^-)$ rest frame.
The Jacobian is calculated as this,

$$\left| \frac{\partial(p_1, \Omega_1, p_2, \Omega_2, p_3, \Omega_3)}{\partial(p_{3\text{body}}, \Omega_{3\text{body}}, \widetilde{M}_{3\text{body}}^2, \widetilde{\Omega}_3, \widetilde{M}_{2\text{body}}^2, \widetilde{\Omega}_2)} \right| = \frac{E_1 E_2 E_3}{p_1^2 p_2^2 p_3^2} \frac{p_{3\text{body}}^2}{(E_1 + E_2 + E_3)} \frac{\widetilde{p}_{2\text{body}}}{\widetilde{M}_{3\text{body}}} \frac{\widetilde{p}_{1\text{body}}}{\widetilde{M}_{2\text{body}}}. \tag{A.5}$$

After the transformation, the differential decay width is written by:

$$\begin{aligned} \frac{d\Gamma_5}{dp_1 d\Omega_1 dp_2 d\Omega_2 dp_3 d\Omega_3} &\left| \frac{\partial(p_1, \Omega_1, p_2, \Omega_2, p_3, \Omega_3)}{\partial(p_{3\text{body}}, \Omega_{3\text{body}}, \widetilde{M}_{3\text{body}}^2, \widetilde{\Omega}_3, \widetilde{M}_{2\text{body}}^2, \widetilde{\Omega}_2)} \right| = \frac{e^4 |G_{ll'}|^2}{3 \cdot 2^{18} \cdot \pi^{10} M} \frac{p_{3\text{body}}^2}{(E_1 + E_2 + E_3)} \\ &\times \frac{\widetilde{p}_{2\text{body}}}{\widetilde{M}_{3\text{body}}} \frac{\widetilde{p}_{1\text{body}}}{\widetilde{M}_{2\text{body}}} [(Q_{LL} T_{\text{SM}} + Q_{RL} T_{RL}^Q + B_{RL} T_{RL}^B + L \leftrightarrow R) + I_\alpha \Re(T_\alpha^I) + I_\beta \Re(T_\beta^I)]. \end{aligned} \tag{A.6}$$

We introduce some expressions;

$$d\Gamma_{\text{SM}} = F T_{\text{SM}} \tag{A.7}$$
$$d\Gamma_{RL}^S = F T_{RL}^S \tag{A.8}$$
$$d\Gamma_{RL}^V = F T_{RL}^V \tag{A.9}$$
$$d\Gamma_{LRRL}^{SV} = F T_{LRRL}^{SV} \tag{A.10}$$
$$d\Gamma_{LLRR}^{SV} = F T_{LLRR}^{SV}. \tag{A.11}$$

Here $F = \frac{e^4|G_{ll'}|^2}{3\cdot 2^{18}\cdot\pi^{10}M}\frac{p_{3\text{body}}^2}{(E_1+E_2+E_3)}\frac{\widetilde{p}_{2\text{body}}}{\widetilde{M}_{3\text{body}}}\frac{\widetilde{p}_{1\text{body}}}{\widetilde{M}_{2\text{body}}}$. From the expressions of Eq. (A.3), we can calculate the constants $b \sim g$ through the calculation of integral of $d\Gamma^S_{RL}, d\Gamma^V_{RL}, d\Gamma^{SV}_{LRRL}$, and $d\Gamma^{SV}_{LLRR}$.

$$\begin{aligned}
d(b) &= d(d) = d\Gamma^Q_{RL}/\Gamma_{\text{SM}} = \frac{1}{3\Gamma_{\text{SM}}}(16d\Gamma^S_{RL} - d\Gamma^V_{RL}),\\
d(c) &= d(e) = d\Gamma^B_{RL}/\Gamma_{\text{SM}} = \frac{4}{3\Gamma_{\text{SM}}}(d\Gamma^V_{RL} - 4d\Gamma^S_{RL}),\\
d(f) &= d\Gamma^I_\alpha/\Gamma_{\text{SM}} = 4d\Gamma^{SV}_{LRRL}/\Gamma_{\text{SM}},\\
d(g) &= d\Gamma^I_\beta/\Gamma_{\text{SM}} = 2d\Gamma^{SV}_{LLRR}/\Gamma_{\text{SM}}.
\end{aligned} \tag{A.12}$$

This calculation is performed by using the similar formula to Eq. (5), for example,

$$b = \frac{\Gamma^Q_{RL}}{\Gamma_{\text{SM}}} = \frac{1}{N_{\text{gen}}}\sum_{\mathbf{x}\in\Omega}\frac{16d\Gamma^S_{RL}(\mathbf{x}) - d\Gamma^V_{RL}(\mathbf{x})}{3d\Gamma_{\text{SM}}(\mathbf{x})} \tag{A.13}$$

We generate each variable $(p_{3\text{body}}, \Omega_{3\text{body}}, \widetilde{M}_{3\text{body}}, \widetilde{\Omega}_3, \widetilde{M}_{2\text{body}}, \widetilde{\Omega}_2)$ randomly so that its variables follow the distribution of $d\Gamma_{\text{SM}}$.

Appendix B. Formula of Branching Fraction with Coupling Constants

The formalisms Eq. (6), (7), (8), and (9) can be rewritten to the formalism expressed through the coupling constants g^i_{jk} as follows.

$$\begin{aligned}
BR^{e^-e^+e^-\bar{\nu}_e\nu_\tau}_{\text{exp}} = BR^{e^-e^+e^-\bar{\nu}_e\nu_\tau}_{\text{SM}}\{&|g^V_{LL}|^2(1+\frac{|g^S_{LL}|^2}{4|g^V_{LL}|^2}) + (0.2501\pm0.0001)|g^S_{RL}|^2 + (0.8465\pm0.1073)|g^V_{RL}|^2\\
&+(2.693\pm0.215)|g^T_{RL}|^2 - (0.1540\pm0.1073)g^S_{RL}g^{T*}_{RL} + (0.4303\pm0.0001)g^S_{LL}g^{V*}_{RR}\\
&+(0.06039\pm0.00004)g^S_{LR}g^{V*}_{RL} + (0.3623\pm0.0002)g^V_{LR}g^{T*}_{RL} + L\leftrightarrow R\}.\\
BR^{\mu^-e^+e^-\bar{\nu}_\mu\nu_\tau}_{\text{exp}} = BR^{\mu^-e^+e^-\bar{\nu}_\mu\nu_\tau}_{\text{SM}}\{&|g^V_{LL}|^2(1+\frac{|g^S_{LL}|^2}{4|g^V_{LL}|^2}) + (0.2506\pm0.0001)|g^S_{RL}|^2 + (0.3484\pm0.1468)|g^V_{RL}|^2\\
&+(1.699\pm0.294)|g^T_{RL}|^2 - (0.6538\pm0.1468)g^S_{RL}g^{T*}_{RL} + (52.20\pm0.01)g^S_{LL}g^{V*}_{RR}\\
&+(45.33\pm0.01)g^S_{LR}g^{V*}_{RL} + (272.0\pm0.1)g^V_{LR}g^{T*}_{RL} + L\leftrightarrow R\}.\\
BR^{e^-\mu^+\mu^-\bar{\nu}_e\nu_\tau}_{\text{exp}} = BR^{e^-\mu^+\mu^-\bar{\nu}_e\nu_\tau}_{\text{SM}}\{&|g^V_{LL}|^2(1+\frac{|g^S_{LL}|^2}{4|g^V_{LL}|^2}) + (0.2536\pm0.0001)|g^S_{RL}|^2 + (0.3802\pm0.0001)|g^V_{RL}|^2\\
&+(1.775\pm0.001)|g^T_{RL}|^2 - (0.6342\pm0.0001)g^S_{RL}g^{T*}_{RL} + (0.2616\pm0.0001)g^S_{LL}g^{V*}_{RR}\\
&+(0.05633\pm0.00001)g^S_{LR}g^{V*}_{RL} + (0.3380\pm0.0001)g^V_{LR}g^{T*}_{RL} + L\leftrightarrow R\}.\\
BR^{\mu^-\mu^+\mu^-\bar{\nu}_\mu\nu_\tau}_{\text{exp}} = BR^{\mu^-\mu^+\mu^-\bar{\nu}_\mu\nu_\tau}_{\text{SM}}\{&|g^V_{LL}|^2(1+\frac{|g^S_{LL}|^2}{4|g^V_{LL}|^2}) + (0.2512\pm0.0001)|g^S_{RL}|^2 + (0.3704\pm0.0001)|g^V_{RL}|^2\\
&+(1.745\pm0.015)|g^T_{RL}|^2 - (0.6344\pm0.0004)g^S_{RL}g^{T*}_{RL} + (98.67\pm0.01)g^S_{LL}g^{V*}_{RR}\\
&-(4.510\pm0.001)g^S_{LR}g^{V*}_{RL} - (27.060\pm0.006)g^V_{LR}g^{T*}_{RL} + L\leftrightarrow R\}.
\end{aligned} \tag{B.1}$$

Appendix C. Particle Identification (PID) in the Selection

To identify electron and muon, we use following method. A likelihood ratio cut $\mathcal{P}_\mu = \mathcal{L}_\mu/(\mathcal{L}_\mu + \mathcal{L}_\pi + \mathcal{L}_K) > 0.7$ is applied to select muons [10]. To identify electrons the likelihood ratio parameter $\mathcal{P}_e = \mathcal{L}_e/(\mathcal{L}_e + \mathcal{L}_{\text{other}})$ is constructed [11]. To select $\tau^- \to e^-e^+e^-\bar{\nu}_e\nu_\tau$ events, we require $\mathcal{P}_{e^-} > 0.7$ and $\mathcal{P}_{e^-}, \mathcal{P}_{e^+} > 0.5$. For the $\tau^- \to \mu^-e^+e^-\bar{\nu}_\mu\nu_\tau$ events $\mathcal{P}_{e^-}, \mathcal{P}_{e^+} > 0.5$; while for the $\tau^- \to e^-\mu^+\mu^-\bar{\nu}_e\nu_\tau$ events $\mathcal{P}_{e^-} > 0.7$.

References

[1] A. Flores-Tlalpa, G. Lopez Castro and P. Roig, JHEP **1604**, 185 (2016) [arXiv:1508.01822 [hep-ph]].
[2] W. Fetscher, H. J. Gerber and K. F. Johnson, Phys. Lett. B 173, 102 (1986).
[3] M. Fael and C. Greub, arXiv:1611.03726 [hep-ph].
[4] G. M. Pruna, A. Signer and Y. Ulrich, arXiv:1611.03617 [hep-ph].
[5] Private communication of Matteo Fael.
[6] S. Jadach, B. F. L. Ward and Z. Was, Comput. Phys. Commun. **130** (2000) 260.
[7] Z. Was, Nucl. Phys. Proc. Suppl. **98**, 96 (2001).
[8] R. Brun *et al*, GEANT 3.21, CERN Report No. DD/EE/84-1 (1984).
[9] M. S. Alam *et al*, Phys. Rev. Lett. 76, 2637 (1996).
[10] A. Abashian *et al.*, Nucl. Instr. Meth. A **491**, 69 (2002).
[11] K. Hanagaki *et al.*, Nucl. Instr. Meth. A **485**, 490 (2002).

IOP Conf. Series: Journal of Physics: Conf. Series **912** (2017) 012003 doi:10.1088/1742-6596/912/1/012003

Hadronic tau decays and the strong coupling

Diogo Boito,[a] Maarten Golterman,[1b] Kim Maltman[c,d] and Santiago Peris[e]

[a]Instituto de Física de São Carlos, Universidade de São Paulo, CP 369 13570-970, São Carlos, SP, Brazil
[b]Department of Physics and Astronomy, San Francisco State University San Francisco, CA 94132, USA
[c]Department of Mathematics and Statistics, York University Toronto, ON Canada M3J 1P3
[d]CSSM, University of Adelaide, Adelaide, SA 5005 Australia
[e]Department of Physics and IFAE-BIST, Universitat Autònoma de Barcelona E-08193 Bellaterra, Barcelona, Spain

E-mail: maarten@sfsu.edu

Abstract. We provide an overview of recent determinations of the strong coupling, α_s, from hadronic τ decays. We contrast two analysis methods, the "truncated-OPE approach" and the "DV-model approach," highlighting the assumptions going into each of these. We argue that a detailed study based on ALEPH data shows the truncated-OPE approach to be quantitatively unreliable, while the DV-model approach passes all tests. New data for hadronic τ decays from Belle and Belle-II could provide more stringent tests of the DV-model approach, and thus potentially lead to a more precise value of α_s from hadronic τ decays.

1. Introduction

Let us begin with an overview of values for the strong coupling, $\alpha_s(m_\tau^2)$ obtained in the most recent analyses of hadronic τ decays. Taking the averages of "FOPT" and "CIPT" values,[2] the most recent values obtained from ALEPH data [2, 3] are

$$\begin{aligned} \alpha_s(m_\tau^2) &= 0.332(12)\ , && [3]\ , \\ &= 0.328(12)\ , && [4]\ , \\ &= 0.301(10)\ , && [5]\ , \\ &= 0.309(9)\ , && [5] \text{ including OPAL data [6, 7]}\ . \end{aligned} \tag{1}$$

While the values of Refs. [3] and [4] are in good agreement, there is a significant discrepancy with the values from Ref. [5]. It is interesting to also quote some values from the PDG [8],

[1] Speaker at conference
[2] FOPT stands for fixed-order perturbation theory, and CIPT for contour-improved perturbation theory; for a review, see for instance Ref. [1]. For comparison we use these averaged values in this section; however, we do not recommend using such averages in general, as this averaging is without theoretical basis.

adjusted to the τ mass:

$$\begin{aligned} \alpha_s(m_\tau^2) &= 0.325(15)\ , && \text{from } \tau \text{ decays}\ , \\ &= 0.315(9)\ , && \text{from lattice QCD}\ , \\ &= 0.314(9)\ , && \text{world average without } \tau\ . \end{aligned} \tag{2}$$

While these values are nominally in agreement with each other, the value from τ decays is on the high side, and this is clearly related to the higher values in Eq. (1), which push up the τ-based PDG average. Two natural questions arise regarding the average from τ decays: (i) is the average actually reliable, and (ii) can we do better? We address these questions in what follows.

As we will see, the differences originate in the way the non-perturbative "contamination" afflicting the determination of α_s from τ decays is treated in the analysis. While the value of the τ mass is large enough that one may attempt to apply QCD perturbation theory, it is sufficiently low that such a determination is not free from non-perturbative effects. Here, we will compare the analysis methods used in Refs. [3, 4, 5], and argue that the different assumptions underlying these methods lead to the different values reported in Eq. (1). We will argue that the method used in Refs. [3, 4] is unreliable, while the method used in Ref. [5] appears to be consistent with the data. However, also Ref. [5] employs an assumption in order to deal with the non-perturbative contamination. While tests based on ALEPH and OPAL data support the validity of this assumption, more precise data, as for instance obtained from Belle and Belle-II, would make it possible to subject also the method of Ref. [5] to more stringent consistency tests.

2. Background

The ratio of the decay width for non-strange hadronic τ decays to the width for $\tau \to \nu_\tau e \overline{\nu}_e$ can be calculated in QCD perturbation theory, and takes the form[3]

$$R_\tau^{ud} \equiv \frac{\Gamma(\tau \to \text{hadrons}_{ud})}{\Gamma(\tau \to \nu_\tau e \overline{\nu}_e)} = 3 S_{\text{EW}} |V_{ud}|^2 \left(1 + \frac{\alpha_s}{\pi} + 5.2 \left(\frac{\alpha_s}{\pi}\right)^2 + \dots\right)\ , \tag{3}$$

where S_{EW} is a short-distance electroweak correction. The perturbative series has been calculated to order α_s^4 [9]. In principle, this result can be used to determine α_s from the vector (V) and axial-vector (A) non-strange decays, for which ALEPH and OPAL obtained relatively precise data. However, at $\alpha_s(m_\tau^2)/\pi \approx 0.1$, the series is rather slowly converging, indicating that QCD perturbation theory may not be all there is at the τ mass. On the experimental side, what one sees is not $\tau \to \nu_\tau$jets, but rather (in the V channel) $\tau \to \nu_\tau \rho^* \to \nu_\tau$pions with $\rho^* = \rho(770)$, $\rho(1450)$, *etc.*, calling into question the relation between observed hadronic τ decays and the perturbative regime.

The finite-energy sum-rule (FESR) approach allows us to look into this more systematically [10]. First, using the optical theorem, R_τ^{ud} can be expressed in terms of the $I = 1$ vacuum polarization $\Pi_{V+A}(q^2)$ as [11]

$$R_\tau^{ud} = 12\pi^2 S_{\text{EW}} |V_{ud}|^2 \int_0^{m_\tau^2} \frac{ds}{m_\tau^2} \left(1 - \frac{s}{m_\tau^2}\right)^2 \left(1 + \frac{2s}{m_\tau^2}\right) \rho_{V+A}(s)\ , \tag{4}$$

where $\rho_{V+A}(s) = \frac{1}{\pi}\text{Im}\ \Pi_{V+A}(s)$ is the non-strange inclusive $V + A$ spectral function measured in hadronic τ decays. Since in this talk we will always consider the case $V + A$ we drop the

[3] In FOPT.

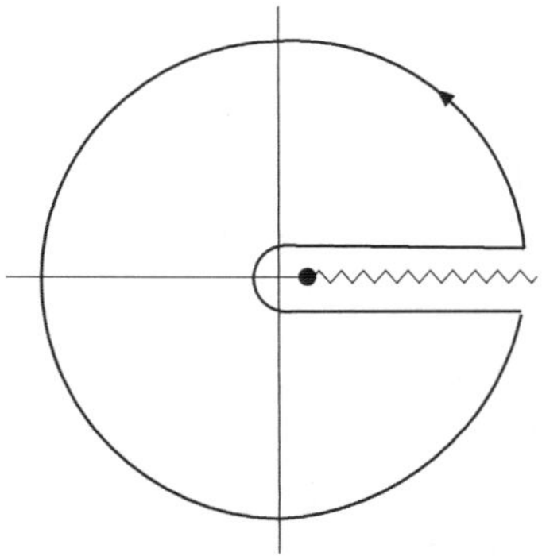

Figure 1. *Contour used in Eq. (5).*

subscript $V+A$ from now on. Second, Cauchy's theorem, applied to the contour shown in Fig. 1, implies that[4]

$$\int_0^{s_0} ds\, w(s)\, \rho(s) = -\frac{1}{2\pi i}\oint_{|z|=s_0} dz\, w(z)\, \Pi(z)\ , \tag{5}$$

where $w(s)$ can be any analytic function. The idea is now to split $\Pi(z=q^2)$ into a perturbative part $\Pi^{\rm pert}$ and non-perturbative parts,

$$\Pi(z) = \Pi^{\rm pert}(z) + \Pi^{\rm OPE}(z) + \Pi^{\rm DV}(z)\ , \tag{6}$$

and substitute this into the contour integral on the right-hand side of Eq. (5). Here $\Pi^{\rm OPE}$ is given by the operator product expansion (OPE),[5]

$$\Pi^{\rm OPE}(q^2) = \frac{C_4}{q^4} - \frac{C_6}{q^6} + \frac{C_8}{q^8} + \dots\ , \tag{7}$$

and $\Pi^{\rm DV}$ represents, by definition, the non-perturbative part not encapsulated by the OPE.[6] Neglecting this latter part, and truncating the OPE, for $w(x)$ a polynomial of degree N the right-hand side then depends on a finite number of parameters, α_s and C_{2k} for $k = 2,\ 3,\ \dots,\ k_{max} = 2N+2$. These parameters can then be fit to the integral on the left-hand side, computed from the data for the spectral function, as a function of s_0.

The physics not represented by the OPE is the physics of resonances, which are visible in the data along the positive real $z = q^2$ axis in the spectral function, *cf.* Sec. 3 below. In other words, the OPE does not provide a reliable representation of the non-perturbative part of $\Pi(z)$ in the region where the circle with radius s_0 crosses the positive real axis in Fig. 1. As we will see below, this leads to the need to introduce a non-trivial *ansatz* for $\Pi^{\rm DV}(z)$, in order to take into account the resonance contributions to $\Pi(q^2)$. Under mild conditions on $\rho^{\rm DV}(s) = \frac{1}{\pi}{\rm Im}\,\Pi^{\rm DV}(s)$, Eq. (5) can be rewritten as [13, 14]

$$\int_0^{s_0} ds\, w(s)\, \rho(s) = -\frac{1}{2\pi i}\oint_{|z|=s_0} dz\, w(z)\, \left(\Pi^{\rm pert}(z) + \Pi^{\rm OPE}(z)\right) - \int_{s_0}^{\infty} ds\, w(s)\, \rho^{\rm DV}(s)\ . \tag{8}$$

If we supply a parametrization for $\rho^{\rm DV}(s)$ in terms of a finite number of parameters, this constitutes the FESR we will apply to determine $\alpha_s(m_\tau^2)$ (one of the parameters of the "theory"

[4] The same equation holds for V and A channels separately.

[5] The coefficients C_{2k} are logarithmically dependent on q^2, but this dependence can be numerically neglected in the application to τ decays [12]. For the non-strange channel, C_2 is negligibly small, being proportional to the squares of the light quark masses.

[6] In the literature, the perturbative part is often defined to be the leading-order term of the OPE. Here we will use "OPE" to refer to power-correction terms (7) only.

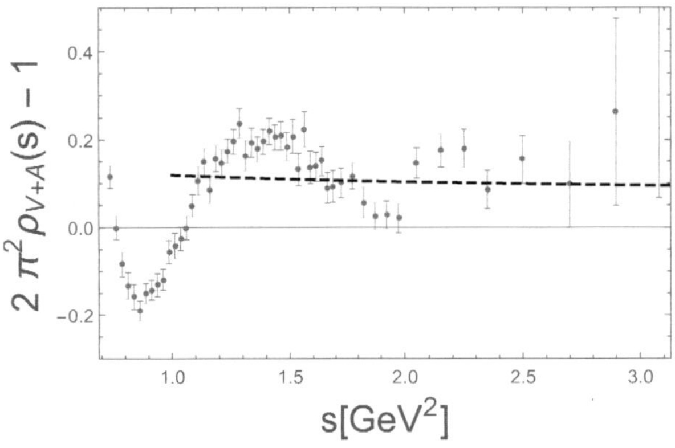

Figure 2. *Blow-up of the ALEPH data (red data points) in the large-s region of the $V + A$ non-strange spectral function (s in* GeV^2*). Black dashed line: perturbation theory (CIPT) contributions to the spectral function. The α_s-independent parton-model contribution to $\rho(s)$ has been subtracted from both.*

part on the right-hand side) from the experimental data, which allow us to compute the integrals on the left-hand side.[7] The representation on the right-hand side is valid for s_0 sufficiently large, and the hope is that it already applies for s_0 (not too far) below m_τ^2.

3. Data

Plots showing the V and A non-strange spectral functions can be found in Fig. 5 of Ref. [3] and Fig. 5 of Ref. [6]. First, these figures suggest that, short of a complete determination of the inclusive spectral functions from Belle and Belle-II data, a determination of the 4π decay modes might already significantly reduce the error on the inclusive spectral functions if combined with the ALEPH data, given the vastly increased statistics of the Belle experiments. Of course, a complete determination of the inclusive spectral function from Belle data is highly desirable as well, given the much higher statistics, and would avoid potential pitfalls with combining data from different experiments.

Second, it is interesting to consider the data from the point of view of what they indicate about non-perturbative effects, in particular, about resonance effects. In Fig. 2, we show the high-s part of the $V + A$ non-strange spectral function measured by ALEPH. What is shown in this figure is $2\pi^2\rho_{V+A} - 1$, *i.e.*, the dynamical QCD contribution to the spectral distribution, subtracting the contribution from QCD without gluons (the parton model). The black, dashed curve shows the theoretical representation of the spectral function without "duality violations," obtained from the right-hand side of Eq. (8) setting $\rho^{\mathrm{DV}} = 0$. The figure shows that the oscillations of the data, due to the presence of resonances, are not small compared to the contribution from QCD perturbation theory, and thus need to be taken into account when fitting the data using Eq. (8).

We will turn to the analysis of the data using the two approaches, the "truncated-OPE" approach and the "DV-model" approach, next.

4. Truncated-OPE approach

In the truncated-OPE approach [15], employed in Refs. [3, 4], one always takes $s_0 = m_\tau^2$, and employs the weights ($x = s/s_0$)

$$\begin{aligned} w_{00}(x) &= (1-x)^2(1+2x) , \\ w_{10}(x) &= (1-x)^3(1+2x) , \\ w_{11}(x) &= (1-x)^3(1+2x)x , \end{aligned} \tag{9}$$

[7] The parameters in Π^{OPE} and Π^{DV} can be viewed as "nuisance" parameters in these fits, although they are of physical interest in their own right.

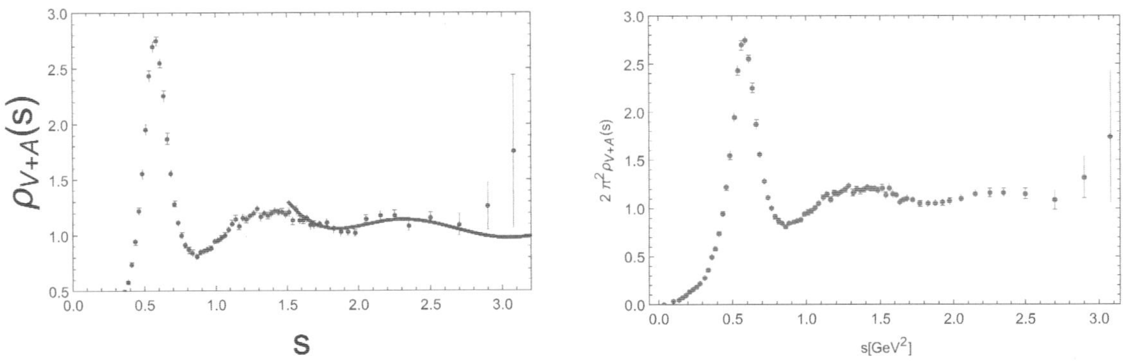

Figure 3. *Left figure: true data, with the model curve for $s \geq 1.55$ GeV^2. Right figure: fake data for $s \geq 1.55$ GeV^2, true data for $s < 1.55$ GeV^2.*

$$
\begin{aligned}
w_{12}(x) &= (1-x)^3(1+2x)x^2 \,, \\
w_{13}(x) &= (1-x)^3(1+2x)x^3 \,.
\end{aligned}
$$

One assumes that only $C_{4,6,8}$ are non-negligible, and that duality violations, represented by $\rho^{\rm DV}$ in Eq. (8) can be neglected. We note that the weights in Eq. (9) all suppress the region near $s = s_0$ by a double or triple zero, which supports (but does not necessarily validate) the assumption that duality violations can be neglected. However, since, from the residue theorem,

$$\frac{1}{2\pi i}\oint dz\, z^n \frac{C_{2k}}{z^k} = C_{2n+2}\,\delta_{k,n+1} \,, \tag{10}$$

we see that, using the weights (9), which have values of n up to 7, we would need to keep terms in the OPE up to C_{16}, since $2k = 2(n+1) = 16$ for $n = 7$. The truncated-OPE approach thus neglects, without theoretical foundation, the OPE coefficients $C_{10,12,14,16}$. Of course, the practical reason for doing this is that at $s_0 = m_\tau^2$, only 5 data integrals are available, so that a maximum of 4 parameters, α_s, C_4, C_6 and C_8, can be fit. The truncated-OPE approach thus has two shortcomings: the truncation of the OPE, and the neglect of resonance oscillations around perturbation theory plus the OPE clearly visible in the $V + A$ spectral function, *cf.* Fig. 2.

This approach can be tested, using a set of fake data based on a model with known values for the parameters [16]. Using a fit from Ref. [5], which describes the V, A and thus $V + A$ spectral functions very well, one can use this fit as a model to generate fake data from a gaussian distribution with the model as central values, and the real-world, ALEPH covariance matrix. Since the model is only valid for $s \geq 1.55$ GeV^2, we generate fake data only in this region, and use the true data for $s < 1.55$ GeV^2. The model has $\alpha_s(m_\tau^2) = 0.312$ (CIPT), and known OPE coefficients (for details, we refer to Ref. [16]).

ALEPH	$\alpha_s(m_\tau^2)$	C_4 (GeV^4)	C_6 (GeV^6)	C_8 (GeV^8)	χ^2/dof
true values	0.312	0.0027	-0.013	0.035	
fake data fit	0.334(4)	-0.0024(4)	0.0007(3)	-0.0008(4)	0.95/1

Table 1. *CIPT fits employing the truncated-OPE approach on the fake data, based on the weights of Eq. (9). By assumption, $C_{10} = C_{12} = C_{14} = C_{16} = 0$. Errors are statistical only.*

In Table 1 we show the results of fitting the fake data with the truncated-OPE approach. We see that the truncated-OPE approach gets it wrong, and leads to a systematic overestimate of the value of $\alpha_s(m_\tau^2)$, despite an excellent value for the χ^2 per degree of freedom. In addition, we also note that there is a large difference between the fitted and true values of the OPE coefficients: clearly, the fit is trying to compensate for the "missing" contributions from the higher-dimension OPE coefficients and/or duality violations that have, by fiat, been set equal to zero.

Reference [4] considered many other choices of sets of weights, in addition to the "ALEPH" weights shown in Eq. (9). In particular, another set of "optimal" weights was considered, given by

$$\begin{aligned} w^{(2,1)}(x) &= 1-3x^2+2x^3 , \\ w^{(2,2)}(x) &= 1-4x^3+3x^4 , \\ w^{(2,3)}(x) &= 1-5x^4+4x^5 , \\ w^{(2,4)}(x) &= 1-6x^5+5x^6 , \\ w^{(2,5)}(x) &= 1-7x^6+6x^7 . \end{aligned} \tag{11}$$

These weights differ from the ALEPH weights of Eq. (9) in that no linear term in x appears in these weights. Therefore, they do not involve C_4, allowing one to fit also C_{10} using the same strategy. This reduces the reliance on truncation of the OPE by one order, although the OPE terms proportional to C_{12}, C_{14} and C_{16} still need to be truncated in this approach.

optimal	$\alpha_s(m_\tau^2)$	C_6 (GeV^6)	C_8 (GeV^8)	C_{10} (GeV^{10})	χ^2/dof
true values	0.312	0.0027	-0.013	0.035	
fake data fit	0.334(4)	0.0008(4)	-0.0008(5)	0.0001(3)	0.92/1

Table 2. *CIPT fits employing the truncated-OPE approach on the fake data, based on the weights of Eq. (11). By assumption, $C_{12} = C_{14} = C_{16} = 0$. Errors are statistical only.*

Table 2 shows the result of the fake-data test. The conclusion is the same as before: while the fit results are completely consistent between Tables 1 and 2, they again disagree with the true model values. A much more detailed analysis of this problem was given in Ref. [16], with the conclusion that the truncated-OPE approach is inherently unreliable, and thus should no longer be used.

5. DV-model approach

Our conclusion is that the truncation of the OPE, employed in the approach described in the previous section, leads to a systematic bias of the value for α_s (as well as the OPE coefficients) obtained in that approach. Already based on earlier investigations raising doubts about the truncated-OPE approach [17, 18], we employed a simpler choice of weights,[8]

$$\begin{aligned} w_0(x) &= 1 , \\ w_2(x) &= 1-x^2 , \\ w_3(x) &= (1-x)^2(1+2x) . \end{aligned} \tag{12}$$

It follows from Eq. (10) that only the coefficients C_6 and C_8 need to be kept in the OPE.[9] However, the first weight does not suppress the region $s \approx s_0$ at all, and the second weight does

[8] Note that the sets of weights considered here share $w_{00}(x) = w^{(2,1)}(x) = w_3(x)$.

[9] This assumes no logarithmic corrections to the OPE coefficients. These are suppressed by powers of α_s, and found to be numerically negligible [12].

this only linearly. It is thus even less likely than in the truncated-OPE approach that duality violations can be ignored. We introduce an *ansatz* for $\rho^{\rm DV}(s)$ in Eq. (8); in each of the channels, V and A, we assume that, for large enough s,

$$\rho^{\rm DV}(s) = e^{-\delta-\gamma s}\sin(\alpha+\beta s)\ , \tag{13}$$

thus introducing 8 new parameters, $\delta_{V,A}$, $\gamma_{V,A}$, $\alpha_{V,A}$ and $\beta_{V,A}$ to the fits. Clearly, we cannot restrict ourselves to $s_0 = m_\tau^2$; instead, we let s_0 vary with a minimum value s_{min}, to be determined by the quality of the fits. It turns out that the optimal value is $s_{min} \approx 1.55$ GeV^2, with good stability for small variations of this value.

Of course, the use of Eq. (13) implies that this approach is model dependent. This is unavoidable, since the physics of resonances is not captured by QCD perturbation theory or the OPE, while it is clearly visible in the data (*cf.* Fig. 2). The ingredients going into the construction of the model are generally thought of as approximate properties of QCD. One starts with assuming Regge behavior of the spectrum for $s > s_{min}$, which, in the large-N_c limit implies that the spectrum in each channel looks like $M^2(n) = M^2(0)+\sigma n$, with $n = 0, 1, 2, \ldots$ and $\sigma \sim 1$ GeV^2. Relaxing the large-N_c limit, these poles pick up an imaginary part leading to a decay width $\Gamma(n) \propto M(n)/N_c$, providing a more realistic picture of the resonance spectrum in each channel. A model which incorporates all these features and satisfies the analyticity constraints of QCD everywhere in the complex plane exists [13, 14, 19, 20, 21], and leads to the *ansatz* (13).

While this approach, the "DV-model" approach, is model dependent, it should be stressed that also the truncated-OPE approach is model dependent in the same sense: it assumes an even simpler model for duality violations, by simply setting $\rho^{\rm DV}(s) = 0$. Since the data clearly show the presence of duality violations, this choice of model is almost certainly worse than the choice (13). Pinching, *i.e.* the suppression of the region $s \approx s_0$ by the choice of weights (9) or (11) may help, but without an explicit model for duality violations this cannot be checked. As we have shown, internal consistency between different sets of weights in the truncated-OPE approach is not sufficient.

Within this framework, many different fits can be performed. One can choose any subset of the weights (12), fit only the V channel or the combined V and A channels, or just the sum $V + A$. A large suite of fits has been considered in Refs. [5, 7, 12], applying this DV-model approach to both the OPAL and ALEPH data. The results of these fits show excellent internal consistency.

Figure 4 shows an example of the simplest possible fit based on the DV-model approach, by considering only the V channel, and only the weight $w_0(x) = 1$. There is excellent agreement between the data and the fits, shown in the left panel. The spectral function following from this fit is shown in the right panel, along with the ALEPH data. Another example of a fit following this approach, now involving all weights in Eq. (12), again considering only the V channel, is shown in Fig. 5.

For many other fits, also involving the A channel, and many consistency checks, we refer to Ref. [5]. The results we obtain, from the ALEPH data, for α_s are

$$\begin{aligned} \alpha_s(m_\tau^2) &= 0.296(10)\ , \quad \text{FOPT}\ , &(14)\\ &= 0.310(14)\ , \quad \text{CIPT}\ . &(15)\end{aligned}$$

The average of these leads to the third value reported in Eq. (1); averaging this with the OPAL-based value obtained in Ref. [7] yields the fourth value reported in this equation.

A test that is of particular interest is to compare the moments with weights (9) between the fits, which allow us to determine all OPE coefficients [5], and the data. These comparisons are shown in Fig. 6. The red points show the data, *i.e.*, the left-hand side of Eq. (8) ("spectral

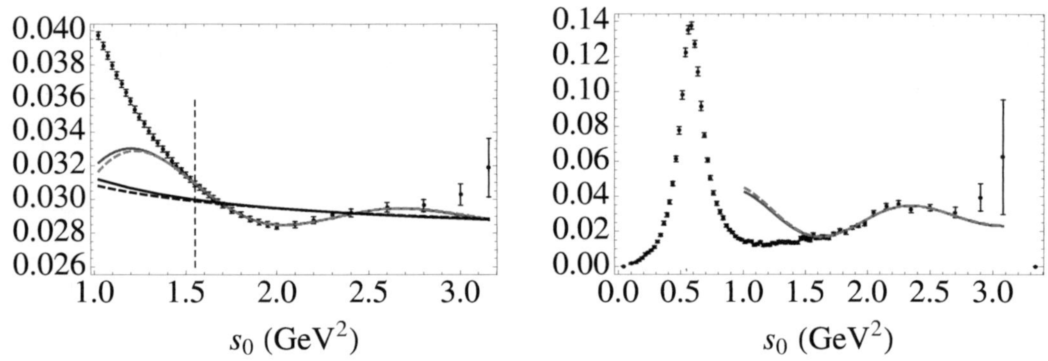

Figure 4. *Left panel: comparison of V-channel, w_0 fits to the data. Right panel: comparison of the theoretical spectral function resulting from this fit with the experimental results. CIPT fits are shown in red (dashed) and FOPT in blue (solid). The (much flatter) black curves in the left panel represent the OPE parts of the fits, i.e., the fit results with the DV parts removed. The vertical dashed line indicates the location of $s_{\rm min}$. From Ref. [5].*

integrals"), while the black curves show the fitted theory, *i.e.*, the right-hand side of Eq. (8) ("OPE+DV integrals"). What is actually shown is the value of these moments evaluated at $s_0 = m_\tau^2$ minus the moment at varying s_0, in order to account for the strong correlations between both the theory integrals at different s_0 and the data integrals at different s_0. We note that there is good agreement for all s_0 shown in the figures.

In Fig. 7 we show the same comparison, but now using the fit results obtained with the truncated-OPE approach, where all moments shown were employed in the fit. The black curves now only show the perturbative plus OPE contribution ("OPE integrals"), because duality violations, represented by the last term on the right-hand side of Eq. (8), are assumed negligible in the truncated-OPE approach. The agreement between data and fitted theory is much less good than for the DV-model approach, for which the corresponding comparisons are shown in Fig. 6. Of course, near $s_0 = m_\tau^2$ data and fitted theory agree, because of the differences which are shown. Differences between theory and experimental integrals are, of course, by definition, zero at $s_0 = m_\tau^2$; but deviation of the theory curves from the data at lower s_0 signals shortcomings in the theoretical representation. For the moments with weights w_{12} and w_{13}, the agreement already deteriorates well above $s_0 = 2$ GeV^2. For more detail, we refer to Ref. [16].

6. Discussion and outlook

The determination of the strong coupling, α_s, from hadronic τ decays is interesting for at least two reasons. First, it provides a value of α_s at the Z mass which is quite competitive in comparison with other methods. The reason for this is that because of the running, relatively modest errors get reduced when the obtained values for α_s are converted to values at the Z mass. Second, because it is at present the only determination with rather high precision at a rather low energy, it provides a nice test of the running of the strong coupling predicted by QCD. With the presently available precision, the QCD prediction is nicely confirmed, as indicated, for instance, by the values shown in Eq. (2).

However, as we have discussed in detail, the determination from τ decays is afflicted by contamination from non-perturbative effects, because of the rather low value of the τ mass. Some of these non-perturbative effects are captured by the OPE, but not all: resonance effects are

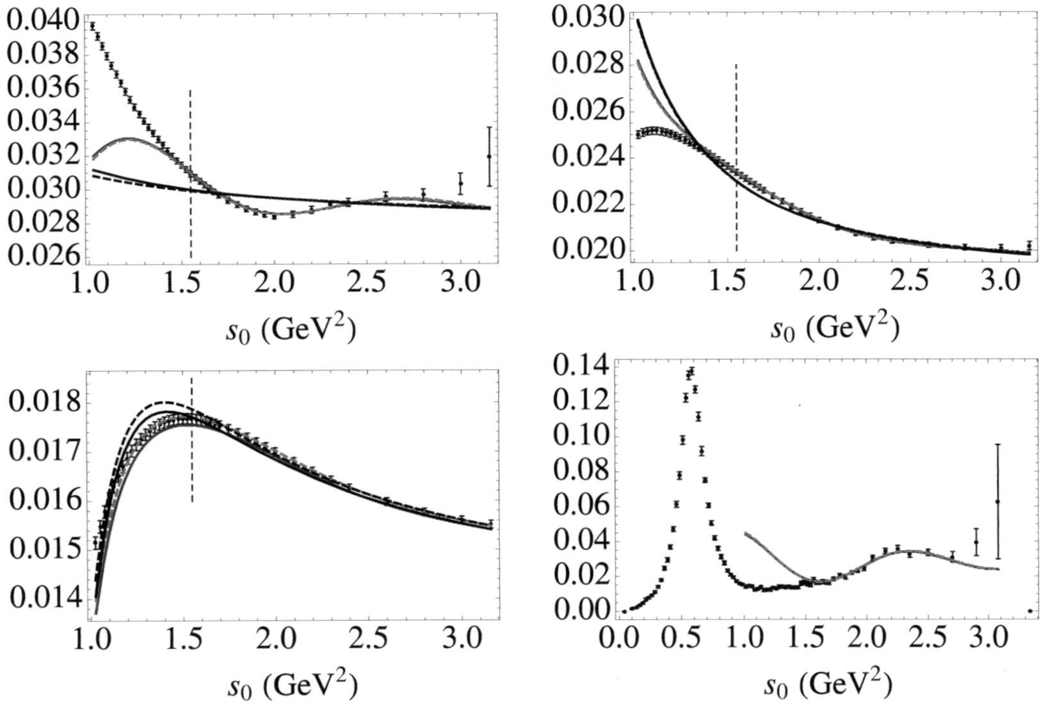

Figure 5. *Comparison of V channel fits using the weights w_0 (upper left panel), w_2 (upper right panel) and w_3 (lower left panel). CIPT fits are shown in red (dashed) and FOPT in blue (solid). Lower right panel: comparison of the theoretical spectral function resulting from this fit with the experimental results. The black curves (which are much flatter for the w_0 case) represent the OPE parts of the fits. The vertical dashed line indicates the location of $s_{\rm min}$. From Ref. [5].*

clearly important, and appear as significant deviations from the quark-gluon picture underlying perturbation theory and the OPE. Another, probably related, issue is the discrepancy between values obtained using CIPT or FOPT, which correspond to two different resummation schemes.[10]

Two methods for extracting α_s from τ decays have been pursued. One of these, the truncated-OPE approach, has now been shown to fail; it cannot, in general, account for all non-perturbative effects that afflict the analysis. The other approach is based on a model for duality violations, and has thus far been shown to be internally consistent, and consistent with the data. However, it is important to test this approach more stringently, and much more precise data for the V channel would provide the possibility to carry out such tests, and, if successful, lead to a possible reduction in the systematic error.[11] With thousands times more τ pairs, inclusive spectral functions obtained from Belle and Belle-II data would almost certainly have a significant impact on our understanding of QCD from hadronic τ decays. Even a precise determination of the $\tau \to \nu_\tau 4\pi$ channels might already have an impact, as the large errors in the large-s region on the currently available V spectral functions are dominated by these channels. In this respect, we believe that it will suffice to focus on the non-strange V channel, since, contrary to popular

[10] Other resummation schemes are possible, and have been considered.
[11] For other ideas on possible tests, see Ref. [22].

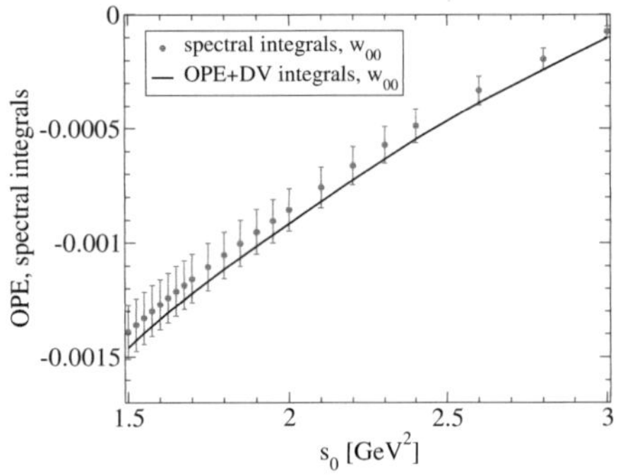

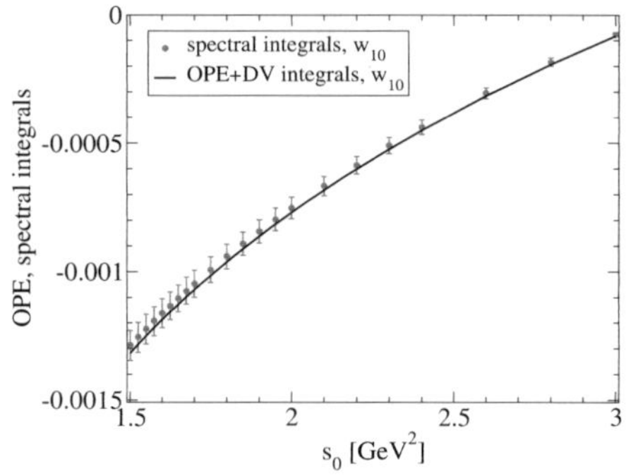

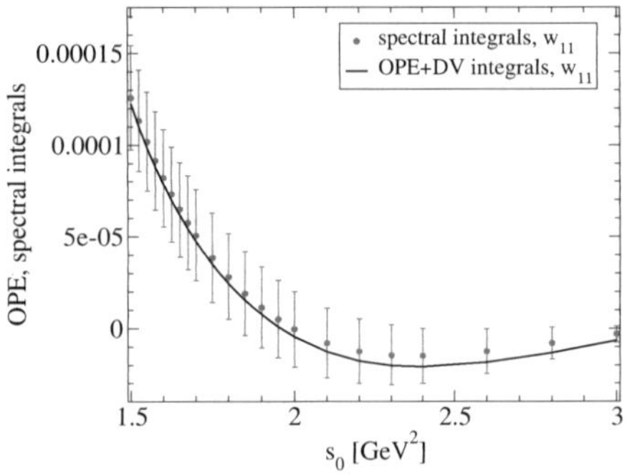

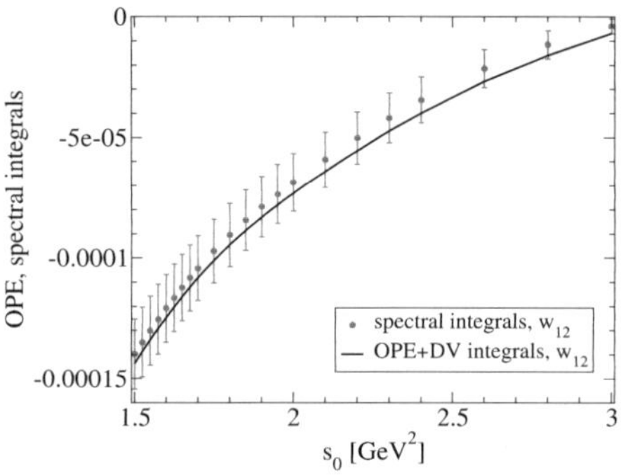

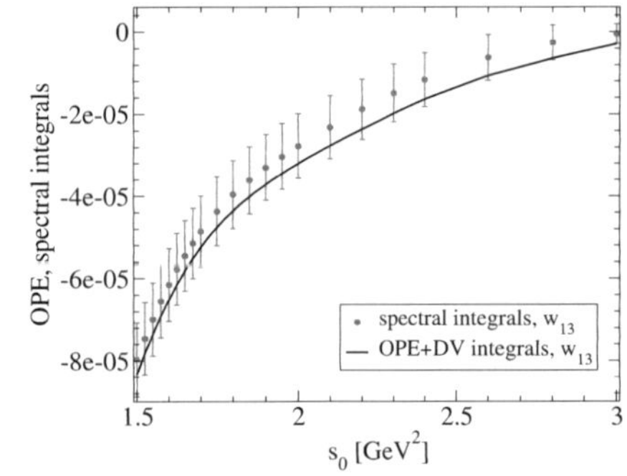

Figure 6. *Comparison between data and fitted theory for the moments of Eq. (9). Fits using the DV-model approach. Shown are differences between the values at $s_0 = m_\tau^2$ minus the values at varying s_0.*

belief, there is no indication that a higher precision can be obtained from $V + A$. It is not unlikely that the asymptotic regime where the right-hand side of Eq. (8) is valid sets in at lower values of s in the V than in the A channel.

It is more difficult to make progress on the optimal resummation scheme. The typical difference between values of $\alpha_s(m_\tau^2)$ obtained from FOPT or CIPT is always about 5%, with CIPT providing the higher value, irrespective of the analysis approach employed. Theoretical considerations based on a renormalon analysis somewhat favor FOPT [23, 24], but other resummation schemes cannot be ruled out [25]. A direct comparison between the value of α_s from τ decays and from the lattice, for instance, tends to favor the CIPT-based value. Again, higher precision determinations based on higher precision data may contribute to a deeper understanding.

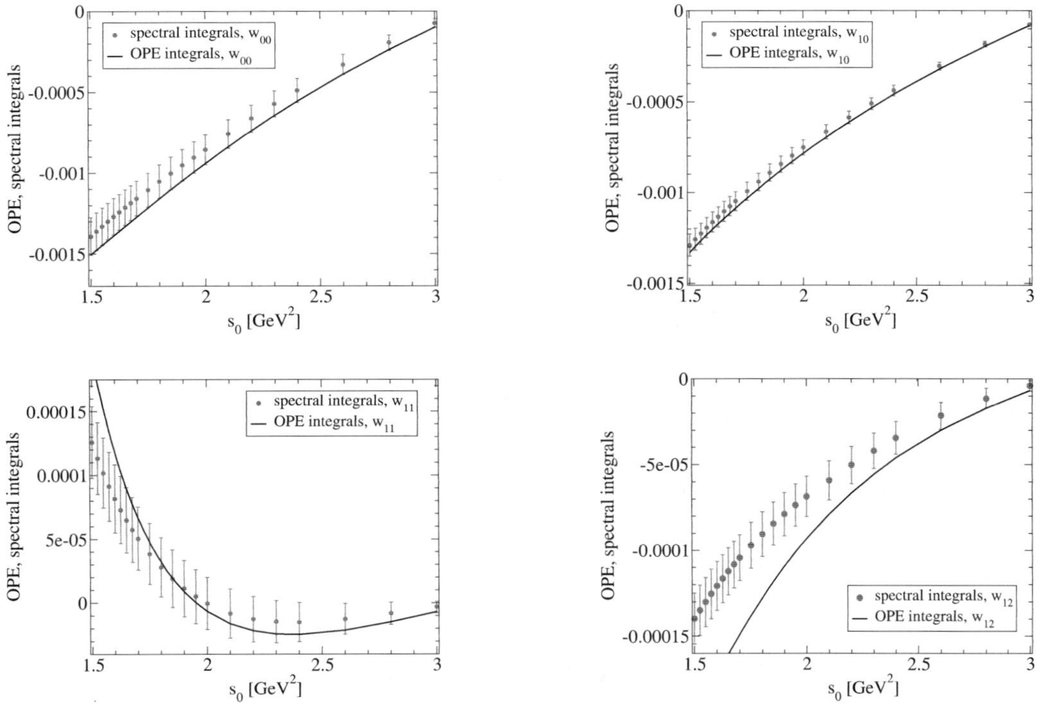

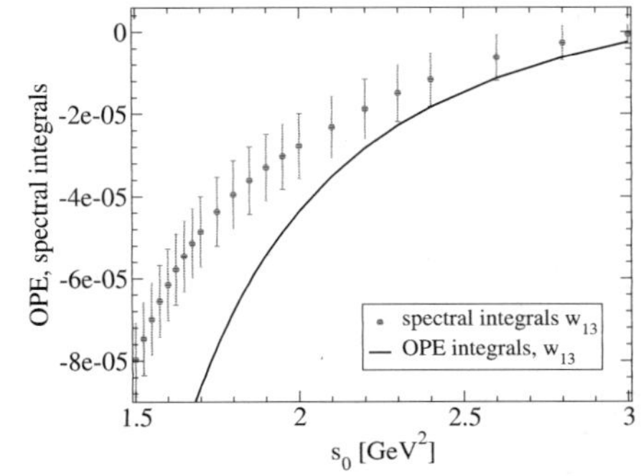

Figure 7. *Comparison between data and fitted theory for the moments of Eq. (9). Fits using the truncated-OPE approach. Shown are differences between the values at* $s_0 = m_\tau^2$ *minus the values at varying* s_0.

Acknowledgments

MG would like to thank the organizers of the Mini Workshop on Tau Physics for the opportunity to present this work, and for warm hospitality. The work of DB is supported by the São Paulo Research Foundation (FAPESP) Grant No. 2015/20689-9 and by CNPq Grant No. 305431/2015-3. This material is based upon work supported by the U.S. Department of Energy, Office of Science, Office of High Energy Physics, under Award Number DE-FG03-92ER40711 (MG). KM is supported by a grant from the Natural Sciences and Engineering Research Council of Canada. SP is supported by CICYTFEDER-FPA2014-55613-P, 2014-SGR-1450.

References

[1] M. Jamin, JHEP **0509**, 058 (2005) [hep-ph/0509001].
[2] R. Barate *et al.* [ALEPH Collaboration], Eur. Phys. J. C **4**, 409 (1998); S. Schael *et al.* [ALEPH Collaboration], Phys. Rept. **421**, 191 (2005) [arXiv:hep-ex/0506072];

[3] M. Davier, A. Hoecker, B. Malaescu, C. Z. Yuan and Z. Zhang, Eur. Phys. J. C **74**, 2803 (2014) [arXiv:1312.1501 [hep-ex]].
[4] A. Pich and A. Rodríguez-Sánchez, Phys. Rev. D **94**, no. 3, 034027 (2016) [arXiv:1605.06830 [hep-ph]].
[5] D. Boito, M. Golterman, K. Maltman, J. Osborne and S. Peris, Phys. Rev. D **91**, 034003 (2015) [arXiv:1410.3528 [hep-ph]].
[6] K. Ackerstaff *et al.* [OPAL Collaboration], Eur. Phys. J. C **7** (1999) 571 [arXiv:hep-ex/9808019].
[7] D. Boito, M. Golterman, M. Jamin, A. Mahdavi, K. Maltman, J. Osborne and S. Peris, Phys. Rev. D **85**, 093015 (2012) [arXiv:1203.3146 [hep-ph]].
[8] C. Patrignani *et al.* [Particle Data Group], Chin. Phys. C **40**, no. 10, 100001 (2016).
[9] P. A. Baikov, K. G. Chetyrkin and J. H. Kühn, Phys. Rev. Lett. **101** (2008) 012002 [arXiv:0801.1821 [hep-ph]].
[10] E. Braaten, S. Narison, and A. Pich, Nucl. Phys. B **373** (1992) 581.
[11] Y.-S. Tsai, Phys. Rev. D **4**, 2821 (1971).
[12] D. Boito, O. Catà, M. Golterman, M. Jamin, K. Maltman, J. Osborne and S. Peris, Phys. Rev. D **84**, 113006 (2011) [arXiv:1110.1127 [hep-ph]].
[13] O. Catà, M. Golterman, S. Peris, JHEP **0508**, 076 (2005) [hep-ph/0506004].
[14] O. Catà, M. Golterman, S. Peris, Phys. Rev. D **77**, 093006 (2008) [arXiv:0803.0246 [hep-ph]].
[15] F. Le Diberder, A. Pich, Phys. Lett. B **289**, 165 (1992).
[16] D. Boito, M. Golterman, K. Maltman and S. Peris, Phys. Rev. D **95**, no. 3, 034024 (2017) [arXiv:1611.03457 [hep-ph]].
[17] K. Maltman and T. Yavin, Phys. Rev. D **78**, 094020 (2008) doi:10.1103/PhysRevD.78.094020 [arXiv:0807.0650 [hep-ph]].
[18] O. Catà, M. Golterman and S. Peris, Phys. Rev. D **79**, 053002 (2009) doi:10.1103/PhysRevD.79.053002 [arXiv:0812.2285 [hep-ph]].
[19] B. Blok, M. A. Shifman and D. X. Zhang, Phys. Rev. D **57**, 2691 (1998) [Erratum-ibid. D **59**, 019901 (1999)] [arXiv:hep-ph/9709333].
[20] I. I. Y. Bigi, M. A. Shifman, N. Uraltsev, A. I. Vainshtein, Phys. Rev. D **59**, 054011 (1999) [hep-ph/9805241].
[21] M. Golterman, S. Peris, B. Phily, E. de Rafael, JHEP **0201**, 024 (2002) [hep-ph/0112042].
[22] I. Caprini, M. Golterman and S. Peris, Phys. Rev. D **90**, 033008 (2014) [arXiv:1407.2577 [hep-ph]]; D. Boito and I. Caprini, Phys. Rev. D **95**, no. 7, 074027 (2017) doi:10.1103/PhysRevD.95.074027 [arXiv:1702.03757 [hep-ph]].
[23] M. Beneke and M. Jamin, JHEP **0809**, 044 (2008) [arXiv:0806.3156 [hep-ph]].
[24] M. Beneke, D. Boito and M. Jamin, JHEP **1301**, 125 (2013) [arXiv:1210.8038 [hep-ph]].
[25] See, for instance, I. Caprini and J. Fischer, Eur. Phys. J. C **64**, 35 (2009) [arXiv:0906.5211 [hep-ph]]; G. Abbas, B. Ananthanarayan, I. Caprini and J. Fischer, Phys. Rev. D **87**, 014008 (2013) [arXiv:1211.4316 [hep-ph]]; I. Caprini, Mod. Phys. Lett. A **28**, 1360003 (2013) [arXiv:1306.0985 [hep-ph]]; G. Abbas, B. Ananthanarayan and I. Caprini, Mod. Phys. Lett. A **28**, 1360004 (2013) [arXiv:1306.1095 [hep-ph]].

IOP Conf. Series: Journal of Physics: Conf. Series **912** (2017) 012004 doi:10.1088/1742-6596/912/1/012004

Two- and three-meson decay modes of the τ-lepton in the Monte Carlo generator TAUOLA

Olga Shekhovtsova

Radzikowskiego str. 152, Cracow 31-342, Poland
Akademicheskaya str. 1, Kharkov 61202, Ukraine

E-mail: shekhovtsova@kipt.kharkov.ua

Abstract. The current status of the two- and three-meson decay study of τ-lepton using the Monte Carlo generator TAUOLA is overviewed.

1. Introduction

The precise experimental data for tau lepton decays collected at B-factories (both Belle [1] and BaBar [2]) provide an opportunity to measure the Standard Model (SM) parameters, such as the strong coupling constant, the quark-mixing matrix, the strange quark mass etc, and for searching new physics, beyond the SM. Leptonic decay modes of the τ-lepton allow to test the universality of the lepton couplings to the gauge bosons [3].

Hadronic decays, in addition to the low energy e^+e^- data, give an information about the hadronization mechanism and resonance dynamics in the energy region of 1-2 GeV where the methods of the perturbative QCD cannot be applied. Unlike the e^+e^- annihilation into hadrons, the τ-lepton decays weakly and thus provides access to both vector and axial-vector currents. In addition, systematical comparison between τ-lepton and e^+e^- data for corresponding hadronic channels allows to test the conservation of the vector current hypothesis [4]. The τ-lepton data are also applied to estimate the hadronic vacuum polarization contribution to the anomalous magnetic momentum of the muon [5]. In addition hadronic decays of the τ-lepton are a tool in high-energy physics to study Higgs boson features [6]. Also hadronic flavour-violating and CP violating decays of τ lepton allow to search for new physics scenario [7].

Being the main hadronic decay modes, the two- and three-meson decays of the τ-lepton play a key role in the studies mentioned above. The current status of the part of the Monte Carlo generator TAUOLA related to the two- and three-meson modes is discussed in this note. The article is organized as follows: Section 2 is devoted to a general presentation of the two- and three-meson currents while Sections 3 and 4 describe the numerical test and fit of the dominant modes to experimental data. Section 5 summarizes this note.

2. Current status of the MC TAUOLA for two- and three-meson states

Since the 90's TAUOLA [8] has been the main MC generator to simulate τ decays. It has been used to simulate decays of the τ lepton by the experimental collaborations CLEO, ALEPH, and at both B-factories BaBar and Belle. The simulation of hadronic decay modes in the original version of TAUOLA was done on the basis of the theoretical results within the Vector Meson Dominance (VMD) approximation. In this approach a hadronic current is modeled by a weighted product of Breit-Wigner functions. This approach was contested in [9]: the investigation of the low energy behaviour of the $\tau \to \pi\pi\pi\nu_\tau$ amplitude demonstrates that the VMD amplitude reproduces only the leading order χPT result but contradicts the next-to-leading-order χPT prediction [10]. Moreover, the model was not sufficient to describe the CLEO $KK\pi$ data [11]. To reproduce the data the CLEO collaboration reshaped the model by introducing two ad-hoc parameters that spoilt the QCD normalization for Wess-Zumino contribution. The parameters are obtained by fitting to data and demonstrated more than 60% deviation from the predicted Wess-Zumino behaviour. However, before making conclusion that the Wess-Zumino anomaly normalization is spoilt it should be checked whether an oversimplified theoretical approximation, like VMD, was applied.

The alternative approach based on the Resonance Chiral Lagrangian (RχL) [12] was proposed in [13, 14]. The computations done within RχL are able to reproduce the low-energy limit of χPT at least up to next-leading-order and demonstrate the right falloff in the high energy region [13, 14]. The hadronic currents for the main two-meson ($\pi\pi$, $K\pi$, KK) [15, 16, 17] and three-meson ($\pi\pi\pi$ and $KK\pi$) [13, 14] decay modes were calculated in the framework of RχL and were implemented in TAUOLA [18]. The set covers more than 88% of the total hadronic τ width.

2.1. Two-meson decay modes in TAUOLA

In the general case, an hadronic current of a two-meson tau-lepton decay mode is described by the two Lorentz structures and corresponding them two hadronic form factors. For a decay $\tau \to h_1(p_1) + h_2(p_2) + \nu_\tau$, the hadronic current reads

$$J^\mu_{P_1P_2} = N(P_1P_2)[c_1(p_1 - p_2)^\mu F^+(s) + (p_1 + p_2)^\mu F^-(s)], \tag{1}$$

where $s = (p_1 + p_2)^2$, $\Delta_{12} = m_1^2 - m_2^2$. The normalization factor $N(P_1P_2)$ depends on the decay channel.

The function F^+ is the vector form factor ($F^V \equiv F^+$) and F^- is related to the scalar form factor F^S: $F^-(s) = (F^S(s) - F^+(s)) \cdot \Delta_{12}/s$. Therefore the two-meson hadronic current is re-written as

$$J^\mu_{P_1P_2} = N(P_2P_2)[(p_1 - p_2 - \frac{\Delta_{12}}{s}(p_1 + p_2))^\mu F^V(s) + \frac{\Delta_{12}}{s}(p_1 + p_2)^\mu F^S(s)]. \tag{2}$$

The form factors F^V and F^S are associated to the $L = 1$ and $L = 0$ waves of the final hadronic system. Fixing the $N(\pi^-\pi^0) = 1$, the other modes normalization form factors are

$$N(K^-K^0) = \frac{1}{\sqrt{2}}, \; N(\pi^-K^0) = \frac{1}{\sqrt{2}}, \; N(\pi^0K^-) = \frac{1}{2}. \tag{3}$$

The expression (2) together with normalizations factors (3) are applied in TAUOLA [18].

The exact appearance of both form factors are specified by the theoretical model describing hadronization. In the isospin symmetry limit, $m_{\pi^-} = m_{\pi^0}$ and $m_{K^-} = m_{K^0}$, the scalar form factor F^S vanishes for the two-pion and two-kaon decay modes and the current is described by the vector form factor F^V only. In the original version of TAUOLA [8] the scalar form factor was not included, as the generator was mainly oriented on the two-pion mode simulation. However, in the case of the $K\pi$ channels the scalar form factor [19] gives an essential contribution and has to be taken into account. For this reason it was added in the simulation in the updated version of the code [18].

Let us move to the theoretical parametrizations for the two-meson form factors applied to TAUOLA and we start with the dominant two-pion mode ($Br(\tau^- \to \pi^-\pi^0\nu_\tau) \simeq 25.52\%$). The scalar two-pion form factor was put to zero and the four theoretical results described below were implemented for the vector two-pion form factor in TAUOLA.

The original version of TAUOLA [8] for the two-pion decay contained only

- Kühn-Santamaría (KS) parametrization [20], *i.e.*

$$F_\pi^V(s) = \frac{1}{1+\beta+\gamma}(BW_\rho(s) + \beta BW_{\rho'}(s) + \gamma BW_{\rho''}(s)) \tag{4}$$
$$BW(s) = \frac{M^2}{M^2 - s - i\sqrt{s}\Gamma_{\pi\pi}(s)},$$

thus the pion form factor was written as a weighted sum of the P-wave Breit -Wigner (BW) functions of the $\rho(770)$, $\rho(1450)$, $\rho(1700)$ vector resonances with their relative strengths 1, α and β. The denominator of BW functions includes only an imaginary part of the two-pion loop contribution through the energy dependent width ($\Gamma_{\pi\pi}(s) = \Gamma_\rho \cdot s/m_\rho^2 \cdot \sigma_\pi^3(s)/\sigma_\pi^3(m_\rho^2)$ with $\sigma_\pi(s) = \sqrt{1-4m_\pi^2/s}$) of the resonances.

To follow state of the art of theoretical calculations for the pion form factor and experimental data analysis on the τ decay into two pions, three other parametrizations of the pion vector form factor were added.

- Gounaris-Sakurai parametrization [21]. It was applied to data analysis by the Belle, ALEPH and CLEO Collaborations.

$$F_\pi^V(s) = \frac{1}{1+\beta+\gamma}(BW_\rho^{GS}(s) + \beta BW_{\rho'}^{GS}(s) + \gamma BW_{\rho''}^{GS}(s)), \tag{5}$$
$$BW^{GS}(s) = \frac{M^2 + dM\Gamma_{\pi\pi}(s)}{M^2 - s + f(s) - i\sqrt{s}\Gamma_{\pi\pi}(s)},$$

For the precise formalue for the function $f(s)$ and parameter d I refer the reader to [22]. Also this parametrization is a weighted sum of BW contributions related to the ρ mesons. As in the case of KS only the two-pion loop contribution is taken into account in the energy dependent width of the ρ-meson states, however, both real (through $f(s)$) and imaginary ($\Gamma_{\pi\pi}$) parts are included. The GS parameters, as the resonance masses, widths and relative strentgs, were fixed to the Belle fit values, Table VII in [22] (the case 'Fit results (fixed $|F(0)|^2$)' was chosen).

- Parametrization based on the RχL approach [15].

$$F_\pi^V(s) = \frac{1 + \sum\limits_{i=\rho,\rho'} \frac{F_{V_i}G_{V_i}}{F^2}\frac{s}{M_i^2-s}}{1 + \left(1 + \sum\limits_{i=\rho,\rho'} \frac{2G_{V_i}^2}{F^2}\frac{s}{M_i^2-s}\right)\frac{2s}{F^2}\left[B_{22}^\pi(s) + \frac{1}{2}B_{22}^K(s)\right]}, \tag{6}$$

This parametrization is obtained directly from the Resonance Lagrangian (see for details [15]). Only two states of the vector ρ resonance contribute to the form factor, however their contribution is included as a weighted sum separately in the nominator and the denominator. Both pion and kaon loops are taken into account in this parametrization through the functions $B_{22}^{\pi;K}$. The relation between $Im(B_{22}^\pi(s))$ and $\Gamma_{\pi\pi}(s)$ can be found in Section 2 of [23]. The model parameters F_{V_i} and G_{V_i} as well as the resonance masses were fitted to the ALEPH (only the $\sqrt{s} < 1.2$ GeV points were taken into account) data [24], the one available at that moment. Only the $\sqrt{s} < 1.2$ GeV points were taken into account, and presented in Table I of [15].

- Combined RχL parametrization [16].

It combines two approaches: dispersion approximation in the low energy region ($s < s_0$) and modified RχL result at the high tail of the spectrum ($s > s_0$).

The low energy part of the spectrum ($s < s_0$) is presented by the Omnès solution of the 3-subtracted dispersion relation [25] and was calculated in [16]:

$$F_\pi^V(s) = \exp\left[\alpha_1 s + \frac{\alpha_2}{2}s^2 + \frac{s^3}{\pi}\int_{4m_\pi^2}^{\infty} ds' \frac{\delta_1^1(s')}{(s')^3(s'-s-i\epsilon)}\right], \tag{7}$$

where the parameters α_1, α_2 are the subtraction constants and δ_1^1 is the $I = 1$, $J = 1$ two-pseudoscalar scattering phase shift. The expresion for δ_1^1 via the simplified version of the form factor is presented in Eq. (16) of [16].

The high energy tail of the spectrum ($s > s_0$) is based on the modified RχL result and includes contributions from the the $\rho(770)$, $\rho(1450)$, $\rho(1700)$:

$$\begin{aligned} F_\pi^V(s) &= \frac{M_\rho^2 + (\beta+\gamma)s}{M_\rho^2 - s + \frac{2s}{F^2}M_\rho^2\left[B_{22}^\pi(s) + \frac{1}{2}B_{22}^K(s)\right]} \\ &- \frac{\beta s}{M_{\rho'}^2 - s + \frac{192\pi s\Gamma_{\rho'}}{M_{\rho'}\sigma_\pi^3(s)}B_{22}^\pi(s)} - \frac{\gamma s}{M_{\rho''}^2 - s + \frac{192\pi s\Gamma_{\rho''}}{M_{\rho''}\sigma_\pi^3(s)}B_{22}^\pi(s)}. \end{aligned}$$

Both pion and kaon loops are taken into account. The s_0 parameter is supposed to satisfy $1.0\text{GeV}^2 < s_0 < 1.5\text{GeV}^2$ [16]. In order to get the numerical values of the model parameters in Eqs. (7) and (8) the fit to the Belle data [22] was done, its results are summarized in Table 1 and Eqs. (29), (30) in [16].

The K^-K^0 decay channel. Up to the normalization factor the vector form factor for the $\tau^- \to K^-K^0\nu_\tau$ coincides with the vector form factor of the two-pion mode. The two parametrizations are currently included in TAUOLA: the KS approach (4) and the RχL one, Eq. (8). More details and discussion can be found in [23].

For the $(K\pi)^-$ modes both scalar and vector form factors are essential. The vector form factor was included within the two parametrizations [17] and Eqs. (17), (18) of [19]. Both parametrizations are based on the RχL approach, however, the final state interaction effects are treated in different ways. More details about these parametrizations can be found in Sections 2.4 and 5.4 of Ref. [18] and in the theoretical papers [17, 19]. The scalar $K\pi$ form factor was calculated on the basis of the private code of M. Jamin that corresponds to the numerical solution of a set of coupled dispersion relations specified in [26].

2.2. Three-meson decay modes in TAUOLA

When the τ lepton decays into three hadrons and a neutrino: $\tau \to h_1(p_1)+h_2(p_2)+h_3(p_3)+\nu_\tau$, Lorentz invariance determines the decomposition of the hadronic current in terms of five Lorentz invariant structures multiplied by hadronic form factors (F_i)

$$\begin{aligned} J^\mu = N\{T_\nu^\mu[c_1(p_2-p_3)^\nu F_1(q^2,s_1,s_2) + c_2(p_3-p_1)^\nu F_2(q^2,s_1,s_2) \\ +c_3(p_1-p_2)^\nu F_3(q^2,s_1,s_2)] + c_4 q^\mu F_4(q^2,s_1,s_2) - \frac{i}{4\pi^2F^2}c_5\epsilon^\mu_{.\,\nu\rho\sigma}p_1^\nu p_2^\rho p_3^\sigma F_5(q^2,s_1,s_2)\}, \end{aligned} \tag{8}$$

where as usual $T_{\mu\nu} = g_{\mu\nu} - q_\mu q_\nu/q^2$ denotes the transverse projector, $q^\mu = (p_1+p_2+p_3)^\mu$ is the total momentum of the hadronic system and the two-pion invariant mass squared is

$s_i = (p_j + p_k)^2$. The normalization factor N is $N = \cos\theta_{Cabibbo}/F$ for modes with an even number of kaons and $N = \sin\theta_{Cabibbo}/F$ for an odd kaon number modes. The TAUOLA value for the c_i coefficients is presented in Table 1 in Ref. [18].

I would like to stress that the decomposition (9) is model independent while the hadronic functions $F_i(q^2, s_1, s_2)$, $i = 1...5$ are model dependent. The vector form factor (F_5) vanishes for the three-pion modes due to the G-parity conservation. Of the three hadronic form factors F_i, $i = 1, 2, 3$ corresponding to the axial-vector part of the hadronic tensor, only two are independent. In strangeness-conserving decays the pseudoscalar form factor F_4 is proportional to m_π^2/q^2 [18], so that is suppressed with respect to F_i, $i = 1, 2, 3$.

As in the case of the two-meson decay modes I start with the parametrizations implemented in TAUOLA for the predominant three-meson modes $\tau \to \pi\pi\pi\nu_\tau$ ($Br(\tau \to \pi\pi\pi\nu_\tau) \simeq 18.67\%$).

Currently the following three-pion form factors are available in TAUOLA.

- CPC version [8].

It corresponds to the original version of TAUOLA and includes only the dominant $a_1 \to \rho\pi$ production mechanism. The form factor is a product of the Breit-Wigner for the a_1 and ρ meson.

- CLEO parametrization [27].

It is based on the Dalitz plot analysis carried out by the CLEO collaboration for $\tau^- \to \pi^0\pi^0\pi^-\nu_\tau$. The analysis included the following intermediate states: $a_1 \to (\rho;\rho')\pi$ (where $(\rho;\rho')\pi$ can be either in S- or in D-wave), $a_1 \to \sigma\pi$, $a_1 \to f_2(1270)\pi$, $a_1 \to f_0(1370)\pi$. As expected the $\rho\pi$ S-wave amplitude with a branching fraction 70% was dominant, the next one was the σ-meson part (about 16%) and the Dalitz plot analysis demonstrated that it could not be neglected. It is interesting to mention also that the $\rho' \equiv \rho(1450)$ state showed up more strongly in the D-wave amplitude than in the S-wave amplitude. All resonances were modelled by Breit-Wigner functions and the hadronic current was written as a weighted sum of their product.

- Modified RχL parametrization [30].

Initially the RχL three-pion current based on [14] was implemented in the TAUOLA code, see Section II.A of [18]. The RχL three-pion current was written as a sum of the chiral contribution corresponding to the direct vertex ($W^- \to (\pi\pi\pi)^-$), single-resonance contributions (*e.g.* $W^- \to (\rho\pi)^-$) and double-resonance contributions (*e.g.* $W^- \to a_1^- \to (\rho\pi)^-$). Only contributions from the intermediate vector and axial-vector were included in the original RχL hadronic currents. More details about the parametrization can be found in Section II.A of [18].

However, the comparison of the RχL predictions with the BaBar spectra (see Section 4), demonstrated a disagreement in the low energy tail of the two-pion spectra, both $\pi^+\pi^-$ and $\pi^-\pi^-$, Fig.1 in [29], and required a modification of the current. We ascribed this disagreement to the scalar resonance absence in the theoretical studies in [14] and to solve the problem an $f_0(500) = \sigma$ scalar resonance contribution was added. I would like to stress that the σ-meson contribution was included phenomenologically, *i.e.* it was not based on any Lagrangian. Our unique restriction was in requiring the same structure for the current as it was for the vector resonance contribution. In other words the hadronic currents had to contain a single-resonance part ($W^- \to \sigma\pi^-$) and a double-resonance part ($W^- \to a_1^- \to \sigma\pi^-$). In addition, the σ-resonance was modelled by a BW function. Hereinafter this approach will be refereed to as 'modified RχL'.

Also effects related with the Coulomb interaction of the final-state pions was included in TAUOLA. The far-field approximation was applied to estimate the Coulomb interaction effects, *i.e.* the final-state pions were treated as stable point-like objects and the three pion interaction was treated as a superposition of the two pion ones.

Channel	Width, [GeV]		
	PDG	Isospin-averaged masses	Phase space with physical masses
$\pi^-\pi^0$	$(5.778 \pm 0.35\%) \cdot 10^{-13}$	$(5.2283 \pm 0.005\%) \cdot 10^{-13}$	$(5.2441 \pm 0.005\%) \cdot 10^{-13}$
$\pi^0 K^-$	$(9.72 \pm 3.5\%) \cdot 10^{-15}$	$(8.3981 \pm 0.005\%) \cdot 10^{-15}$	$(8.5810 \pm 0.005\%) \cdot 10^{-15}$
$\pi^-\bar{K}^0$	$(1.9 \pm 5\%) \cdot 10^{-14}$	$(1.6798 \pm 0.006\%) \cdot 10^{-14}$	$(1.6512 \pm 0.006\%) \cdot 10^{-14}$
K^-K^0	$(3.60 \pm 10\%) \cdot 10^{-15}$	$(2.0864 \pm 0.007\%) \cdot 10^{-15}$	$(2.0864 \pm 0.007\%) \cdot 10^{-15}$
$\pi^-\pi^-\pi^+$	$(2.11 \pm 0.8\%) \cdot 10^{-13}$	$(2.1013 \pm 0.016\%) \cdot 10^{-13}$	$(2.0800 \pm 0.017\%) \cdot 10^{-13}$
$\pi^0\pi^0\pi^-$	$(2.10 \pm 1.2\%) \cdot 10^{-13}$	$(2.1013 \pm 0.016\%) \cdot 10^{-13}$	$(2.1256 \pm 0.017\%) \cdot 10^{-13}$
$K^-\pi^-K^+$	$(3.17 \pm 4\%) \cdot 10^{-15}$	$(3.7379 \pm 0.024\%) \cdot 10^{-15}$	$(3.8460 \pm 0.024\%) \cdot 10^{-15}$
$K^0\pi^-\bar{K}^0$	$(3.9 \pm 24\%) \cdot 10^{-15}$	$(3.7385 \pm 0.024\%) \cdot 10^{-15}$	$(3.5917 \pm 0.024\%) \cdot 10^{-15}$
$K^-\pi^0K^0$	$(3.60 \pm 12.6\%) \cdot 10^{-15}$	$(2.7367 \pm 0.025\%) \cdot 10^{-15}$	$(2.7711 \pm 0.024\%) \cdot 10^{-15}$

Table 1. The τ decay partial widths. For each channel, the PDG value [31] is compared with the MC prediction. The third column includes results with isospin averaged masses, whereas for the last column physical masses were used.

The $K\pi K$ decay modes, *i.e.* $K^-\pi^-K^+$, $K^0\pi^-\bar{K}^0$ and $K^-\pi^0K^0$ final states. As for the $\pi\pi\pi$ decay mode the MC TAUOLA contains three parametrizations: CPC, based on the CLEO analysis and RχL one. To see the details of these parametrizations I refer the reader to Refs. [8], [11] and Sections II.B and II.C of [18], respectively.

The Table 1 summarizes the TAUOLA update within the RχL framework and contains the RχL predictions for the decay width for every updated channel.

As one can see a difference between Table 1 and PDG values is 5-30% depending on the channel. To improve it a fit to available experimental data was done, see Section 4.

For the rest of three-meson decay modes, $\eta\pi^-\pi^0$ was computed in [28] ($K\pi\pi$ has not been published yet) in the framework of RχL, only the CPC parametrization [8] is included in TAUOLA.

3. Numerical tests and technical points

Since Ref. [8] has been published, numerical tests of TAUOLA MC functioning have not been repeated in a systematic way, despite the technical precision requirements are much higher now and reach sub-per mil level. Therefore the TAUOLA update required to revisit the numerical stability of the generator and of the multiple integration. The following steps were done:

- check of the phase space integration;
- numerical tests for every RχL current including a comparison with semi-analytical results.

The details on this study can be found in Section IV of [18] on the sample of the $\tau^- \to \pi^-\pi^-\pi^+\nu_\tau$ decay mode and for the other channels they are collected in Ref. [33].

From the technical point of view the RχL version of the hadronic form-factors are chosen invoking CALL INIRCHL(1) prior to TAUOLA initialization whereas the old version (either CLEO or VMD) corresponds to CALL INIRCHL(0). The numerical values of the model parameters as well as the flags to choose different types of RχL parametrizations for 3π (simple or modified RχL, to run with or without $\rho(1450)$, to switch on/off the Coulomb interaction), $K\pi$ (to run with or without the scalar form factor) are fixed in the file: *value_parameters.f*.

4. Fit to experimental data for $\tau \to \pi^-\pi^0\nu_\tau$ and $\tau \to \pi^-\pi^-\pi^+\nu_\tau$ modes.

The predicted values of the partial widths in Table 1 from the previous section corresponded to the 'default' numerical values of the RχL parameters specified in Table IV of [18]. However, the model uncertainty varies from 10% to 30% depending on a RχL parameter, thus a fit to experimental distributions had to be carried out in order to adjust numerical values of the model parameters. In this Section the fit result to the Belle and BaBar data for the predominant two-meson ($\tau \to \pi^-\pi^0\nu_\tau$) and three-meson ($\tau \to \pi^-\pi^-\pi^+\nu_\tau$) channels are collected.

In all parametrizations (4)-(7) for the two-pion mode, except for the RχL one, the pion form factor is given by interfering amplitudes from the known isovector meson resonances $\rho(770)$, $\rho'(1450)$ and $\rho''(1700)$ with relative strengths 1, β and γ. Although one could expect from the quark model that β and γ are real, the parameters were allowed to be complex (following the Belle, CLEO and ALEPH analysis) and their phases were left free in the fits.

The fit results to the Belle pion form factor [22] are presented in Figs. 1. The best fit occurs with the GS pion form factor ($\chi^2 = 95.65$), as could be expected, since this parametrization corresponds to the one applied for the Belle analysis, and the worst one is with the RχL one ($\chi^2 = 156.93$), which is not able to reproduce the high energy tail.

The two main differences of the RχL parametrization compared to the others are: 1) the ρ''-meson absence, 2) a real value of the $\rho'(1450)$-meson strength. To check the influence of the $\rho''(1700)$ on the RχL result, the $\rho''(1700)$ was added in the same way as it was done for $\rho'(1450)$. However, this inclusion did not improved the result (see Fig. 2 of [23]). That brings to a conclusion: the fault in reproducing the high energy tail is related to the real value of the $F_{V_i}G_{V_i}/F^2$ parameters which is associated with the β and γ parameters in the other parametrizations. In an effective field theory, like RχL, complex values come only from loops, *e.g.* missing loop contributions in the ρ' and ρ'' propagators, most probably related for the four-pion loops, could be responsible for the disagreement and the complex value of the β and γ parameters might mimic missing multi-particle loop contributions. The same conclusion was reached in Ref. [15] where it was stressed that the two-pseudoscalar loops could not incorporate all the inelasticity needed to describe the data and other multiparticle intermediate states could play a role.

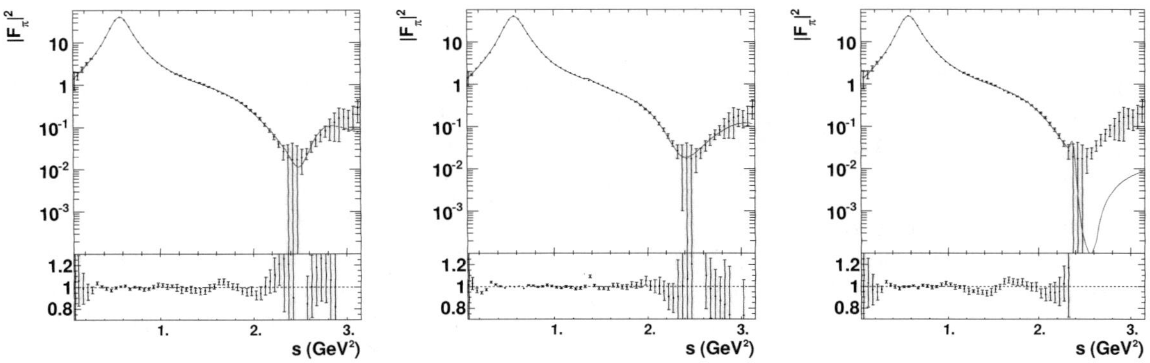

Figure 1. The pion form factor fit to Belle data [22]: the GS parametrization (left panel), the combined RχL parametrization (central panel), the RχL parametrization with $\rho''(1700)$ resonance (right panel). At the bottom: the ratio of a theoretical prediction to the data is given.

For the $\tau \to \pi^-\pi^-\pi^+\nu_\tau$ decay the one-dimensional distributions of the two- and three-pion invariant mass spectra calculated on the base of the modified RχL parametrization have been fitted in [30] to the BaBar preliminary data [32]. The first comparison of the RχL

	M_ρ	$M_{\rho'}$	$\Gamma_{\rho'}$	M_{a_1}	M_σ	Γ_σ	F	F_V
Min	0.767	1.35	0.30	0.99	0.400	0.400	0.088	0.11
Max	0.780	1.50	0.50	1.25	0.550	0.700	0.094	0.25
Fit	0.771849	1.350000	0.448379	1.091865	0.487512	0.700000	0.091337	0.168652

	F_A	$\beta_{\rho'}$	α_σ	β_σ	γ_σ	δ_σ	R_σ
Min	0.1	-0.37	-10.	-10.	-10.	-10.	-10.
Max	0.2	-0.17	10.	10.	10.	10.	10.
Fit	0.131425	-0.318551	-8.795938	9.763701	1.264263	0.656762	1.866913

Table 2. Numerical ranges of the $R\chi L$ parameters used to fit the BaBar data for three pion mode [32]. The approximate uncertainty estimates are collected in Table III of Ref. [30]. The $M_{\rho'}$ and Γ_σ best fit values are observed to be at the boundary of the physically motivated range of variation that we allowed for them.

prediction, specified in [18], showed a disagreement in the low energy tail of both $\pi^-\pi^-$ and $\pi^+\pi^-$ spectra [30], whereas the modified RχL approach was in line with the BaBar data in the whole energy spectra, Fig. 2.

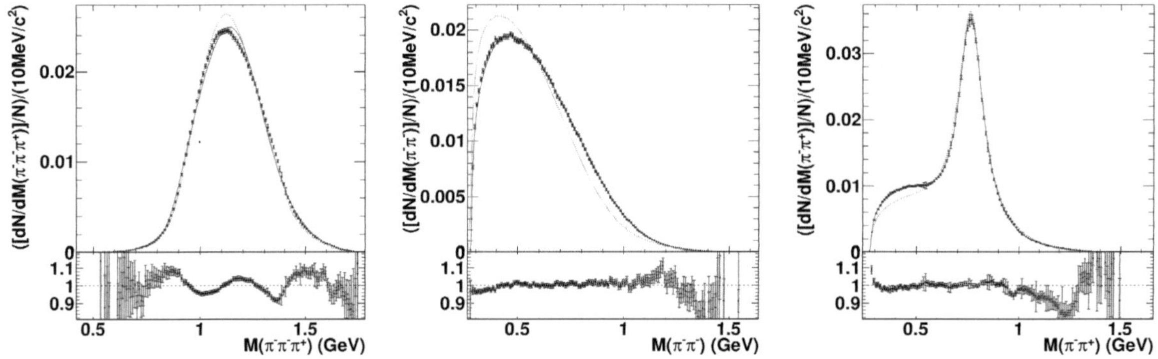

Figure 2. The $\tau^- \to \pi^-\pi^-\pi^+\nu_\tau$ decay invariant mass distribution of the three-pion system (left panel) and the two-pion system (central and right panels). The BaBar data [32] are represented by the data points, with the results from the modified RχL current as described in the text (blue line) and the old fit from Cleo [27] (red-dashed line) overlaid.

The study of the Coulomb interaction of the final pions, which might be important near the production threshold and might mimic the missing effects responsible for the disagreement between the RχL prediction [14] and the BaBar data, demonstrated that the Coulomb interaction alone without the σ meson could not describe the low energy tail of the experimental spectra.

Based on the fitting of the modified RχL spectra to the preliminary BaBar data the numerical values of the RχL parameters were adjusted, Table 2.

For the discussion of the systematic and statistical uncertainties in the fit, technical details and the computing strategy of the fitting procedure I refer to [30].

Finally, using the obtained numerical values of the model parameters, the $\pi^0\pi^0\pi^-$ partial width was estimated: $\Gamma(\tau^- \to \pi^0\pi^0\pi^-\nu_\tau) = (2.1211 \pm 0.0016\%)$ that is only 1% higher than the central PDG value [31] and is within the errors cited by PDG.

5. Conclusion

A set of the theoretical parametrizations for the two- and three-meson decay modes of the τ-lepton installed in the MC event generator was reviewed in this note.

Also it was discussed the results of fits for the two- and three-pion decay modes to the Belle and preliminary BaBar data. In the case of the two-pion decay modes the three parametrizations were chosen for the pion form factor: a) the Gounaris-Sakurai model, b) the RχL approach, c) a framework combining the resonance approach together with the dispersion approximation. All three models reproduced well the energy spectrum up to $1.05 - 1.15$ GeV while the theoretical result on the basis of the RχT was not able reproduce the experimental spectrum for the higher energy region. For the $\pi^-\pi^-\pi^+$ decay mode the fit to the preliminary BaBar data was done using the modified RχL hadronic current. The result can be summarized as following: the theoretical distributions were in line with the BaBar data for both two- and three-pion invariant mass distributions in the whole energy spectra. A small discrepancy between theoretical spectra and experimental data can be explained by missing resonances in the model, such as the axial-vector resonance $a_1'(1600)$, the scalar resonance $f_0(980)$ and the tensor resonance $f_2(1270)$. The first study on the $f_2(1270)$ resonance inclusion within the RχL approach in the three-pion decay mode of the τ-lepton was recently presented in [34].

The last point to be mentioned in this note is the second-class current study in the τ-lepton decays. Presently the second-class current decays are not observable yet and their measurement is a key point of the τ physics study at Belle-II. However, according to theoretical predictions (including the RχL prediction [35]), the main second-class current decay mode $\tau^- \to \pi^-\eta\nu_\tau$ should be finally discovered at the forthcoming Belle-II. Therefore, the future upgrade of TAUOLA should include also the second-class current simulation.

Acknowledgments
It is a pleasure to thank the organizers for nice and fruitful atmosphere in which the Workshop took place. In particular, I would like to thank my many-year collaborator Dr. P. Roig for the invitation and great organization of the meeting.

[1] S. Ryu *et al.* [Belle Collaboration], 2014 *Phys. Rev. D* **89** 072009.
[2] B. Aubert *et al.* [BaBar Collaboration], 2008 *Phys. Rev. Lett.* **100** 011801.
[3] B. Aubert *et al.* [BaBar Collaboration], 2010 *Phys. Rev. Lett.* **105** 051602.
[4] F. Jegerlehner and A. Nyffeler, 2009 *Phys. Rept.* **477** 1.
[5] M. Davier, A. Höecker, B. Malaescu and Z. Zhang, 2011 *Eur. Phys. J. C* **71** 1515. Erratum: [Eur. Phys. J. C **72** (2012) 1874].
[6] G. Aad *et al.* [ATLAS and CMS Collaborations], 2016 *J. High Energy Phys.* JHEP08(2016)045.
[7] A. Celis, V. Cirigliano and E. Passemar, 2014 *Phys. Rev. D* **89** 013008.
[8] S. Jadach, J. H. Kühn and Z. Was, Comput. Phys. Commun. **64** (1990) 275. S. Jadach, Z. Was, R. Decker and J. H. Kühn, 1993 *Comput. Phys. Commun.* **76** 361.
[9] D. Gómez Dumm, A. Pich and J. Portolés, 2004 *Phys. Rev. D* **69** 073002.
[10] G. Colangelo, M. Finkemeier and R. Urech, Phys. Rev. D **54** (1996) 4403.
[11] T. E. Coan *et al.* [CLEO Collaboration], 2004 *Phys. Rev. Lett.* **92** (2004) 232001.
[12] G. Ecker, J. Gasser, A. Pich and E. de Rafael, 1989 *Nucl. Phys. B* **321** *311.*
[13] D. G. Dumm, P. Roig, A. Pich and J. Portolés, 2010 Phys. Rev. D **81** *034031.*
[14] D. G. Dumm, P. Roig, A. Pich and J. Portolés, 2010 Phys. Lett. B **685** *158.*
[15] J. J. Sanz-Cillero and A. Pich, 2003 Eur. Phys. J. C **27** *587.*
[16] D. Gómez Dumm and P. Roig, 2013 Eur. Phys. J. C **73** *2528.*
[17] M. Jamin, A. Pich and J. Portolés, 2008 Phys. Lett. B **664** *78.*
[18] O. Shekhovtsova, T. Przedzinski, P. Roig and Z. Was, 2012 Phys. Rev. D **86** *113008.*
[19] D. R. Boito, R. Escribano and M. Jamin, 2009 Eur. Phys. J. C **59** *821.*
[20] J. H. Kühn and A. Santamaría, Z. Phys. C **48** *(1990) 445.*
[21] G. J. Gounaris and J. J. Sakurai, Phys. Rev. Lett. **21** *(1968) 244.*
[22] M. Fujikawa et al. [BELLE Collaboration], 2008 Phys. Rev. D **78** *072006.*

[23] *O. Shekhovtsova, 2015 JETP Lett.* **102** *329 [Pisma Zh. Eksp. Teor. Fiz.* **102** *(2015) no.6, 366].*
[24] *R. Barate et al. [ALEPH Collaboration], 1997 Z. Phys. C* **76** *15.*
[25] *F. Guerrero and A. Pich, 1997 Phys. Lett. B* **412** *382.*
[26] *M. Jamin, J. A. Oller and A. Pich, 2002 Nucl. Phys. B* **622** *279.*
[27] *D. M. Asner et al. [CLEO Collaboration], 2000 Phys. Rev. D* **61** *012002.*
[28] *D. Gómez Dumm and P. Roig, Phys. Rev. D* **86** *(2012) 076009.*
[29] *O. Shekhovtsova, I. M. Nugent, T. Przedzinski, P. Roig and Z. Was, 2014 Nucl. Phys. Proc. Suppl.* **253 - 255** *73.*
[30] *O. Shekhovtsova, T. Przedzinski, P. Roig and Z. Was, Phys. Rev. D* **86** *(2012) 113008. I. M. Nugent, T. Przedzinski, P. Roig, O. Shekhovtsova and Z. Was, 2013 Phys. Rev. D* **88** *093012.*
[31] *K. Nakamura et al. [Particle Data Group], 2010 J. Phys. G* **37** *075021.*
[32] *I. M. Nugent [BaBar Collaboration], 2014 Nucl. Phys. Proc. Suppl.* **253-255** *38.*
[33] *http://annapurna.ifj.edu.pl/~wasm/RχL/RχL.html .*
[34] *J. J. Sanz-Cillero and O. Shekhovtsova 2017 Refining the scalar and tensor contributions in* $\tau \to \pi\pi\pi\nu_\tau$ *decays. Preprint arXiv:1707.01137 [hep-ph].*
[35] *R. Escribano, S. Gonzàlez-Solis and P. Roig, Phys. Rev. D* **94** *(2016) no.3, 034008. E. A. Garcés, M. H. Villanueva, G. L. Castro and P. Roig, arXiv:1708.07802 [hep-ph].*

Recent upgrades and results from the CMS experiment

J A Murillo Quijada[1] on behalf of the CMS collaboration

[1] Department of Physics at CINVESTAV-IPN, Av. Instituto Politécnico Nacional 2508, Col. San Pedro Zacatenco, 07360, México D.F. (+52) (01) 55 5747 3800

E-mail: javier.murillo@cern.ch

Abstract.
In 2016 the LHC restarted its operation with proton-proton collisions energies equal to $\sqrt{s} = 13$ TeV. During 2016 the CMS detector collected an amount of data equal to ~37 fb^{-1}, significantly surpassing any amount of data collected in previous LHC campaigns. At the end of 2016 the 'Extended Year-End Technical Stop' (EYETS) started, a period where specific upgrades were performed over the detector to improve its efficiency when the data taking process restarted in 2017. Most of the upgrades are part of preparation process towards the High Luminosity LHC (HL-LHC) when the LHC will deliver up to 7 times the LHC design luminosity to the experiments. These upgrades will have significant contributions during 2019-2020 (LS2) and 2024-2016 (LS3). In this document the main upgrades that took place during EYETS are shown, also the current and future status of the LHC and the CMS experiment are discussed. Main preliminary results obtained by analyzing Run 2 data (2015 and 2016) are also shown.

1. The Large Hadron Collider (LHC) and the CMS experiment

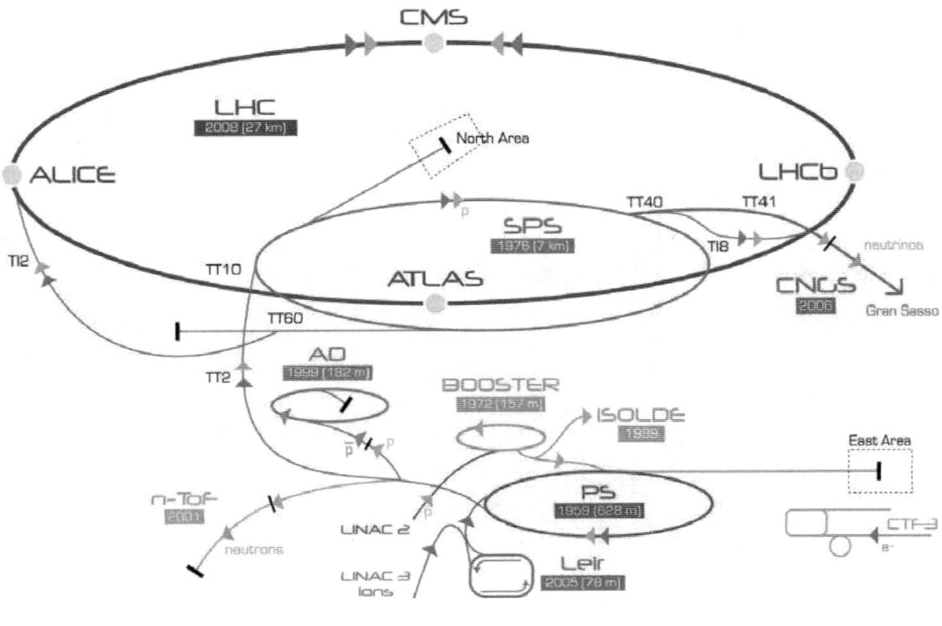

Figure 1. Diagram showing the acceleration complex at CERN that provides proton or ion beams to the LHC. From [2].

The successful operation of the LHC, and the Compact Muon Solenoid (CMS), experiment is the product of the hard work of thousands of people involved in their design, construction and careful

usage. The studies and results presented in this contribution could not have been possible without the support of this collaboration. The CERN laboratory directs the operation of the LHC which consists of ∼27 km circumference that extends to both sides of the border between France and Switzerland. This massive accelerator runs at unprecedented collisions energies, being able to operate at a design luminosity of 10^{34} cm^{-2} s^{-1} [1].
CMS along with ATLAS is one of the largest detectors located around the LHC ring (∼28.7m long and diameter of 15m). These two experiments are 'tagged' as *multipurpose detectors* due to the huge variety of physics analyses that can be performed with them. Results corresponding to 2015 and 2016 data taking processes from proton-proton collisions at $\sqrt{s}$ = 13 TeV are described in section 4. Proton bunches (made up of $\sim 10^{11}$ constituents) are produced, split and accelerated sequentially through different accelerators before injection to the LHC as it is illustrated in Figure 1. The acceleration process starts with the LINAC accelerator where protons are accelerated to 31% of the speed of light, the resulting beam is introduced to the Proton Synchrotron Booster (PSB), which increases the energy of the protons to 1.4 GeV and their speed to 91.6% of the speed of light. Protons are then accelerated by the Proton Synchrotron (PS) to 25 GeV and then injected to the Super Proton Synchrotron (SPS), where they are accelerated to 450 GeV. Finally the protons are injected to the LHC ring, being accelerated to an energy up to 7 TeV, reaching speed over 99.999999% of the speed of light, being able to produce proton-proton head-on collisions at up to $\sqrt{s}$ = 14 TeV. At the end of this acceleration chain the proton bunches travel 25ns from each other [1]. The laboratory also hosts additional smaller experiments that are within the scope of interest of nuclear and particle physics such as NA61-SHINE, NA62, COMPASS, CNGS, incorporated to the SPS and TOTEM and LHCf incorporated to the LHC [1].

Rather than being a perfect circumference the LHC is made up of eight insertion straight sections of which four are dedicated to the experiments, one is used for radio frequency cavities, used to accelerate proton bunches, two for beam cleaning and one for beam dumping. The accelerator also comprises eight arcs (sectors) joining these straight insertions. A total of 154 dipole magnets are in each of the arcs and are independently connected in series in the same cryostat with its own power source [1].

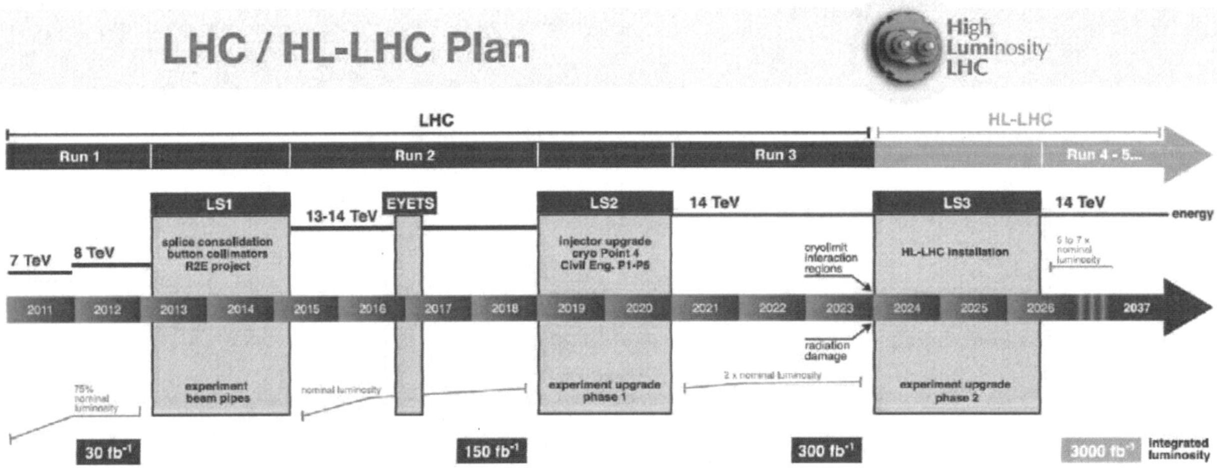

Figure 2. Diagram showing the plan ahead considered for the LHC towards the HL-LHC. From [3] [4].

CERN has adopted the European Strategy for Particle Physics, which has as high priority the full exploitation of the LHC. This gave rise to the High Luminosity LHC (HL-LHC), project, which consists of a set of huge upgrades over the LHC machine and all its detectors. The current timeline towards the HL-LHC project is shown in Figure 2. The HL-LHC final installation is planned to start in ∼2024-2026, after the end of Run 3, where the LHC will be operating at its design head-on proton-proton collision energy of 14 TeV and will deliver a nominal luminosity about two times larger than the LHC design luminosity. On the other hand, the HL-LHC when completed, is planned to operate (from ~2026) at 14 TeV as well but deliver between 5 to 7 times the design luminosity of the LHC, allowing the gathering of at least an order of magnitude more data than what will be achieved by joining together all the data collected during Run I, Run II and Run III (2011 to 2024) [3] [4].

The new upgrades and key innovative technologies that are planned for HL-LHC are for example the installation of 2 new 300-meters service tunnels and two accessing shafts at the ATLAS and CMS detectors sites. Also the installation of 12 powerful quadrupole magnets for the ATLAS and CMS detectors in order to increase the focusing of each of the beams before collision. A set of 4 pairs of bending magnets will be installed in the LHC tunnel, these ones are shorter and more powerful than the existing ones. This will free up some space for the installation of up to 20 new collimators and the replacement of at least 60, so the protection of the LHC machine is reinforced. Additionally, a set of 16 ultra-precise superconducting "crab" cavities for both ATLAS and CMS will be used to tilt the beams before collision, so the luminosity value increases substantially [3] [4].

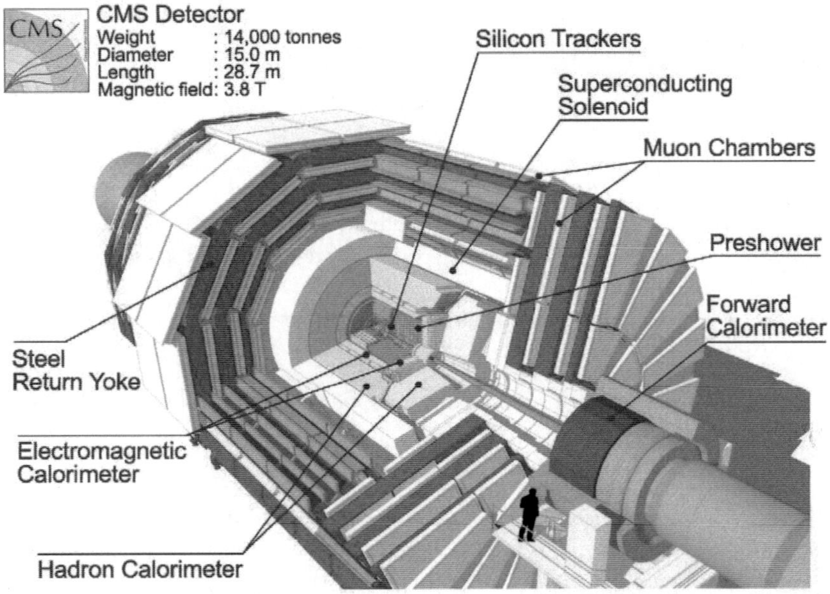

Figure 3. Illustration of the CMS detector and all its main systems. From [6].

The CMS detector is located close to the French village of Cessy, between the Jura mountains and the Geneva Lake, it is designed to study the products from proton-proton, lead-lead and lead-proton collisions. The detector has a huge and powerful solenoid in its inner section (13m long and diameter of 6m) with a nominal high-magnetic-field of 3.9T that surrounds a tracking system composed by a Silicon Pixel and Silicon Strip Tracker in its inner sections and a calorimetry system composed by a Scintillating Crystal Electromagnetic Calorimeter and a Brass-Scintilator Hadronic Calorimeter [5]. It is the heaviest detector at CERN with ~14000 tonnes. An overall layout of the detector is shown if Figure 3, where all the detector main sub-systems are highlighted. The detector has specialized muon detectors in the region outside the solenoid with detection devices based in three main technologies, aluminium Drift Tubes (DT) in the barrel region, Cathode Strip Chambers (CSC) in the endcap region and Resistive Plate Chambers (RPC) in both barrel and endcap regions. The first two provide precise position measurement and trigger, whilst the third one offers precise timing information as well as a second an independent trigger. Forward sampling calorimeters extend the detection range to regions very close to the beam direction. CMS has highly efficient trigger and data acquisition systems to record interesting events. The collaboration uses highly distributed and efficient computing infrastructure, over which specialized software can execute numerous tasks such as reconstruction of events, analysis of the collected data as well as the production of simulated data [5] [7].

2. EYETS: Extended Year-End Technical Stop

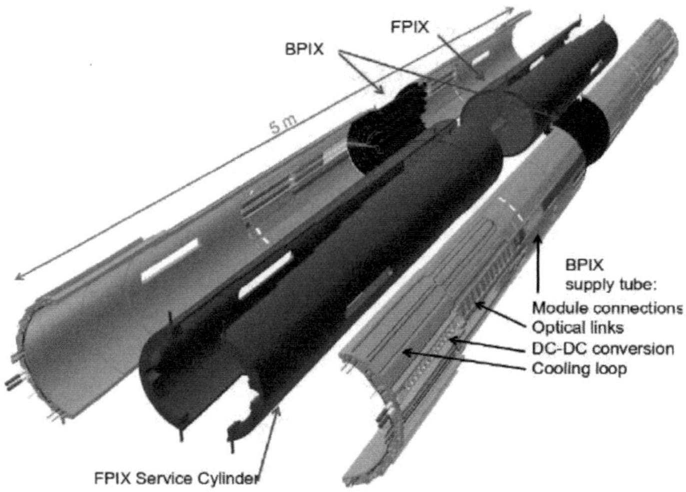

Figure 4. Illustration of the new pixel system for the CMS detector. It is located at closer distance than its previous counterpart with four layer in its forward and barrel sections. From [8].

After the successful data taking process during 2016 for proton-proton collisions and also proton-ion collisions, which are discussed in the following section, the LHC and all its detectors experienced a Technical Stop that lasted from December 2016 to May 2017. During this period the CMS detector suffered significant upgrades crucial for the coming increase in luminosity during the rest of Run 2, and also for Run 3 and HL-LHC phases.

The biggest upgrade for the CMS detector consisted in the replacement of its pixel detector. A diagram of this system is shown in Figure 4. This sub-detector is at closest distance from the beam line ($\sim$ 29mm) than any other device, so a replacement was necessary due to radiation damage, and also due to the need of a more efficient tracking system, which determines the trajectories of charged particles. This new pixel system is designed to keep high tracking performance at luminosities up to 2 times the design luminosity of the LHC and in an environment where the number of primary collisions that are produced per proton bunch crossing (pileup) exceeds the value of 50. The new pixel detector has 4 barrel and forward layers instead of 3 that were present in the previous detector [8].

Another important upgrade was done over the Hadronic Endcap and Forward calorimeters. Over the first one a partial upgrade was performed, which will be continued during the long stop that will take place during 2019 and 2020 (LS2) as it is indicated in Figure 2. This upgrade consists in replacing a set of hybrid photodiodes for silicon photo-detectors which are expected to perform better in a higher luminosity environment [9]. For the case of the Hadronic Forward calorimeter, its photomultiplier boxes were split in to two different channels, to be able to discriminate more efficiently between signal and backgrounds, which is crucial in this forward region of the detector [10]. Also, during this EYETS period the installation of an additional set ot muon detectors GE1/1 that use gas electron multipliers (GEM) technology in the first endcap muon station was started. This installation will continue also during LS2. This system, when successfully installed, is expected to improve substantially the forward muon triggering and reconstruction. Figure 5 shows with red color the location of this new system in the forward region where the background signals are higher [11].

As mentioned in the previous section, the CMS detector uses three main systems for muon detection in its muon spectrometer (muon chambers). For monitoring and for detector optimal performance pur-

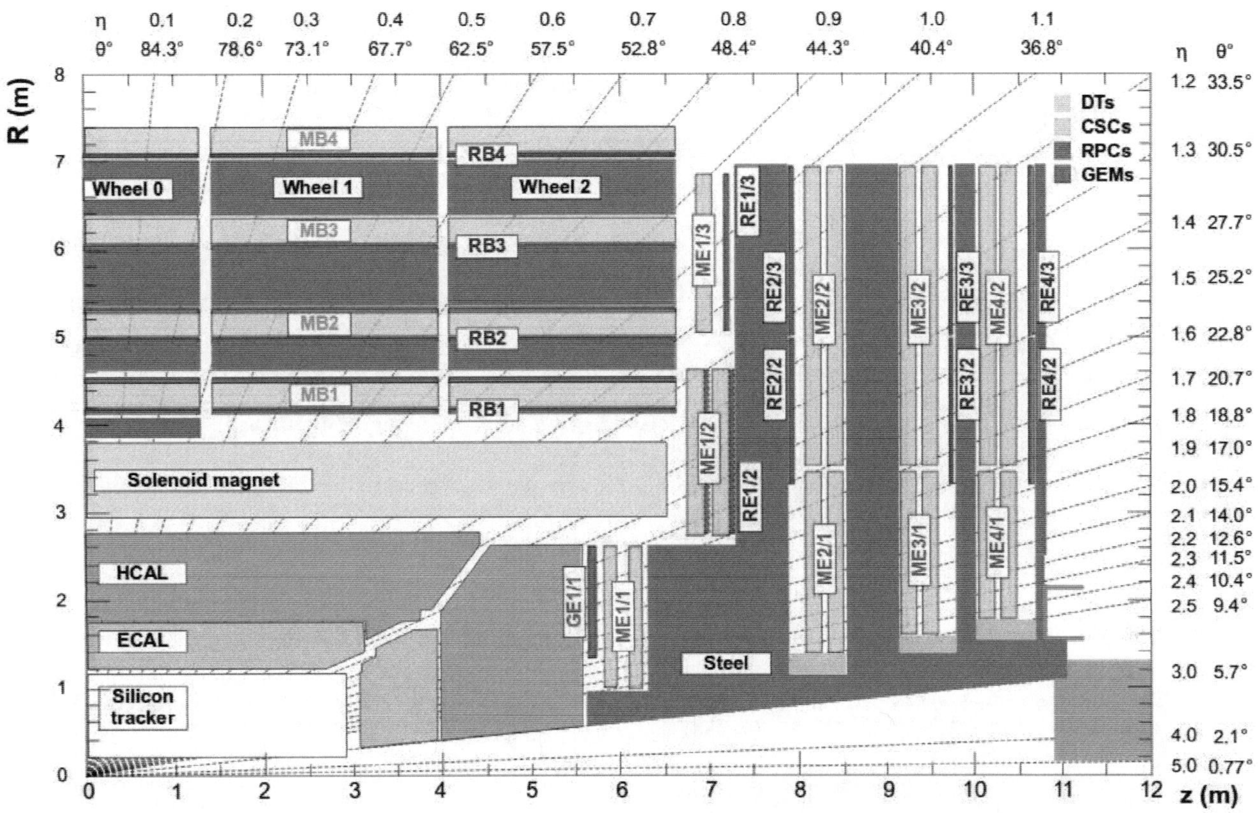

Figure 5. Longitudinal view of one of the quadrants of the CMS detector, the location of the GE1/1 ring is indicated with red color in the forward region. From [11].

poses, México collaborates strongly with RPCs system. The mexican team contributes with analysis of the RPC detectors currents dependence with the luminosity and the effect of installation of shielding layers in the muon detector barrel and endcap. Also studies are done to determine the optimal high voltage value used in these chambers for data collection [12]. Additionally, México has a leading role in this sub-system upgrade coordination, its central system, R&D activities and consolidation studies towards possible future upgrades. Finally, this team as well provides a crucial contribution in the development of the software that delivers the average chambers currents and integrated charge per data run. The total amount of charge (integrated charge) detected by these chambers has been analyzed as well by this team.

3. 2016 Data taking campaign

As it is indicated in Figure 2, Run 2 extends from 2015 to the end of 2018. Proton-proton head-on collisions have occurred so far at $\sqrt{s}$ = 13 TeV during this phase. The integrated luminosity target over this period for both ATLAS and CMS detectors is $\sim$100fb^{-1}; over three times more data than in Run 1, that extended from 2011 to 2015. CMS and the LHC have been upgraded so that the monitored peak instantaneous luminosity value has surpassed the LHC design (planned) luminosity during 2015 and 2016 data taking campaigns. This is the reason why the ongoing upgrade work is crucial (section 1, 2), as the LHC is producing a much higher luminosity environment than planned. This sets the need in all the experiments of highly as possible radiation-hard and efficient detectors. In terms of amount of collected data, both experiments ATLAS and CMS planned to record $\sim$25fb^{-1} of data during 2016. As indicated in Figure 6, both experiments surpassed this goal and ended up recording $\sim$37.8fb^{-1} (for proton-proton head on collisions). Figure 7 shows the amount of data delivered during

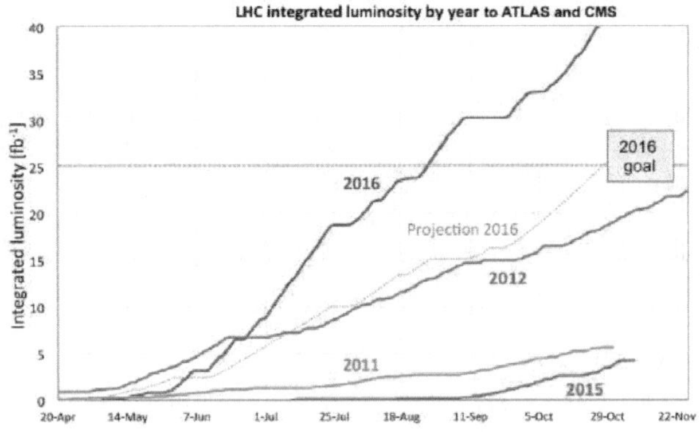

Figure 6. Cumulative proton-proton luminosity delivered by the LHC versus time for the different data-taking campaigns: 2011, 2012, 2015 and 2016. From [13].

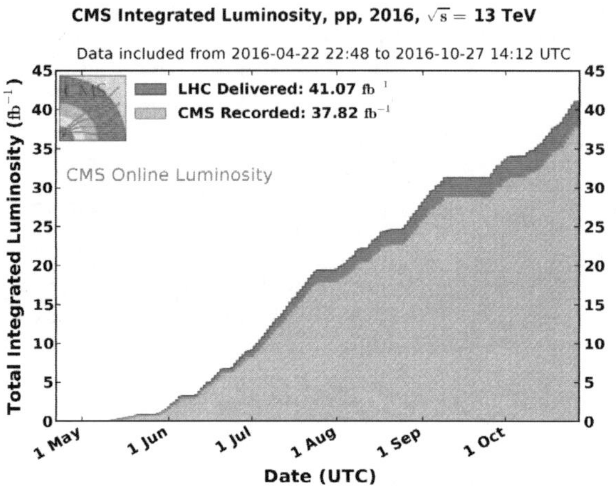

Figure 7. Cumulative proton-proton luminosity delivered by the LHC versus time in 2016 compared with the data recorded by CMS. From [13].

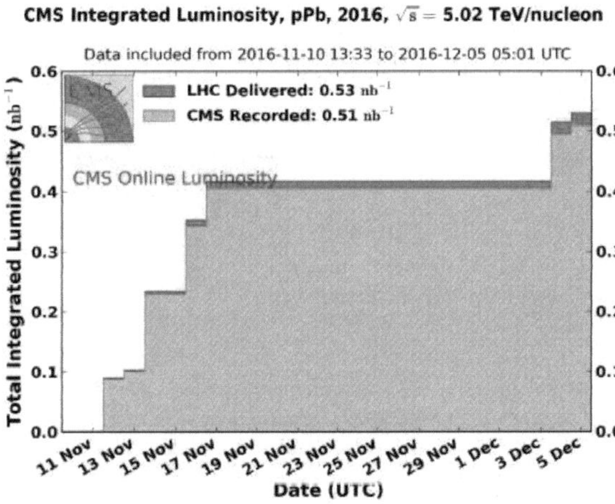

Figure 8. Cumulative proton-ion luminosity delivered by the LHC versus time in 2016 at $\sqrt{s}$ = 5.02 TeV, compared with the data recorded by CMS. From [13].

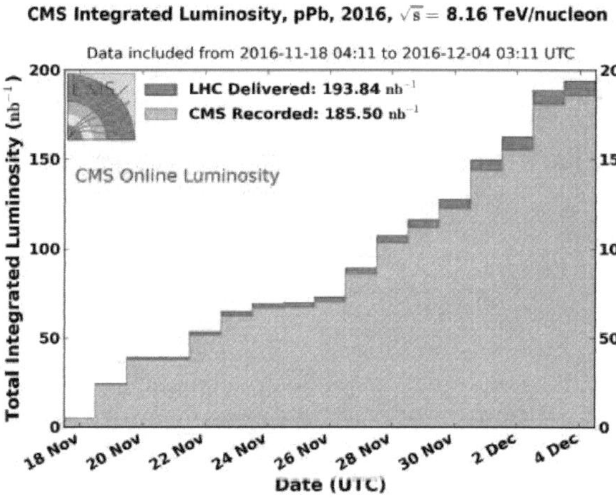

Figure 9. Cumulative proton-ion luminosity delivered by the LHC versus time in 2016 at $\sqrt{s}$ = 8.16 TeV, compared with the data recorded by CMS. From [13].

2016 by the LHC with time compared with the amount of data recovered by CMS.

A proton-ion (lead nucleus) campaign took place at the LHC during November 2016. Two paralell data taking processes occurred, a short one at a centre-of-mass energy of $\sqrt{s} = 5.02$ TeV and another larger one at $\sqrt{s} = 8.16$ TeV as it is indicated in Figures 8 and 9. For the latter, the target was to record $\sim$50nb^{-1} and $\sim$185.5nb^{-1} were actually recorded, surpassing for more than 3 times the planned targed for these collisions. These data will be used for forward physics and ultraperipheral collisions studies, where the production of J/ψ, Υ and ρ particles from photon-induced decays will be measured.

4. CMS Results from proton-proton collisions at $\sqrt{s} = 13$ TeV

The CMS experiment just started to operate with a proton-proton collision energy of 13 TeV in 2015 (section 1). It previously operated at 8 TeV during Run-I (2011 to 2015). So the analysis of these new type of data from higher energy became the center of interest at CERN due to the increased production rate for all processes under study. The experiment experienced a short data taking campaign at this upgraded energy during 2015 ($\sim$3.81fb^{-1} for proton-proton collisions), compared with the $\sim$37.8fb^{-1} recorded in 2016 (section 3). In this section some preliminary results corresponding to proton-proton collisions at $\sqrt{s} = 13$ TeV recorded during 2015 and 2016 are shown.

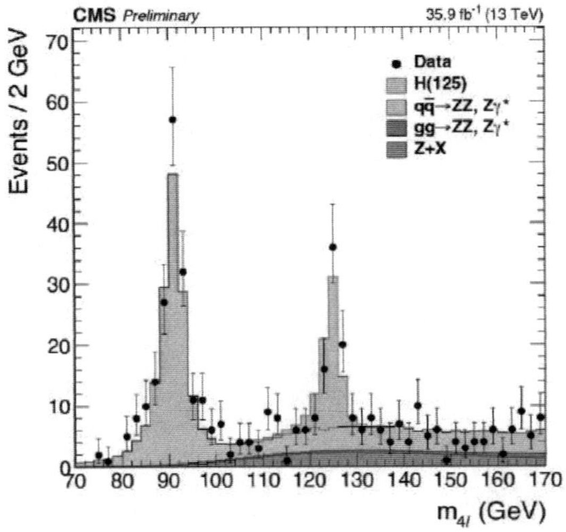

Figure 10. Mass distribution for higgs candidates decaying to 4-leptons. The well modelled component in pink color corresponds to signal events. From [15].

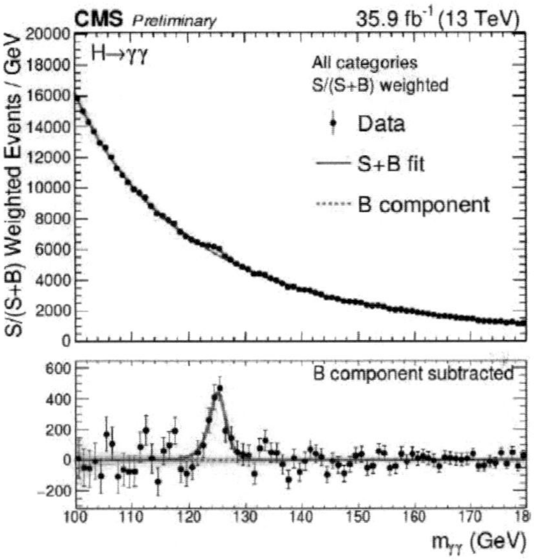

Figure 11. Mass distribution for higgs candidates decaying to two photons ($\gamma\gamma$). The pad below shows clearly the signal peak after subtracting the background component. From [15].

Measurements and observations performed during LHC Run 1 led a legacy. For example during this period a particle consistent with the Standard Model of particle physics (SM), higgs boson was observed using two independent channels ($\mathrm{H} \to \gamma\gamma$ and $\mathrm{H} \to \mathrm{ZZ} \to 4\ell$) and two independent experiments (ATLAS and CMS). The final combination of the results considering these two channels and the two experiments gives a mass of $\mathrm{m}_H = 125.09 \pm 0.21$ (stat.) $\pm$ 0.11 (syst.) GeV, where the two uncertainties stand for statistical and systematic uncertainties respectively [14]. Also the observation and studies of this observed particle during Run 1 led to the conclusion that all its production and decay rates are consistent with the SM predictions. The top quark was as well studied during this period, precision measurements were performed over SM parameters, and also there was a continuous search of new physics, not finding in all these investigations any deviation from the SM. It is the main priority of all research teams to reproduce all these studies and searches at higher collision energy, from where much higher statistics and precision can be obtained easily. Some preliminary results towards this end are shown in what follows.

4.1. Observation of the Higgs boson at 13 TeV

Using the 2015 and 2016 proton-proton collisions datasets, the higgs-like particle was re-observed at $\sqrt{s} = 13$ TeV, but this time the signal can be extracted much easier. For example Figure 10 shows the mass distribution of Higgs candidates decaying to 4 leptons (known as the 'golden channel'). These produced leptons can be any of the following combinations: $\mu\mu\mu\mu$, $\mu\mu ee$ and $eeee$. Due to its relatively

IOP Conf. Series: Journal of Physics: Conf. Series **912** (2017) 012005 doi:10.1088/1742-6596/912/1/012005

large signal-to-background ratio, It is one of the most important channels for study of this particle properties. Here the component of the distribution with pink color with higher value at ~125.0 GeV corresponds to signal decays, showing much higher significance than in Run 1 [14] [15]. Also a preliminary CMS-only measurement of the higgs mass using this distribution gives $m_H = 125.26 \pm 0.20$(stat.) $\pm$ 0.08(syst.) GeV, reaching already higher precision than the Run 1 combination [15]. The width of this particle has been measured as $\Gamma_H < 1.10$ GeV at 95% confidence level (C.L.) and its total production rate (relativelly to the SM prediction) has been found to be $\mu_{comb} = 1.05^{+0.19}_{-0.17}$, not finding any deviation from the SM and reaching already higher precision from the Run 1 measurement. 7 channels for the production of this particle have been considered for the measurement of this production rate [15].

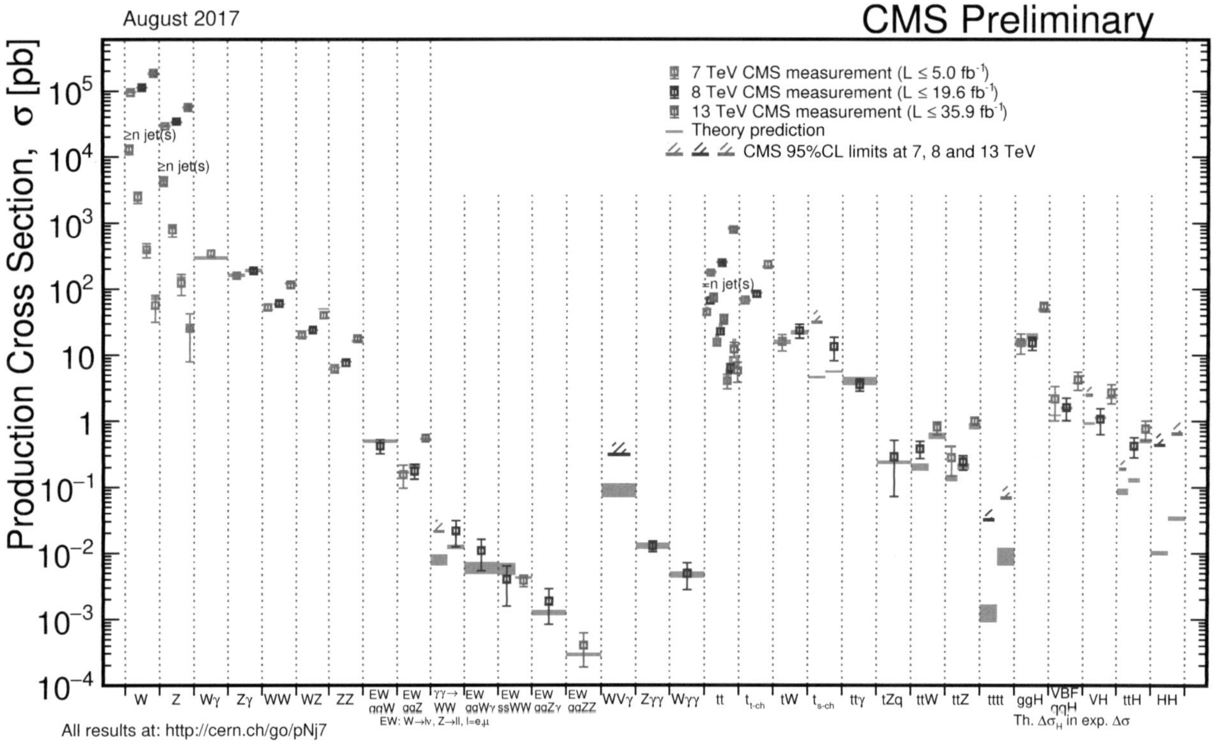

Figure 12. Plot showing the SM prediction for production cross-section for different particle systems versus the correspoding measurements at $\sqrt{s} = 7$, 8 and 13 TeV. No deviation from the SM has been found so far. From [16].

Additionally, the observation of this higgs-like particle has been reproduced in the H$\rightarrow \gamma\gamma$ channel as can been seen in Figure 11, here the background component is higher than with the H$\rightarrow 4\ell$ channel, but the resonance can be clearly seen at ~125.0 GeV in the pad below after substracting the background. A mass measurement using this channel and updated dataset is in progress [15]. Preliminary fiducial cross-section measurements were performed as well using these two channels, not finding any deviation from the SM predictions [15].

4.2. Standard Model cross-check

Production of system of particles involved with Standard Model processes has been measured as it is shown in Figure 12. In this plot are shown the predicted values from the SM of production cross sections for different products compared with has been measured experimentally at proton-proton collision energies of $\sqrt{s} = 7$, 8, 13 TeV, respectively. As the plot shows, no single deviation from the

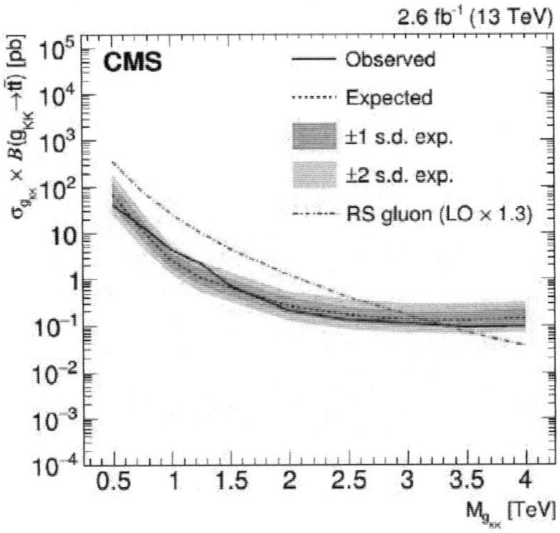

Figure 13. Measured upper 95% C.L. values for the g_{KK} cross-section $\sigma_{g_{KK}}$, the point where the experimental distribution intersects the theoretical curve and goes over, sets the exclusion limit. From [18].

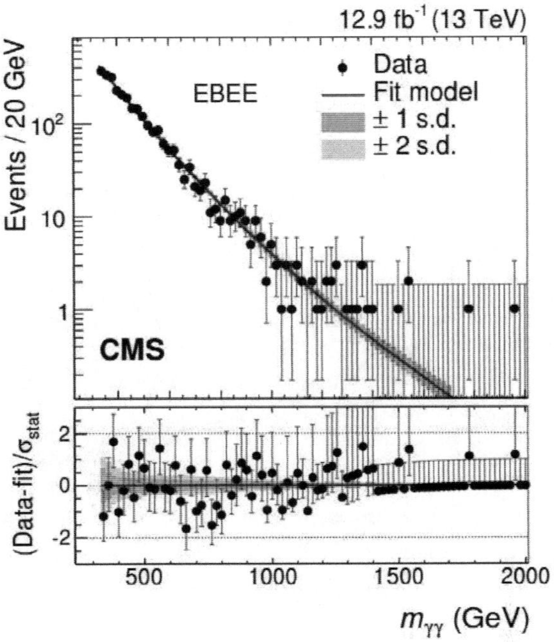

Figure 14. Distribution for the reconstructed invariant mass of the $\gamma\gamma$ system. No resonance or excess out of the uncertainties is observed. From [19].

SM predictions has been found. For the case of the production of 4 top quarks (tttt), upper limits at 95% C.L. were set, which are over the prediction and then not in contradiction with the SM [16]. As can be seen in the figure, the production cross-sections for the different products increases with the center of mass energy of the collision.

4.3. Search of Beyond the Standard Model Physics

Unstable Beyond the SM standard model particles decaying to measurable products have been search extensively. For example there are Beyond the SM Models (BSM), that introduce massive particles that decay into a $t\bar{t}$ pair. For these studies the search has been focused over Z' and a Kaluza-Klein (KK) excitations of gluons g_{KK}. The first one originated in a Sequential SM (SSM), theory, which is the SM + 1 additional massive boson with the same coupling constants [18]. On the other hand g_{KK}'s are originated by extending the Randall-Sundrum (RS) model [18]. Z' particle has been searched considering masses up to $\sim$ 4 TeV and relative width (to its mass) values $\Gamma_{Z'} = 1\%$, 10% and 30%, while g_{KK} has been searched with a relative decay width of $\sim$ 17% [18]. Exclusion limits with this new dataset are improved for all models with respect previous results (at $\sqrt{s} = 8$ TeV) . The Z' particle with a relative width $\Gamma_{Z'} = 30\%$ is excluded up to a mass value of 4.0 TeV, while the exclusion limit is 3.9 TeV for a relative width of 10%, where the previous limit was 2.9 TeV. As it is shown in Figure 13, the upper exclusion limit for g_{KK} is 3.3 TeV, the previous limit here was 2.8 TeV [18].

As indicated in Figure 14, a massive higgs resonance has been as well investigated. This extended search is based in the assumption that the production mechanism is completely due to gluon-gluon fusion (ggF). The decay channel used for this study is H$\to \gamma\gamma$. As the figure indicates, no resonance has been measured in this study. Similarly, no excess corresponding to an additional higgs particle has been seen when measuring the upper limit of the production cross-section (at 95% C.L) and scanning over masses between 0.5 and 4.5 TeV and widths relatively to the mass of 1.4×10^{-4} and 5.6×10^{-2}. On the other hand, exclusion limits have been set for the mass of Randall-Sundrum gravitons that may decay into two photons [19].

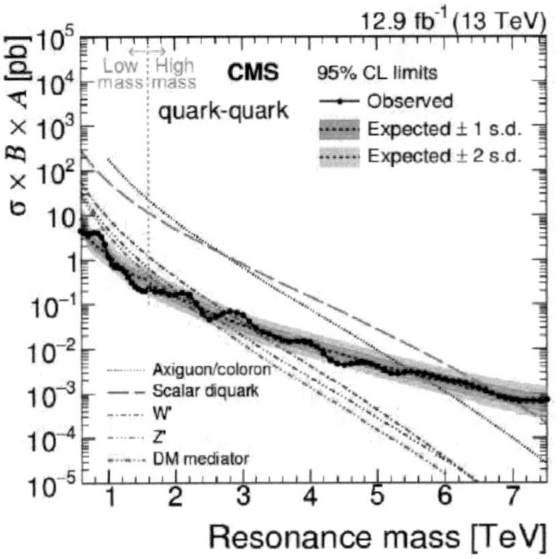

Figure 15. Measured upper 95% C.L. values for BSM particles cross-sections σ decaying to a dijet pair. From [20].

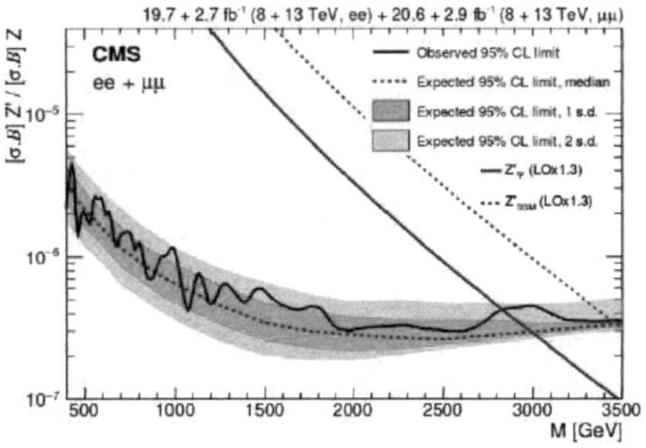

Figure 16. Measured upper 95% C.L. values for BSM particles cross-sections σ decaying to a dilepton pair. From [21].

Similar searches have been performed over massive particles decaying to a pair of jets (dijet) originated from gluons and/or quarks and also BSM particles decaying to a couple of leptons. Figure 15 shows experimental upper limits for cross-section at 95% C.L. where no sign of excess or signal from a BSM particle is seen. In this plot exclusion limits are set for 5 different models introducing massive particles decaying to a dijet system [20]. Similarly, Figure 16 shows that no excess has been observed for BSM particles decaying to a pair of leptons. Here a exclusion limit at 95% C.L. of 3.87 TeV has been set for a SSM Z' particle and a limit of 2.82 TeV for Z'$_\psi$ from a 'superstring model' [21], where the previous limits were 2.90 and 2.57 TeV [21]. Additionally, limits were as well set for gravitons G_{RS} decaying to a pair of leptons, and again, no sign of new physics is observed.

4.4. Supersymmetry search

Another search of BSM physics has been focused over Super-Symmetry (SUSY), theory, a strong candidate for dark matter making up ~25% of the universe. In models based on SUSY, new particles are introduced such that all fermionic (bosonic) degrees of freedom in the SM are paired with corresponding bosonic (fermionic) degrees of freedom in an extended theory [23]. As it is displayed in Figure 17 exclusion limits have been set for particles predicted by SUSY, for example the gluino $\tilde{g}$ and the neutralino $\tilde{\chi}_1^0$. The plot in this figure shows a bidimensional region that excludes simultaneously low mass regions for these two particles. $m_{\tilde{g}}$ is excluded at up to $\sim$ 1.9 TeV. Similar search has been performed over charginos $\chi_1^{\pm}$ along with neutralinos as it is shown in Figure 18. So far, exclusion limits have been extended substantially with respect to what was achieved with the Run 1 dataset, and no deviation or signal BSM has been detected. Similar searches have been performed over stop quarks $\tilde{t}$, not finding any excess or BSM signal [23].

4.5. Top and B-physics

Properties of the top quark and composite states made up of a bottom b-quark have been measured. As shown in Figure 19 (in red color) a quite precise measurement of the top quark mass m_t has been achieved by just using the 2015 dataset ($\sim$ 2.2 fb^{-1}). Still the final CMS combination from Run

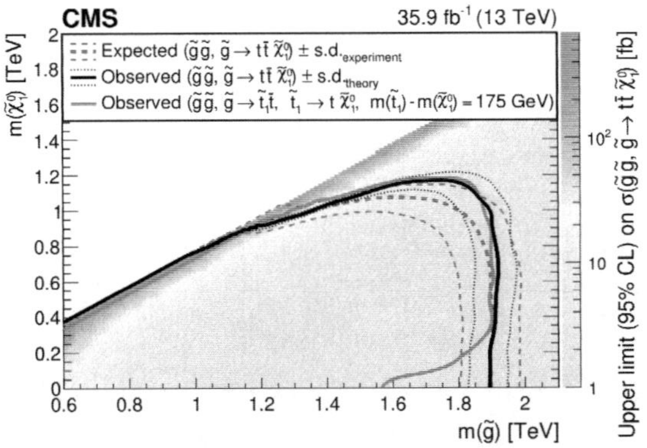

Figure 17. Bidimensional distribution showing the exclusion limits at 95% C.L. for the gluino $\tilde{g}$ and neutralino $\tilde{\chi}_1^0$ supersymmetric particles. From [22].

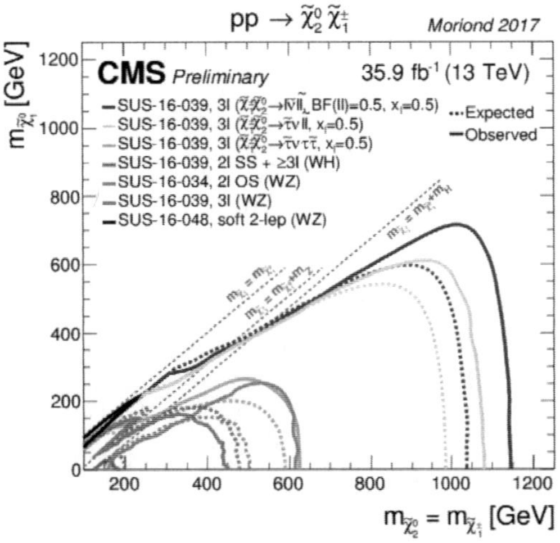

Figure 18. Bidimensional distribution showing the exclusion limits at 95% C.L. for the chargino ($\tilde{\chi}_2^0$, $\tilde{\chi}_{1\pm}$) and neutralino $\tilde{\chi}_1^0$ supersymmetric particles. From [23].

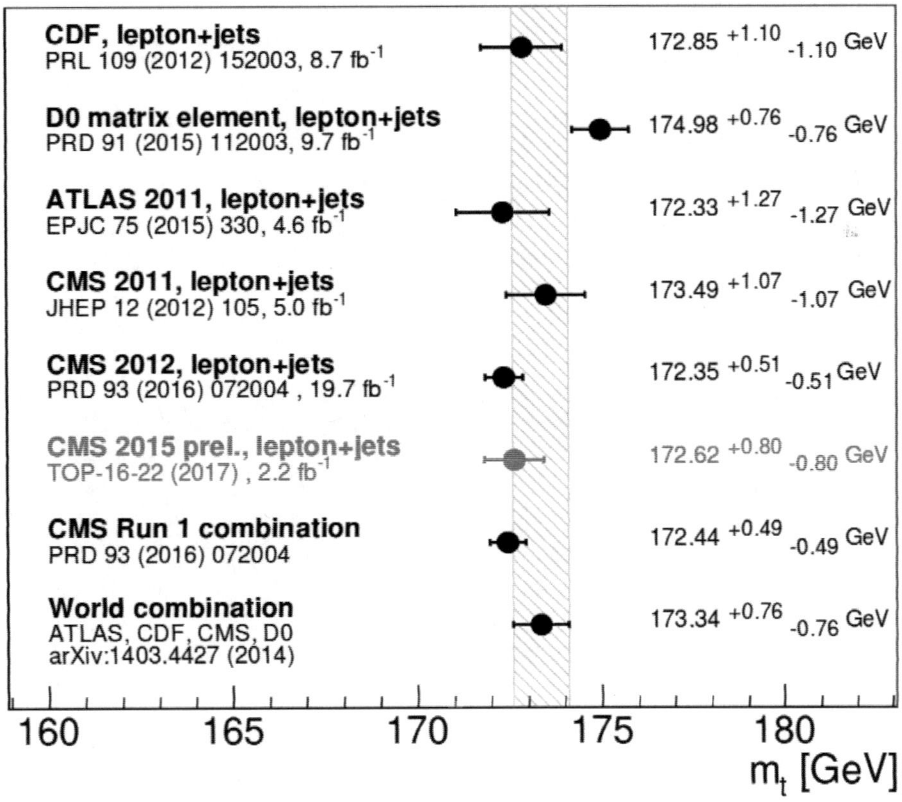

Figure 19. History of top mass measurements from CDF, D, ATLAS and CMS experiment. CMS Run 1 combination is more precise already than the current world combination. From [24].

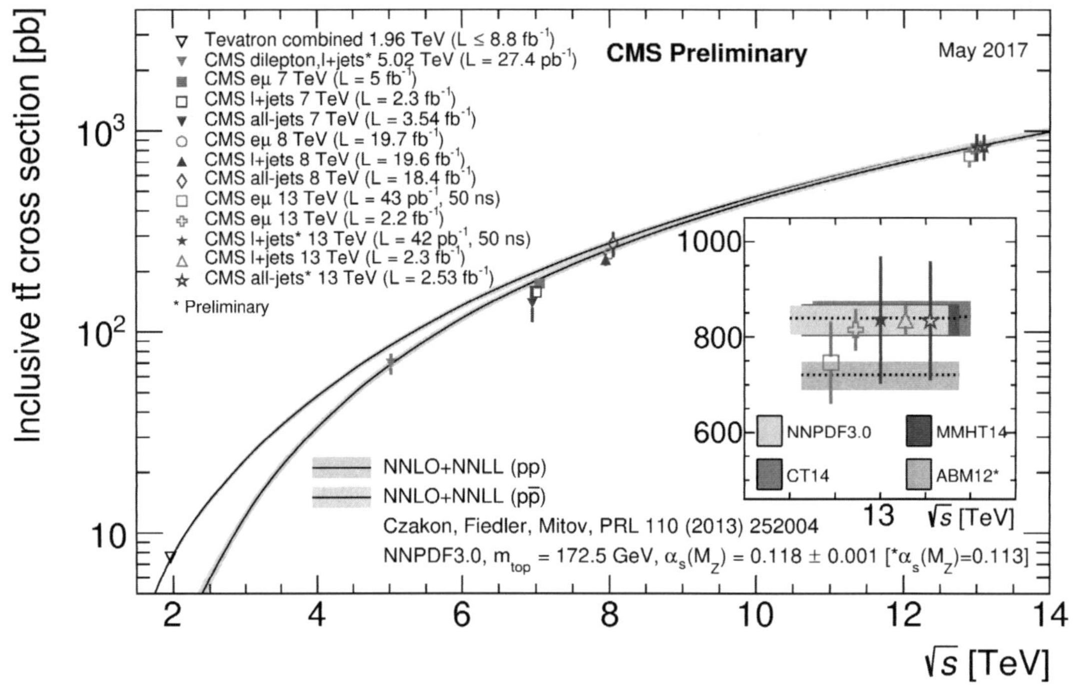

Figure 20. History of inclusive $t\bar{t}$ cross section measurements. Most recent results correspond to the 5 analyses listed at the bottom of the legend, considering dilepton, semileptonic and all jets channels at $\sqrt{s} = 13$ TeV. From [16] [17].

1 dataset, is still more precise with a measured value of $m_t = 172.44^{+0.49}_{-0.49}$ GeV, this Run 1 result is even more precise than the current world combination, computed in 2014, as indicated in the plot [24].

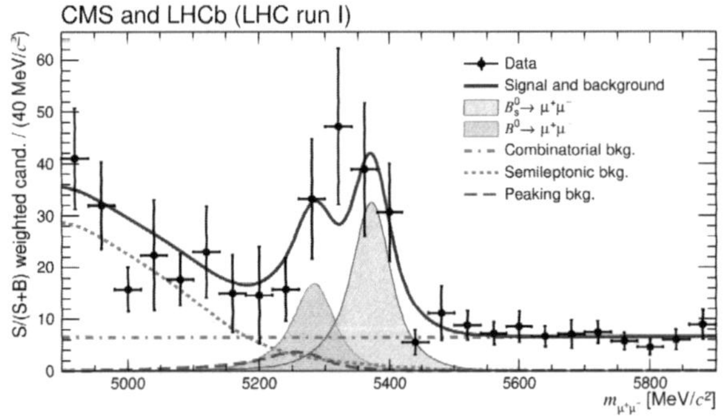

Figure 21. $\mu^+\mu^-$ invariant mass distribution of the Run 1 CMS and LHCb combination analysis to measure the decays of B^0_s and B^0 mesons to a pair of muons. From [25].

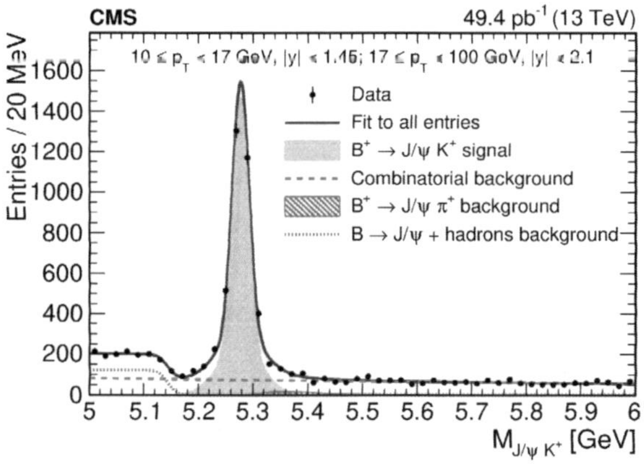

Figure 22. $J/\psi K^+$ candidates invariant mass distribution corresponding to B^+ decays in the Run 2. From [25].

Also the $t\bar{t}$ pair production cross section (corresponding to proton-proton collisions at $\sqrt{s} = 13$ TeV) has been measured. These measurements have been performed using the $t\bar{t}$ pair decay channels: dilepton, semileptonic (ℓ + jets) and all jets, not finding any deviation from the predicted values from the SM as shown in Figure 20. Still there are a lot measurements in progress as the measurement of $t\bar{t}$ spin correlation, W-boson polarisation, $t\bar{t}$ charge asymmetry, top-quark width between other important parameters measured during Run 1 [24].

An interesting parameter that could be sensitive to indentification of new physics is the measurement of the branching fraction of the rare decay $B^0_s \rightarrow \mu^+\mu^-$ and $B^0 \rightarrow \mu^+\mu^-$. The most precise measurements over these parameters were obtained by combining the Run 1 proton-proton collisions datasets from the CMS and LHCb experiments (at $\sqrt{s}$ = 7 and 8 TeV) [25]. This published analysis initiated a phase of precision in the measurement of these two rates. It has been estimated that the production of these decays during Run 2 will be doubled with respect to Run 1. So far the measurements are in agreement with the SM expectation, but the experimental precision reached during Run 1 was still not comparable with the theoretical uncertainty. This is the reason why this is one of the most interesting measurements to reproduce during Run 2 with proton-proton collisions at 13 TeV [25]. Figure 21 shows the invariant mass of the $\mu^+\mu^-$ system, here the resonances corresponding to the B^0_s and B^0 mesons decays at $\sim$ 5.36 and $\sim$ 5.27 GeV can be clearly discriminated from each other [25]. Additionally, other analyses have been completed at 13 TeV, like the measurement of the total and differential inclusive B^+ hadron cross sections. The invariant mass of reconstructed B^+ candidates is shown in Figure 22. The results show a resonable agreement with the theoretical calculations within the uncertainties [26]. In a similar way as with top physics, there are several ongoing measurements of the production and decays of b-mesons, production of bottomium and charmonium states, spectroscopy measurements and search of CP violation and exotic states like tetraquarks and pentaquarks (made up 4 and 5 valence quarks respectively) between others [27].

5. Conclusions

An intensive campaign of upgrade, data-taking, monitoring and analysis work is on-going within the CMS collaboration at CERN in the current Run 2 phase of the LHC. A crucial upgrade process (EYETS) took place at the CMS detector and the LHC from December 2016 to May 2017, being the most significant and important modification, the replacement of the CMS pixel detector in its innermost section ~3cm away from the collision point. In terms of amount of recorded data from proton-proton head-on collisions at $\sqrt{s}$ = 13 TeV, the initial target for 2016 was surpassed by a factor of $\sim \frac{3}{2}$, being as well the case for proton-ion collisions at $\sqrt{s}$ = 8.16 TeV where the planned target was surpassed by a factor of ~3.7. CERN has decided to follow the European Strategy and its High Luminosity LHC plan, whose objetive is the full explotation of the LHC by performing a set of huge upgrades to the LHC itself and its detectors in order to increase the nominal luminosity by a factor of 5-7 times the LHC design luminosity. In terms of physics analysis using the recorded datasets with increased collision energies in 2015, 2016 and 2017, the CMS experiment has started to reproduce all the investigations and measurements that were performed during LHC Run 1. Preliminary results cross-checked the observation of the higgs-like particle discovered during Run 1, and discarded the existence of another higgs resonance ~750 GeV in the H$\rightarrow \gamma\gamma$ channel. There are good prospects to reproduce searches of new physics and precision measurements with this new dataset, as at the moment the amount of data recorded during Run 2 surpasses by more than a factor a 2 what was recorded during the full Run 1 campaign. This will deliver more precision in the measurements and hopefuly some sign of new physics that clearly, as shown throughout this document, has not shown up.

References

[1] L. Evans and P. Bryant, LHC Technical Report, LHC Machine. 2008, JINST **3**, s08001
[2] European Organization for Nuclear Research (CERN), acceleration complex, OPEN-PHO-CHART-2013-001-1, URL: https://cds.cern.ch/record/1621583
[3] HL-LHC High Luminosity Large Hadron Collider: The HL-LHC project, URL: http://hilumilhc.web.cern.ch/about/hl-lhc-project,
[4] Manfred Krammer, The update of the European strategy for particle physics, Published 6 December 2013, 2013, The Royal Swedish Academy of Sciences URL: http://iopscience.iop.org/1402-4896/2013/T158/014019
[5] The CMS experiment at the CERN LHC, Published 14 August 2008, Journal of Instrumentation, Volume **3**, August 2008

[6] CMS Collaboration, CMS detector design illustration, URL: http://cms.web.cern.ch/news/cms-detector-design
[7] CMS Collaboration, Observation of a new boson at a mass of 125 GeV with the CMS experiment at the LHC, **arXiv:1207.7235 [hep-ex]**
[8] CMS Technical Design Report for the Pixel Detector Upgrade, CERN-LHCC-2012-016; **CMS-TDR-11**
[9] The upgrade of the CMS hadron calorimeter with silicon photomultipliers / JINST, Volume **12**, January 2017
[10] CMS Collaboration, CMS Hadron Forward Calorimeter Phase I Upgrade / 2015 J. Phys. Conf. Ser. 587 012007
[11] Colaleo, A (INFN Bari); Safonov, A (Texas A & M); Sharma, A (CERN); Tytgat, M (Gent University), CMS Technical Design Report for the Muon Endcap GEM Upgrade, CERN-LHCC-2015-012; **CMS-TDR-013**
[12] CMS Resistive Plate Chamber overview, from the present system to the upgrade phase, **arXiv:1209.1941**
[13] URL: https://twiki.cern.ch/twiki/bin/view/CMSPublic/LumiPublicResults, CMS Collaboration, CMS PAS LUM-13-001
[14] ATLAS, CMS Collaborations, Combined Measurement of the Higgs Boson Mass in pp Collisions at s=7 and 8 TeV with the ATLAS and CMS Experiments, **arXiv:1503.07589 [hep-ex]**
[15] CMS Collaboration, Results for Moriond Conference 2017, URL: https://cms.cern/news/cms-new-results-Moriond-2017
[16] CMS Collaborarion, Public Standard Model Group Results, URL: https://twiki.cern.ch/twiki/bin/view/CMSPublic/PhysicsResultsCombined
[17] Czakon, Fiedler, Mitov, The total top quark pair production cross-section at hadron colliders through O(α_{S^4}), PRL 110 (2013) 252004, **arXiv:1303.6254 [hep-ph]**
[18] CMS Collaboration, Search for $t\bar{t}$ resonances in highly boosted lepton+jets and fully hadronic final states in proton-proton collisions at $\sqrt{s}$ = 13 TeV, **arXiv:1704.03366 [hep-ex]**
[19] CMS Collaboration, Search for high-mass diphoton resonances in proton-proton collisions at 13 TeV and combination with 8 TeV search, **arXiv:1609.02507 [hep-ex]**
[20] CMS Collaboration, Search for dijet resonances in proton-proton collisions at $\sqrt{s}$ = 13 TeV and constraints on dark matter and other models, **arXiv:1611.03568 [hep-ex]**
[21] CMS Collaboration, Search for narrow resonances in dilepton mass spectra in proton-proton collisions at $\sqrt{s}$ = 13 TeV and combination with 8 TeV data, **arXiv:1609.05391 [hep-ex]**
[22] CMS Collaboration, Search for supersymmetry in pp collisions at $\sqrt{s}$ = 13 TeV in the single-lepton final state using the sum of masses of large-radius jets, **arXiv:1705.04673 [hep-ex]**
[23] CMS Collaboration, Search for supersymmetry in pp collisions at sqrt(s) = 13 TeV in the single-lepton final state using the sum of masses of large-radius jets, **arXiv:1705.04673 [hep-ex]**
[24] CMS Collaboration, Top Physics Summary for LHCP conference URL:https://twiki.cern.ch/twiki/bin/view/CMSPublic/PhysicsResultsTOPSummaryFigures
[25] CMS and LHCb Collaborations, Observation of the rare $B^0_s \rightarrow \mu^+\mu^-$ decay from the combined analysis of CMS and LHCb data, **arXiv:1411.4413 [hep-ex]**
[26] CMS Collaboration, Measurement of the total and differential inclusive B^+ hadron cross sections in pp collisions at $\sqrt{s}$ = 13 TeV, **arXiv:1609.00873 [hep-ex]**
[27] CMS Collaboration, CMS B-physics publications, URL: http://cms-results.web.cern.ch/cms-results/public-results/publications/BPH/index.html
[28] ATLAS, CMS Collaborations, Combined Measurement of the Higgs Boson Mass in pp Collisions at s=7 and 8 TeV with the ATLAS and CMS Experiments, Phys. Rev. Lett. **114**, 191803 (2015), **arXiv:1503.07589 [hep-ex]**
[29] CKMfitter Group - J. Charles et al., Eur. Phys. J. C**41**, 1-131 (2005), **hep-ph-0406184**

IOP Conf. Series: Journal of Physics: Conf. Series **912** (2017) 012006 doi:10.1088/1742-6596/912/1/012006

Prospects for measurement of $R(D)$ and $R(D^*)$ in Belle II.

Jorge Martínez-Ortega
CINVESTAV, Physics Department, Mexico City, Mexico.
E-mail: jmartinez@fis.cinvestav.mx

Abstract. The ratios $R(D) = \mathcal{B}(\bar{B} \to D\tau^-\bar{\nu}_\tau)/\mathcal{B}(\bar{B} \to D\ell^-\bar{\nu}_\ell)$ and $R(D^*) = \mathcal{B}(\bar{B} \to D^*\tau^-\bar{\nu}_\tau)/\mathcal{B}(\bar{B} \to D^*\ell^-\bar{\nu}_\ell)$, measured by the Babar, Belle and LHCb experiments, show a 3.9σ deviation with respect to the Standard Model Prediction. If it is confirmed, it would be a clear evidence for Physics Beyond the SM. Here we present a description of the Belle II experiments, and the plans to measure the ratios mentioned.

1. Introduction

The Babar and Belle experiments, known as B factories, collected more than 1.5 ab^{-1} of e^+e^- collisions, mainly at $\Upsilon(4S)$ resonance. These samples allowed to discover CP violation on B mesons, and to confirm to high precision the Cabbibo-Kobayashi-Maskawa mechanism for quark mixing in the Standard Model. However, besides the success of the SM, there is room to New Physics, as the $R(D)$ and $R(D^*)$ anomalies show. Semileptonic decays are an excellent tool to look for evidence of New Physics beyond the Standard Model (NP). In particular, lepton universality tests can help to probe the SM predictions. In the SM, the Branching fractions $\mathcal{B}(\bar{B} \to D^{(*)}\tau^-\bar{\nu}_\tau)$ and $\mathcal{B}(\bar{B} \to D^{(*)}\ell^-\bar{\nu}_\ell)$ are different only due to the mass of the charged lepton involved. Figure 1 show the tree level diagram of the possible decays. Assuming leptonic universality, the theoretical prediction for the ratios $R(D)$ and $R(D^*)$ [3, 4], are given by

$$R(D) \quad = \frac{\mathcal{B}(\bar{B}\to D\tau^-\bar{\nu}_\tau)}{\mathcal{B}(\bar{B}\to D\mu^-\bar{\nu}_\mu)} \quad = 0.299 \pm 0.011, \tag{1}$$

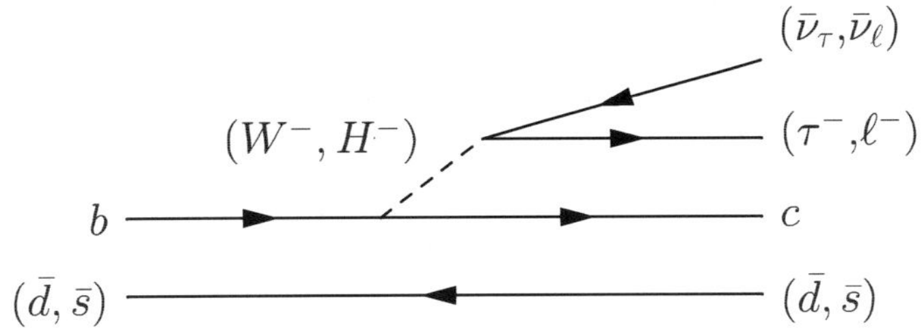

Figure 1. Tree level Feynman diagram showing the possible scenarios for the decays $\bar{B} \to D^{(*)}\tau^-\bar{\nu}_\tau$ and $\bar{B} \to D^{(*)}\ell^-\bar{\nu}_\ell$.

IOP Conf. Series: Journal of Physics: Conf. Series **912** (2017) 012006 doi:10.1088/1742-6596/912/1/012006

Table 1. Summary of experimental measurements of semitauonic B decays [5].

Experiment	Tag method	τ mode	$R(D)$	$R(D^*)$	ρ
Belle 07	inclusive	$e\nu\nu$, $\pi\nu$	0.38 ± 0.11	0.34 ± 0.08	–
Belle 10	inclusive	$\ell\nu\nu$, $\pi\nu$			
BaBar 12	hadronic	$\ell\nu\nu$	$0.440 \pm 0.058 \pm 0.042$	$0.332 \pm 0.024 \pm 0.018$	-0.27
Belle 15	hadronic	$\ell\nu\nu$	$0.375 \pm 0.064 \pm 0.026$	$0.293 \pm 0.038 \pm 0.015$	-0.32
Belle 16	semileptonic	$\ell\nu\nu$	–	$0.302 \pm 0.030 \pm 0.011$	–
Belle 17	hadronic	$\pi\nu$, $\rho\nu$	–	$0.270 \pm 0.035 \pm 0.027$	–
LHCb 16	–	$\ell\nu\nu$	–	$0.336 \pm 0.027 \pm 0.030$	–
HFAG	–	–	$0.407 \pm 0.039 \pm 0.024$	$0.304 \pm 0.013 \pm 0.007$	-0.20
SM	–	–	0.300 ± 0.008	0.252 ± 0.003	–

$$R(D^*) = \frac{\mathcal{B}(\bar{B}\to D^*\tau^-\bar{\nu}_\tau)}{\mathcal{B}(\bar{B}\to D^*\mu^-\bar{\nu}_\mu)} = 0.252 \pm 0.003. \quad (2)$$

However, the current experimental measurements shown a 3.9σ deviation from the SM prediction. We show in the figure the different measurements from BaBar, Belle and LHCb. A new measurement from a Belle II experiment, could help to resolve or confirm this discrepancy.

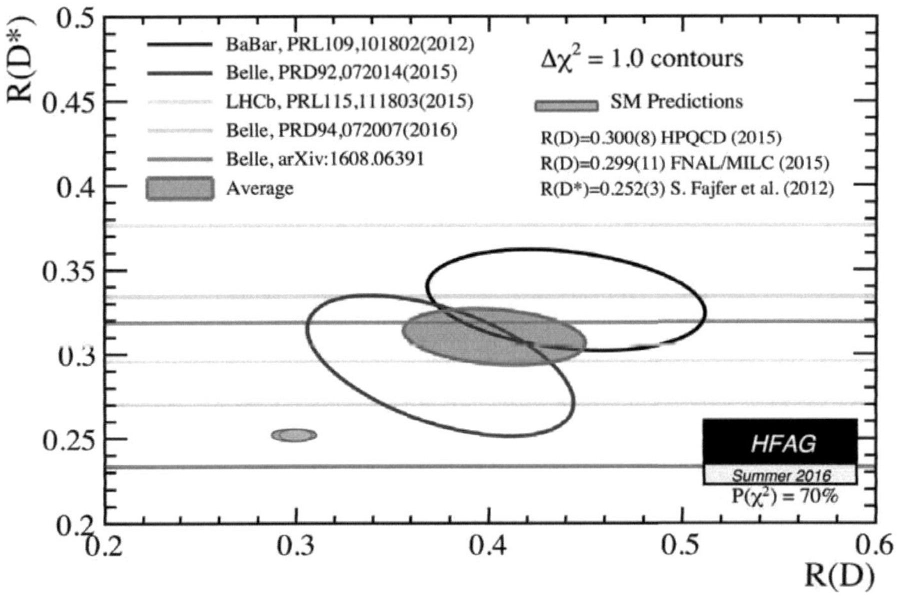

Figure 2. Measured values and SM prediction of $R(D)$ and $R(D^*)$

2. Belle II and SuperKEKB

The Belle II experiment is being built at SuperKEKB factory, in Tsukuba, Japan [6, 7]. It is expected to start taking data at the beginning of 2018, and to collect 50ab^{-1} of integrated luminosity by year 2024. The LHCb experiment is currently running, providing excellent results, and it will be continuing doing so for the coming years, creating a healthy competition with Belle

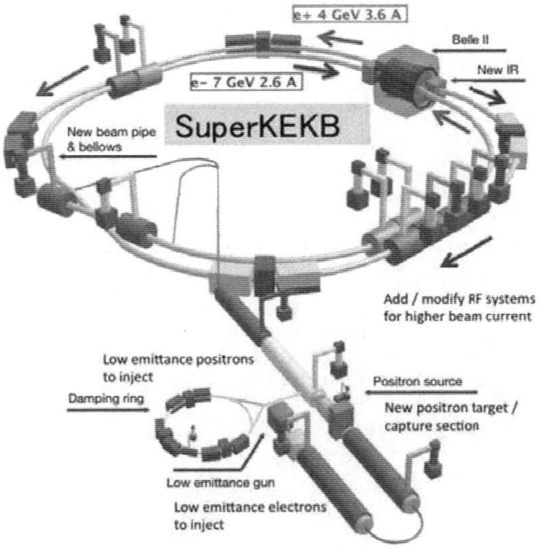

Figure 3. The SUperKEKB accelerator.

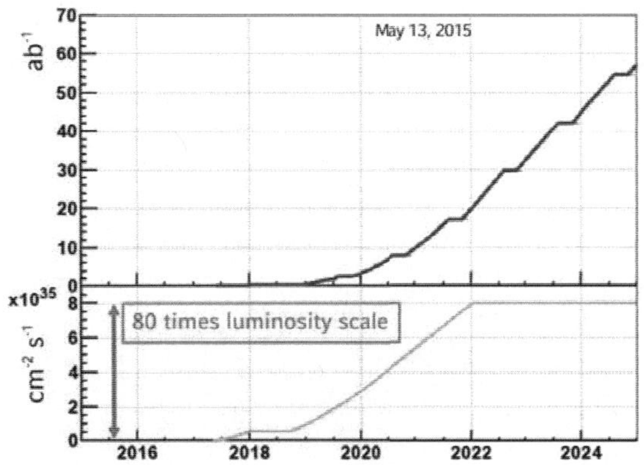

Figure 4. Instantaneous and integrated luminosity projections for SuperKEKB.

II. However, the Belle II experiment is a solid-angle detector working on a clean environment of electron-positron collisions allowing the possibility of the full reconstruction of final states, taking into account the kinematical constraints. These features give the capability of studying in great detail decays such as $\mathcal{B}(\bar{B} \to D^{(*)}\tau^-\bar{\nu}_\tau)$, where the neutrino energy is missed. Below we list some of the characteristics of SuperKEKB and the Belle II detector.

- SuperKEKB is an asymmetric energy beam collider, which provides a boost to the $\Upsilon(4S)$ system with respect to the laboratory reference frame. This boost will be $\beta\gamma = 0.28$ (4.0 GeV for positrons, and 7.5 GeV for electrons).
- The beryllium beam pipe has a 1.0 cm radius, this allows to have a close interaction point with the first layer of the silicon detector layer, at 1.3 cm.
- The new vertex detector is comprised of two devices, the silicon Pixel Detector (PXD) and 36 Silicon Vertex Detector (SVD), with altogether six layers around a 10mm radius Be beam pipe. The first layers at r = 14mm and r = 22mm will use pixelated sensors of the DEPFET type. The remaining four layers at radii of 38mm, 80mm, 115mm, and 140mm will be equipped 40 with double-sided silicon strip sensors. .
- A large Central Drift Chamber (CDC) with thousands of wires, will allow very good resolution at track reconstruction and dE/dx measurement.
- Two Chreenkov detectors, the time of propagation detector (TOP) and the aerogel ring-imaging Cherenkov detector (ARICH) in the endcap, will provide particle identification (PID), allowing very good separation of kaons and pions.
- The Electromagnetic calorimeter (ECL) is located inside a superconducting solenoid coil that provides a 1.5 T magnetic field. The ECL uses Thallium activated Cesium Iodide CsI(Tl) crystals in the barrel region, and pure CsI crystals for the endcaps regions. It is equipped with faster readout electronics with respect to the Belle experiment, in order to manage larger amount of background events.
- Outside the solenoid coil, there is an iron flux-return instrumented to detect K_L^0 mesons and to identify muons (KLM). The KLM consists mainly of Resistive Plate Chambers (RPC) in

IOP Conf. Series: Journal of Physics: Conf. Series **912** (2017) 012006 doi:10.1088/1742-6596/912/1/012006

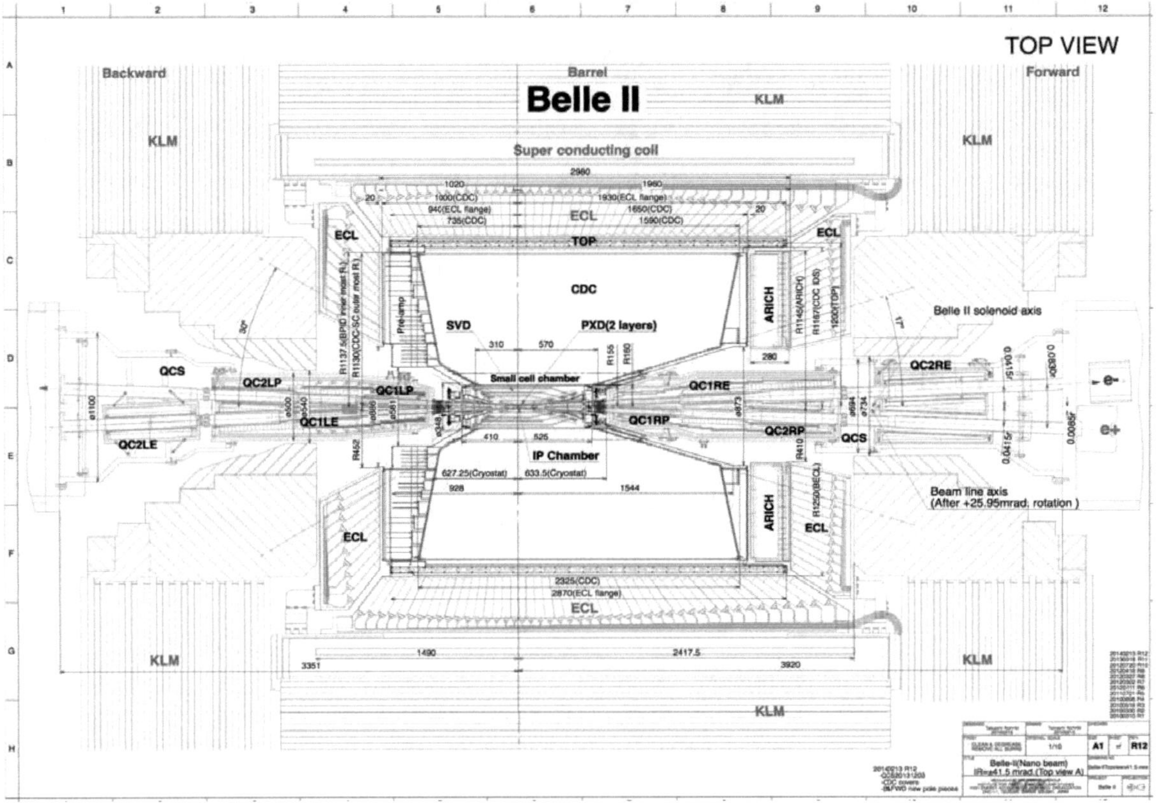

Figure 5. Belle II top view.

the barrel regions, and scintillators in the endcaps regions. K_L^0 mesons shower hadronically in the absorptive iron plates, and can be detected in the KLM active detectors.

3. The method for the measurement.

The strategy of the $R(D^{(*)})$ measurement is to take a ratio of the yields of the $B \to D^{(*)}\tau\nu_\tau$ and the $B \to D^{(*)}\ell\nu_\ell$ events. Three different methods can be used un a $B-$factory experiment: hadronic tag, semileptonic tag, and untagged. The Full Event Interpretation functionality in Belle II Software [8], allows to perform physics analysis taking into account the Missing Energy from neutrinos.

In the hadronic tag analysis the most important variables to separate signal events are related to the missing momentum such as the extra energy on ECL, E_{extra}, and the missing mass squared, M^2_{miss},

$$M^2_{miss} = (p_{e^+e^-} - p_{B_{sig}} - p_{B_{comp}})^2. \tag{3}$$

The missing mass squared is used to separate $B \to D^{(*)}\tau\nu_\tau$ signals from $B \to D^{(*)}\ell\nu_\ell$. The $B \to D^{(*)}\tau\nu_\tau$ are distributed in the higher missing squared mass, and then a multivariate analysis can be used to separate signal events from background processes. One of the advantages of this analysis is the capability of the τ polarization, $P_\tau(D^*)$, measurement using kinematics of the two-body τ decays. $P_\tau(D^*)$ is measured using the distribution of $\cos\theta_{hel}$, which is the cosine of the angle between the momentum of the τ-daughter meson and the direction opposite the momentum of the $\tau\nu_\tau$ system.

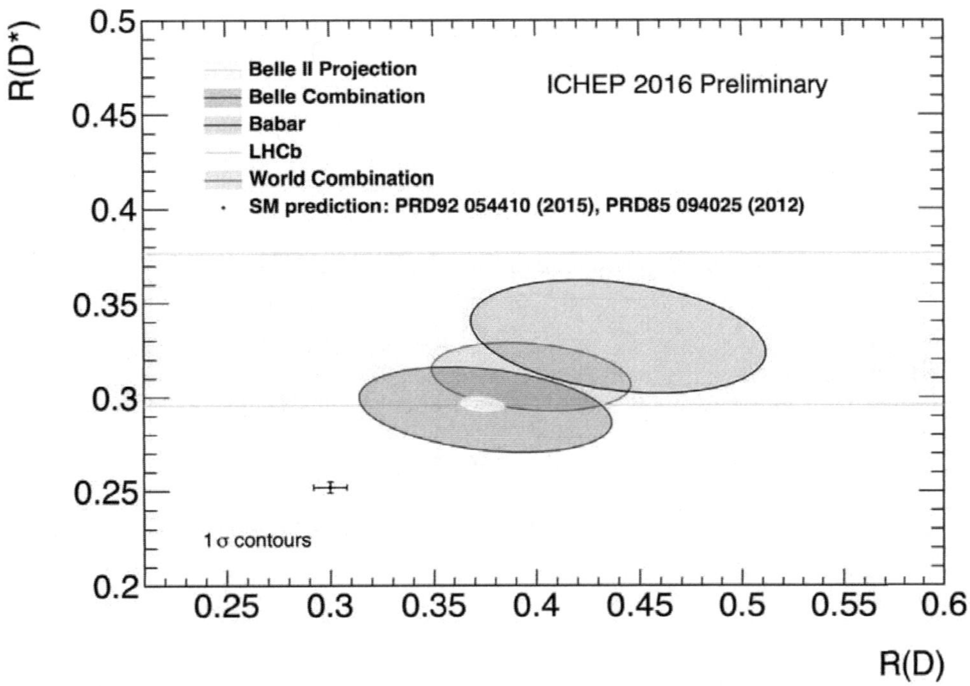

Figure 6. Expected Belle II constraints on the $R(D)$ vs $R(D^*)$ plane, compared to existing experimental constraints from Belle.

For the analysis with semileptonic tagging, the energy deposited in the calorimeter is used to separate $B \to D^*\tau\nu_\tau$ and $B \to D^*\ell\nu_\ell$ from other background, and a multivariate analysis using the angles between the original B meson and the D meson and the lepton can be used to discriminate between $B \to D^*\tau\nu_\tau$ and $B \to D^*\ell\nu_\ell$ events.

In the untagged method, the most important variable to extract signal candidates is the beam-constrained mass of the accompanying B-meson. In this analysis the products of the signal decay firstly are selected and the rest of the event is used to build various shape or topological parameters that discriminate B-meson decays from other hadronic modes. The signal efficiency of the untagged method can be higher than in other methods.

4. Belle II projections

Based on the existing results from Belle and expected improvements at Belle II, the Belle II precisions in the $R(D^{(*)})$ and $P_\tau(D^*)$ measurements are estimated and shown in Table 2 [9]. In Figure 6, the expected precisions at Belle II are compared to the current results and the SM expectations. The $R(D^{(*)})$ precision will be comparable to the current theoretical uncertainty in the SM expectations.

5. Final Remarks

Belle II is ready to collect the legacy of the first generation of B-factories and to continue on the path set by them. Its main purpose will be searching for NP signatures at the intensity frontier. Solving or measuring a confirmation the $R(D^{(*)})$ anomalies, will help to probe the SM to a high degree of precision.

Table 2. Expected precision on $R(D^{(*)})$ and $P_\tau(D^*)$ at Belle II. The first and the second values are the expected statistical and the systematic errors, respectively. These expectations arevshown as the relative (absolute) values for $R(D^{(*)})(P_\tau(D^*))$.

	5 ab^{-1}	50 ab^{-1}
$R(D)$	$(6.0 \pm 3.9)\%$	$(2.0 \pm 2.5)\%$
$R(D^*)$	$(3.0 \pm 2.5)\%$	$(1.0 \pm 2.0)\%$
$P_\tau(D^*)$	0.18 ± 0.08	0.06 ± 0.04

Acknowledgments
JMO would like to thank to Red-FAE-CONACyT and the CINVESTAV Physics Department for the funding provided to attend to the XXXI Annual Meeting of Particles and Fields Division, and to CONACyT for the funding provided by means of project FOINS-296.

References
[1] Elisa Guido. Belle II physics prospects. *PoS*, FPCP2017:036, 2017.
[2] J Bennett. Belle ii physics prospects, status and schedule. *Journal of Physics: Conference Series*, 770(1):012044, 2016.
[3] Dante Bigi and Paolo Gambino. Revisiting $B \to D\ell\nu$. *Phys. Rev.*, D94(9):094008, 2016.
[4] Florian U. Bernlochner, Zoltan Ligeti, Michele Papucci, and Dean J. Robinson. Combined analysis of semileptonic B decays to D and D^*: $R(D^{(*)})$, $|V_{cb}|$, and new physics. *Phys. Rev.*, D95(11):115008, 2017.
[5] HFAG Collaboration. Average of r(d) and r(d*) for fpcp 2017, 2017.
[6] James Kahn (on behalf of the Belle II collaboration). The belle ii experiment. *CERN Proceedings*, 1(0):45, 2017.
[7] Ivan Heredia de la Cruz. The Belle II experiment: fundamental physics at the flavor frontier. *J. Phys. Conf. Ser.*, 761(1):012017, 2016.
[8] Thomas Keck. The Full Event Interpretation for Belle II. Master's thesis, KIT, Karlsruhe, 2014-11.
[9] Gianluca Inguglia. Studies of dark sector & B decays involving τ at Belle and Belle II. *PoS*, ICHEP2016:131, 2016.

IOP Conf. Series: Journal of Physics: Conf. Series **912** (2017) 012007 doi:10.1088/1742-6596/912/1/012007

The $\tau^- \to \eta^{(\prime)}\pi^-\nu_\tau\gamma$ decays as backgrounds in the search for second class currents

Adolfo Guevara

Departamento de Física, Cinvestav IPN, Apdo. Postal 14-740, 07000 Ciudad de México, México

E-mail: aguevara@fis.cinvestav.mx

Abstract. As constructed, the Standard Model does not include genuine Second Class Currents, however these can be induced through the breaking of isospin or charge conjugation. The experimental limits from B-Factories are getting closer to the predictions made for induced Second Class Currents through isospin breaking in the processes $\tau \to \pi\eta^{(\prime)}\nu_\tau$, therefore a more careful analysis of the background in these decays becomes necessary. In this work we analyze the $\tau \to \pi\eta^{(\prime)}\gamma\nu_\tau$ decays as background of the non-radiative process. We find that the radiative processes are very important background for the non-radiative ones whenever the photon escapes detection. Photons cannot be completely excluded since they need to be kept for the identification of the eta and eta' in the final state. However, we find that making appropriate cuts in the energy of the photon, this process can be disregarded as an important background.

1. Introduction

In an effort to understand better the interactions between particles, especially of those that interact strongly, T.D. Lee and C.N. Yang proposed a new operator [1] to study the conservation of its eigenvalues. This operator is defined as the product of a π radians rotation around the I_2 direction in isospin space, times the charge conjugation operator

$$G = Ce^{i\pi I_2}. \tag{1}$$

They found that, for states with zero baryon number and strangeness, $N_B = S = 0$, the operator G has only two eigenvalues, $G = \pm 1$. Therefore, for such states G can be regarded as a parity operator. And thus, hadronic currents with no change in strangeness, charm or bottomness, $\Delta S = \Delta C = \Delta B = 0$, will have a definite G-parity. Depending on their G-parity, these hadronic currents can be classified into first and second class as shown in Table 1 .

Table 1. Definition of first and second class (hadronic) currents according to their parity, spin and G-parity.

	S	P	V	A
1st	$+1$	-1	$+1$	-1
2nd	-1	$+1$	-1	$+1$

The Standard Model (SM) as constructed by Glashow, Weinberg and Salam [2] (and extended to include quarks as done by Glashow, Iliopoulos and Maiani [3]) involves only quark vector

IOP Conf. Series: Journal of Physics: Conf. Series **912** (2017) 012007 doi:10.1088/1742-6596/912/1/012007

currents with $G = +1$ for those with $\Delta S = \Delta C = \Delta B = 0$. Thus, a measurement of a Second Class Current (SCC) could mean evidence for physics beyond the SM (BSM). However, such signal could be also reproduced by an isospin-breaking current, which are present within the SM. Therefore, SM induced SCC must be very well characterized in the search for BSM interactions. Several channels for detection of SCC have been suggested, however the ones suggested by Leroy and Pestieau [4], namely the $\tau^\pm \to \eta^{(\prime)}\pi^\pm\nu_\tau$ decays, are the cleanest channels in which to search for genuine SCC. Since the experimental limits for η [6] and η' [5] for such process are getting close to the theoretical prediction [7], all possible background processes for the measurement of these decays need to be in very good control in order for the experiments to give more precise and reliable results. It so happens that processes in the SM with isospin breaking are suppressed by factors involving differences of light quark masses. In these decays, there is also a kinematical suppression factor (compared to the most common decays into pions), which in total gives

$$\left(\frac{m_{\pi^0}}{m_\eta}\cdot\frac{m_d - m_u}{m_s}\right)^2 \sim 10^{-5}. \tag{2}$$

On the other hand, the radiative process $\tau^- \to \eta^{(\prime)}\pi^-\gamma\nu_\tau$ involves the effective vertex $W^*\pi\eta^{(\prime)}\gamma$ which is not a G eigenstate, since the photon does not have a definite isospin value. However, these decays will have an α_{EM} factor, which is of the order of the isospin suppression. Then, if in the radiative process the photon escapes detection, it could give a fake signal of a genuine SCC detection. Therefore, it becomes utterly necessary to give a prediction of the radiative decay to overcome the difficulty of its misidentification. To do so, we first estimate the Branching Fraction of the radiative process considering only the model dependent contribution stemming from the effective vertex $W^*\pi\eta^{(\prime)}\gamma$ using the Meson Dominance Model (MDM), which gives a simple description of the problem. We then give an estimation of the bremsstrahlung contribution to obtain a lower bound in the photon energy and safely neglect this contribution within Resonance Chiral Theory (RχT) by neglecting the momentum of the photon in the form factors. Then we compute the form factors for the radiative process using RχT to get the photon energy spectrum and compare it to that obtained through the MDM to give an upper bound in the photon energy so that the radiative decay can be safely neglected in the search for SCC. Other studies of background in the search for SCC through the $\tau \to \eta^{(\prime)}\pi\nu_\tau$ decays are being studied, see for example [8].

2. Amplitude and form factors

In order to compute the required observables we need first the amplitude of the process. By choosing the four-momenta convention $\tau(P) \to \pi^-(p)\eta^{(\prime)}(p_0)\nu_\tau(p')\gamma(k,\epsilon)$, one finds that the most general expression for this decay is

$$\mathcal{M} = \frac{eG_F V_{ud}^*}{\sqrt{2}}\epsilon^{*\mu}\left[\bar{u}(p')\gamma^\nu(1-\gamma_5)\left(V_{\mu\nu}^{IB} + V_{\mu\nu} + A_{\mu\nu}\right)u(P)\right], \tag{3}$$

where

$$V_{\mu\nu}^{IB} = \frac{H_\nu(p,p_0)}{-2P\cdot k}(M_\tau + \not{P} - \not{k})\gamma_\mu + V_{\mu\nu}^{SI} \tag{4}$$

is the bremsstrahlung contribution, *i.e.*, that is suppressed by the isospin and the α_{EM} factors. The remaining terms are structure dependent stemming from the $W^*\pi\eta\gamma$ effective vertex, which have no Dirac structure. In the previous equation, $H_\nu(p,p_0) = \langle\eta^{(\prime)}\pi^-|\bar{d}\gamma_\nu u|0\rangle$ is the hadronic current and $V_{\mu\nu}^{SI}$ is the corresponding part for the radiation off the π^-; their general form can be seen in [9].

Now, gauge invariance can be applied to reduce the number of independent form factors, which gives the vector hadronic current $V_{\mu\nu}$ in terms of four form factors and, likewise, the axial hadronic current is given in terms of four axial form factors (using Schouten's identity),

$$\begin{aligned} V_{\mu\nu} =& v_1(p\cdot k\, g_{\mu\nu} - p_\mu k_\nu) + v_2(g_{\mu\nu}\, p_0\cdot k - p_{0\mu}k_\nu) \\ &+ v_3(p_\mu\, p_0\cdot k - p_{0\mu}\, p\cdot k)p_\nu \\ &+ v_4(p_\mu\, p_0\cdot k - p_{0\mu}\, p\cdot k)p_{0\nu}, \end{aligned} \tag{5a}$$

$$\begin{aligned} A_{\mu\nu} =& i\varepsilon_{\mu\nu\rho\sigma}(a_1 p_0^\rho k^\sigma + a_2 k^\rho W^\sigma) \\ &+ i\varepsilon_{\mu\rho\sigma\tau}k^\rho p^\sigma p_0^\tau \left[a_3 W_\nu + a_4(p_0+k)_\nu\right], \end{aligned} \tag{5b}$$

where $W = P - p' = p + p_0 + k$. However, this decomposition is not unique and the form factors v_i and a_i will be determined from the specific model for the hadronic interactions. We will first compute them using the Meson Dominance Model to give an estimation of the SD contribution to the Branching Fraction.

3. Estimation of the bremsstrahlung contribution

Although the bremsstrahlung contribution to the process will be suppressed by α_{EM} and the isospin factor, the infrared behavior of the photon could at some point surpass such suppressions. Therefore it is necessary to estimate the effect of the $V_{\mu\nu}^{IB}$ term at low photon energies. To do so we rely on Low's theorem [10]. This theorem proofs that the amplitude can be written as a power series in the photon's four momentum,

$$\mathcal{M}_\gamma = \frac{A}{k} + B + \mathcal{O}(k). \tag{6}$$

Here, A and B are two quantities that are given in terms of the non-radiative amplitude $\mathcal{M}_0$, in which the four momentum of the photon is neglected on the form factors. Since we are only interested in the low energy behavior of the photon's momentum the amplitude can be approximated by

$$\mathcal{M}_\gamma = -e\mathcal{M}_0\left(\frac{P\cdot\epsilon}{P\cdot k} - \frac{p\cdot\epsilon}{p\cdot k}\right) + \cdots, \tag{7}$$

being that higher orders in k will emerge solely by keeping the energy dependence in the form factors of the $W^*\pi\eta^{(\prime)}$ effective vertex. The non-radiative amplitude can be computed from the expressions in ref. [11]. Thus, by setting a cut for the energy of the photon of 10 MeV the

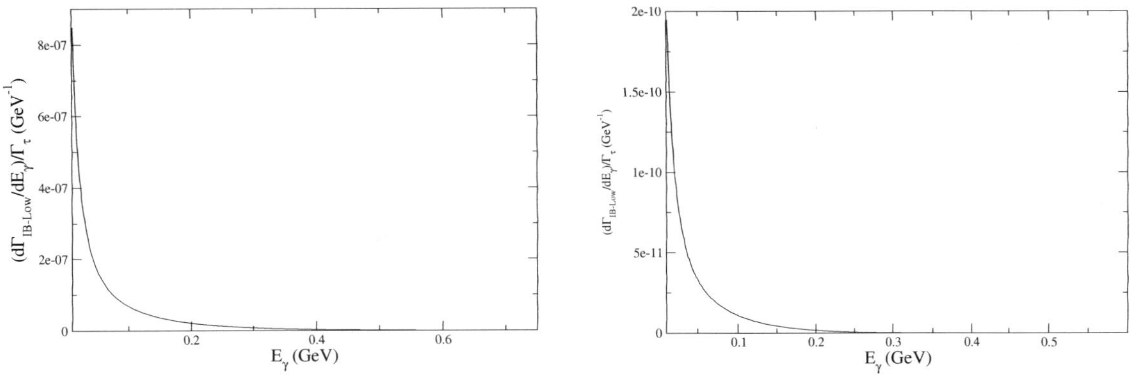

Figure 1. Photon energy spectra for the leading terms of the bremsstrahlung amplitude for $\tau\to\pi\eta\gamma\nu_\tau$ (left) and $\tau\to\pi\eta'\gamma\nu_\tau$ (right).

Branching Ratios obtained are $\mathcal{B}(\tau \to \pi\eta\gamma\nu_\tau) \sim 2.5 \times 10^{-8}$ and $\mathcal{B}(\tau \to \pi\eta'\gamma\nu_\tau) \sim 4.6 \times 10^{-12}$, where the corresponding energy spectra are shown in Figures 1. Therefore, applying this (realistic) bound on the photon's energy one will safely neglect the bremsstrahlung contribution.

4. Form factors within the Meson Dominance Model

In the Meson Dominance Model (MDM), one assumes that the weak and electromagnetic couplings are dominated by the exchange of a few light mesons and their excitations. This approach is useful provided one is able to determine the relevant couplings through data fitting from other independent sources or through model assumptions. The form factors are given by the diagrams of Figure 2, where the vertices are obtained with the rules given by

$$V'^{\mu}(r) \to V^{\alpha}(s)P(t) : ig_{V'VP}\varepsilon^{\mu\alpha\rho\sigma}s_\rho t_\sigma, \tag{8a}$$
$$V^{\mu}(r) \to \gamma^{\alpha}(s)P(t) : ig_{V\gamma P}\varepsilon^{\mu\alpha\rho\sigma}s_\rho t_\sigma, \tag{8b}$$
$$V^{\mu}(r) \to \gamma^{\alpha}(s)P(t) : ig_{V\gamma P}\varepsilon^{\mu\alpha\rho\sigma}s_\rho t_\sigma, \tag{8c}$$
$$A^{\mu}(r) \to V^{\alpha}(s)P(t) : ig_{VAP}(r \cdot s\, g_{\mu\alpha} - r_\alpha s_\mu), \tag{8d}$$
$$V^{\mu}(r) \to \gamma^{\alpha}(s)S(t) : ig_{V\gamma S}(r \cdot s\, g_{\mu\alpha} - r_\alpha s_\mu). \tag{8e}$$

The couplings are phenomenologically obtained from different independent decays. All such determinations of the coupling constants can be seen in ref. [9]. From all diagrams of Fig. 2,

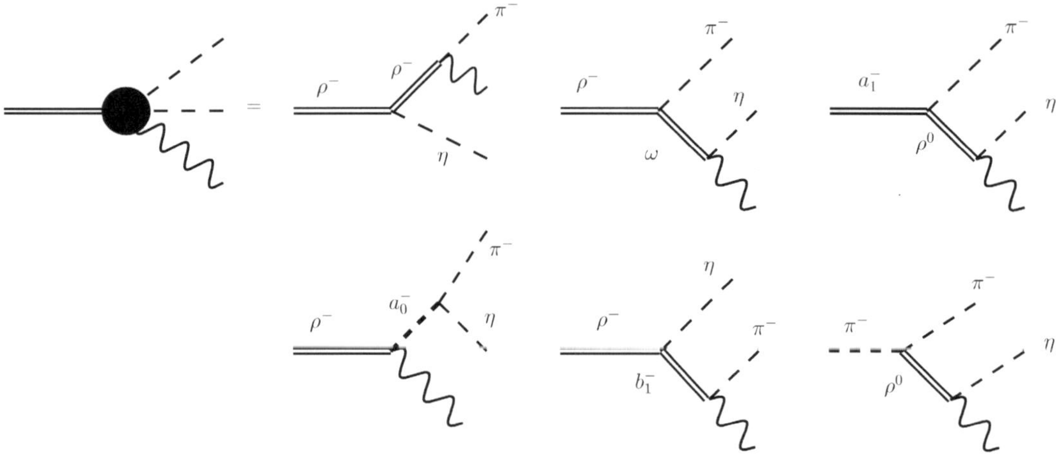

Figure 2. Contribution to the form factors in the Meson Dominance Model

two are not taken into account. First, the contribution of the diagram with the b_1 exchange will be neglected due to the smallness of the Branching Ratios $\mathcal{B}(b_1 \to \pi\gamma) = (1.6 \pm 0.4) \times 10^{-3}$ and $\mathcal{B}(b_1 \to \rho\eta) < 10\%$. Second, the diagram with the pion pole is very suppressed since the pion is far off its mass shell. The second approximation cannot be taken within RχT, since it breaks chiral symmetry. With this we find the expressions of the form factors to be

$$v_1^{MDM} = iC_\rho\left[-\frac{g_{\rho^-\rho^-\eta}g_{\rho^-\pi^-\gamma}}{D_\rho\left[(p+k)^2\right]}p\cdot p_0+\frac{g_{\rho^-\omega\pi^-}g_{\omega\eta\gamma}}{D_\omega\left[(p_0+k)^2\right]}p_0\cdot(p_0+k)+\frac{g_{\rho^-a_0^-\gamma}g_{a_0^-\pi^-\eta}}{D_{a_0}\left[(p+p_0)^2\right]}\right], \quad (9)$$

$$v_2^{MDM} = iC_\rho\left[\frac{g_{\rho^-\rho^-\eta}g_{\rho^-\pi^-\gamma}}{D_\rho\left[(p+k)^2\right]}p\cdot(p+k)-\frac{g_{\rho^-\omega\pi^-}g_{\omega\eta\gamma}}{D_\omega\left[(p_0+k)^2\right]}p\cdot p_0+\frac{g_{\rho^-a_0^-\gamma}g_{a_0^-\pi^-\eta}}{D_{a_0}\left[(p+p_0)^2\right]}\right], \quad (10)$$

$$v_3^{MDM} = iC_\rho\left[-\frac{g_{\rho^-\rho^-\eta}g_{\rho^-\pi^-\gamma}}{D_\rho\left[(p+k)^2\right]}\right], \quad (11)$$

$$v_4^{MDM} = iC_\rho\left[\frac{g_{\rho^-\omega\pi^-}g_{\omega\eta\gamma}}{D_\omega\left[(p_0+k)^2\right]}\right], \quad (12)$$

$$a_1^{MDM} = C_{a_1}\left[\frac{g_{\rho^-a_0^-\gamma}g_{a_0^-\pi^-\eta}}{D_{a_0}\left[(p+p_0)^2\right]}\right](p_0+k)\cdot W, \quad (13)$$

$$a_2^{MDM} = 0, \quad (14)$$

$$a_3^{MDM} = 0, \quad (15)$$

$$a_4^{MDM} = -\frac{a_1^{MDM}}{(p_0+k)\cdot W}. \quad (16)$$

The definition of the constant C_X is given in ref [9] and D_X is the denominator of the propagator of resonance X whose width may depend on the energy of the resonance.

5. Form factors within the Resonance Chiral Theory

Chiral Perturbation Theory (χPT) [12] is the effective field theory of low-energy Quantum Chromodynamics (QCD) built upon its chiral symmetry. However, the validity of this theory stops when the lowest meson resonance, namely the ρ, becomes an active degree of freedom ($\sim$ 700 GeV). So, in order to extend the validity of the approach to higher energies the meson resonances must be included as active degrees of freedom. This is done by relying on the Large N_C approximation of QCD in a manner compatible with chiral symmetry. This extension is called Resonance Chiral Theory (RχT [13]) and by including the lightest meson resonances it is able to extend its applicability to the GeV scale. It also incorporates information about high energy QCD by imposing relations from short distances to form factors, significantly reducing the number of free parameters of the model. It is worth to mention that no gauge symmetry principle is imposed to the resonances.

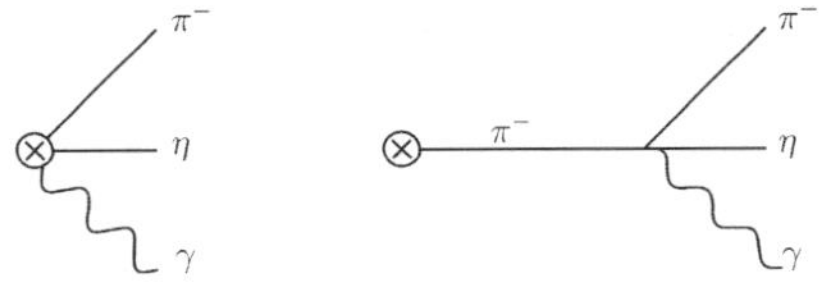

Figure 3. Contribution to the form factors from Chiral Perturbation Theory (no resonances).

Thus, we obtain that the contributions to the form factors are given by the diagrams shown in Figures 3, 4 and 5. To compute every vertex of each diagram a larger base of operators must be considered than that in [13], since this gives only the even-intrinsic parity sector of the lowest chiral order. The operator base are obtained form references [14] and [15] for the odd-intrinsic and even-intrinsic parity sectors, respectively. Due to the extension of the form factors obtained with this model, they will not be shown here. The complete expressions are shown

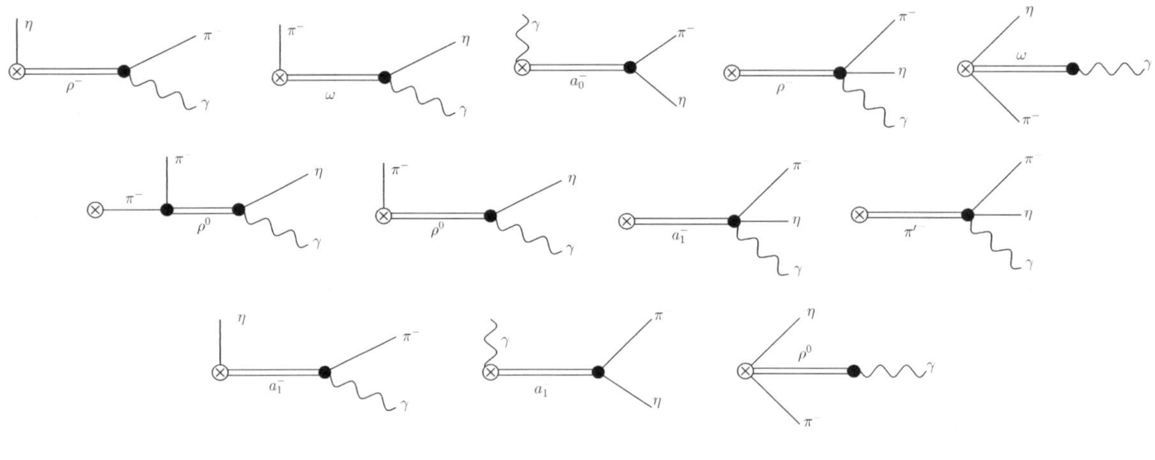

Figure 4. Contribution to the vector (first line) and axial (last two lines) form factors with one resonance exchange in RχT.

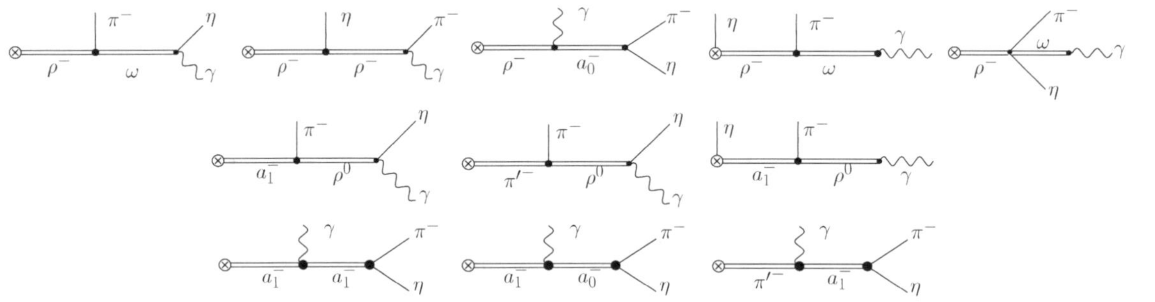

Figure 5. Contribution to the vector (first line) and axial (last two lines) form factors with two resonances exchange in RχT.

in ref [9], along with the short-distance conditions [14, 16] that fix some relations among the coupling constants and phenomenological determinations of given couplings. However, it must be noticed that not all the parameters of the theory can be determined through short distance or phenomenological relations, their absolute values are all bounded by some constraints stemming from the same order expansion constants in plain χPT at NNLO [15]. Take, for example, the coupling strength of the terms in the lagrangian with only a vector resonance, λ_i^V [14]. This coupling must be restricted by the relation

$$\lambda_i^V \sim 3C_i^R \frac{M_V^2}{F} \sim 0.05 \text{ GeV}^{-1}, \tag{17}$$

and it can be taken as an upper bound for the absolute value of the coupling since the relation (C_i^R is a NNLO constant of χPT) $C_i^R \sim \frac{1}{F^2(4\pi)^4}$ is linked to $L_i^R \sim \frac{1}{(4\pi)^2} \sim 5 \times 10^{-3}$, which is the size of the largest NLO couplings, $|L_i^R|$. In a similar way, all bounds for the coupling constants are obtained [9]. The uncertainties on the observables are obtained by randomly varying the values of the coupling constants[1] within the bounds given by the chiral couplings at NNLO.

[1] Two different treatments have been followed in generating the values of the coupling constants: In the first, we generate randomly points within the bounds with uniform probability; in the second, we generate the points

6. Suppressing background in the search for SCC

In studying the $\tau \to \pi\eta^{(\prime)}\gamma\nu_\tau$ decays the question arises of whether one should consider the photon at all. Now, most of the ways in which the η and η' are detected involve photons, and therefore the analysis of the radiative process is utterly necessary to avoid a false signal of genuine SCC. The way to do so is by restricting the energy of the photon. So, to safely neglect the radiative process without affecting the detection of the non-radiative decay we must give a region of the phase space in which the former one is suppressed. Since the difference between both is the photon, it is natural to disregard the radiative process by discarding some values of the photon energy. The way to do this is, again, by relying on Low's theorem, which tells us that structure dependent contributions to the amplitude come from $\mathcal{O}(k)$ terms. Therefore, an upper bound to the energy of the photon must be applied to highly suppress the radiative process (along with the lower bound to neglect the bremsstrahlung contribution).

First, we will analyze the MDM prediction. What we find is that the Branching Fraction is, as

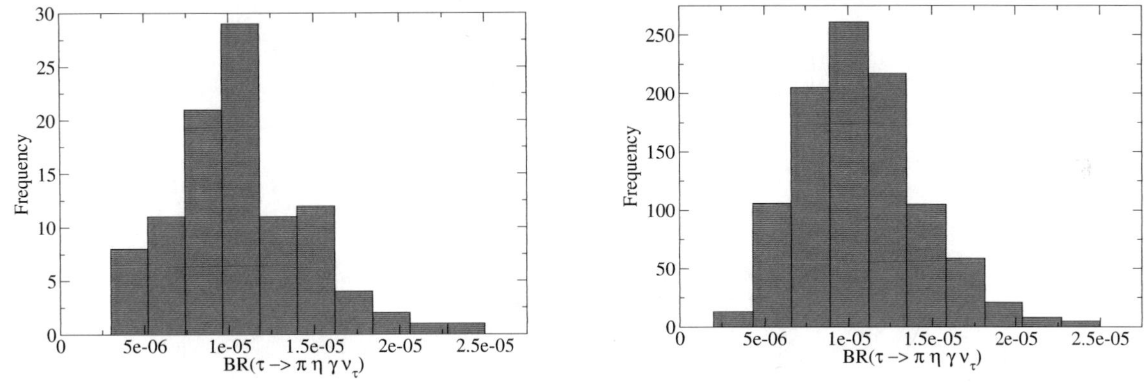

Figure 6. Histogram for $\mathcal{B}(\tau \to \pi\eta\gamma\nu_\tau)$ for 100 (left) and 1000 (right) random points in parameter space in MDM.

supposed, of the order of the non-radiative decay $\left(\mathcal{B}(\tau \to \pi\eta\nu_\tau) \sim 10^{-5}[11]\right)$ as can be seen in Figs 6. Also, from Figs 7 we can see that a good upper bound for the energy of the photon, E_γ might be 100 MeV, which is verified by the results in Fig 8. It must be noticed that despite the peak around (1.25±0.05) GeV in the invariant mass $m_{\eta\pi}$, no marked dynamics is responsible for this effect. Therefore, by reevaluating the histogram with 1000 points in parameter space using the cut in E_γ we obtain the bound $\mathcal{B}(\tau \to \pi\eta\gamma\nu_\tau) \leq 0.6 \times 10^{-7}$.

The procedure for the $\tau \to \pi\eta'\gamma\nu_\tau$ decay is completely analogous. The spectra and the histograms will not be shown here, but can be seen in ref [9]. By using a sample of 1000 points in parameter space and the same upper bound on E_γ we find that $\mathcal{B}(\tau \to \pi\eta\gamma\nu_\tau) \leq 0.2 \times 10^{-8}$. It must be noticed that the MDM prediction is manly given by the a_1^- exchange diagram; if all other contributions are neglected ∼80% of the process is given by this contribution in the η channel, while for the η' it essentially saturates the amplitude.

Using the form factors of RχT one finds the spectra shown in Figs. 9, which also points that a cut in E_γ of 100 MeV will give a strong enough suppression and reproduce the behavior in the $[1.15, 1.35]$ GeV region in the $m_{\eta\pi}$ spectrum obtained with MDM. A remarkable result comes about when the two-resonance-exchange diagrams are neglected, there is little difference between

following a gaussian distribution, being the width equal to the bounds given. Here we will give the numerical values using the second method and present spectra using the first method; for the MDM parameters and the first method, see ref [17] for further discussion on the comparison between the two methods.

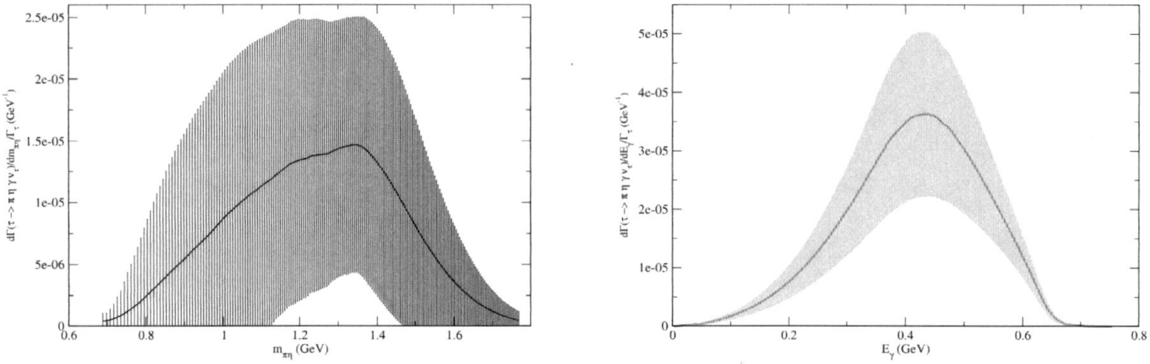

Figure 7. Normalized spectra for $\mathcal{B}(\tau \to \pi\eta\gamma\nu_\tau)$ in MDM for the invariant mass of the $\pi^-\eta$ system (left) and photon energy (right).

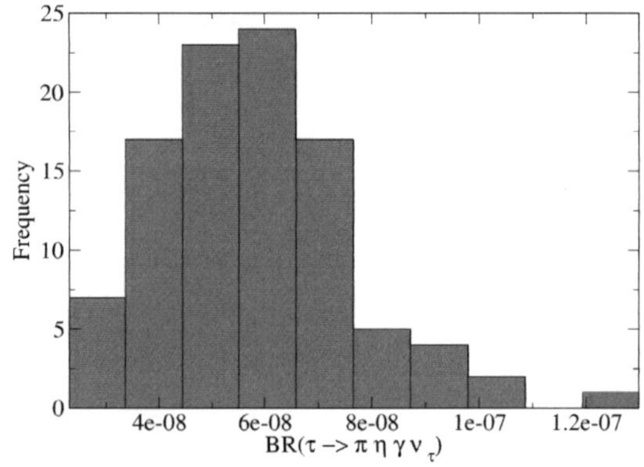

Figure 8. Histogram for $\mathcal{B}(\tau \to \pi\eta\gamma\nu_\tau)$ using 100 points in MDM parameter space rejecting photons with energies E_γ >100 MeV.

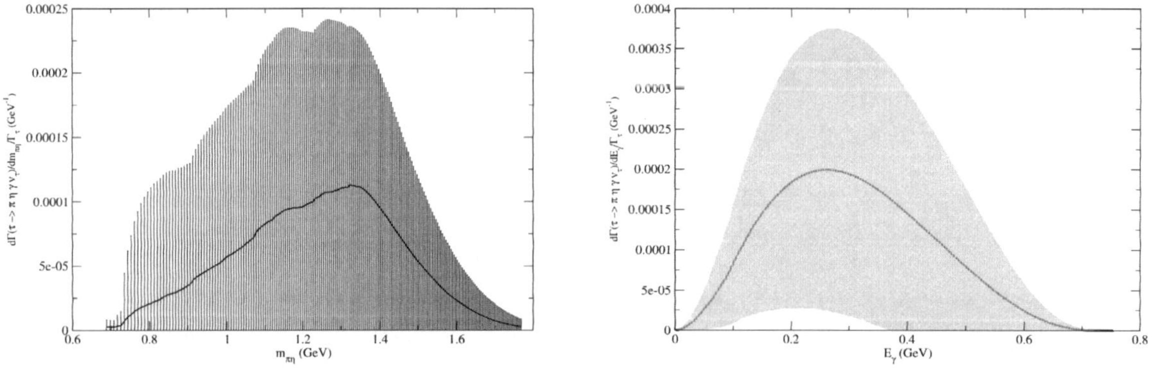

Figure 9. Normalized spectra for $\mathcal{B}(\tau \to \pi\eta\gamma\nu_\tau)$ in RχT for the invariant mass of the $\pi^-\eta$ system (left) and photon energy (right).

this and the full result, as can be seen in Table 2. This is an important feature, for our result can be implemented in TAUOLA without the two-resonance contributions, which will reduce computation time [18].

Table 2. Branching fractions for different kinematical constraints and parameter space points.

	complete $\tau \to \pi\eta\gamma\nu_\tau$	without 2R $\tau \to \pi\eta\gamma\nu_\tau$	complete $\tau \to \pi\eta'\gamma\nu_\tau$	without 2R $\tau \to \pi\eta'\gamma\nu_\tau$
100 points	$(2.3 \pm 0.9) \cdot 10^{-5}$	$(2 \pm 2)10^{-5}$	$(2.3 \pm 3.5) \cdot 10^{-6}$	$(2.1 \pm 1.8) \cdot 10^{-6}$
1000 points	$(3.0 \pm 0.6) \cdot 10^{-5}$	$(2.3 \pm 0.5) \cdot 10^{-5}$	$(2.2 \pm 0.4) \cdot 10^{-6}$	$(2.0 \pm 0.4) \cdot 10^{-6}$
$E_\gamma < 100$ MeV	$(1.2 \pm 0.6) \cdot 10^{-6}$	$(1.0 \pm 0.3) \cdot 10^{-6}$	$(2 \pm 1) \cdot 10^{-7}$	$(2 \pm 1) \cdot 10^{-7}$

7. Conclusions

Belle-II will start collecting data in the very near future and because the experimental limits are now of the order of the theoretical prediction, this brings an excellent opportunity to look for SCC in τ decays. Encouraged by the prediction of MDM, we computed the Branching Fractions of the processes of interest in RχT. Our results are shown in Table 3 for the full phase space and after imposing the cut on the energy of the photon. We also show there the results of the induced SCC through isospin breaking from reference [11].

Table 3. Our prediction of the Branching Ratios $\tau^- \to \pi^-\eta^{(\prime)}\gamma\nu_\tau$ decays without cuts and imposing a rejection of photons with $E_\gamma > 100$ MeV.

Bkg	BR (no cuts)	BR ($E_\gamma^{\text{cut}} > 100$ MeV)	BR SCC signal
η	$(3.0 \pm 0.6) \cdot 10^{-5}$	$(1.2 \pm 0.6) \cdot 10^{-6}$	$\sim 1.7 \cdot 10^{-5}$
η'	$(2.2 \pm 0.4) \cdot 10^{-6}$	$(2 \pm 1) \cdot 10^{-7}$	$[10^{-7}, 10^{-6}]$

The results of Table 3 show that the $\tau \to \pi\eta\gamma\nu_\tau$ decay will have a negligible contribution to the background by imposing the bounds told on the energy of the photon; however, in the case of the $\tau \to \pi\eta'\gamma\nu_\tau$ decay it is not the case, since the large uncertainty in the prediction of the non-radiative process shows that the radiative process might still be an important background in the search for genuine SCC in the η' channel. The only way to surpass this problem would be to obtain a far more precise result of the isospin breaking prediction for the decay for the η'.

With this, we have pointed out for the first time the importance of the process studied as an important background on the search for SCC, where we found that the G-parity violation gives a comparable suppression factor to α_{EM}, the electromagnetic one. We also found that the bremsstrahlung contribution is highly suppressed imposing a fairly lower bound on the energy of the photon E_γ.

We also found that the diagrams with two resonance exchange can be neglected in the computation of the form factors. This is a very useful result, since the form factors obtained through RχT will be implemented in Monte Carlo generators. This will greatly reduce the time of computation and also will facilitate the implementation of the code by using the simplified expressions involving the no-resonance and the one-resonance-exchange diagrams. This can easily be seen in the complete expressions for the form factor available on ref [9].

Acknowledgments

We have benefited from discussions on this topic with Jean Pestieau, Jorge Portolés, and Juan José Sanz-Cillero. Financial support from projects 296 (Fronteras de la Ciencia), 236394, 250628 (Ciencia Básica) and Sistema Nacional de Investigadores (SNI) (Conacyt, México) is acknowledged.

References

[1] T.D. Lee and C.N. Yang, Nouvo Cim. **3** (1956) 749.
[2] S. L. Glashow, Nucl. Phys. **22** (1961) 579, S. Weinberg, Phys. Rev. Lett. **19** (1967) 1264, A. Salam, Conf. Proc. **C** 68-05-19 (1968) 367.
[3] S. L. Glashow, J. Iliopoulos and L. Maiani, Phys. Rev. D**2** (1970) 1285.
[4] C. Leroy and J. Pestieau, Phys. Lett. B**72** (1978) 398.
[5] P. del amo Sánchez et al [BaBar collaboration], Phys. Rev. D**83**, 032002 (2011).
[6] B. Aubert *et al.* [BaBar Collaboration], Phys. Rev. D **77** (2008) 112002.
[7] C. A. Domínguez, Phys. Rev. D **20**, 802 (1979); S. Tisserant and T. N. Truong, Phys. Lett. B **115** (1982) 264; A. Pich, Phys. Lett. B **196**, 561 (1987); A. Bramon, S. Narison and A. Pich, Phys. Lett. B **196**, 543 (1987); E. Berger and H. Lipkin, Phys. Lett. B**189**, 226 (1987); Y. Meurice, Mod. Phys. Lett. A **2**, 699 (1987); Phys. Rev. D **36**, 2780 (1987); C. K. Zachos and Y. Meurice, Mod. Phys. Lett. A **2**, 247 (1987); S. Fajfer and R. J. Oakes, Phys. Lett. B **213**, 376 (1988); H. Pietschmann and H. Rupertsberger, Phys. Rev. D **40**, 3115 (1989); J. O. Eeg and O. Lie-Svendsen, Phys. Lett. B **200**, 182 (1988); M. Suzuki, Phys. Rev. D**36**, 950 (1987); E. Braaten, R. J. Oakes and S. M. Tse, Int. J. Mod. Phys. A **5**, 2737 (1990); J. L. Díaz- Cruz and G. López Castro, Mod. Phys. Lett. A **6**, 1605 (1991); H. Neufeld and H. Rupertsberger, Z. Phys. C **68** (1995) 91.
[8] G. Hernández-Tomé, G. López-Castro and P. Roig, E-print: arXiv:1707.03037
[9] A. Guevara, G. López Castro and P. Roig, Phys. Rev. D**95** (2016) 054015.
[10] F. E. Low, Phys. Rev. **110** (1958) 974.
[11] R. Escribano, S. Gonzàlez-Solís and P. Roig, Phys. Rev. D**94** (2016) 034008.
[12] J. Gasser and H. Leutwyler, Annals Phys. **158** (1984) 142; Nucl. Phys. B **250** 465.
[13] G. Ecker, J. Gasser, A. Pich, E. De Rafael, Nucl. Phys. B**321** (1989) 311; G. Ecker, J. Gasser, H. Leutwyler, A. Pich, E. De Rafael, Phys. Lett. B**223** (1989) 425.
[14] K. Kampf and J. Novotný, Phys. Rev. D**84** (2011) 014036.
[15] V. Cirigliano, G. Ecker, M. Eidenmüller, R. Kaiser, A. Pich and J. Portolés, Nucl. Phys. B**753** (2006) 139.
[16] p. Roig, J.J. Sanz-Cillero, Phys. Lett. B**733** (2014) 158.
[17] Adolfo Guevara, PhD Thesis, *Low-energy meson phenomenology with Resonance Chiral Lagrangians.* (2017), Cinvestav, Mexico City, Mexico. E-print: arXiv:1708.00554
[18] Stanislaw Jadach *et al*, Comput.Phys.Commun. **64** (1990) 275-299; Comput.Phys.Commun. 76 (1993) 361

Forward Physics and the glue at small x

Martin Hentschinski
Departamento de Actuaria, Física y Matemáticas
Universidad de las Américas Puebla
Santa Catarina Martir, 72820
Puebla, Mexico

E-mail: martin.hentschinski@udlap.mx

Abstract.

1. Introduction

The Regge- or high energy limit of perturbative Quantum Chromodynamics (QCD) refers to the study of scattering amplitudes in the limit of infinitely large center of mass energy $\sqrt{s} \to \infty$. To treat such processes using perturbative methods, it is further necessary to insist on the presence of a hard scale Q^2, which is substantially bigger than the typical hadronic scale Λ_{QCD} of the order of a few 100 MeVs; in particular such a hard scale sets usually the scale of the strong coupling constant $\alpha_s = \alpha_s(Q^2)$. Asymptotic freedom provides then a small strong coupling constant $\alpha_s(Q^2) \ll 1$ and perturbative methods can be applied. The perturbative Regge limit of QCD is therefore characterized by the presence of the following hierarchy of scales

$$\sqrt{s} \gg Q \gg \Lambda_{\text{QCD}} \tag{1}$$

A classical example of such a process is provided by Deep Inelastic Scattering (DIS) of electrons on protons at large, see also Fig. 1. The hadronic part of this process is characterized by two important quantities: Q^2 the hard scale or resolution scale corresponding to the virtuality of the photon exchanged between electron and the probed proton and Bjorken x, the ratio of Q^2 and $2p{\cdot}q$ *i.e.* $x = Q^2/(2p{\cdot}q)$. The latter quantity is closely related to the square of the center of mass energy of the virtual photon-proton scattering process, $W^2 = 2p \cdot q - Q^2$; in particular in the limit $W^2 \gg Q^2$, both quantities will coincide and the small x limit of Deep Inelastic Scattering directly corresponds to the perturbative high energy limit of this process. To be more precise the perturbative high energy limit will be from now on always referred to as the limit $x \to 0$ of some quantity x at *fixed* hard scale Q^2, where x is always given by the ratio of the square of the hard scale Q^2 and the squared center of mass energy s (or a closely related quantity). While the structure-less electron provides in DIS processes a perfect probe to investigate the proton at low x, the perturbative high energy limit of QCD can be also explored in hadron-hadron collisions, in particular at the LHC, especially suitable for such explorations due to its high center of mass energy. The low x limit can be probed in this case using production processes in the forward direction of one of the protons, see [1] for a detailed discussion.

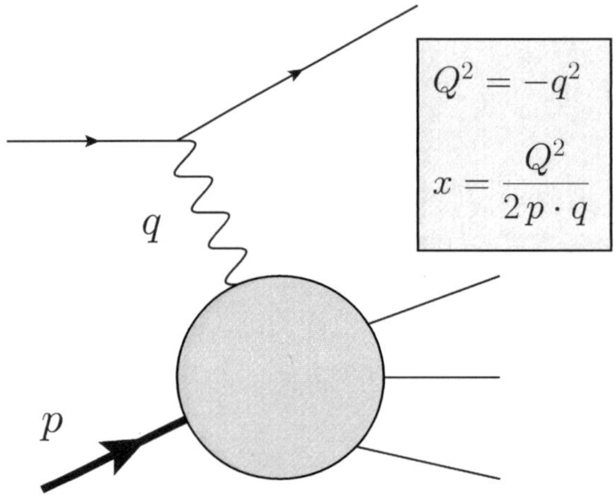

Figure 1. Deep Inelastic Scattering of electrons on protons

From a theory point of view the starting point for the exploration of the perturbative high energy limit of QCD is provided by the Balitsky-Fadin-Kuraev-Lipatov (BFKL) evolution equation [2]. If $x \ll 1$ large logarithms $\ln 1/x$ can compensate for the smallness of the strong coupling constant $\alpha_s(Q^2) \ll 1$ and the naïve expansive in $\alpha_s(Q^2)$ must break down. It is therefore necessary to re-arrange the perturbative expansion and to resum logarithmic enhanced terms $\left(\alpha_s(Q^2)\ln\frac{1}{x}\right)^n$ to all orders in the strong coupling α_s. The perturbative expansion is then no longer organized into contributions which are of leading (LO), next-to-leading order (NLO), etc. in α_s but in terms of leading logarithmic (LL) $\left(\alpha_s(Q^2)\ln\frac{1}{x}\right)^n$, next-to-leading logarithmic (NLL) $\alpha_s(Q^2)\left(\alpha_s(Q^2)\ln\frac{1}{x}\right)^n$ etc. terms. At the same time one is entitled to exploit the small-ness of the variable $x = Q^2/s$ and to make use of an expansion of scattering amplitudes in x or correspondingly in inverse powers of the squared center of mass energy s. In contrast to the conventional expansion in $1/Q^2$ in the Bjorken limit of QCD ($Q^2 \to \infty$ with x fixed), known under the term collinear factorization, the above mentioned expansion in $1/s$ is commonly referred to has high-energy factorization. Using such an expansion of scattering amplitudes in $1/s$ as well as by resumming logarithmic enhanced terms to all orders Kuraev, Fadin and Lipatov, and Balitsky and Lipatov arrived at a evolution equation for the gluon Green's function, named the BFKL evolution equation. It predicts a strong power-like rise of the total cross-section with energy. The observation of such a strong rise in the low x region of the proton structure function F_2 at the HERA experiments at DESY provided then a first triumph of BFKL evolution. Nevertheless the LL BFKL resummation has a number of deficits: it is limited to fixed strong coupling *i.e.* perturbative corrections corresponding to the running of the coupling are not included at this level of precision, and – as any LL resummation – it is subject to large scale uncertainties. This uncertainties where meant to be fixed by the determination of the NLL corrections to the BFKL equation [3]. While those corrections turned out to be large and negative and therefore initially put into question the validity of the BFKL approach, it was soon realized [4] that the largeness of the corrections is directly related to the presence of large collinear logarithms in the next-to-leading BFKL kernel as well as large running coupling corrections *i.e.* terms proportional to the first perturbative term of the QCD beta-functions β_0. Below in Sec. 2 we will describe the details of a BFKL fit to combined HERA data which takes into account resummation of both contributions within the NLO BFKL kernel.

Even though NLO BFKL evolution provides a very good description of currently available low x data, there exist unitarity bounds for hadronic cross-sections which indicate that the power-like rise with increasing x cannot continue forever; instead the observed rise must eventually slow down and saturate. A dynamical mechanism which achieves perturbative unitarity of QCD cross-sections at low x is known under the term 'gluon saturation' as originally proposed by Gribov, Levin and Ryskin [5]. The framework of the Color Glass Condensate (CGC) provides then an effective action approach to gluon saturation [6]. It gives a description of the high energy hadron or nucleus as a weakly-coupled yet non-perturbative system of gluons with high occupancy number, characterized by a semi-hard scale Q_s, called the saturation scale. As a result one obtains a non-linear generalization of the BFKL evolution equation [7–9], which includes multiple-scattering effects; the latter provide the necessary slow down and saturation of the growth with energy, see [10] and references therein

The outline of the following discussion is as follows: In Section 2 we discuss the details of a NLO BFKL fit to combined HERA data. In Section 3 we describe an application of this fit to exclusive vector meson production at the LHC while Section 4 gives an outlook on processes which can possible serve to search for gluon saturation effects.

2. The NLO collinear improved BFKL unintegrated gluon density

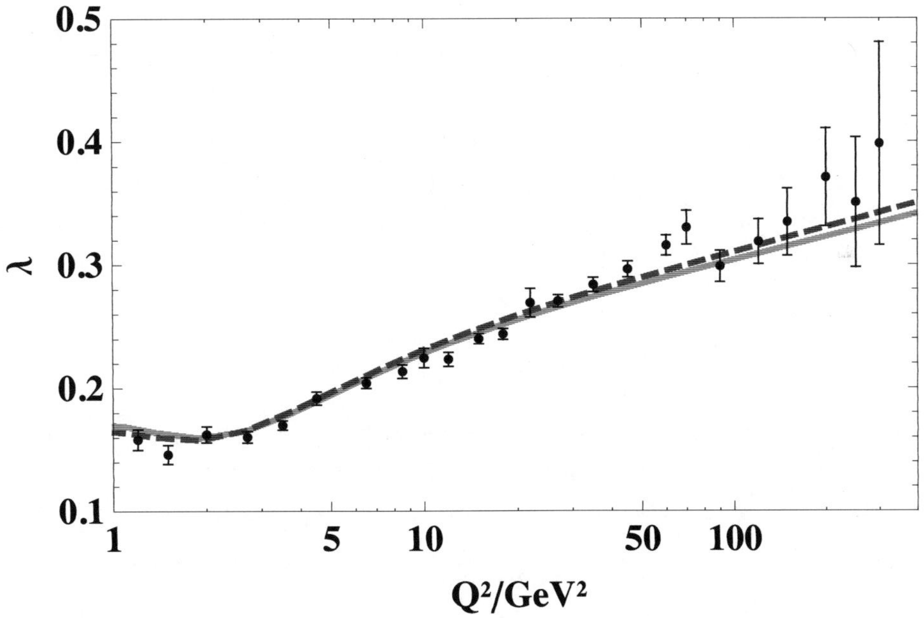

Figure 2. The effective Pomeron intercept $\lambda(Q^2)$ obtained from combined HERA data, using the following parametrization of the proton structure function $F_2(x, Q^2) = c(Q^2)x^{-\lambda(Q^2)}$. The theory predictions corresponds to the two different NLO BFKL fits performed in [13]

The fit achieved in [12, 13] uses a model for the the proton impact factor which serves to parametrize our ignorance about the non-perturbative structure of the proton, absorbed into the initial conditions of the evolution. The employed model is given by

$$\Phi_p\left(\frac{\boldsymbol{q}^2}{Q_0^2}, \delta\right) = \frac{\mathcal{C}}{\pi\Gamma(\delta)}\left(\frac{\boldsymbol{q}^2}{Q_0^2}\right)^{\delta} e^{-\frac{\boldsymbol{q}^2}{Q_0^2}}. \tag{2}$$

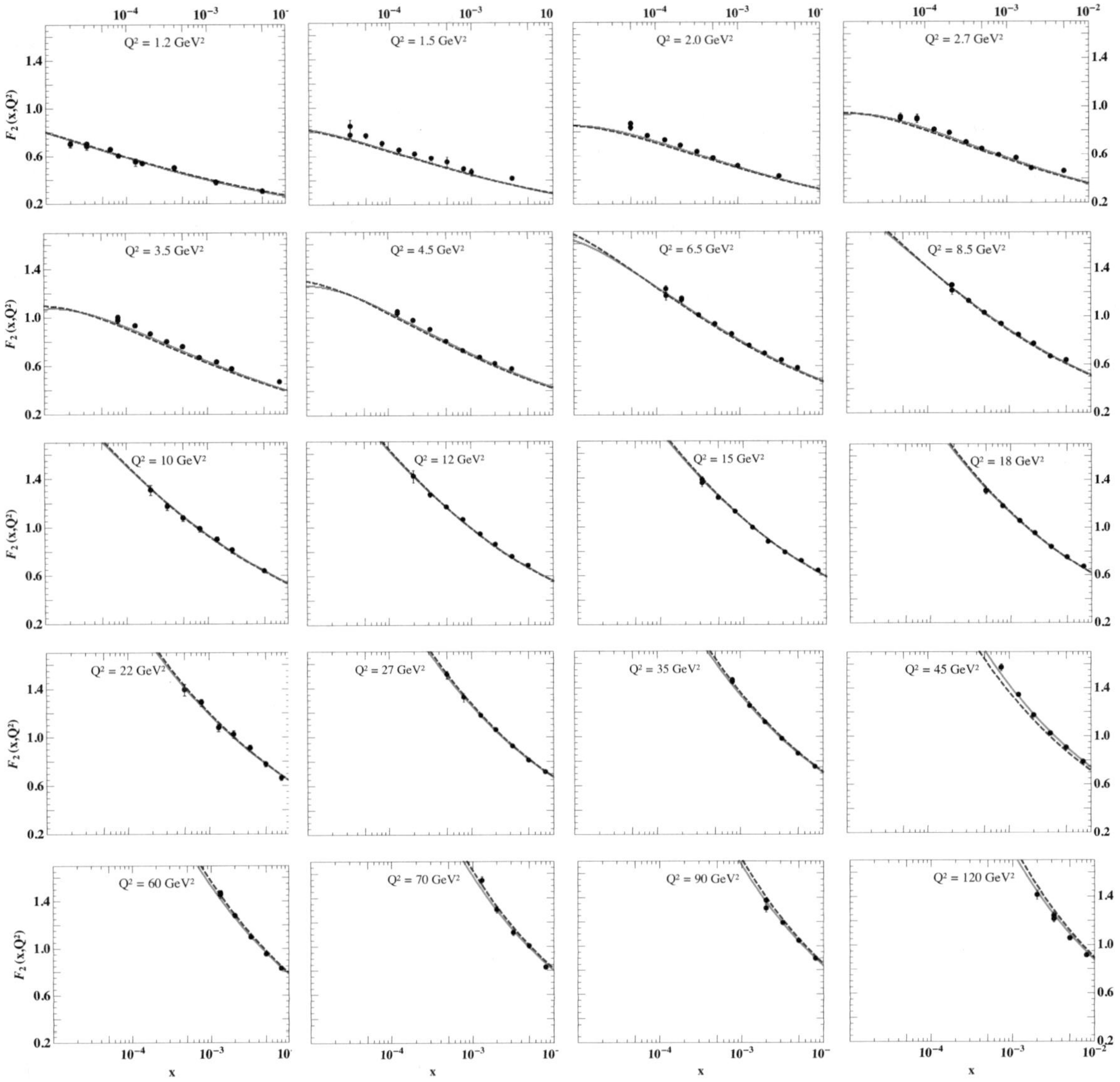

Figure 3. Comparison of the results of the BFKL fit obtained in [13] to combined HERA data.

	virt. photon impact factor	Q_0/GeV	δ	$\mathcal{C}$
fit 1	leading order (LO)	0.28	8.4	1.50
fit 2	LO with kinematic improvements	0.28	6.5	2.35

Table 1. *Parameters of the proton impact factor obtained in [13] through a fit to combined HERA data*

It introduces 2 free parameters plus an overall normalization factor and provides a Poisson-distribution peaked at $\boldsymbol{q}^2 = \delta Q_0^2$. Depending on the precise form of the virtual photon impact factor, two sets of parameters have been determined, summarized in Tab. 1, where for the second fit the leading order virtual photon impact factor has been supplemented with DGLAP inspired kinematic corrections [14]; both fits have been performed for $n_f = 4$ mass-less flavors. The results of the fit are shown in Fig. 2 and Fig. 3 demonstrating excellent agreement with data. In [15] the results of this fit have been used to introduce a NLL BFKL unintegrated gluon density as the following convolution of proton impact factor and BFKL Green's function

$$G(x, \boldsymbol{k}^2, Q_0^2) = \int \frac{d\boldsymbol{q}^2}{\boldsymbol{q}^2} \mathcal{F}^{\rm DIS}(x, \boldsymbol{k}^2, \boldsymbol{q}^2) \Phi_p \left(\frac{\boldsymbol{q}^2}{Q_0^2} \right) . \tag{3}$$

In Mellin space conjugate to transverse momentum space this unintegrated gluon density can be written as

$$G\left(x, \boldsymbol{k}^2, M\right) = \frac{1}{\boldsymbol{k}^2} \int_{\frac{1}{2}-i\infty}^{\frac{1}{2}+i\infty} \frac{d\gamma}{2\pi i} \, \hat{g}\left(x, \frac{M^2}{Q_0^2}, \frac{\overline{M}^2}{M^2}, \gamma\right) \left(\frac{\boldsymbol{k}^2}{Q_0^2}\right)^{\gamma} \tag{4}$$

where M is a characteristic hard scale of the process which in the case of the DIS fit has been identified with the virtuality of the photon and $\overline{M}$ is a corresponding scale which enters the running coupling constant. $\hat{g}$ is finally an operator in γ space and defined as

$$\hat{g}\left(x, \frac{M^2}{Q_0^2}, \frac{\overline{M}^2}{M^2}, \gamma\right) = \frac{\mathcal{C} \cdot \Gamma(\delta - \gamma)}{\pi \Gamma(\delta)} \cdot \left(\frac{1}{x}\right)^{\chi\left(\gamma, \frac{\overline{M}^2}{M^2}\right)} \cdot \left\{1 + \frac{\bar{\alpha}_s^2 \beta_0 \chi_0\left(\gamma\right)}{8N_c} \log\left(\frac{1}{x}\right) \left[-\psi\left(\delta - \gamma\right) + \log \frac{M^2}{Q_0^2} - \partial_\gamma \right] \right\}, \tag{5}$$

where $\bar{\alpha}_s = \alpha_s N_c/\pi$ with N_c the number of colors and $\chi(\gamma, \overline{M}^2/M^2)$ is the next-to-leading logarithmic (NLL) BFKL kernel after collinear improvements; in addition large terms proportional to the first coefficient of the QCD beta function, $\beta_0 = 11N_c/3 - 2n_f/3$ have been resumed through employing a Brodsky-Lepage-Mackenzie (BLM) optimal scale setting scheme [16]. The NLL kernel with collinear improvements reads

$$\chi\left(\gamma, \frac{\overline{M}^2}{M^2}\right) = \bar{\alpha}_s \chi_0\left(\gamma\right) + \bar{\alpha}_s^2 \tilde{\chi}_1\left(\gamma\right) - \frac{1}{2} \bar{\alpha}_s^2 \chi_0'\left(\gamma\right) \chi_0\left(\gamma\right) + \chi_{RG}(\bar{\alpha}_s, \gamma, \tilde{a}, \tilde{b}) - \frac{\bar{\alpha}_s^2 \beta_0}{8N_c} \chi_0(\gamma) \log \frac{\overline{M}^2}{M^2}. \tag{6}$$

with the leading-order BFKL eigenvalue,

$$\chi_0(\gamma) = 2\psi(1) - \psi(\gamma) - \psi(1-\gamma) . \tag{7}$$

The term responsible for the resummation of collinear enhanced terms reads

$$\chi_{RG}(\bar{\alpha}_s, \gamma, a, b) = \bar{\alpha}_s(1 + a\bar{\alpha}_s)\left(\psi(\gamma) - \psi(\gamma - b\bar{\alpha}_s)\right) - \frac{\bar{\alpha}_s^2}{2} \psi''(1-\gamma) - \frac{b\bar{\alpha}_s^2 \cdot \pi^2}{\sin^2\left(\pi\gamma\right)} + \frac{1}{2} \sum_{m=0}^{\infty} \left(\gamma - 1 - m + b\bar{\alpha}_s - \frac{2\bar{\alpha}_s(1 + a\bar{\alpha}_s)}{1 - \gamma + m} + \sqrt{(\gamma - 1 - m + b\bar{\alpha}_s)^2 + 4\bar{\alpha}_s(1 + a\bar{\alpha}_s)}\right). \tag{8}$$

For details on the derivation of this term we refer to the discussion in [12], see also [4, 17]. Employing BLM optimal scale setting and the momentum space (MOM) physical renormalization scheme based on a symmetric triple gluon vertex [18] with $Y \simeq 2.343907$ and gauge parameter $\xi = 3$ one obtains the following next-to-leading order BFKL eigenvalue

$$\begin{aligned} \tilde{\chi}_1(\gamma) &= \tilde{\mathcal{S}}\chi_0(\gamma) + \frac{3}{2}\zeta(3) + \frac{\Psi''(\gamma) + \Psi''(1-\gamma) - \phi(\gamma) - \phi(1-\gamma)}{4} \\ &- \frac{\pi^2 \cos(\pi\gamma)}{4\sin^2(\pi\gamma)(1-2\gamma)}\left[3 + \left(1 + \frac{n_f}{N_c^3}\right)\frac{2+3\gamma(1-\gamma)}{(3-2\gamma)(1+2\gamma)}\right] \\ &+ \frac{1}{8}\left[\frac{3}{2}(Y-1)\xi + \left(1 - \frac{Y}{3}\right)\xi^2 + \frac{17Y}{2} - \frac{\xi^3}{6}\right]\chi_0(\gamma), \end{aligned} \tag{9}$$

where $\tilde{\mathcal{S}} = \frac{(4-\pi^2)}{12}$, see also the discussion in [19]. The coefficients $\tilde{a}, \tilde{b}$ which enter the collinear resummation term Eq. (8) are obtained as the coefficients of the $1/\gamma$ and $1/\gamma^2$ poles of the NLO eigenvalue. In the case of Eq. (9) one has

$$\begin{aligned} \tilde{a} &= -\frac{13}{36}\frac{n_f}{N_c^3} - \frac{55}{36} + \frac{3Y-3}{16}\xi + \frac{3-Y}{24}\xi^2 - \frac{1}{48}\xi^3 + \frac{17}{16}Y \\ \tilde{b} &= -\frac{n_f}{6N_c^3} - \frac{11}{12}. \end{aligned} \tag{10}$$

Employing BLM optimal scale setting, the running coupling constant becomes dependent on the Mellin-variable γ and reads

$$\tilde{\alpha}_s\left(\overline{M}\cdot Q_0, \gamma\right) = \frac{4N_c}{\beta_0\left[\log\left(\frac{\overline{M}\cdot Q_0}{\Lambda^2}\right) + \frac{1}{2}\chi_0(\gamma) - \frac{5}{3} + 2\left(1 + \frac{2}{3}Y\right)\right]}, \tag{11}$$

In addition, in order to access the region of small photon virtualities, in [12,13], a parametrization of the running coupling introduced by Webber in Ref. [20] has been used,

$$\alpha_s\left(\mu^2\right) = \frac{4\pi}{\beta_0 \ln\frac{\mu^2}{\Lambda^2}} + f\left(\frac{\mu^2}{\Lambda^2}\right), \quad f\left(\frac{\mu^2}{\Lambda^2}\right) = \frac{4\pi}{\beta_0}\frac{125\left(1 + 4\frac{\mu^2}{\Lambda^2}\right)}{\left(1 - \frac{\mu^2}{\Lambda^2}\right)\left(4 + \frac{\mu^2}{\Lambda^2}\right)^4}, \tag{12}$$

with $\Lambda = 0.21$ GeV. At low scales this modified running coupling is consistent with global data of power corrections to perturbative observables, while for larger values it coincides with the conventional perturbative running coupling constant. For further details we refer the interested reader to [12,13] and references therein.

3. Exclusive J/Ψ and Υ photo-production cross-sections

A very interesting process to further study the gluon at very low values of x at the LHC is provided by exclusive photo-production of vector mesons J/Ψ and Υ on a proton. In particular the description of the rise with center-of-mass energy W of the $\gamma p \to Vp$ cross-section ($V = J/\Psi, \Upsilon$), combining measurements at HERA and the LHC, is of interest. At the LHC the cross-sections for the process $\gamma p \to Vp$ can be extracted from ultra-peripheral proton-proton (pp) and proton-lead (pPb) collisions, which allow to test the gluon distribution in the proton down to very small values of the proton momentum fraction $x > 4 \cdot 10^{-6}$. For this processes the mass of the heavy quarks, *i.e.* charm (J/Ψ) and bottom (Υ), provide the hard mass-scale which allows for a description within perturbative QCD.

Currently there exist various studies of the data collected during the 7/8 TeV run [21–27]. In following we discuss whether the above obtained perturbative BFKL fit is capable to describe the rise of the $\gamma p \to Vp$ cross-section with energy, following the treatment in [28].

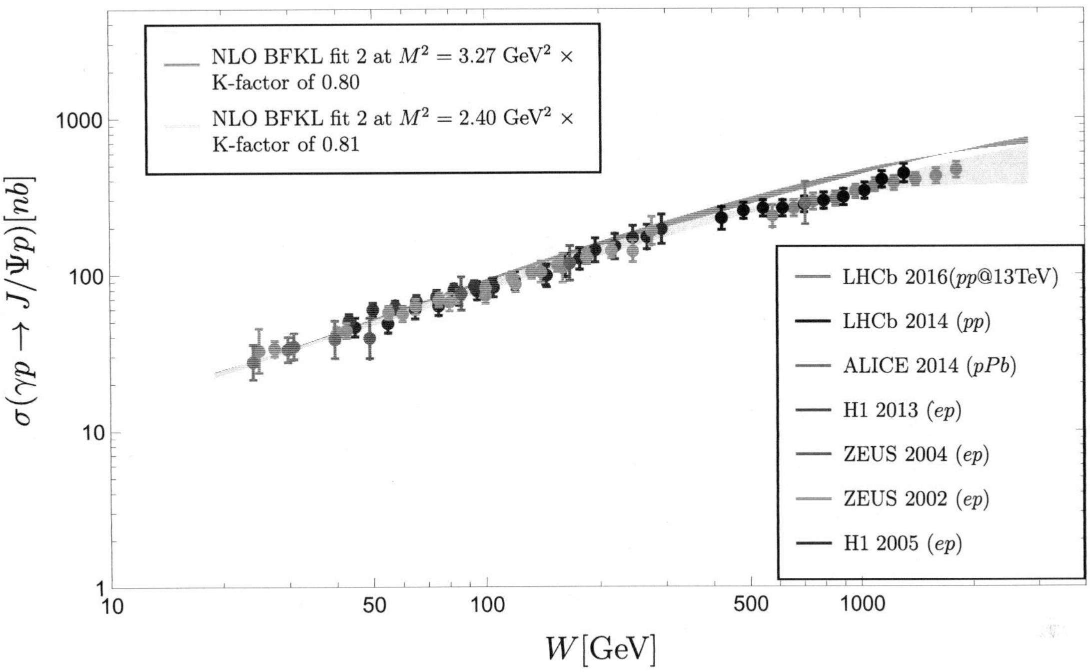

Figure 4. *Energy dependence of the J/Ψ photo-production cross-section as provided by the BFKL fit 1 (up) and 2 (down). The uncertainty band reflects a variation of the scale $\overline{M}^2 \to \{\overline{M}^2/2, \overline{M}^2 \cdot 2\}$. We also show photo-production data measured at HERA by ZEUS [29] and H1 [30] as well as LHC data obtained from ALICE [31] and LHCb (W^+ solutions) [32].*

To use the above unintegrated gluon density for the description of the process $\gamma p \to Vp$, we still require the impact factor for the transition $\gamma \to V$ which we extract from a description based on a factorization of the amplitude in the high energy limit into light-front wave function and and dipole amplitude. In the dilute limit, the factorization into light-front wave function and dipole amplitude becomes equivalent to the factorization into impact factor and unintegrated gluon density and it is therefore possible to recover the required impact factor from these results. Our starting point is the following expression for the imaginary part of the vector meson photo-production scattering amplitude [33, 34]

$$\Im\mathrm{m}\mathcal{A}_T^{\gamma^* p\to Vp}(W,t) = 2\int d^2\boldsymbol{r}\int d^2\boldsymbol{b}\int_0^1 \frac{dz}{4\pi}\,(\Psi_V^*\Psi)_T\, e^{-i[\boldsymbol{b}-(1-z)\boldsymbol{r}]\cdot\boldsymbol{\Delta}}\mathcal{N}(x,r,b)\,, \tag{13}$$

where $\mathcal{N}(x,r,b)$ is the dipole amplitude and T denotes transverse polarization of the virtual photon respectively and $t = -\boldsymbol{\Delta}^2$. The overlap between the photon and the vector meson light-front wave function reads

$$(\Psi_V^*\Psi)_T(r) = \hat{e}_f e\frac{N_c}{\pi z(1-z)}\left\{m_f^2 K_0(\epsilon r)\phi_T(r,z) - \left[z^2+(1-z)^2\right]\epsilon K_1(\epsilon r)\partial_r\phi_T(r,z)\right\}, \tag{14}$$

where $\epsilon^2 = m_f^2$ for real photons and $r = \sqrt{\boldsymbol{r}^2}$, while $f = c, b$ denotes the flavor of the heavy quark, with charge $\hat{e}_f = 2/3,\ 1/3$, corresponding to J/ψ and Υ mesons respectively. For the scalar parts of the wave functions $\phi_{T,L}(r,z)$, we follow closely [24] and employ the boosted Gaussian wave-functions with the Brodsky-Huang-Lepage prescription [35]. For the ground state vector meson

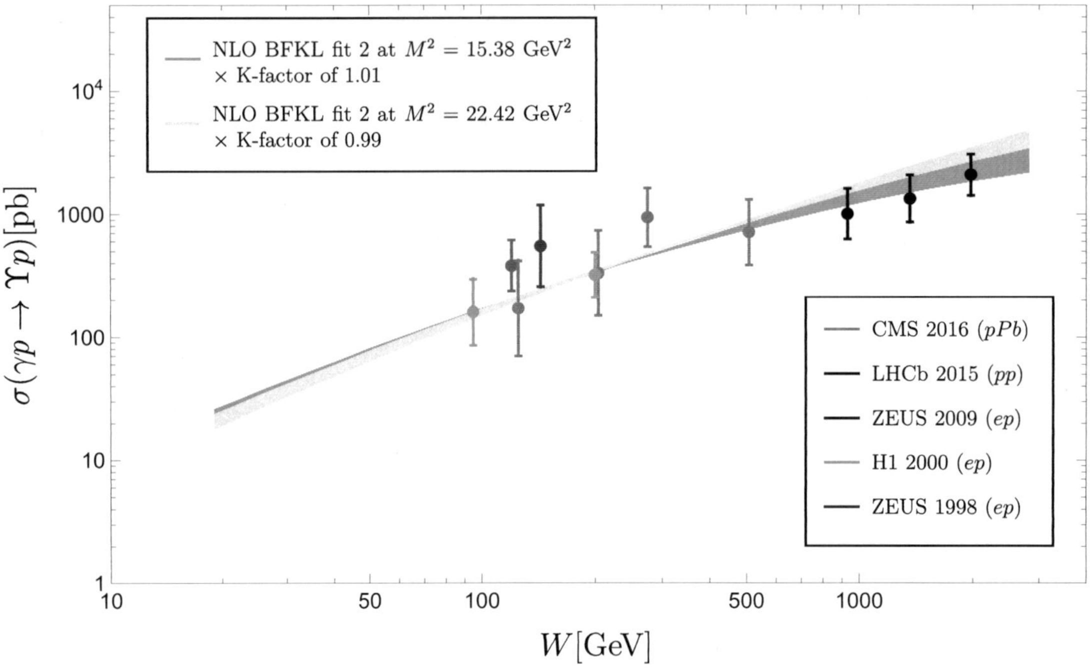

Figure 5. *Energy dependence of the Υ photo-production cross-section as provided by the BFKL fit 1 (up) and 2 (down). The uncertainty band reflects a variation of the scale $\overline{M}^2 \to \{\overline{M}^2/2, \overline{M}^2\cdot 2\}$. We also show HERA data measured by H1 [36] and ZEUS [37] and LHC data by LHCb [38] and CMS [39].*

Meson	m_f/GeV	$\mathcal{N}_T$	$\mathcal{R}^2$/GeV^{-2}	M_V/GeV	$8\mathcal{R}^{-\epsilon}$/GeV^2	$\frac{1}{4}M_V^2$/GeV^2
J/ψ	$m_c = 1.27$	0.596	2.45	3.097	3.27	2.40
Υ	$m_b = 4.2$	0.481	0.57	9.460	15.38	22.42

Table 2. Parameters of the boosted Gaussian vector meson wave functions for J/ψ and Υ obtained in [24, 26]. The last two columns give the two possible hard scales used in the BFKL analysis.

(1s) the scalar function $\phi_{T,L}(r, z)$, has the following general form [34, 40],

$$\phi^{1s}_{T,L}(r, z) = \mathcal{N}_{T,L} z(1-z) \exp\left(-\frac{m_f^2 \mathcal{R}_{1s}^2}{8z(1-z)} - \frac{2z(1-z)r^2}{\mathcal{R}_{1s}^2} + \frac{m_f^2 \mathcal{R}_{1s}^2}{2}\right). \tag{15}$$

The free parameters N_T and $\mathcal{R}_{1s}$ of this model have been determined in various studies from the normalization condition of the wave function and the decay width of the vector mesons. In the following we use the most recent available values *i.e.* [24] (for the J/Ψ) and [26] (for the Υ) The results are summarized in Tab. 2. In the forward limit $t = 0$, the entire dependence of the integrand on the impact parameter $\boldsymbol{b}$ is contained in the dipole amplitude which results into the following inclusive dipole cross-section,

$$2\int d^2\boldsymbol{b}\mathcal{N}(x, r, b) = \sigma_0 N(x, \boldsymbol{r}). \tag{16}$$

The relation between the latter and an unintegrated gluon density has been worked in [41] and is given by

$$\sigma_0 N(\boldsymbol{r}, x) = \frac{4\pi}{N_c} \int \frac{d^2\boldsymbol{k}}{\boldsymbol{k}^2} \left(1 - e^{i\boldsymbol{k}\cdot\boldsymbol{r}}\right) \alpha_s G(x, \boldsymbol{k}^2) \,. \tag{17}$$

This expression can then be used to calculate the BFKL impact factor from the light-front wave function overlap Eq. (14). In particular we find

$$\begin{aligned} \Im\mathrm{m}\mathcal{A}_T^{\gamma^* p \to Vp}(W, 0) &= \int d^2\boldsymbol{r} \int_0^1 \frac{dz}{4\pi} \,(\Psi_V^* \Psi)_T(r) \, \cdot \sigma_0 N\,(x, r) \\ &= \alpha_s(\overline{M} \cdot Q_0) \int_{\frac{1}{2}-i\infty}^{\frac{1}{2}+i\infty} \frac{d\gamma}{2\pi i} \int_0^1 \frac{dz}{4\pi} \; \hat{g}\left(x, \frac{M^2}{Q_0^2}, \frac{\overline{M}^2}{M^2}, Q_0, \gamma\right) \cdot \Phi_{V,T}(\gamma, z, M) \cdot \left(\frac{M^2}{Q_0^2}\right)^{\gamma} . \end{aligned} \tag{18}$$

In the above expression, M and $\overline{M}$ are the mass-scales introduced in Eq. (5). The scale of the strong coupling α_s in Eq. (17), (18) has been set in accordance with the conventions used in the HERA fit[1] [13]. From Eq. (18) we obtain

$$\begin{aligned} \Phi_{V,T}(\gamma, z, M) &= \frac{4\pi}{N_c} \int d^2\boldsymbol{r} \int \frac{d^2\boldsymbol{k}}{(\boldsymbol{k}^2)^2} \left(1 - e^{i\boldsymbol{k}\cdot\boldsymbol{r}}\right) \left(\frac{\boldsymbol{k}^2}{M^2}\right)^{\gamma} (\Psi_V^* \Psi)_T\,(r) \\ &= e\hat{e}_f 8\pi^2 \mathcal{N}_T \frac{\Gamma(\gamma)\Gamma(1-\gamma)}{m_f^2} \left(\frac{m_f^2 \mathcal{R}^2}{8z(1-z)}\right)^2 e^{-\frac{m_f^2 \mathcal{R}^2}{8z(1-z)}} e^{\frac{m_f \mathcal{R}^2}{2}} \left(\frac{8z(1-z)}{M^2 \mathcal{R}^2}\right)^{\gamma} \\ &\left[U\left(2-\gamma, 1, \frac{\epsilon^2 \mathcal{R}^2}{8z(1-z)}\right) + [z^2 + (1-z)^2] \frac{\epsilon^2 (2-\gamma)}{2 \cdot m_f^2} U\left(3-\gamma, 2, \frac{\epsilon^2 \mathcal{R}^2}{8z(1-z)}\right)\right], \end{aligned} \tag{19}$$

where $U(a, b, z)$ is a hypergeometric function of the second kind or Kummer's function.

Even though the real part of the scattering amplitude is suppressed by powers of α_s in the high energy limit, it can still provide a sizable correction to the cross-section and should be therefore included. In the high energy limit it is possible to obtain this real part from the imaginary part using dispersion relation. One has

$$\frac{\Re\mathrm{e}\mathcal{A}(W^2, t)}{\Im\mathrm{m}\mathcal{A}(W^2, t)} = \tan\frac{\lambda\pi}{2}, \qquad \text{with} \qquad \lambda = \frac{d\ln\mathcal{A}(W^2, t)}{d\ln W^2} \,. \tag{20}$$

Eq. (20) is frequently used in the literature in the study of photo-production of vector mesons. Within our framework we write first the imaginary part of the scattering amplitude as a double

[1] A precise determination of the scale of this running coupling would require the complete NLO corrections to the impact factor which are currently not available

Mellin transform

$$\Im\mathrm{m}\mathcal{A}_T^{\gamma p\to Vp}(W^2,0) = \alpha_s(\overline{M}\cdot Q_0)\cdot \int\limits_{\delta-i\infty}^{\delta+i\infty}\frac{d\omega}{2\pi i}\left(\frac{1}{x}\right)^{\omega}\int\limits_{\frac{1}{2}-i\infty}^{\frac{1}{2}+i\infty}\frac{d\gamma}{2\pi i}\left(\frac{M^2}{Q_0^2}\right)^{\gamma}$$

$$\int\limits_0^1 \frac{dz}{4\pi}\Phi_{V,T}(\gamma,z,M)\frac{\mathcal{C}\cdot\Gamma(\delta-\gamma)}{\pi\Gamma(\delta)}\cdot\Bigg\{\frac{1}{\omega-\chi\left(\gamma,\frac{\overline{M}^2}{M^2}\right)}$$
$$+\frac{\bar{\alpha}_s^2\beta_0\chi_0(\gamma)/(8N_c)}{\left[\omega-\chi\left(\gamma,\frac{\overline{M}^2}{M^2}\right)\right]^2}\left[-\psi(\delta-\gamma)-\frac{d\ln\left[\Phi_{V,T}(\gamma,z,M)\right]}{d\gamma}\right]\Bigg\}, \tag{21}$$

where the ω-contour runs to the right of all singularities and $x = \frac{M_V^2}{W^2-m_p^2}$, with $m_p = .938$ GeV the proton mass; note that $\mathcal{A}_L = 0$ due to $Q = 0$. To determine the real part we identify λ with the Mellin variable conjugate to the x, $\lambda = \omega$. The complete amplitude is then obtained through multiplying the integrand of Eq. (21) by a factor $\left(i + \tan\frac{\omega\pi}{2}\right)$. The Mellin transform w.r.t. ω is then easily evaluated through taking residues at the single and double pole at $\omega = \chi(\gamma, \overline{M}^2.M^2)$, while residues at $\omega = -1, -3, \ldots$ are sub-leading in the high energy/low x limit and therefore neglected. As a consequence we obtain – in contrast to the bulk of phenomenological studies in the literature – an energy dependent ratio of real and imaginary parts. Particular for small values of W, the real part provides a relative large correction, up to 41% in the case of the J/Ψ and 70% in the case of the Υ, see also the discussion in [42]. On the other hand, since this ratio is decreasing with increasing W, we find that this energy-dependent ratio leads to a slow-down of the growth with energy in the high energy region.

While we calculated so far the differential cross-section at momentum transfer $t = 0$, experimental data which we wish to analyze, are usually given for cross-sections integrated over t. It is therefore necessary to model the t-dependence and to relate in this way the differential cross-section at $t = 0$ to the integrated cross-section. Here we follow the prescription given in [21, 22], who assume an exponential drop-off with $|t|$, $\sigma \sim \exp\left[-|t|B_D(W)\right]$ with an energy dependent t slope parameter B_D, which can be motivated by Regge theory,

$$B_D(W) = \left[b_0 + 4\alpha'\ln\frac{W}{W_0}\right]\mathrm{GeV}^{-2}. \tag{22}$$

For the numerical values we use $\alpha' = 0.06$ GeV^{-2}, $W_0 = 90$ GeV and $b_0^{J/\Psi} = 4.9$ GeV^{-2} in the case of the J/Ψ, while $b_0^{\Upsilon} = 4.63$ GeV^{-2} for Υ production, as proposed in [21, 22]. The total cross-section for vector meson production is therefore obtained as

$$\sigma^{\gamma p\to Vp}(W^2) = \frac{1}{B_D(W)}\frac{d\sigma}{dt}(\gamma p\to Vp)\bigg|_{t=0}. \tag{23}$$

In Fig. 4 and Fig. 5 we compare perturbative BFKL evolution driven by the BFKL gluon with $x = M_V^2/(W^2 - m_p^2)$ (where M_V is the mass of the vector meson and m_p the proton mass) to experimental data. Even though we require in the case of the J/Ψ a K-factor of the order of 0.8 to address an off-set in the overall normalization, we find that the W-dependence of the total exclusive vector meson photo-production cross-section is both for J/Ψ and Υ production very well described in terms of collinear improved BFKL evolution. As a new feature we included in the present comparison 2016 13 TeV LHCb data. In accordance with the 2014 LHCb data, they

IOP Conf. Series: Journal of Physics: Conf. Series **912** (2017) 012008 doi:10.1088/1742-6596/912/1/012008

seem to prefer the choice $M^2 = (M_{J/\Psi}/2)^2$ for the hard scale. The good agreement with the data seems to suggest that at currently accessible collider data non-linear effects are not yet dominant. Nevertheless care is needed in the interpretation of these results: at first the uncertainty band of the BFKL prediction increases in the region of largest W value (in particular in the case of the J/Ψ which is characterized by a smaller hard scale). This indicates the existence of a potential instability of the current implementation of NLO BFKL evolution in this region of phase space. Moreover, non-linear effects are for exclusive vector meson production only manifest through evolution, but not on the level of the observable. This suggests that it is necessary to identify observables which are sensitive to saturation effects already at the level of the observables to allow for improved saturation searches

4. Searching for high density effects

Since the CGC-framework assumes high occupation number of the gluonic target field, the latter can be no longer treated perturbatively. The resulting formalism represents therefore the DIS scattering process in the low x limit as the scattering of the virtual photon on a strong background color field, which represents the target. Defining light-cone vectors $n, \bar{n}$ through the four momenta of virtual photon and target, *i.e.* $n \sim p$ and $n \cdot \bar{n} = 1$ with $v^- \equiv n \cdot v, v^+ \equiv \bar{n} \cdot v$ for a generic four vector v, theory calculations are performed using quark and gluon propagators in the presence of the background field $A^{+,c}$. In particular one has for the propagators

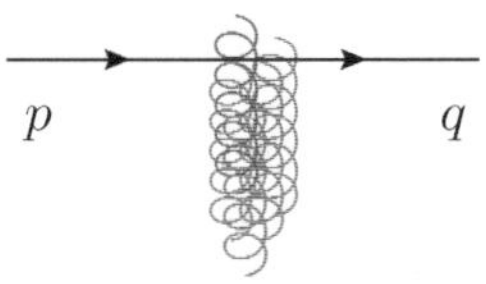

$$= (2\pi)^d \delta^{(d)}(p-q)\tilde{S}_F^{(0)}(p) + \tilde{S}_F^{(0)}(p) \quad [\text{diagram}] \quad \tilde{S}_F^{(0)}(q),$$

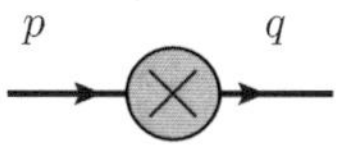

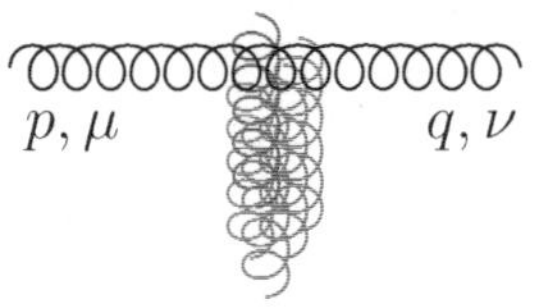

$$= (2\pi)^d \delta^{(d)}(p-q)\tilde{G}_{\mu\nu}^{(0)}(p) + \tilde{G}_{\mu\alpha}^{(0)}(p) \quad [\text{diagram}] \quad \tilde{G}_{\alpha\nu}^{(0)}(q), \tag{24}$$

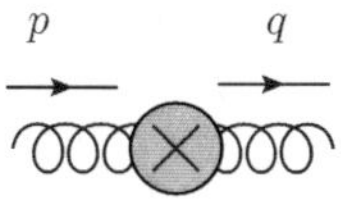

with the free propagators

$$\tilde{S}_F^{(0)}(p) = \frac{i\not{p} + m}{p^2 - m^2 + i0}, \qquad \tilde{G}_{\mu\nu}^{(0)}(p) = \frac{i d_{\mu\nu}(p)}{p^2 + i0}, \qquad d_{\mu\nu}(p) = -g_{\mu\nu} + \frac{n_\mu^- p_\nu + p_\mu n_\nu^-}{n^- \cdot p}, \tag{25}$$

and the interaction terms

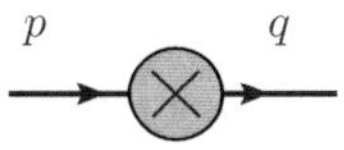

$$= 2\pi\delta(p^- - q^-)\not{n}^- \int d^{d-2}\boldsymbol{z} e^{-i\boldsymbol{z}\cdot(\boldsymbol{p}-\boldsymbol{q})} \Big\{\theta(p^-)\big[V(\boldsymbol{z}) - 1\big] - \theta(-p^-)\big[V^\dagger(\boldsymbol{z}) - 1\big]\Big\}, \tag{26}$$

$$= -2\pi\delta(p^- - q^-)2p^- \int d^{d-2}\boldsymbol{z} e^{-i\boldsymbol{z}\cdot(\boldsymbol{p}-\boldsymbol{q})} \cdot \Big\{\theta(p^-)\big[U(\boldsymbol{z}) - 1\big] - \theta(-p^-)\big[U^\dagger(\boldsymbol{z}) - 1\big]\Big\}. \tag{27}$$

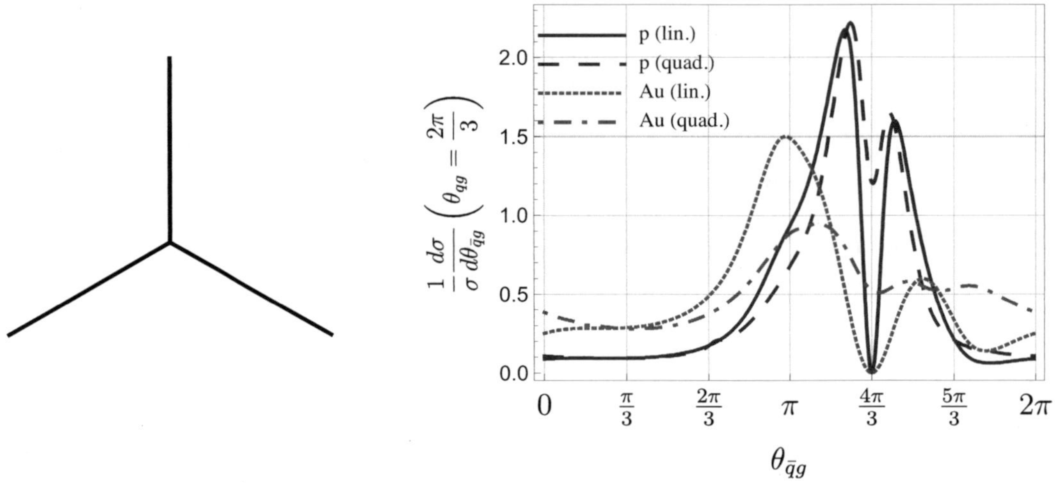

Figure 6. *Left:* Mercedes-Benz star configuration: all transverse momenta have equal absolute value, with relative angle $\Delta = 2\pi/3$. *Right:* Normalized cross-section for three parton production in the presence of large gluon densities against $\Delta\theta_{\bar{q}g}$ with $\Delta_{qg} = 2\pi/3$ for proton and gold up to linear and quadratic order in $N^{(2)}$. For the study the photon momentum fractions of the final state particles have been fixed to $z_1 = z_2 = 0.2$, $z_3 = 0.6$ while $|\boldsymbol{p}_1| = |\boldsymbol{p}_2| = |\boldsymbol{p}_3| = 2$ GeV and $Q^2 = 9$ GeV^2 (plot originally published in [43]).

The latter contain Wilson lines in the fundamental (V) and adjoint (U) representation which resum the interaction with the background field with

$$V(\boldsymbol{z}) \equiv V_{ij}(\boldsymbol{z}) \equiv \mathrm{P}\exp ig \int_{-\infty}^{\infty} dx^- A^{+,c}(x^-, \boldsymbol{z}) t^c \tag{28}$$

and $U^{ba}(z_t) = 2\,\mathrm{tr}(t^a V(z_t) t^b V^\dagger(z_t))$; P denotes path ordering of the gluon field in the exponent while t^c is a SU(N_c) generator in the fundamental representation.

Current x values reached at today's collider experiments are most likely still too large to provide definite evidence for the presence of saturation effects. A first possibility to increase gluon densities and therefore the sensitivity to saturation is to consider scattering on large nuclei, where gluon densities are expected to be enhanced by a factor $A^{\frac{1}{3}}$, with A the mass number of the nucleon. A second possibility is to consider exclusive final states: while for inclusive observables the presence of high gluon densities leads mainly to a modification of the x dependence of gluon densities [11], exclusive production processes can provide matrix elements which differ for high gluon densities from their low density counter parts. They are therefore more sensitive to effects related to high gluon densities and provide a tool for their exploration. Such a study has been carried out in [43, 44], where the differential cross-section for the process

$$\gamma^*(q) + \text{target}\ (p) \to q(p_1) + \bar{q}(p_2) + g(p_3) + X, \tag{29}$$

has been calculated for first time within the CGC formalism. For details on the calculation we refer to [43, 44]. The results of a first numerical study are shown in Fig. 6. With the differential cross-section for the process Eq. (29) expressed in terms of dipole and quadrupole operators of Wilson lines Eq. (28),

$$S^{(2)}_{(\boldsymbol{x}_1\boldsymbol{x}_2)} \equiv \frac{1}{N_c}\mathrm{tr}\left[V(\boldsymbol{x}_1)V^\dagger(\boldsymbol{x}_2)\right], \quad S^{(4)}_{(\boldsymbol{x}_1\boldsymbol{x}_2\boldsymbol{x}_3\boldsymbol{x}_4)} \equiv \frac{1}{N_c}\mathrm{tr}\left[V(\boldsymbol{x}_1)V^\dagger(\boldsymbol{x}_2)V(\boldsymbol{x}_3)V^\dagger(\boldsymbol{x}_4)\right], \tag{30}$$

this study uses the large N_c and Gaussian approximation [45] to express the quadrupole in terms of the dipole operator. For the dipole profile itself a model motivated by recent rcBK[2] fits to HERA data [46] has been used,

$$S^{(2)}_{(\boldsymbol{x}_1\boldsymbol{x}_2)} = \int d^2\boldsymbol{l}\, e^{-i\boldsymbol{l}\cdot\boldsymbol{x}_{12}}\, \Phi(\boldsymbol{l}^2) = 2\left(\frac{Q_0|\boldsymbol{x}_{12}|}{2}\right)^{\rho-1} \frac{K_{\rho-1}(Q_0|\boldsymbol{x}_{12}|)}{\Gamma(\rho-1)},$$
$$\text{where} \qquad \Phi(\boldsymbol{l}^2) = \frac{\rho-1}{Q_0^2\pi}\left(\frac{Q_0^2}{Q_0^2+\boldsymbol{l}^2}\right)^{\rho}, \tag{31}$$

and Q_0 is a scale proportional to the saturation scale. To justify the dilute limit associated with the Gaussian approximation, we study the cross-section at large photon virtuality $Q^2 = 9$ GeV^2 and expand the differential cross-section up to quadratic order in $N^{(2)} = 1 - S^{(2)}$. We further use $\rho = 2.3$ and $Q_0^{\text{proton}} = 0.69$ GeV, as motivated by inclusive DIS fits at $x = 0.2 \times 10^{-3}$. Since we want to search for effects due to large gluon densities, we further simulate scattering on a gold nucleus through the re-scaling $Q_0^{\text{Au}} = A^{1/6} \cdot Q_0^{\text{proton}} = 1.67$ GeV. At linear order in $N^{(2)}$, the cross-section is proportional to the Fourier transform of the dipole, $\Phi\left((\boldsymbol{p}_1+\boldsymbol{p}_2+\boldsymbol{p}_3)^2\right)$ and therefore gives direct access to the gluon distribution in the target. At this order the cross-section is – up to the detailed shape and x-dependence of the dipole factor Eq. (31) – identical to a corresponding BFKL calculation. The configuration dominant in the 'collinear' limit corresponds on the other hand to the vanishing of the sum of transverse momenta $\boldsymbol{p}_1 + \boldsymbol{p}_2 + \boldsymbol{p}_3 = 0$, *i.e.* the transverse momentum transfer between projectile and target is zero. To study deviations from this configuration we first chose $|\boldsymbol{p}_1| = |\boldsymbol{p}_2| = |\boldsymbol{p}_3|$. For this choice the 'collinear' limit $\boldsymbol{p}_1 + \boldsymbol{p}_2 + \boldsymbol{p}_3 = 0$ is then given by the Mercedes-Benz star configuration, see Fig. 6. We observe vanishing of the partonic cross section at these 'collinear' configurations, Fig. 6, with a strong double peak. This vanishing is explained by the vanishing of the partonic matrix element at leading order in $N^{(2)}$ for zero momentum transfer between projectile and target, which itself is expected from gauge invariance. For a complete study at hadronic level, we expect the double peak to turn into a single peak. While in the case of the proton we find (besides the double-peak structure) mainly a smearing of the collinear peak at $\Delta_{\bar{q}g} = 4\pi/3$, the effect of a larger gluon saturation scale for a gold nucleus can be clearly seen: while the shape of the linear gold curve is similar to the proton curve, it is wider and depleted with respect to the former. Including further quadratic terms in $N^{(2)}$, we start to probe effects due to high gluon densities at the level of the matrix element beyond the mere modification of Eq. (31) and beyond the corresponding BFKL matrix element. In the case of the proton, these corrections are small, as expected for $x = 0.2 \cdot 10^{-3}$ and a relatively large photon virtuality $Q^2 = 9$ GeV^2. For a highly saturated gold nucleus, they are on the other hand sizable. While this apparently puts doubts on the reliability of the current expansion up to quadratic order in $N^{(2)}$, it clearly demonstrates the sensitivity of this processes to high and saturated gluon densities and identifies it as suitable tool in searching for such effects already at currently accessible collider energies.

References

[1] K. Akiba *et al.* [LHC Forward Physics Working Group], J. Phys. G **43** (2016) 110201 doi:10.1088/0954-3899/43/11/110201 [arXiv:1611.05079 [hep-ph]].

[2] L. N. Lipatov, Sov. J. Nucl. Phys. **23** (1976) 338, E. A. Kuraev, L. N. Lipatov, V. S. Fadin, Phys. Lett. B **60** (1975) 50, Sov. Phys. JETP **44** (1976) 443, Sov. Phys. JETP **45** (1977) 199. I. I. Balitsky, L. N. Lipatov, Sov. J. Nucl. Phys. **28** (1978) 822.

[3] V. S. Fadin, L. N. Lipatov, Phys. Lett. B **429** (1998) 127, M. Ciafaloni, G. Camici, Phys. Lett. B **430** (1998) 349.

[4] G. P. Salam, JHEP **9807** (1998) 019 [hep-ph/9806482].

[2] The running coupling Balitsky-Kovchegov equation (rcBK) provides a non-linear generalization of the BFKL equation as discussed above, see [8, 46] for details

[5] L. V. Gribov, E. M. Levin and M. G. Ryskin, Phys. Rept. **100**, 1 (1983).

[6] L. D. McLerran and R. Venugopalan, Phys. Rev. D **49**, 2233 (1994), Phys. Rev. D **49**, 3352 (1994).

[7] J. Jalilian-Marian, A. Kovner, L. D. McLerran and H. Weigert, Phys. Rev. D **55**, 5414 (1997); J. Jalilian-Marian, A. Kovner, A. Leonidov and H. Weigert, Nucl. Phys. B **504**, 415 (1997), Phys. Rev. D **59**, 014014 (1999), Phys. Rev. D **59**, 014015 (1999), Phys. Rev. D **59**, 034007 (1999), A. Kovner, J. G. Milhano and H. Weigert, Phys. Rev. D **62**, 114005 (2000); A. Kovner and J. G. Milhano, Phys. Rev. D **61**, 014012 (2000); E. Iancu, A. Leonidov and L. D. McLerran, Nucl. Phys. A **692**, 583 (2001), Phys. Lett. B **510**, 133 (2001); E. Ferreiro, E. Iancu, A. Leonidov and L. McLerran, Nucl. Phys. A **703**, 489 (2002); H. Weigert, Nucl. Phys. **A703**, 823-860 (2002); J. -P. Blaizot, E. Iancu, H. Weigert, Nucl. Phys. **A713**, 441-469 (2003).

[8] I. Balitsky, Nucl. Phys. B **463**, 99 (1996); Y. V. Kovchegov, Phys. Rev. D **60**, 034008 (1999), Phys. Rev. D **61**, 074018 (2000).

[9] Y. V. Kovchegov and E. Levin, Nucl. Phys. B **577**, 221 (2000) [hep-ph/9911523]; M. Hentschinski, H. Weigert and A. Schafer, Phys. Rev. D **73**, 051501 (2006) [hep-ph/0509272]; E. Levin and A. Prygarin, Phys. Rev. C **78**, 065202 (2008) [arXiv:0804.4747 [hep-ph]].

[10] E. Iancu and R. Venugopalan, In *Hwa, R.C. (ed.) et al.: Quark gluon plasma* 249-3363; H. Weigert, Prog. Part. Nucl. Phys. **55**, 461 (2005); J. Jalilian-Marian, Y. V. Kovchegov, Prog. Part. Nucl. Phys. **56**, 104-231 (2006); F. Gelis, E. Iancu, J. Jalilian-Marian, R. Venugopalan, Ann. Rev. Nucl. Part. Sci. **60**, 463-489 (2010); J. L. Albacete and C. Marquet, Prog. Part. Nucl. Phys. **76**, 1 (2014).

[11] J. Bartels and K. Kutak, Eur. Phys. J. C **53**, 533 (2008) [arXiv:0710.3060 [hep-ph]]; L. Motyka and M. Sadzikowski, Acta Phys. Polon. B **45**, no. 11, 2079 (2014) [arXiv:1411.7774 [hep-ph]].

[12] M. Hentschinski, A. Sabio Vera and C. Salas, Phys. Rev. Lett. **110** (2013) no.4, 041601 [arXiv:1209.1353 [hep-ph]].

[13] M. Hentschinski, A. Sabio Vera and C. Salas, Phys. Rev. D **87** (2013) no.7, 076005 [arXiv:1301.5283 [hep-ph]].

[14] J. Kwiecinski, A. D. Martin and A. M. Stasto, Phys. Rev. D **56**, 3991 (1997) [hep-ph/9703445]; A. Bialas, H. Navelet and R. B. Peschanski, Nucl. Phys. B **603**, 218 (2001) [hep-ph/0101179].

[15] G. Chachamis, M. Deàk, M. Hentschinski, G. Rodrigo and A. Sabio Vera, JHEP **1509**, 123 (2015) [arXiv:1507.05778 [hep-ph]].

[16] S. J. Brodsky, G. P. Lepage and P. B. Mackenzie, Phys. Rev. D **28** (1983) 228.

[17] A. Sabio Vera, Nucl. Phys. B **722** (2005) 65 [hep-ph/0505128].

[18] W. Celmaster and R. J. Gonsalves, Phys. Rev. D **20** (1979) 1420.

[19] S. J. Brodsky, V. S. Fadin, V. T. Kim, L. N. Lipatov and G. B. Pivovarov, JETP Lett. **76** (2002) 249 [Pisma Zh. Eksp. Teor. Fiz. **76** (2002) 306] [hep-ph/0207297]; JETP Lett. **70** (1999) 155 [hep-ph/9901229].

[20] B. R. Webber, JHEP **9810** (1998) 012 [hep-ph/9805484].

[21] S. P. Jones, A. D. Martin, M. G. Ryskin and T. Teubner, J. Phys. G **41**, 055009 (2014) [arXiv:1312.6795 [hep-ph]].

[22] S. P. Jones, A. D. Martin, M. G. Ryskin and T. Teubner, JHEP **1311**, 085 (2013) [arXiv:1307.7099].

[23] V. P. Goncalves, L. A. S. Martins and W. K. Sauter, Eur. Phys. J. C **76** (2016) no.2, 97 [arXiv:1511.00494 [hep-ph]].

[24] N. Armesto and A. H. Rezaeian, Phys. Rev. D **90**, no. 5, 054003 (2014) [arXiv:1402.4831 [hep-ph]].

[25] V. P. Goncalves, B. D. Moreira and F. S. Navarra, Phys. Rev. C **90**, no. 1, 015203 (2014) [arXiv:1405.6977 [hep-ph]].

[26] V. P. Goncalves, B. D. Moreira and F. S. Navarra, Phys. Lett. B **742**, 172 (2015) [arXiv:1408.1344 [hep-ph]].

[27] W. Schäfer and A. Szczurek, Phys. Rev. D **76**, 094014 (2007) [arXiv:0705.2887 [hep-ph]]; A. Cisek, W. Schäfer and A. Szczurek, JHEP **1504**, 159 (2015) [arXiv:1405.2253 [hep-ph]].

[28] I. Bautista, A. Fernandez Tellez and M. Hentschinski, Phys. Rev. D **94** (2016) no.5, 054002 doi:10.1103/PhysRevD.94.054002 [arXiv:1607.05203 [hep-ph]].

[29] S. Chekanov *et al.* [ZEUS Collaboration], Eur. Phys. J. C **24**, 345 (2002) [hep-ex/0201043]; Nucl. Phys. B **695**, 3 (2004) [hep-ex/0404008].

[30] C. Alexa *et al.* [H1 Collaboration], Eur. Phys. J. C **73**, no. 6, 2466 (2013) [arXiv:1304.5162 [hep-ex]]; Eur. Phys. J. C **46**, 585 (2006) [hep-ex/0510016].

[31] B. B. Abelev *et al.* [ALICE Collaboration], Phys. Rev. Lett. **113**, no. 23, 232504 (2014) [arXiv:1406.7819 [nucl-ex]].

[32] R. Aaij *et al.* [LHCb Collaboration], J. Phys. G **40**, 045001 (2013) [arXiv:1301.7084 [hep-ex]]; J. Phys. G **41**, 055002 (2014) [arXiv:1401.3288 [hep-ex]].

[33] H. Kowalski, L. Motyka and G. Watt, Phys. Rev. D **74**, 074016 (2006) [hep-ph/0606272].

[34] B. E. Cox, J. R. Forshaw and R. Sandapen, JHEP **0906**, 034 (2009) [arXiv:0905.0102 [hep-ph]].

[35] S. J. Brodsky, T. Huang and G. P. Lepage, "The Hadronic Wave Function in Quantum Chromodynamics," SLAC-PUB-2540.

[36] C. Adloff *et al.* [H1 Collaboration], Phys. Lett. B **483**, 23 (2000) [hep-ex/0003020].

[37] J. Breitweg *et al.* [ZEUS Collaboration], Phys. Lett. B **437** (1998) 432 [hep-ex/9807020]; Phys. Lett. B **680**, 4 (2009) [arXiv:0903.4205 [hep-ex]].
[38] R. Aaij *et al.* [LHCb Collaboration], JHEP **1509**, 084 (2015) [arXiv:1505.08139 [hep-ex]].
[39] CMS Collaboration [CMS Collaboration], "Measurement of exclusive Y photoproduction in pPb collisions at $\sqrt{s_{\mathrm{NN}}} = 5.02$ TeV," CMS-PAS-FSQ-13-009.
[40] J. Nemchik, N. N. Nikolaev and B. G. Zakharov, Phys. Lett. B **341**, 228 (1994) [hep-ph/9405355], J. Nemchik, N. N. Nikolaev, E. Predazzi and B. G. Zakharov, Z. Phys. C **75**, 71 (1997) [hep-ph/9605231], Z. Phys. C **75**, 71 (1997) [hep-ph/9605231].
[41] K. Kutak and A. M. Stasto, Eur. Phys. J. C **41**, 343 (2005) [hep-ph/0408117].
[42] S. P. Baranov, Phys. Rev. D **76**, 034021 (2007).
[43] A. Ayala, M. Hentschinski, J. Jalilian-Marian and M. E. Tejeda-Yeomans, arXiv:1604.08526 [hep-ph].
[44] A. Ayala, M. Hentschinski, J. Jalilian-Marian and M. E. Tejeda-Yeomans, Nucl. Phys. B **920** (2017) 232 doi:10.1016/j.nuclphysb.2017.03.028 [arXiv:1701.07143 [hep-ph]].
[45] F. Dominguez, C. Marquet, B. -W. Xiao and F. Yuan, Phys. Rev. D **83**, 105005 (2011).
[46] J. L. Albacete, N. Armesto, J. G. Milhano, P. Quiroga-Arias and C. A. Salgado, Eur. Phys. J. C **71**, 1705 (2011);

IOP Conf. Series: Journal of Physics: Conf. Series **912** (2017) 012009 doi:10.1088/1742-6596/912/1/012009

Partial restoration of chiral symmetry in cold nuclear matter: the ϕ-meson case

J J Cobos-Martínez[1,2,5], K. Tsushima[2], G. Krein[3] and A. W. Thomas[4]

[1] Cátedra CONACyT, Departamento de Física, Centro de Investigación y de Estudios Avanzados del Instituto Politécnico Nacional, Apartado Postal 14-740, 07000, Ciudad de México, México.

[2] Laboratório de Física Teórica e Computacional-LFTC, Universidade Cruzeiro do Sul, 01506-000, São Paulo, SP, Brazil

[3] Instituto de Física Teórica, Universidade Estadual Paulista, Rua Dr. Bento Teobaldo Ferraz, 271-Bloco II, 01140-070, São Paulo, SP, Brazil

[4] ARC Centre of Excellence for Particle Physics at the Terascale and CSSM, Department of Physics, University of Adelaide, Adelaide SA 5005, Australia

E-mail: jcobos@fis.cinvestav.mx

Abstract.
The work presented at this workshop is divided into two parts. In the first part, the mass and decay width of the ϕ-meson in cold nuclear matter are computed in an effective Lagrangian approach. The medium dependence of these properties are obtained by evaluating kaon-antikaon loop contributions to the ϕ-meson self-energy, employing medium-modified kaon masses calculated using the quark-meson coupling model. The loop integral is regularized with a dipole form factor, and the sensitivity of the results to the choice of cutoff mass in the form factor is investigated. At normal nuclear matter density, we find a downward shift of the ϕ mass by a few percent, while the decay width is enhanced by an order of magnitude. Our results support the literature which suggest that one should observe a small downward mass shift and a large broadening of the decay width. In the second part, we present ϕ-meson–nucleus bound state energies and absorption widths for four selected nuclei, calculated by solving the Klein-Gordon equation with complex optical potentials. The attractive potential for the ϕ-meson in the nuclear medium originates from the in-medium enhanced $K\overline{K}$ loop in the ϕ-meson self-energy. The results suggest that the ϕ-meson should form bound states with all the nuclei considered. However, the identification of the signal for these predicted bound states will need careful investigation because of their sizable absorption widths.

1. Introduction

The properties of light vector mesons at finite baryon density, such as their masses and decay widths, have attracted considerable experimental and theoretical interest over the last few decades [1–5]. This has been in part due to their potential to carry information on the partial restoration of chiral symmetry, and the possible role of QCD of van der Waals forces in the binding of quarkonia to nuclei. In particular, there is special interest on the ϕ-meson, the main reasons being: i) despite its nearly pure $s\bar{s}$ content, the ϕ-meson does interact strongly with a nucleus, composed predominantly of light u and d quarks, through the excitation of

[5] Speaker

below-threshold virtual kaon and anti-kaon states that might have their properties changed in medium [6–10]; (ii) the ϕN interaction in vacuum [11–14] and a possible in-medium mass shift of the ϕ are related to the strangeness content of the nucleon [15], which may have implications beyond the physics of the strong interaction [16–18]; (iii) medium modifications of ϕ-meson properties have been proposed [19] as a possible source for the anomalous nuclear mass number A-dependence observed in ϕ-meson production from nuclear targets [20]; (iv) furthermore, as the ϕ-meson is a nearly pure $s\bar{s}$ state and gluonic interactions are flavor blind, studying it serves to test theories of the multiple-gluon exchange interactions, including long range QCD van der Waals forces [21], which are believed to play a role in the binding of the J/Ψ and other exotic heavy-quarkonia to matter [22–33].

Heavy-ion collisions and photon- or proton-induced reactions on nuclear targets have been used to extract information on the in-medium properties of hadrons. Several experiments have focused on the light vector ρ, ω, and ϕ mesons, since their mean-free paths can be comparable with the size of a nucleus after being produced inside the nucleus. However, a unified consensus has not yet been reached among the different experiments—see Refs. [1–5] for comprehensive reviews on the current status.

For the ϕ-meson, although the precise values are different, a large in-medium broadening of its decay width has been reported by most of the experiments performed, while only a few of them find evidence for a substantial mass shift [20; 34–38]. In 2007 the KEK-E325 collaboration reported a 3.4% mass reduction of the ϕ-meson [34] and an in-medium decay width of $\approx$ 14.5 MeV at normal nuclear matter density $\rho_0 = 0.15$ fm^{-3}. These conclusions were based upon the measurement of the invariant mass spectra of e^+e^- pairs in 12 GeV p+A reactions, with copper and carbon being used as targets [34]. Even though this result may indicate a signal for partial restoration of chiral symmetry in nuclear matter, it is not possible to draw a definite conclusion solely from this. In fact, recently, a large in-medium ϕ-meson decay width (>30 MeV) has been extracted at various experimental facilities without observing any mass shift [20; 35–38],

It is therefore clear that the search for evidence of a light vector meson mass shift in nuclear matter is indeed a complicated issue and further experimental efforts [32; 39] are required in order to understand the phenomenon better. Indeed, the J-PARC E16 collaboration [32; 39] intends to perform a more systematic study for the mass shift of vector mesons with higher statistics than the above-mentioned experiment at KEK-E325.

However, either complementary or alternative experimental methods are desired. The study of the ϕ-meson–nucleus bound states is complementary to the invariant mass measurements, such as Ref. [34], where only a small fraction of the produced ϕ-mesons decay inside the nucleus and may be expected to provide extra information on the ϕ-meson properties at finite baryon density. Along these lines, and motivated by the 3.4% mass reduction reported by the KEK-E325 experiment [34], the E29 collaboration at J-PARC has recently put forward a proposal [40; 41] to study the in-medium mass modification of the ϕ-meson via the possible formation of ϕ-meson–nucleus bound states [31; 42]. Furthermore, there is also a proposal at JLab, following the 12 GeV upgrade, to study the binding of ϕ and η mesons to ^{4}He [43]. This new experimental approach [31; 32; 42; 43] for the measurement of the ϕ-meson mass shift in nuclei, will produce a slowly moving ϕ-meson [31; 32; 42; 43], where the maximum nuclear matter effect can be probed. In this way, one may indeed anticipate the formation of a ϕ-meson–nucleus bound state, where the ϕ-meson is trapped inside the nucleus.

Meson-nucleus systems bound by attractive strong interactions are very interesting objects [4; 5]. First, they are strongly interacting exotic many-body systems and to study them serves, for example, to understand better the multi-gluon exchange interactions, including QCD "van der Waals" forces [21], which are believed to play a role in the binding of the J/Ψ and other exotic heavy-quarkonia to matter (a nucleus) [22–33]. Second, they provide unique laboratories for the study of hadron properties at finite density, which may not only lead to a deeper understanding of the strong interaction [1–5] but of the structure of finite nuclei as well [44; 45].

A downward mass shift of the ϕ-meson in a nucleus is directly connected with the possible existence of an attractive potential between the ϕ-meson and the nucleus where it has been produced, the strength of which is expected to be of the same order as that of the mass shift.

Concerning the theoretical evaluation of the ϕ-meson mass shift, various authors predict a downward shift of the in-medium ϕ-meson mass and a broadening of its decay width, many of them focusing on the self-energy of the ϕ-meson due to the kaon-antikaon loop. Ko et al. [46] used a density-dependent kaon mass determined from chiral perturbation theory and found that at normal nuclear matter density, ρ_0, the ϕ-meson mass decreases very little, by at most 2%, and a $\Gamma_\phi \approx 25$ MeV width which broadens drastically for large densities. Hatsuda and Lee calculated the in-medium ϕ-meson mass based on the QCD sum rule approach [47; 48], and predict a 1.5%-3% decrease at ρ_0. Other investigations also predict a small downward mass shift and a large broadening of the ϕ-meson width at ρ_0: Ref. [49] reports a negative mass shift of $< 1\%$ and a decay width of 45 MeV; Ref. [50] predicts a decay width of 22 MeV but does not report a result on the mass shift; and Ref. [51] gives a rather small negative mass shift of $\approx 0.81\%$ and a decay width of 30 MeV. More recently, Ref. [52] reported a downward mass shift of $< 2\%$ and a large broadening width of 45 MeV at ρ_0; and finally, in Ref. [53], extending the work of Refs. [50; 51], the authors reported a negative mass shift of 3.4% and a large decay width of 70 MeV at ρ_0. The reason for these differences may lie in the different approaches used to estimate the kaon-antikaon loop contributions to the ϕ-meson self-energy and this might have consequences for the formation of ϕ-meson–nucleus bound states.

From a practical point of view, the important question is whether this attraction, if it exists, is sufficient to bind the ϕ-meson to a nucleus. A simple argument can be given as follows. One knows that for an attractive spherical well of radius R and depth V_0, the condition for the existence of a nonrelativistic s-wave bound state of a particle of mass m is $V_0 > \frac{\pi^2\hbar^2}{8mR^2}$. Using $m = m^*_\phi$, where m^*_ϕ is the ϕ-meson mass at normal nuclear matter density found in Ref. [34] and $R = 5$ fm (the radius of a heavy nucleus), one obtains $V_0 > 2$ MeV. Therefore, the prospects of capturing a ϕ-meson seem quite favorable, provided that the ϕ-meson can be produced almost at rest in the nucleus. Real nuclei, however, have a surface and this estimate can be quite misleading [4]. A full calculation using realistic density profiles of nuclei is required for a more reliable estimate.

The work presented at this workshop was carried out in collaboration with professors K. Tushima, G. Krein, and A. W. Thomas and has been published in Refs. [54; 55]. In Ref. [54] we studied the ϕ-meson mass shift and decay width in nuclear matter, based on an effective Lagrangian approach, by evaluating the $K\overline{K}$ loop contribution in the ϕ-meson self-energy, with the in-medium K and $\overline{K}$ masses explicitly calculated by the quark-meson coupling (QMC) model [56]. This initial study has been extended in Ref. [54] to some selected nuclei by computing the ϕ-meson–nucleus bound complex potential in the local density approximation and solving the Klein-Gordon equation in order to obtain the bound state energies and absorption widths. The nuclear density distributions for all nuclei studied, except for ^{4}He, are explicitly calculated using the QMC model [57].

2. ϕ-meson mass and decay width in nuclear matter

The ϕ-meson property modifications in nuclear matter, such as its mass and decay width, are strongly correlated to its coupling to the $K\overline{K}$ channel, which is the dominant decay channel in vacuum. Thus, one expects that a significant fraction of the density dependence of the ϕ-meson self-energy in nuclear matter arises from the in-medium modification of the $K\overline{K}$ intermediate state in the ϕ-meson self-energy.

We briefly review the computation [54] of the ϕ-meson self-energy in vacuum and in nuclear matter using an effective Lagrangian approach [58]. The interaction Lagrangian $\mathcal{L}_{\rm Int}$ involves

$\phi K\overline{K}$ and $\phi\phi K\overline{K}$ couplings dictated by a local gauge symmetry principle:

$$\mathcal{L}_{\rm Int} = \mathcal{L}_{\phi K\overline{K}} + \mathcal{L}_{\phi\phi K\overline{K}}, \tag{1}$$

where (we use the convention $K = \begin{pmatrix} K^+ \\ K^0 \end{pmatrix}$, $\overline{K} = \begin{pmatrix} K^- & \overline{K}^0 \end{pmatrix}$ for isospin doublets)

$$\mathcal{L}_{\phi K\overline{K}} = \mathrm{i}g_\phi \phi^\mu \left[\overline{K}(\partial_\mu K) - (\partial_\mu \overline{K})K\right], \tag{2}$$

$$\mathcal{L}_{\phi\phi K\overline{K}} = g_\phi^2 \phi^\mu \phi_\mu \overline{K}K. \tag{3}$$

We note that the use of the effective interaction Lagrangian of Eq. (1) without the term given in Eq. (3) may be considered as being motivated by the hidden gauge approach in which there are no four-point vertices, such as Eq. (3), that involve two pseudoscalar mesons and two vector mesons [59; 60]. This is in contrast to the approach of using the minimal substitution to introduce vector mesons as gauge particles where such four-point vertices do appear. However, these two methods have been shown to be consistent if both the vector and axial vector mesons are included [61–64]. Therefore, we present results with and without such an interaction [54; 55]. We consider first the contribution from the $\phi K\overline{K}$ coupling, given by Eq. (2), to the scalar part of the ϕ-meson self-energy, $\Pi_\phi(p)$.

For a ϕ-meson at rest the scalar self-energy is given by

$$\mathrm{i}\Pi_\phi(p) = -\frac{8}{3}g_\phi^2 \int \frac{\mathrm{d}^4 q}{(2\pi)^4}\, \vec{q}^{\,2} D_K(q) D_K(q-p), \tag{4}$$

where $D_K(q) = \left(q^2 - m_K^2 + \mathrm{i}\epsilon\right)^{-1}$ is the kaon propagator; $p = (p^0 = m_\phi, \vec{0})$ is the ϕ-meson four-momentum vector at rest, with m_ϕ the ϕ-meson mass; $m_K (= m_{\overline{K}})$ is the kaon mass; and g_ϕ is the coupling constant. When $m_\phi < 2m_K$ the self-energy $\Pi_\phi(p)$ is real. However, when $m_\phi > 2m_K$, which is the case here, $\Pi_\phi(p)$ acquires an imaginary part.

The integral in Eq. (4) is divergent but it will be regulated using a phenomenological form factor, with cutoff parameter Λ_K, as in Refs. [54; 65]. The sensitivity of the results to the cutoff value is analyzed below.

The mass and decay width of the ϕ-meson in vacuum (m_ϕ and Γ_ϕ), as well as in nuclear matter (m_ϕ^* and Γ_ϕ^*), are determined self-consistently [54] from

$$m_\phi^2 = \left(m_\phi^0\right)^2 + \mathrm{Re}\,\Pi_\phi(m_\phi^2), \tag{5}$$

$$\Gamma_\phi = -\frac{1}{m_\phi}\mathrm{Im}\,\Pi_\phi(m_\phi^2). \tag{6}$$

The coupling constant g_ϕ is determined [54] from Eqs. (4) and (6), and the experimental value for the $\phi \to K\overline{K}$ decay width in vacuum, corresponding to the branching ratio of 83.1% of the total decay width [66].

The nuclear density dependence of the ϕ-meson mass and decay width is driven by the intermediate $K\overline{K}$ state interactions with the nuclear medium. This effect enters through m_K^* in the kaon propagators in Eq. (4). The in-medium mass m_K^* is calculated within the QMC model [54], which has proven to be very successful in studying the properties of hadrons in nuclear matter and finite nuclei. For a more complete discussion of the model see Refs. [6; 56; 67]. Here we just make a few necessary comments. In order to calculate the in-medium properties of K and $\overline{K}$, we consider infinitely large, uniformly symmetric, spin-isospin-saturated nuclear matter in its rest frame, where all the scalar and vector mean field potentials, which are responsible for the nuclear many-body interactions, become constant in the Hartree approximation [54]. In

Figure 1 (left panel) we present the resulting in-medium kaon Lorentz scalar mass (=antikaon Lorentz scalar mass), calculated using the QMC model, as a function of the baryon density. The kaon effective mass at normal nuclear matter density $\rho_0 = 0.15$ fm^{-3} has decreased by about 13%. We also recall, in connection with the calculation of the in-medium $K\overline{K}$ loop contributions to the ϕ-meson self-energy, that the isoscalar-vector ω mean field potentials arise both for the kaon and antikaon. However, they have opposite signs and cancel each other. Equivalently, they can be eliminated by a variable shift in the loop calculation [6; 56; 67] of the ϕ-meson self-energy, and therefore we do not show them here.

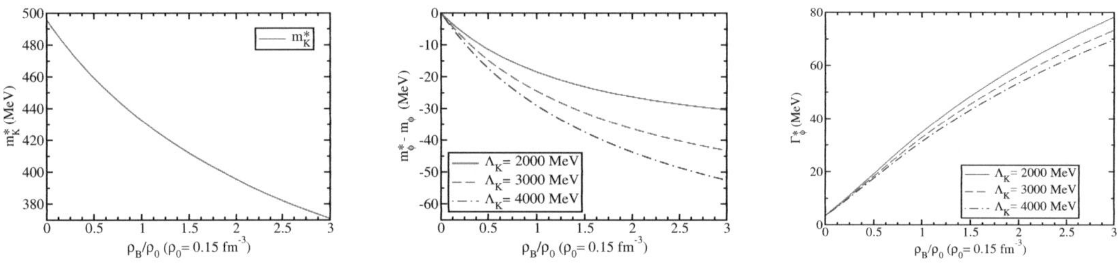

Figure 1. Left panel: In-medium kaon mass (=antikaon) Lorentz scalar mass m^*_K; centre and right panels: ϕ-meson mass shift and decay width, respectively, in symmetric nuclear matter for three values of the cutoff parameter Λ_K.

In Figure 1, we present the ϕ-meson mass shift (centre panel) and decay width (right panel) as a function of the nuclear matter density, ρ_B, for three values of the cutoff parameter Λ_K. As can be seen, the effect of the in-medium kaon and antikaon mass change yields a negative mass shift for the ϕ-meson. This is because the reduction in the kaon and antikaon masses enhances the $K\overline{K}$ loop contribution in nuclear matter relative to that in vacuum. For the largest value of the nuclear matter density, the downward mass shift turns out to be a few percent at most for all values of Λ_K. On the other hand, we see that Γ^*_ϕ is very sensitive to the change in the kaon and antikaon masses, increasing rapidly with increasing nuclear matter density, up to a factor of ~ 20 enhancement for the largest value of ρ_B. In Table 1 we present the values for m^*_ϕ and Γ^*_ϕ at normal nuclear matter density ρ_0. We see that the negative kaon and antikaon mass shift of 13% [54] induces a downward mass shift of the ϕ-meson of just $\approx 2\%$, while the broadening of the ϕ-meson decay width is an order-of-magnitude larger than its vacuum value.

In Ref. [54] we evaluated the impact of adding the $\phi\phi K\overline{K}$ interaction of Eq. (3) on the in-medium ϕ-meson mass and decay width. We found that one still gets a downward shift of the in-medium ϕ-meson mass as well as a significant broadening of the decay width when this interaction is added. In both cases, though, the absolute values are slightly different from those shown in Figure 1. In Table 1 we present the values for m^*_ϕ and Γ^*_ϕ at ρ_0 obtained by adding the gauged Lagrangian of Eq. (3). In both cases, for the mass and decay width in nuclear matter, the effect of adding Eq. (3) can be compensated by the use of a larger cutoff Λ_K.

The results described above support those which suggest that one should observe a small downward mass shift and a large broadening of the decay width of the ϕ-meson in a nuclear medium. Furthermore, they open experimental possibilities for studying the binding and absorption of ϕ-meson in nuclei. Although the mass shift found in this study may be large enough to bind the ϕ-meson to a nucleus, the broadening of its decay width will make it difficult to observe a signal for the ϕ-meson–nucleus bound state formation experimentally. We explore this further in the second part of this talk.

IOP Conf. Series: Journal of Physics: Conf. Series **912** (2017) 012009 doi:10.1088/1742-6596/912/1/012009

	$\Lambda_K = 2000$	$\Lambda_K = 3000$	$\Lambda_K = 4000$
m^*_ϕ	1000.9 (1009.5)	994.9 (1004.3)	990.4 (1000.6)
Γ^*_ϕ	34.8 (37.8)	32.8 (36.0)	31.3 (34.7)

Table 1. ϕ-meson mass and width at normal nuclear matter density ρ_0 with and without the gauged Lagrangian of Eq. (3). The values in parentheses were computed by adding the gauged Lagrangian of Eq. (3). All quantities are given in MeV.

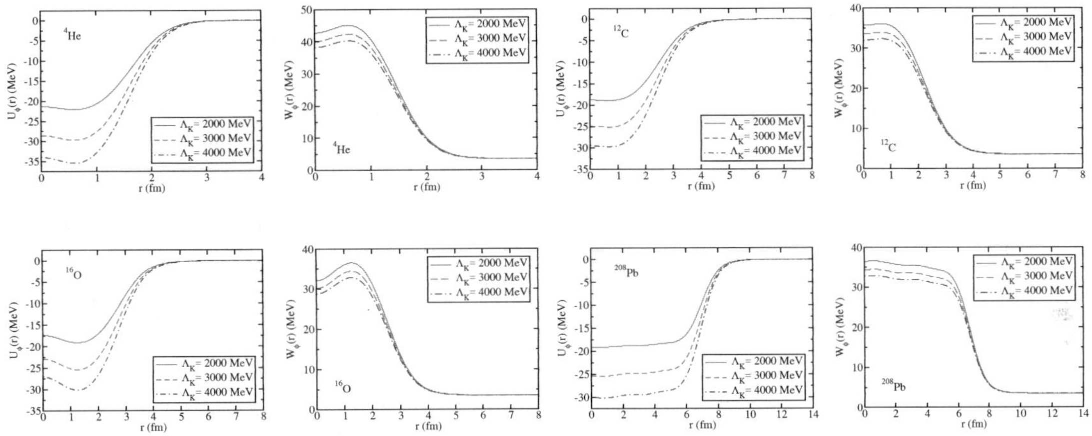

Figure 2. Real $[U_\phi(r)]$ and imaginary $[W_\phi(r)]$ parts of the ϕ-meson-nucleus potentials in the nuclei ^{12}C, ^{16}O, ^{40}Ca , and ^{208}Pb, for three values of the cutoff parameter Λ_K.

3. ϕ-meson–nuclear bound states

In this part we discuss the situation where the ϕ-meson is placed in a nucleus. The nuclear density distributions for the nuclei ^{12}C, ^{16}O, ^{40}Ca , and ^{208}Pb are obtained using the QMC model [57]. For ^{4}He, we use the parametrization for the density distribution obtained in Ref. [68]. Then, using a local density approximation we calculate the ϕ-meson complex potentials for a nucleus A, which can be written as

$$V_{\phi A}(r) = U_\phi(r) - \frac{\mathrm{i}}{2} W_\phi(r), \tag{7}$$

where r is the distance from the center of the nucleus and $U_\phi(r) = \Delta m_\phi(\rho_B(r)) \equiv m^*_\phi(\rho_B(r)) - m_\phi$, and $W_\phi(r) = \Gamma_\phi(\rho_B(r))$ are, respectively, the ϕ-meson mass shift and decay width in a nucleus A, with $\rho_B(r)$ the baryon density distribution for the particular nucleus.

Figure 2 shows the ϕ-meson potentials calculated for the selected nuclei, for three values of the cutoff parameter Λ_K. One can see that the depth of the real part of the potential is sensitive to the cutoff parameter, varying from -20 MeV to -35 MeV for ^{4}He and from -20 MeV to -30 MeV for ^{208}Pb. On the other hand, one can see that the imaginary part does not vary much with Λ_K. These latter observations may well have consequences for the feasibility of experimental observation of the expected bound states.

Using the ϕ-meson potentials obtained in this manner, we next calculate the ϕ-meson–nucleus bound state energies and absorption widths for the selected nuclei. Before proceeding, a few comments on the use of Eq. (8) are in order. In this study, we consider the situation where the ϕ-meson is produced nearly at rest. Then, it should be a very good approximation to neglect the possible energy difference between the longitudinal and transverse components of the ϕ-meson

		$\Lambda_K = 2000$		$\Lambda_K = 3000$		$\Lambda_K = 4000$	
		E	$\Gamma/2$	E	$\Gamma/2$	E	$\Gamma/2$
$^{4}_{\phi}$He	1s	n (-0.8)	n	n (-1.4)	n	-1.0 (-3.2)	8.3
$^{12}_{\phi}$C	1s	-2.1 (-4.2)	10.6	-6.4 (-7.7)	11.1	-9.8 (-10.7)	11.2
$^{16}_{\phi}$O	1s	-4.0 (-5.9)	12.3	-8.9 (-10.0)	12.5	-12.6 (-13.4)	12.4
	1p	n (n)	n	n (n)	n	n (-1.5)	n
$^{208}_{\phi}$Pb	1s	-15.0 (-15.5)	17.4	-21.1 (-21.4)	16.6	-25.8 (-26.0)	16.0
	1p	-11.4 (-12.1)	16.7	-17.4 (-17.8)	16.0	-21.9 (-22.2)	15.5
	1d	-6.9 (-8.1)	15.7	-12.7 (-13.4)	15.2	-17.1 (-17.6)	14.8
	2s	-5.2 (-6.6)	15.1	-10.9 (-11.7)	14.8	-15.2 (-15.8)	14.5
	2p	n (-1.9)	n	-4.8 (-6.1)	13.5	-8.9 (-9.8)	13.4
	2d	n (n)	n	n (-0.7)	n	-2.2 (-3.7)	11.9

Table 2. ϕ-meson–nucleus single-particle energies, E, and half widths, $\Gamma/2$, obtained with and without the imaginary part of the potential of Eq. (7), for three values of the cutoff parameter Λ_K. When only the real part is included, where the corresponding single-particle energy E is given inside parenthesis, $\Gamma = 0$ for all nuclei. "n" indicates that no bound state is found. All quantities are given in MeV.

wave function ψ^{μ}_{ϕ}. After imposing the Lorentz condition, $\partial_\mu \psi^{\mu}_{\phi} = 0$, to solve the Proca equation becomes equivalent to solving the Klein-Gordon equation

$$\left(-\nabla^2 + \mu^2 + 2\mu V(\vec{r})\right) \phi(\vec{r}) = \mathcal{E}^2 \phi(\vec{r}), \tag{8}$$

where $\mu = m_\phi m_A/(m_\phi + m_A)$ is the reduced mass of the ϕ-meson-nucleus system with m_ϕ (m_A) the mass of the ϕ-meson (nucleus A) in vacuum, and $V(\vec{r})$ is the complex ϕ-meson-nucleus potential of Eq. (7). We solve the Klein-Gordon equation using the momentum space methods [69]. The calculated bound state energies (E) and absorption widths (Γ), which are related to the complex energy eigenvalue $\mathcal{E}$ by $E = \Re\mathcal{E} - \mu$ and $\Gamma = -2\Im\mathcal{E}$, are listed in Table 2 for three values of the cutoff parameter Λ_K, with and without the imaginary part of the potential.

We first discuss the case in which the imaginary part of the ϕ-meson–nucleus potential is set to zero. The results are given within parentheses in Table 2. From the values shown, we see that the ϕ-meson is expected to form bound states with all the nuclei selected, for all values of the cutoff parameter Λ_K. The bound state energy is obviously dependent on Λ_K, increasing as Λ_K increases. Next, we discuss the results obtained when the imaginary part of the potential is included. Adding the absorptive part of the potential changes the situation appreciably. From the results presented in Table 2 we note that for the largest value of the cutoff parameter Λ_K, which yields the deepest attractive potentials (see Figure 2), the ϕ-meson is expected to form bound states with all the selected nuclei, including the lightest one, the ^{4}He nucleus. However, in this case, whether or not the bound states can be observed experimentally is sensitive to the value of the cutoff parameter Λ_K. One also observes that the width of the bound state is insensitive to the values of Λ_K for all nuclei. Furthermore, since the so-called dispersive effect of the absorptive potential is repulsive, the bound states disappear completely in some cases, even though they were found when the absorptive part was set to zero. This feature is obvious for the ^{4}He nucleus, making it especially relevant to the future experiments, planned at J-PARC and JLab using light and medium-heavy nuclei [31; 32; 42; 43].

4. Summary and discussion

We have calculated the ϕ-meson mass and width in nuclear matter within an effective Lagrangian approach up to three times of normal nuclear matter density. Essential to our results are the in-medium kaon masses, which are calculated in the QMC model, where the scalar and vector meson mean fields couple directly to the light u and d quarks (antiquarks) in the K ($\overline{K}$) mesons. At normal nuclear matter density, allowing for a very large variation of the cutoff parameter Λ_K, although we have found a sizable negative mass shift of 13% in the kaon mass, this induces only a few percent downward shift of the ϕ-meson mass. On the other hand, it induces an order-of-magnitude broadening of the decay width.

We have also calculated the ϕ-meson–nucleus bound state energies and absorption widths for various nuclei. The ϕ-meson–nucleus potentials were calculated using a local density approximation, with the inclusion of the $K\overline{K}$ loop in the ϕ-meson self-energy. The nuclear density distributions, as well as the in-medium K and $\overline{K}$ meson masses, were consistently calculated by employing the quark-meson coupling model. Using the ϕ-meson–nucleus complex potentials found, we have solved the Klein-Gordon equation in momentum space, and obtained ϕ-meson–nucleus bound state energies and absorption widths.

Furthermore, we have studied the sensitivity of our results to the cutoff parameter Λ_K in the form factor at the $\phi K\overline{K}$ vertex appearing in the ϕ-meson self-energy. We expect that the ϕ-meson should form bound states for all four nuclei selected, provided that the ϕ-meson is produced in (nearly) recoilless kinematics. This feature, is even more obvious in the (artificial) case where the absorptive part of the potential is ignored. Given the similarity of the binding energies and widths reported here, the signal for the formation of the ϕ-meson–nucleus bound states may be difficult to identify experimentally. Therefore, the feasibility of observation of the ϕ-meson–nucleus bound states needs further investigation, including explicit reaction cross section estimates.

Acknowledgements
This work was partially supported by Conselho Nacional de Desenvolvimento Científico e Tecnológico - CNPq, Grants No. 152348/2016-6 (J.J.C-M.), No. 400826/2014-3 and No. 308088/2015-8 (K.T.), No. 305894/2009-9 (G.K.), and No. 313800/2014-6 (A.W.T.), and Fundação de Amparo à Pesquisa do Estado de São Paulo-FAPESP, Grants No. 2015/17234-0 (K.T.) and No. 2013/01907-0 (G.K.). This research was also supported by the University of Adelaide and by the Australian Research Council through the ARC Centre of Excellence for Particle Physics at the Terascale (CE110001104), and through Grant No. DP151103101 (A.W.T.). J.J.C-M. also acknowledges the support given by Red-FAE-CONACyT (México) to attend the workshop where this work was presented.

References

[1] Hayano R S and Hatsuda T 2010 *Rev. Mod. Phys.* **82** 2949 (*Preprint* 0812.1702)
[2] Leupold S, Metag V and Mosel U 2010 *Int. J. Mod. Phys.* **E19** 147–224 (*Preprint* 0907.2388)
[3] Hosaka A, Hyodo T, Sudoh K, Yamaguchi Y and Yasui S 2016 (*Preprint* 1606.08685)
[4] Krein G, Thomas A W and Tsushima K 2017 (*Preprint* 1706.02688)
[5] Metag V, Nanova M and Paryev E Ya 2017 (*Preprint* 1706.09654)
[6] Tsushima K, Saito K, Thomas A W and Wright S V 1998 *Phys. Lett.* **B429** 239–246 [Erratum: Phys. Lett.B436,453(1998)] (*Preprint* nucl-th/9712044)
[7] Laue F *et al.* (KaoS) 1999 *Phys. Rev. Lett.* **82** 1640–1643 (*Preprint* nucl-ex/9901005)
[8] Schaffner-Bielich J, Koch V and Effenberger M 2000 *Nucl. Phys.* **A669** 153–172 (*Preprint* nucl-th/9907095)
[9] Akaishi Y and Yamazaki T 2002 *Phys. Rev.* **C65** 044005
[10] Fuchs C 2006 *Prog. Part. Nucl. Phys.* **56** 1–103 (*Preprint* nucl-th/0507017)

[11] Titov A I, Oh Y s and Yang S N 1997 *Phys. Rev. Lett.* **79** 1634–1637 (*Preprint* nucl-th/9702015)

[12] Titov A I, Oh Y s, Yang S N and Morii T 1998 *Phys. Rev.* **C58** 2429–2449 (*Preprint* nucl-th/9804043)

[13] Oh Y s, Titov A I, Yang S N and Morii T 1999 *Phys. Lett.* **B462** 23–28 (*Preprint* nucl-th/9905044)

[14] Oh Y s and Bhang H C 2001 *Phys. Rev.* **C64** 055207 (*Preprint* nucl-th/0104068)

[15] Gubler P and Ohtani K 2014 *Phys. Rev.* **D90** 094002 (*Preprint* 1404.7701)

[16] Bottino A, Donato F, Fornengo N and Scopel S 2002 *Astropart. Phys.* **18** 205–211 (*Preprint* hep-ph/0111229)

[17] Ellis J R, Olive K A and Savage C 2008 *Phys. Rev.* **D77** 065026 (*Preprint* 0801.3656)

[18] Giedt J, Thomas A W and Young R D 2009 *Phys. Rev. Lett.* **103** 201802 (*Preprint* 0907.4177)

[19] Sibirtsev A, Hammer H W, Meissner U G and Thomas A W 2006 *Eur. Phys. J.* **A29** 209–220 (*Preprint* nucl-th/0606044)

[20] Ishikawa T *et al.* 2005 *Phys. Lett.* **B608** 215–222 (*Preprint* nucl-ex/0411016)

[21] Appelquist T and Fischler W 1978 *Phys. Lett.* **77B** 405–410

[22] Brodsky S J, Schmidt I A and de Teramond G F 1990 *Phys. Rev. Lett.* **64** 1011

[23] Luke M E, Manohar A V and Savage M J 1992 *Phys. Lett.* **B288** 355–359 (*Preprint* hep-ph/9204219)

[24] Sibirtsev A, Tsushima K, Saito K and Thomas A W 2000 *Phys. Lett.* **B484** 23–29 (*Preprint* nucl-th/9904015)

[25] Gao H, Lee T S H and Marinov V 2001 *Phys. Rev.* **C63** 022201 (*Preprint* nucl-th/0010042)

[26] Beane S R, Chang E, Cohen S D, Detmold W, Lin H W, Orginos K, Parreo A and Savage M J 2015 *Phys. Rev.* **D91** 114503 (*Preprint* 1410.7069)

[27] Brambilla N, Krein G, Tarrs Castell J and Vairo A 2016 *Phys. Rev.* **D93** 054002 (*Preprint* 1510.05895)

[28] Gao H, Huang H, Liu T, Ping J, Wang F and Zhao Z 2017 *Phys. Rev.* **C95** 055202 (*Preprint* 1701.03210)

[29] Kawama D (J-PARC E16) 2013 *PoS* **Hadron2013** 178

[30] Kawama D *et al.* (J-PARC E16) 2014 *JPS Conf. Proc.* **1** 013074

[31] Ohnishi H *et al.* 2014 *Acta Phys. Polon.* **B45** 819–826

[32] Aoki K (J-PARC E16) 2015 *Proceedings, 10th Workshop on Particle Correlations and Femtoscopy (WPCF 2014): Gyngys, Hungary, August 25-29, 2014* (*Preprint* 1502.00703) URL https://inspirehep.net/record/1342838/files/arXiv:1502.00703.pdf

[33] Morino Y *et al.* (J-PARC E16) 2015 *JPS Conf. Proc.* **8** 022009

[34] Muto R *et al.* (KEK-PS-E325) 2007 *Phys. Rev. Lett.* **98** 042501 (*Preprint* nucl-ex/0511019)

[35] Mibe T *et al.* (CLAS) 2007 *Phys. Rev.* **C76** 052202 (*Preprint* nucl-ex/0703013)

[36] Qian X *et al.* (CLAS) 2009 *Phys. Lett.* **B680** 417–422 (*Preprint* 0907.2668)

[37] Wood M H *et al.* (CLAS) 2010 *Phys. Rev. Lett.* **105** 112301 (*Preprint* 1006.3361)

[38] Polyanskiy A *et al.* 2011 *Phys. Lett.* **B695** 74–77 (*Preprint* 1008.0232)

[39] Yokkaichi S *et al.* 2006 URL http://rarfaxp.riken.go.jp/~yokkaich/paper/jparc-proposal-0604.pdf

[40] Buhler P *et al.* 2009 URL http://j-parc.jp/researcher/Hadron/en/pac_0907/pdf/Ohnishi.pdf

[41] Buhler P *et al.* 2010 URL http://j-parc.jp/researcher/Hadron/en/pac_1007/pdf/KEK_J-PARC-PAC2010-02.pdf

[42] Buhler P *et al.* 2010 *Prog. Theor. Phys. Suppl.* **186** 337–342

[43] Paolone M *et al.* 2014 URL https://www.jlab.org/exp_prog/PACpage/PAC42/PAC42_FINAL_Report.pdf

[44] Stone J R, Guichon P A M, Reinhard P G and Thomas A W 2016 *Phys. Rev. Lett.* **116** 092501 (*Preprint* 1601.08131)

[45] Guichon P A M, Matevosyan H H, Sandulescu N and Thomas A W 2006 *Nucl. Phys.* **A772** 1–19 (*Preprint* nucl-th/0603044)

[46] Ko C M, Levai P, Qiu X J and Li C T 1992 *Phys. Rev.* **C45** 1400–1402

[47] Hatsuda T and Lee S H 1992 *Phys. Rev.* **C46** R34

[48] Hatsuda T, Shiomi H and Kuwabara H 1996 *Prog. Theor. Phys.* **95** 1009–1028 (*Preprint* nucl-th/9603043)

[49] Klingl F, Waas T and Weise W 1998 *Phys. Lett.* **B431** 254–262 (*Preprint* hep-ph/9709210)

[50] Oset E and Ramos A 2001 *Nucl. Phys.* **A679** 616–628 (*Preprint* nucl-th/0005046)

[51] Cabrera D and Vicente Vacas M J 2003 *Phys. Rev.* **C67** 045203 (*Preprint* nucl-th/0205075)

[52] Gubler P and Weise W 2015 *Phys. Lett.* **B751** 396–401 (*Preprint* 1507.03769)

[53] Cabrera D, Hiller Blin A N and Vicente Vacas M J 2017 *Phys. Rev.* **C95** 015201 (*Preprint* 1609.03880)

[54] Cobos-Martnez J J, Tsushima K, Krein G and Thomas A W 2017 *Phys. Lett.* **B771** 113–118 (*Preprint* 1703.05367)

[55] Cobos-Martnez J J, Tsushima K, Krein G and Thomas A W 2017 (*Preprint* 1705.06653)

[56] Saito K, Tsushima K and Thomas A W 2007 *Prog. Part. Nucl. Phys.* **58** 1–167 (*Preprint* hep-ph/0506314)

[57] Saito K, Tsushima K and Thomas A W 1996 *Nucl. Phys.* **A609** 339–363 (*Preprint* nucl-th/9606020)

[58] Klingl F, Kaiser N and Weise W 1996 *Z. Phys.* **A356** 193–206 (*Preprint* hep-ph/9607431)

[59] Lin Z w, Ko C M and Zhang B 2000 *Phys. Rev.* **C61** 024904 (*Preprint* nucl-th/9905003)

[60] Lee S H, Song C and Yabu H 1995 *Phys. Lett.* **B341** 407–412 (*Preprint* hep-ph/9408266)

[61] Yamawaki K 1987 *Phys. Rev.* **D35** 412

[62] Meissner U G and Zahed I 1987 *Z. Phys.* **A327** 5–15

[63] Meissner U G 1988 *Phys. Rept.* **161** 213

[64] Saito S and Yamawaki K (eds) 1987 *LOW-ENERGY EFFECTIVE THEORY OF QCD. PROCEEDINGS, INTERNATIONAL WORKSHOP, NAGOYA, JAPAN, APRIL 13-14, 1987*

[65] Krein G, Thomas A W and Tsushima K 2011 *Phys. Lett.* **B697** 136–141 (*Preprint* 1007.2220)

[66] Olive K A *et al.* (Particle Data Group) 2014 *Chin. Phys.* **C38** 090001

[67] Guichon P A M 1989 *Nucl. Phys.* **A497** 265C–270C

[68] Saito K, Tsushima K and Thomas A W 1997 *Phys. Rev.* **C56** 566–569 (*Preprint* nucl-th/9703011)

[69] Kwan Y R and Tabakin F 1978 *Phys. Rev.* **C18** 932–943

IOP Conf. Series: Journal of Physics: Conf. Series **912** (2017) 012010 doi:10.1088/1742-6596/912/1/012010

On the iterative solution of the gap equation in the Nambu-Jona-Lasinio model

A Martinez and A Raya

Instituto de Física y Matemáticas, Universidad Michoacana de San Nicolás de Hidalgo. Edificio C-3, Ciudad Universitaria. Francisco J. Múgica s/n, Col. Felícitas del Río. Morelia 58040, Michoacan, Mexico.

E-mail: angelo@ifm.umich.mx, raya@ifm.umich.mx

Abstract. In this work we revise the standard iterative procedure to find the solution of the gap equation in the Nambu–Jona-Lasinio model within the most popular regularization schemes available in literature in the super-strong coupling regime. We observe that whereas for the hard cut-off regularization schemes, the procedure smoothly converges to the physically relevant solution, Pauli-Villars and Proper-Time regularization schemes become chaotic in the sense of discrete dynamical systems. We call for the need of an appropriate interpretation of the non-convergence of this procedure to the solution of the gap equation.

1. Introduction

The Nambu-Jona-Lasinio (NJL) model [1, 2] was one of the earliest attempts to describe strong interactions among protons and neutrons. Spontaneous breaking of chiral symmetry is presented in the model through an analogy with the phenomenon of superconductivity in condensed matter physics. The model has a single coupling constant for two kinds of 4-Fermi interactions and is non-renormalizable, which demands for a regulator. Nevertheless, up to date, the model continues being used as an effective description of quarks dynamics either as originally proposed, where fermion fields are now regarded as quark fields, or with slightly modifications [3, 4, 5] to account for confinement. NJL model has been widely used to sketch the quantum chromodynamics (QCD) phase diagram (see, for instance, Ref. [6] for a review) and its magnetized extension [7] from a complementary point of view of lattice [8] and other quantum field theoretical approaches [9, 10].

Several regularization schemes have been used in literature within the context of NJL model. Perhaps the simplest regulator one can think of is to establish a hard cut-off in the four-momentum integrals, or over the spatial momentum integrals, which thus allows a straightforward incorporation of temperature and density effects within the Matsubara formalism [11]. Pauli-Villars regularization has also been used in NJL studies due to the advantages of the scheme. Nevertheless, external magnetic field effects are best included through the Schwinger Proper-Time representation of the fermion propagator. This observation has been exploited to address the problem of magnetic catalysis [12] and inverse magnetic catalysis observed in lattice simulations [13, 14] and confirmed by several approaches [7, 8, 15]. Under this environment, the effective coupling of the model can reach very large values, and therefore, exploring the validity of these regularization schemes in this super-strong coupling regime is a natural question to address.

In this contribution, we revise the commonly used iterative strategy to solve the gap equation for the NJL model in vacuum under each of the above mentioned regulation schemes. For this purpose, we have organized the remaining of the manuscript as follows: In Sect. 2 we present the model and the gap equation in each regularization scheme. Section 3 is devoted to illustrate the iterative procedure to self consistently find the dynamically generated mass as a function of the parameters of the model. A summary and outlook is presented in Sect. 4.

2. NJL model and the gap equation

The Lagrangian for the NJL model proposed in Ref. [1, 2] has the form

$$\mathcal{L} = \overline{\psi}(i\partial\!\!\!/ - m_q)\psi + G\left\{\left(\overline{\psi}\psi\right)^2 + \left(\overline{\psi}i\gamma^5\vec{\tau}\psi\right)^2\right\}, \tag{1}$$

where in the modern applications, ψ is the quark field, m_q is the current quark mass, $\vec{\tau}$ are the Pauli matrices acting on isospin space and G is the coupling constant. Within the Hartree-Fock approximation, the gap equation is written as

$$m = m_q - 2G\left\langle\bar{\psi}\psi\right\rangle, \tag{2}$$

where m is the dynamically generated mass and $-\left\langle\bar{\psi}\psi\right\rangle$ is the chiral condensate, defined as

$$-\left\langle\bar{\psi}\psi\right\rangle = \int \frac{d^4k}{(2\pi)^4}\mathrm{Tr}[iS(k)], \tag{3}$$

with the dressed quark propagator

$$S(k) = \frac{(k\!\!\!/ + m)}{k^2 - m^2}. \tag{4}$$

It should be noted that the dynamical mass predicted by the NJL model is momentum independent; in vacuum, the dynamical mass depends only on the coupling constant strength once a regulator has been appropriately chosen. As mentioned before, the NJL model is non-renormalizable and thus the integrals in the gap equation Eq. (2), require a regularization. For the purposes of this work, we apply 3D-cut–off (3D), 4D-cut–off (4D), Pauli-Villars (PV) and Proper-Time (PT) regularizations. Within each prescription, the gap equation acquires the form

$$m = \frac{GN_fN_c}{\pi^2}m\left[\Lambda_{3D}\sqrt{\Lambda_{3D}^2+m^2} - m^2\ln\left(\frac{\Lambda_{3D}+\sqrt{\Lambda_{3D}^2+m^2}}{m}\right)\right], \tag{5}$$

$$m = \frac{GN_fN_c}{2\pi^2}m\left[\Lambda_{4D}^2 - m^2\ln\left(\frac{\Lambda_{4D}^2+m^2}{m^2}\right)\right], \tag{6}$$

$$m = \frac{GN_fN_c}{2\pi^2}m\left[\Lambda_{PV}^2 - m^2 + m^2\ln\left[\frac{m^2}{\Lambda_{PV}^2}\right]\right], \tag{7}$$

$$m = \frac{GN_fN_c}{2\pi^2}m\int_{\Lambda_{PT}}^{\infty} ds\frac{e^{-m^2s}}{s^2}. \tag{8}$$

In the above expressions, N_f and N_c are the number of families we are considering and the number of colors, respectively. Finally, for the purposes of the analysis in this contribution, we transform the gap equations Eqs. (5), (6), (7) and (8) into their dimensionless form

$$M = G'M\left\{\sqrt{1+M^2} - M^2\ln\left[\frac{1+\sqrt{1+M^2}}{M^2}\right]\right\}, \qquad (3D\,\text{cut-off}) \qquad (9)$$

$$M = G'M\left\{1 - M^2\ln\left[1+M^{-2}\right]\right\}, \qquad (4D\,\text{cut-off}\,) \qquad (10)$$

$$M = G'M\left\{1 - M^2 + M^2\ln\left[M^2\right]\right\}, \qquad (\text{Pauli-Villars}) \qquad (11)$$

$$M = G'M\int_1^{\infty} d\tau\frac{e^{-M^2\tau}}{\tau^2}, \qquad (\text{Proper-Time}) \qquad (12)$$

where we performed the substitution $M = m/\Lambda$ for 3D, 4D and PV regularizations and $M^2 = m^2\Lambda$ for PT regularization, whereas G' absorbs numerical factors to render all equations in similar footing.

3. Solving with iterations

To introduce the iterations to the gap equations, we define.

$$y = m,$$
$$y = G'f(m), \qquad (13)$$

where $f(m)$ is the RHS of any of the gap equations in Eqs. (5), (6), (7) and (8). Plotting these two equations for a fixed value $G' = 1$, for example, we get Fig. 1. Next, we perform an iteration of these two equations, first using an starting value m_0 then plugging into $f(m)$ to get $m_1 = f(m_0)$ and so on. Graphically, this iteration procedure can be depicted by a cobweb plot, in the language of discrete dynamical systems [16], as shown in Fig. 2.

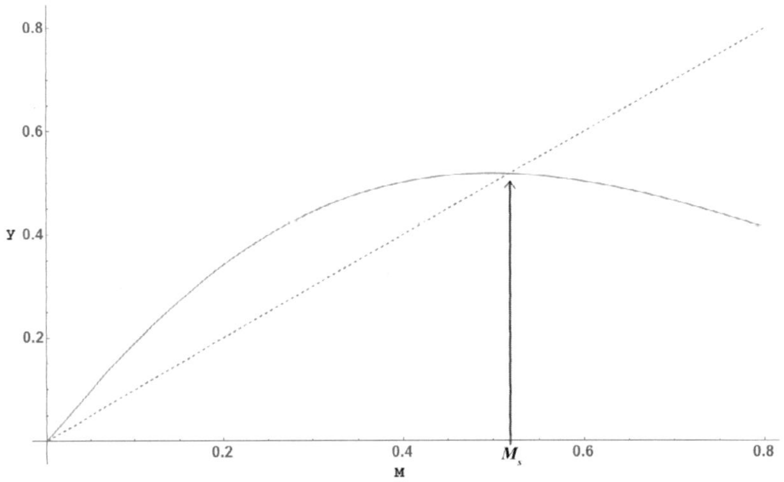

Figure 1. Plot of the system of Eqs. (13), were we have used the RHS of Eq. (8) in place of $G'f(m)$. The solution of the gap equation is the intersection of the two curves, M_s.

For each value of the coupling constant G', the curves in Fig. 1 change as well as their intersection and thus, the iteration procedure converges to a different point, and the generated mass for that particular value of G' changes as well. Below we discuss results corresponding to each of the gap Eqs. (5), (6), (7) and (8).

In Fig. 3 we show the last 200 steps of an iterative procedure with a total of 500 steps for each value of G'. We observe that for $G' < 1$, the procedure converges to $M_s = 0$, which is consistent

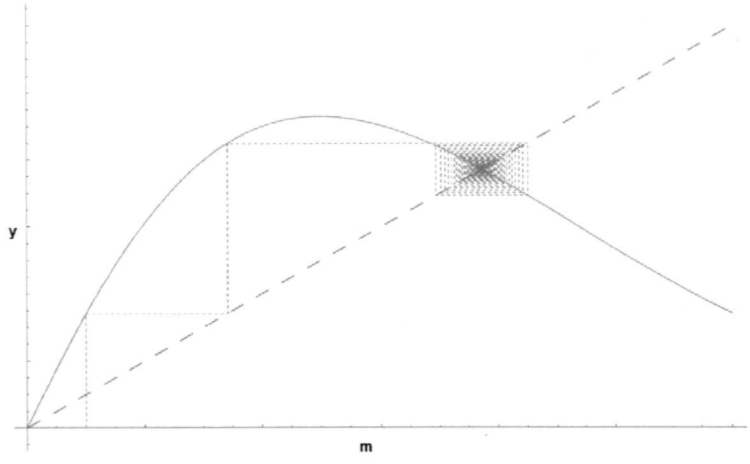

Figure 2. Graphic representation of the iteration process (coweb plot). In this case, we can see that the sequence of iterations tends to converge to the intersection of the curves.

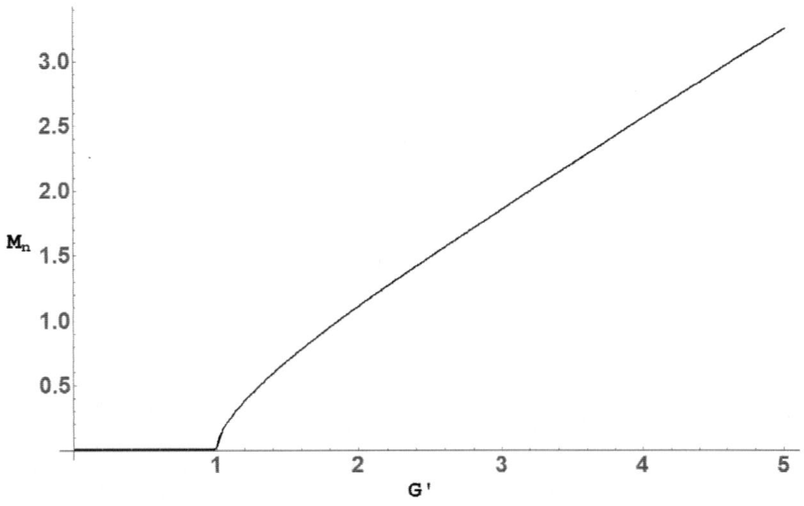

Figure 3. Iterations plot for Eq. (9). In the vertical axes, the point at where the iterations converges. At the horizontal axes the various values of the coupling constant G'.

with the known feature of NJL model: Unless the coupling exceeds a critical value, the gap equation would have only the trivial solution $m = 0$. The critical value is therefore $G'_c = 1$. It is said that at this critical value, the gap equation bifurcates away from the trivial solution, because above this critical value, the iterative procedure always converges to $M_s = m > 0$, that grows smoothly as G' increases. Such a behavior is expected and, in fact, corroborated in Fig. 4, where a similar procedure was performed on the gap Eq. (10). We emphasize that the behavior is independent of the initial choice m_0 and settles the best with a larger number of iteration steps. The critical value remains the same, though the growth of m with G' is slightly different.

From Figs. 5 and 6, corresponding to the PV (11) and PT (12) regularizations, we observe that iterations fail to converge to a fixed value M_s after a second critical coupling is reached at.

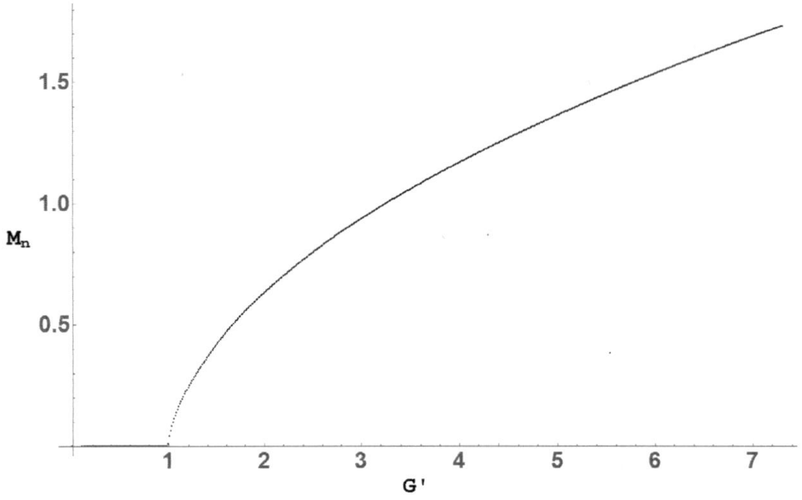

Figure 4. Iterations plot for Eq. (10). In the vertical axes, the point at where the iterations converges. At the horizontal axes the various values of the coupling constant G'.

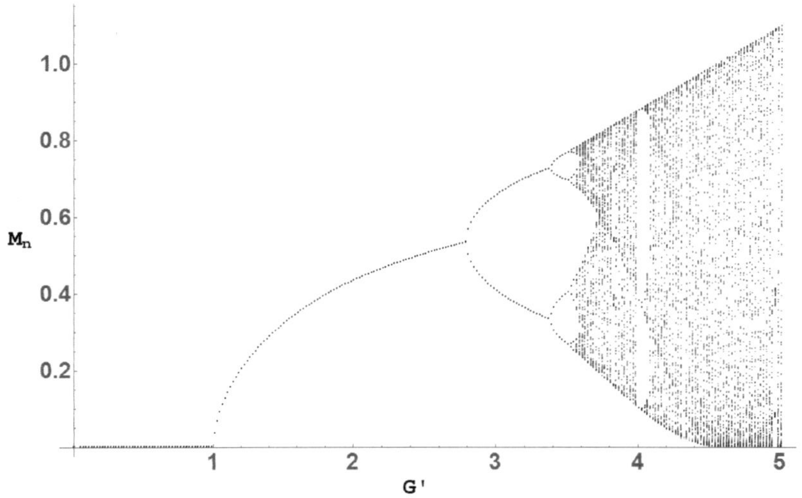

Figure 5. Iterations plot for Eq. (11). In the vertical axes, the point at where the iterations converges. In this case for $G' \approx 2.8$ and above, the iterations stop converging and start to bounce between two values of M_n. For $G' > 3.6$ the iterations go into a chaotic regime. At the horizontal axes the various values of the coupling constant G'.

This happens for $G' \approx 2.8$ for PV and $G' \approx 3.9$ for PT. In both cases, a bifurcation develops, meaning that the iterations do not converge and they keep bouncing between two values, as shown in Fig. 7. Furthermore, as the coupling gets stronger, each branch bifurcates once again and so on until the iterations become chaotic. The iteration diagram resembles the *logistic map* of discrete dynamical systems [17].

4. Summary and Outlook
In this contribution we have confirmed that applying an iteration scheme to solve the gap equation in the NJL model is still an effective method to explore the dependence of the generated

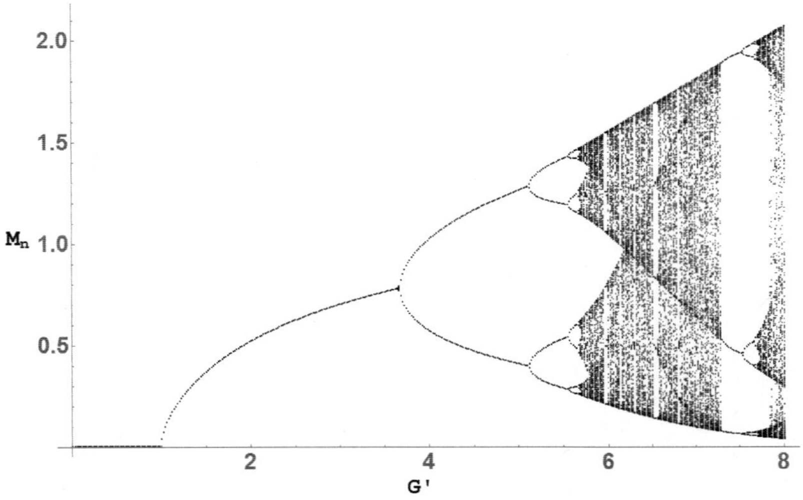

Figure 6. Iterations plot for Eq. (12). In the vertical axes, the point at where the iterations converges. In this case for $G' \approx 3.9$ and above the iterations stop converging and start to oscillate between two values of M_n. Then for $G' > 5.75$ the iterations go into a chaotic regime. At the horizontal axes the various values of the coupling constant G'.

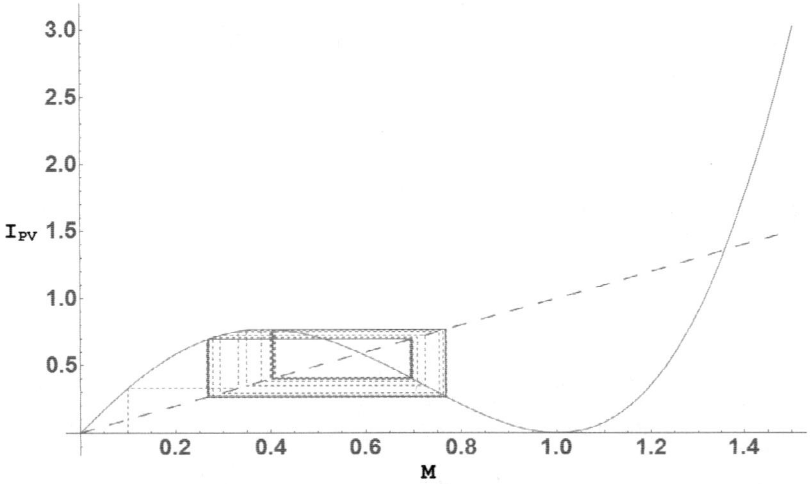

Figure 7. Example of iterations that do not converge and start bouncing between four values of M_n at fixed $G' = 3.51$ in PV regularization. In this case, the bouncing among the four values of M_n correspond to the second bifurcation in Fig. 5.

mass on the coupling constant and other relevant parameters, at least in a broad range of values of G' which contains the typical values used to describe hadron physics phenomenology. For $3D$ and $4D$ cut-off regularization schemes, the iterative succession converges quickly for every value of the coupling constant, even at the super-strong coupling regime. On the other hand, applying an iterative procedure to solve Eqs. (7) or (8) can result in ill solutions for super-strong couplings. The procedure fails to converge after a second critical value for the coupling constant is reached at: iterative steps bounce between two values, four values for larger coupling and so on, till the behavior becomes completely chaotic, as shown in Figs. 5 and 6. This observation implies that special care should be taken when choosing the coupling in any of these two regularization schemes, which are preferred, for instance, to insert the influence of an external magnetic that could render the coupling super-strong.

Furthermore, unless a neat physical interpretation of the partial iterative results can be given, as the one suggested in Ref. [18] as a resummation of bubble insertions into the self-energy, the chaotic behavior might hint a failure of the procedure. This and other considerations are under scrutiny and shall be presented elsewhere [19].

Acknowledgments
We acknowledge valuable discussions with Alejandro Ayala, Gastao Krein, Bruno El-Bennic and M Loewe. We also acknowledge Consejo Nacional de Ciencia y Tecnología (México) for financial support under grant 256494.

References

[1] Y. Nambu and G. Jona-Lasinio. *Dynamical Model of Elementary Particles Based on an Analogy with Superconductivity. I. Phys. Rev.*, 122:345–358, Apr 1961.
[2] Y. Nambu and G. Jona-Lasinio. *Dynamical Model of Elementary Particles Based on an Analogy with Superconductivity. II. Phys. Rev.*, 124:246–254, Oct 1961.
[3] H. L. L. Roberts, A. Bashir, L. X. Gutiérrez-Guerrero, C. D. Roberts, and D. J. Wilson. *π and ρ mesons, and their diquark partners, from a contact interaction. Phys. Rev. C*, 83:065206, Jun 2011.
[4] Gutiérrez-Guerrero, L. X. and Bashir, A. and Cloët, I. C. and Roberts, C. D. *Pion form factor from a contact interaction. Phys. Rev. C*, 81:065202, Jun 2010.
[5] Gustavo A. Contrera, A. Gabriela Grunfeld, and David Blaschke. Supporting the search for the CEP location with nonlocal PNJL models constrained by Lattice QCD. *Eur. Phys. J.*, A52(8):231, 2016.
[6] Michael Buballa. NJL model analysis of quark matter at large density. *Phys. Rept.*, 407:205–376, 2005.
[7] Alejandro Ayala, C. A. Dominguez, L. A. Hernández, M. Loewe, Alfredo Raya, J. C. Rojas, and C. Villavicencio. *Thermomagnetic properties of the strong coupling in the local Nambu˘Jona-Lasinio model. Phys. Rev. D*, 94:054019, Sep 2016.
[8] Ricardo L. S. Farias, Varese S. Timóteo, Sidney S. Avancini, Marcus B. Pinto, and Gastão Krein. Thermomagnetic effects in quark matter: Nambu-jona-lasinio model constrained by lattice qcd. *The European Physical Journal A*, 53(5):101, May 2017.
[9] P. Watson and H. Reinhardt. Quark gap equation in an external magnetic field. *Phys. Rev.*, 89(4):045008, February 2014.
[10] N. Mueller, J. A. Bonnet, and C. S. Fischer. Dynamical quark mass generation in a strong external magnetic field. *Phys. Rev.*, 89(9):094023, May 2014.
[11] Joseph I. Kapusta and Charles Gale. *Finite-Temperature Field Theory: Principles and Applications.* Cambridge Monographs on Mathematical Physics. Cambridge University Press, 2 edition, 2006.
[12] Igor A. Shovkovy. *Magnetic Catalysis: A Review*, pages 13–49. Springer Berlin Heidelberg, Berlin, Heidelberg, 2013.
[13] Gergely Endrodi. Critical point in the QCD phase diagram for extremely strong background magnetic fields. *JHEP*, 07:173, 2015.
[14] G. S. Bali, F. Bruckmann, G. Endrődi, Z. Fodor, S. D. Katz, and A. Schäfer. Qcd quark condensate in external magnetic fields. *Phys. Rev. D*, 86:071502, Oct 2012.
[15] Aftab Ahmad and Alfredo Raya. Inverse magnetic catalysis and confinement within a contact interaction model for quarks. *J. Phys.*, G43(6):065002, 2016.
[16] S.H. Strogatz. *Nonlinear Dynamics And Chaos.* Studies in nonlinearity. Sarat Book House, 2007.
[17] R. M. May. Simple mathematical models with very complicated dynamics. *Nature*, 261:459–467, June 1976.
[18] Ken-Ichi Aoki, Shinnosuke Onai, and Daisuke Sato. *Analysis of Spontaneous Mass Generation by Iterative Method in the Nambu-Jona-Lasinio Model and Gauge Theories.* In *Proceedings, KMI-GCOE Workshop on Strong Coupling Gauge Theories in the LHC Perspective (SCGT 12): Nagoya, Japan, December 4-7, 2012*, pages 439–442, 2014.
[19] Angelo Martinez and Alfredo Raya. Work in progress.

On flavor symmetries of phenomenologically viable string compactifications

Saúl Ramos-Sánchez
Instituto de Física, Universidad Nacional Autónoma de México, POB 20-364, Cd.Mx. 01000, México
E-mail: ramos@fisica.unam.mx

Abstract. Heterotic orbifolds can explain the origin of flavor symmetries and the flavor representations of matter fields in particle physics as a result of the geometric properties of the associated string states in the compact space. After a review of the method to obtain flavor symmetries in these models, we determine the most frequent non-Abelian flavor symmetries appearing in promising Abelian heterotic orbifolds. Interestingly, these symmetries correspond only to D_4, $\Delta(54)$ and products of these symmetries and Abelian factors. A large set of promising models exhibits purely Abelian flavor symmetries. We finally explore the phenomenological potential of a sample model endowed with $\Delta(54)$ assuming certain *ad hoc* flavon expectation values.

1. Introduction

One of the goals of flavor phenomenology is to discover the underlying structure in particle physics that may solve some questions left unanswered in the SM, such as the origin of the family replication, the patterns of quark and lepton mixing matrices, the origin of CP violation and the absence of flavor-changing neutral currents (FCNC). The field theoretic approach consists in first freely choosing a non–Abelian discrete symmetry within $\mathrm{SU}(3)^5_{flavor}$, the largest global symmetry of the SM in the absence of Yukawa couplings, and then introducing a number of *ad hoc* matter fields, some of them with unjustified expectation values (VEVs), to fulfill different basic phenomenological constraints, such as the quark and lepton masses and mixings. Once these restrictions are met, this bottom-up approach delivers a number of consequences, which frequently include interesting new physics. There are plenty of useful symmetries which have been thoroughly studied (see e.g. [1, 2] for a review), and it is hard to learn which of them corresponds to the actual description of our Universe.

Looking for the origin of such symmetries might at least reduce the number of possibilities. In particular, given the constraining environment of string theory, one may wonder what kind of flavor symmetries can emerge in string compactifications that reproduce many properties of the standard model (SM) or its minimal supersymmetric extension (MSSM). The first general studies of this question were [3, 4], in the context of orbifold compactifications of the heterotic string perhaps due to their geometric simplicity. Around those studies, there has been some progress in understanding the qualities of flavor symmetries arising in some phenomenologically viable heterotic orbifolds [5, 6, 7], their enhancements [8] and their generalizations in models endowed with magnetic fluxes [9]. Recently, there has also been progress in the study of flavor symmetries from promising orientifold D-brane models [10, 11].

In this paper, we focus on the $E_8 \times E_8$ heterotic string compactified on symmetric, toroidal $\mathbb{Z}_N$ and $\mathbb{Z}_N \times \mathbb{Z}_M$ orbifolds, which are the simplest compactifications since, among other features, the resulting space corresponds to a special point in the Calabi-Yau moduli space, where the underlying CFT of string theory is valid. Thus, obtaining the structure of the couplings in the effective theory can be done by computing correlation functions of asymptotic string states and vertex operators [12, 13, 14, 15]. These computations lead to a set of selection rules that determine which couplings among the effective fields are non-vanishing [16, 17, 18, 19, 20, 21, 22], and that can be expressed in terms of discrete symmetries. As we shall review in section 2, based on [3, 4], these symmetries are the core of the flavor symmetries from heterotic orbifolds.

One question we address in this paper is what symmetries actually emerge from orbifold compactifications that fulfill a minimal set of necessary conditions that guarantee their phenomenological viability. Here, to be considered phenomenologically promising, an orbifold model must yield the SM gauge group, such that the hypercharge generator be non-anomalous and (with normalization) compatible with grand unification, three generations of quarks and leptons, at least a couple of Higgs superfields, H_u and H_d, and only vectorlike exotics w.r.t. the SM gauge group. These models are identified from a set of several millions of consistent orbifold compactifications, what renders the task very time-consuming. Fortunately, this search becomes accessible thanks to tools such as the `orbifolder` [23], which automatizes the computation of matter spectra and the selection of the promising models.

The `orbifolder` allows one to perform a search of viable orbifold models by scanning randomly the parameter space and comparing the spectra of the generated models. In section 3, we apply this technique to obtain a sample of randomly generated viable orbifold models, similar to those presented in [24, tab A.1], in order to identify the most favored flavor symmetries. Our statistical results, that coincide with those of [24] except for few cases ($\mathbb{Z}_7$ and $\mathbb{Z}_2 \times \mathbb{Z}_4$), help us remark that the most common non-Abelian symmetries arising in phenomenologically promising models are the dihedral group D_4, $\Delta(54)$ and products of these symmetries with Abelian discrete symmetries.

Orbifold compactifications do not only constrain the symmetry groups that can be used as flavor symmetries, but also the number of effective matter fields appearing in the models and their flavor representations. Therefore, in contrast to field-theoretic models where flavor representations can be chosen at convenience, in string models one is restricted to use the field representations given by the theory. It is possible to go even beyond this statement: if all moduli fields determining the size and shape of the compactification space are fixed, even the dynamics of possible flavon fields and hence their VEVs are fully set by the theory. Unfortunately, despite recent progress in this direction [25, 26], it is still too early to define whether moduli can be stabilized in this context. Consequently, the details of the phenomenology that can be extracted from these constructions must rely on admissible (compatible with the small-field limit of SUGRA), yet *ad hoc*, VEVs of moduli and would-be flavon fields. This is the approach we follow in section 4, where we review the main phenomenological results of our previous work [7], based on the $\Delta(54)$ flavor symmetry, which is the favored symmetry in $\mathbb{Z}_3 \times \mathbb{Z}_3$ orbifolds.

2. Geometry of flavor in heterotic orbifolds

We follow the discussion of [3, 4], that leads to the building blocks of the flavor symmetries of heterotic orbifolds.

2.1. Elements of heterotic orbifolds

Here, we introduce the basic formalism of Abelian orbifolds, by using a $\mathbb{Z}_N \times \mathbb{Z}_M$ as the standard. $\mathbb{Z}_N$ orbifolds follow readily by ignoring all the elements related to the second symmetry. $\mathbb{Z}_N \times \mathbb{Z}_M$ heterotic orbifolds, with $N, M \in \mathbb{Z}$ are characterized by the quotient of a six-dimensional torus $\mathbb{T}^6$ divided by the joint action of two Abelian isometries of $\mathbb{T}^6$. Since $\mathbb{Z}_N \times \mathbb{Z}_M$ must be an

Table 1: All admissible $\mathbb{Z}_N$ and $\mathbb{Z}_N{\times}\mathbb{Z}_M$ orbifolds in six dimensions. We provide the corresponding twist vectors and the number of allowed geometries (ignoring roto-translations) according to [27].

$\mathbb{Z}_N$	twist vector	# geometries	$\mathbb{Z}_N{\times}\mathbb{Z}_M$	twist vectors		#geometries
$\mathbb{Z}_3$	$\frac{1}{3}(1,1,-2)$	1	$\mathbb{Z}_2{\times}\mathbb{Z}_2$	$\frac{1}{2}(0,1,-1)$	$\frac{1}{2}(1,0,-1)$	12
$\mathbb{Z}_4$	$\frac{1}{4}(1,1,-2)$	3	$\mathbb{Z}_2{\times}\mathbb{Z}_4$	$\frac{1}{2}(0,1,-1)$	$\frac{1}{4}(1,0,-1)$	10
$\mathbb{Z}_6$-I	$\frac{1}{6}(1,1,-2)$	2	$\mathbb{Z}_2{\times}\mathbb{Z}_6$-I	$\frac{1}{2}(0,1,-1)$	$\frac{1}{6}(1,0,-1)$	2
$\mathbb{Z}_6$-II	$\frac{1}{6}(1,2,-3)$	4	$\mathbb{Z}_2{\times}\mathbb{Z}_6$-II	$\frac{1}{2}(0,1,-1)$	$\frac{1}{6}(1,1,-2)$	4
$\mathbb{Z}_7$	$\frac{1}{7}(1,2,-3)$	1	$\mathbb{Z}_3{\times}\mathbb{Z}_3$	$\frac{1}{3}(0,1,-1)$	$\frac{1}{3}(1,0,-1)$	5
$\mathbb{Z}_8$-I	$\frac{1}{8}(1,2,-3)$	3	$\mathbb{Z}_3{\times}\mathbb{Z}_6$	$\frac{1}{3}(0,1,-1)$	$\frac{1}{6}(1,0,-1)$	2
$\mathbb{Z}_8$-II	$\frac{1}{8}(1,3,-4)$	2	$\mathbb{Z}_4{\times}\mathbb{Z}_4$	$\frac{1}{4}(0,1,-1)$	$\frac{1}{4}(1,0,-1)$	5
$\mathbb{Z}_{12}$-I	$\frac{1}{12}(1,4,-5)$	2	$\mathbb{Z}_6{\times}\mathbb{Z}_6$	$\frac{1}{6}(0,1,-1)$	$\frac{1}{6}(1,0,-1)$	1
$\mathbb{Z}_{12}$-II	$\frac{1}{12}(1,5,-6)$	1				

isometry of $\mathbb{T}^6$, the geometry of the torus defines which symmetries, up to deformations, can be moded out. If the generators of $\mathbb{Z}_N$ and $\mathbb{Z}_M$, ϑ and ω respectively, are considered as rotations in six dimensions (i.e. ignoring roto-translations), all admissible choices of N, M and the number of inequivalent torus geometries are displayed in table 1.

The geometry of $\mathbb{T}^6$ is encoded in its six-dimensional lattice Γ, whose basis vectors are $\{e_1, \ldots, e_6\}$, which build, *in the simplest cases*, a space with metric given by the Cartan matrix of a semi-simple Lie algebra. It is convenient to express this geometry in terms of the complex coordinates z_1, z_2, z_3, on which the orbifold generators act as

$$\begin{aligned} \vartheta: &\quad z_i \to z_i e^{2\pi \mathrm{i} v_i}, &\quad 0 \le |v_i| < 1,\ i = 1,2,3\,, \\ \omega: &\quad z_i \to z_i e^{2\pi \mathrm{i} w_i}, &\quad 0 \le |w_i| < 1, \end{aligned} \tag{1}$$

where the so-called *twist vectors* $\boldsymbol{v} = (v_1, v_2, v_3)$ and $\boldsymbol{w} = (w_1, w_2, w_3)$ are subject to the $\mathcal{N} = 1$ conditions $\pm v_1 \pm v_2 \pm v_3 = 0$ and $\pm w_1 \pm w_2 \pm w_3 = 0$. In table 1 we list our choice of the twist vectors for all admissible $\mathbb{Z}_N$ and $\mathbb{Z}_N{\times}\mathbb{Z}_M$ orbifolds.

The action of the orbifold on $\mathbb{T}^6$ is not free; that is, some points are left invariant or fixed (up to translations $n_\alpha e_\alpha$, $n_\alpha \in \mathbb{Z}$, in the torus). For example, since ϑ y ω are only rotations, the origin is always a fixed point in the orbifold. Besides this trivial fixed point, there are non-trivial fixed points away from the origin. The fixed points turn out to be curvature singularities of the compact space, but do not lead to undesirable gravitational effects in the four-dimensional uncompactified space, which is flat at first approximation.

Together with torus translations, the orbifold rotations build the *space group* $S = (\mathbb{Z}_N \times \mathbb{Z}_M) \ltimes \Gamma$, whose elements are $g = (\vartheta^n \omega^m, n_\alpha e_\alpha)$, with $0 \le n < N$, $0 \le m < M$ and $n_\alpha \in \mathbb{Z}$. The action of the space group on the complex coordinates is given by

$$S:\ z \to gz = \vartheta^n \omega^m z + n_\alpha e_\alpha\,. \tag{2}$$

It is possible to associate each fixed point z_f with a space group element, called *constructing element* g_f, such that $z_f = g_f z_f$. We notice that for each choice of (n, m) there are different choices of $\{n_\alpha\}$ leading to different fixed points. The fixed points are truly inequivalent if their corresponding constructing elements belong to different conjugacy classes within S. So, at the end, for each *sector* (n, m) there is a finite number of fixed points.

Once the generic geometrical aspects of the compactification in six dimensions have been set, these must be embedded into the gauge degrees of freedom of the heterotic strings. We consider here the $\mathcal{N}=1$ $E_8\times E_8$ heterotic string. Modular invariance of the partition function demands that each orbifold twist be embedded either as a rotation or as a shift vector in the 16 dimensions of $E_8\times E_8$, and that the $\mathbb{T}^6$ translations be connected to a so-called Wilson line. We denote the shift vectors as V, W for the embedding of ϑ, ω, and the Wilson lines as A_α, $\alpha = 1,\ldots,6$, as the embedding of e_α. Constructing elements in S are then embedded into the gauge degrees of freedom as

$$g = (\vartheta^n\omega^m, n_\alpha e_\alpha) \quad \hookrightarrow \quad V_g \equiv nV + mW + n_\alpha A_\alpha\,. \tag{3}$$

The gauge embedding is subject to some constraints. First, both V and W must be consistent with a $\mathbb{Z}_N\times\mathbb{Z}_M$ action. This amounts to requiring e.g. that NV must lie in the root lattice of $E_8\times E_8$, Λ. Analogous conditions must be imposed to W. Secondly, Wilson lines must be consistent with the torus geometry and the orbifold action on it. The fact that the Γ basis vectors e_α are in general related by the action of ϑ and ω translates to relations among different A_α. Furthermore, from these considerations, just as shift vectors, Wilson lines A_α have an order N_α, such that $N_\alpha A_\alpha \in \Lambda$ (without summation over α). Finally, modular invariance additionally imposes in $\mathbb{Z}_N\times\mathbb{Z}_M$ heterotic orbifolds that [28]

$$\begin{aligned}
N\,(V^2 - v^2) &= 0 \mod 2\,, & N_\alpha\,(V\cdot A_\alpha) &= 0 \mod 2\,, \quad \alpha = 1,\ldots,6\,,\\
M\,(W^2 - w^2) &= 0 \mod 2\,, & N_\alpha\,(W\cdot A_\alpha) &= 0 \mod 2\,,\\
M\,(V\cdot W - v\cdot w) &= 0 \mod 2\,, & N_\alpha\,A_\alpha^2 &= 0 \mod 2\,,\\
 & & \gcd(N_\alpha, N_\beta)\,(A_\alpha\cdot A_\beta) &= 0 \mod 2\,, \quad \alpha\neq\beta\,.
\end{aligned} \tag{4}$$

If all the previous requirements are fulfilled, all ingredients can be used to compactify a heterotic string. Note that $A_\alpha = 0\ \forall\alpha$ is always a possibility.

We now turn to the matter fields Φ in orbifold compactifications. These correspond to closed string states $|\Phi\rangle$ that are invariant under the orbifold action and that must be massless because the mass of massive strings is some factor of the Planck scale, M_{pl}, and thus too high to appear in the effective theory at low energies. Closed strings comprise left-movers, that equip physical states with gauge quantum numbers, and right-movers providing the so-called H-momentum [20], which is basically the momentum of the state in the compact dimensions.

In heterotic orbifolds, bulk or *untwisted fields* correspond to the orbifold-invariant states arising directly from the ten-dimensional closed strings of the uncompactified heterotic string, whose field limit is ten-dimensional $\mathcal{N}=1$ supergravity endowed with an $E_8\times E_8$ Yang-Mills theory. Thus, the four-dimensional gauge superfields, generating the unbroken gauge group $\mathcal{G}_{4D} \subset E_8\times E_8$, and some four-dimensional matter states with non-trivial gauge quantum numbers under $\mathcal{G}_{4D}$ live in the bulk of a heterotic orbifold.

Additionally, there are the so–called *twisted fields*, which arise from strings that are closed only due to the action of the orbifold. Twisted fields are always localized at singularities of the orbifold and are thus related to a constructing element g and its embedding into the gauge degrees of freedom V_g defined in (3). Since it is V_g what defines the gauge quantum numbers of the localized strings, we notice that for each sector (n,m) all states have identical gauge numbers unless $A_\alpha \neq 0$ for some α, i.e. unless there are non-trivial Wilson lines.

2.2. Coupling selection rules

Couplings among string states are subject to a set of constraints called *string selection rules* [16, 17, 18, 19, 20, 21, 22], due to symmetries of the underlying CFT of the compactified string theory. The main restrictions are

① *Gauge invariance*: the sum of all gauge quantum numbers must be trivial.

② *H-momentum conservation*: all H-momenta of the states must add up to zero.

③ *Space-group invariance*: the product of constructing elements must be trivial, which for r interacting strings means

$$\prod_{f=1}^{r}\left(\theta_{(f)},\, n_\alpha^{(f)} e_\alpha\right) \stackrel{!}{=} \left(\mathbb{1},\, \bigcup_f (\mathbb{1}-\theta_{(f)})\Gamma\right), \tag{5}$$

where $g_f = (\theta_{(f)},\, n_\alpha^{(f)} e_\alpha)$, $\theta_{(f)} = \vartheta^{n^{(f)}}\omega^{m^{(f)}}$, corresponds to the constructing element of a twisted state localized at z_f, and $\cup_f(\mathbb{1}-\theta_{(f)})\Gamma$ is known as the *invariant sublattice* of fixed points. The space group selection rule (5) indicates if r closed strings can interact or not taking as criterion their localizations in the compact space.

These selection rules establish for which combination of string states there is a non-zero correlation function, and thus a non-vanishing coupling for the associated effective fields. We can easily see that the only selection rule that is relevant for flavor symmetries is the space-group invariance. Notice that the rotational part of that selection rule implies

$$\sum_{f=1}^{r} n^{(f)} = 0 \mod N\,, \qquad \sum_{f=1}^{r} m^{(f)} = 0 \mod M\,, \tag{6}$$

which is a way to express invariance under a $\mathbb{Z}_N \times \mathbb{Z}_M$ symmetry for fields with charges $(n^{(f)}, m^{(f)})$. In explicit examples, it can be shown, as we do in the next section, that the translational part also implies the existence of additional orbifold and geometry dependent Abelian symmetries with charges $n_\alpha^{(f)}$. That is, in the four-dimensional model emerging from an Abelian heterotic orbifold, space-group invariance amounts to including additional Abelian symmetries and assign thus appropriate discrete charges to each field in the model.

2.3. *Non-Abelian flavor symmetries*
As guiding examples of the origin of flavor symmetries, let us consider the cases illustrated in figure 1. The first figure corresponds to a $\mathbb{Z}_2$ symmetry that, instead of being moded out of the whole $\mathbb{T}^2$, divides only one compact dimension, i.e. S^1. This scenario is realized in heterotic

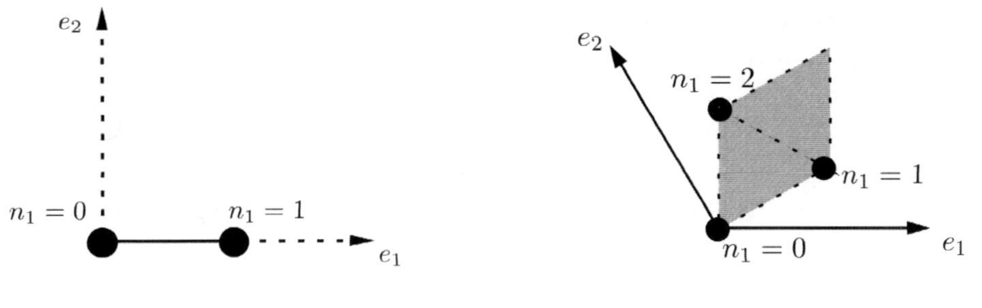

(a) Fixed points and charges of $S^1/\mathbb{Z}_2$ (b) Fixed points and charges of $\mathbb{T}^2/\mathbb{Z}_3$

Figure 1: Geometrical origin of the D_4 and $\Delta(54)$ flavor symmetries in orbifolds. When unaffected by Wilson lines, the fixed points of an $S^1/\mathbb{Z}_2$ ($\mathbb{T}^2/\mathbb{Z}_3$) orbifold realize an S_2 (S_3) permutation symmetry. String selection rules impose an additional $\mathbb{Z}_2\times\mathbb{Z}_2$ ($\mathbb{Z}_3\times\mathbb{Z}_3$) symmetry based on the localization charges of twisted states. The resulting flavor symmetry is $S_2 \ltimes \mathbb{Z}_2^2 = D_4$ ($S_3 \ltimes \mathbb{Z}_3^2 = \Delta(54)$).

orbifolds when $\mathbb{T}^6$ can be factorized as $\mathbb{T}^2 \times \mathbb{T}^4$ and there is a Wilson line associated to one of the compact directions (e_2 in the depicted case). The $\mathbb{Z}_2$ acts as a reflection on the points of S^1, so that the elements of $\mathbb{Z}_2$ that act on z_1 are expressed as $\{\vartheta^0 = 1, \vartheta^1 = -1\}$. Thus, the action of ϑ on the Γ basis vector reads $e_1 \to -e_1$. With this at hand, we observe that two fixed points occur in the sector ϑ^n with $n = 1$ and are given by the constructing elements $g_0 = (\vartheta, 0)$ and $g_1 = (\vartheta, e_1)$, corresponding to $z_0 = 0$ and $z_1 = \frac{1}{2}e_1$.

From the rotational part of space-group invariance, we realize that the only non-vanishing couplings are those satisfying $\vartheta^r = 1$, i.e. when an even number of strings from the ϑ sector interact. This implies that states are charged with a $\mathbb{Z}_2$ charge that can be either $n = 0$ for untwisted sector states or $n = 1$ for states in the ϑ sector. On the other hand, the translational component of space-group invariance requires verifying first what the invariant sublattice is. Since $(1-\vartheta)e_1 = 2e_1$, then the invariant sublattice is given by any integer multiple of $2e_1$. Consider now a coupling between, say, two twisted states, related to the constructing elements $(\vartheta, n_1^{(1)}e_1)$ and $(\vartheta, n_1^{(2)}e_1)$. The product of these states yields $(1, (n_1^{(1)} - n_1^{(2)})e_1)$, which is equivalent to $(1, (n_1^{(1)} + n_1^{(2)})e_1)$ because they differ by a contribution of the invariant sublattice. Hence, the selection rule implies that $n_1^{(1)} + n_1^{(2)}$ must be even, or $\sum_f n_1^{(f)} = 0 \mod 2$ in general, i.e. another $\mathbb{Z}_2$ that associates the charge $n_1^{(f)}$ to the twisted states. We find thus that couplings between two twisted states are allowed only if both states lie at the same fixed point. We have found a $\mathbb{Z}_2 \times \mathbb{Z}_2$ symmetry whose generators can be expressed in the space of the ϑ fixed points as the matrices $R = \mathrm{diag}(-1, -1)$ and $T = \mathrm{diag}(1, -1)$.

Finally, in the absence of Wilson lines associated with the direction e_1, both singularities are indistinguishable from the point of view of a four-dimensional observer because the gauge quantum numbers of states living at both singularities are equal. Therefore, there is a permutation symmetry S_2 of the states at the singularities, whose generator can be expressed as

$$P = \begin{pmatrix} 0 & 1 \\ 1 & 0 \end{pmatrix} . \tag{7}$$

Noting that conjugating with P any element of $\mathbb{Z}_2 \times \mathbb{Z}_2$ generated by R and T yields another element of this group, we see that $\mathbb{Z}_2 \times \mathbb{Z}_2$ is a normal subgroup of the full discrete symmetry, which can thus be written as $S_2 \ltimes (\mathbb{Z}_2 \times \mathbb{Z}_2)$, isomorphic to D_4. This is the flavor symmetry of the fields arising from an $S^1/\mathbb{Z}_2$ sector of a heterotic orbifold.

As we discuss in full detail in [7], similar considerations yield the flavor symmetry $\Delta(54)$ in the $\mathbb{T}^2/\mathbb{Z}_3$ orbifold, depicted in figure 1(b). In this case, $\vartheta e_1 \to e_2$ and $\vartheta e_2 \to -e_1 - e_2$. Further, the constructing elements associated with the three fixed points of the ϑ sector are $(\vartheta, n_1 e_1)$, $n_1 = 0, 1, 2$. With the invariant sublattice given by the basis $\{3e_1, 3e_2\}$, one can show that space-group invariance implies that only couplings of a multiple of three states are allowed and that $\sum_f n_1^{(f)} = 0 \mod 3$ must be satisfied. The related Abelian symmetry is then $\mathbb{Z}_3 \times \mathbb{Z}_3$, generated by $R = \mathrm{diag}(\rho, \rho, \rho)$ and $T = \mathrm{diag}(1, \rho, \rho^2)$, with $\rho = e^{2\pi i/3}$. Furthermore, in the absence of Wilson lines related to $\mathbb{T}^2$, there is a permutation symmetry S_3, whose generators are

$$\sigma = \begin{pmatrix} 0 & 0 & 1 \\ 1 & 0 & 0 \\ 0 & 1 & 0 \end{pmatrix} , \qquad \tau = \begin{pmatrix} 0 & 1 & 0 \\ 1 & 0 & 0 \\ 0 & 0 & 1 \end{pmatrix} . \tag{8}$$

The multiplicative closure of all elements reveals that the emerging flavor symmetry is $S_3 \ltimes (\mathbb{Z}_3 \times \mathbb{Z}_3)$, isomorphic to $\Delta(54)$.

Another interesting aspect is that the flavor representations of matter fields are also determined by this structure. States associated to the fixed points of $S^1/\mathbb{Z}_2$ can only build doublets $\mathbf{2}$ of D_4. In $\mathbb{T}^2/\mathbb{Z}_3$, the only matter representations in the ϑ sector are triplets $\mathbf{3}_{11}$

of $\Delta(54)$ (in the notation of [29]); in the $\vartheta^2 = \vartheta^{-1}$ sector the representations are conjugate, i.e. $\mathbf{3}_{12}$. Bulk states are in all cases trivial singlets of the corresponding flavor symmetries. No further representations appear unless states form condensates.

The permutation symmetries are broken completely if there are non-vanishing Wilson lines associated to the compact directions. Thus, the flavor symmetries suffer an explicit breakdown to $\mathbb{Z}_2\times\mathbb{Z}_2$ or $\mathbb{Z}_3\times\mathbb{Z}_3$ in $S^1/\mathbb{Z}_2$ or $\mathbb{T}^2/\mathbb{Z}_3$, respectively, when affected by non-trivial Wilson lines. Flavor symmetries can also be broken spontaneously to non-Abelian subgroups if some fields localized at the singularities develop VEVs. The details depend on the VEV structure and the number of flavon fields.

This discussion has been explicitly developed for all possible sub-orbifolds (in less than six dimensions) appearing in Abelian toroidal heterotic orbifolds [3], resulting in a reduced number of family symmetries. The findings include, besides D_4 and $\Delta(54)$, only the symmetries $(D_4 \times D_4)/\mathbb{Z}_2$, $(D_4 \times \mathbb{Z}_4)/\mathbb{Z}_2$, $(D_4 \times \mathbb{Z}_8)/\mathbb{Z}_2$ and $S_7 \ltimes \mathbb{Z}_7^6$. As we shall shortly see, not all of these symmetries are realized in phenomenologically viable models.

3. Favored flavor symmetries in $\mathbb{Z}_N$ and $\mathbb{Z}_N\times\mathbb{Z}_M$ orbifolds

The tools explained in the previous section can be applied to all Abelian orbifolds. Therefore, given a set of consistent shift vectors and Wilson lines, it is possible first to obtain the spectrum and then to determine the flavor symmetry by inspecting which Wilson lines are trivial and what is the orbifold action on the Γ basis vectors associated to them.

For example, $\mathbb{Z}_3\times\mathbb{Z}_3$ orbifold models in their simplest geometry ((1,1) in the nomenclature of [27]) admit three independent Wilson lines, each associated to one of the z^i complex planes with basis vectors e_{2i-1} and e_{2i}. The global flavor symmetry in the absence of Wilson lines is $(S_3 \times S_3 \times S_3) \ltimes \mathbb{Z}_3^5$, somewhat similar to the naive expectation of $\Delta(54)^3$. Now, if one of the Wilson lines is non-zero, then the flavor symmetry is $(S_3 \times S_3) \ltimes \mathbb{Z}_3^5$ due to the explicit breaking of one S_3. Two non-vanishing Wilson lines yield $S_3 \ltimes \mathbb{Z}_3^5$, which can be rewritten as $\Delta(54) \times \mathbb{Z}_3^3$. (This symmetry shall play an important role in the next section.) And three non-trivial Wilson lines lead to a purely Abelian $\mathbb{Z}_3^5$ flavor symmetry.

There are other more complicated cases, such as $\mathbb{Z}_6$-II orbifolds, whose structure leads to the possibility of three non-trivial Wilson lines: one related to the $\mathbb{T}^2$ of the z^2 plane, and two associated to the e_5 and e_6 directions. From the orbifold action determined by the twist $v = (1, 2, -3)/6$, we see that the orbifold corresponds to the factorization $\mathbb{T}^2/\mathbb{Z}_6 \times \mathbb{T}^2/\mathbb{Z}_3 \times S^1/\mathbb{Z}_2 \times S^1/\mathbb{Z}_2$, where the Wilson lines can affect the last three factors. In this case, it is not enough to know the number of non-zero Wilson lines of a given model to figure out the flavor symmetry. One needs also the direction with which it is associated. If a model is furnished with one non-trivial Wilson line, then the original flavor symmetry $(S_3 \times S_2 \times S_2) \ltimes (\mathbb{Z}_3^2 \times \mathbb{Z}_2^3)$ is broken down to one of two possibilities, $\Delta(54) \times D_4 \times \mathbb{Z}_2$ or $(S_2 \times S_2) \ltimes (\mathbb{Z}_3^2 \times \mathbb{Z}_2^3)$, isomorphic to $(D_4 \times D_4)/\mathbb{Z}_2 \times \mathbb{Z}_3^2$. Two non-vanishing Wilson lines render the flavor symmetries $\Delta(54) \times \mathbb{Z}_2^3$ or $D_4 \times \mathbb{Z}_3 \times \mathbb{Z}_3 \times \mathbb{Z}_2$. And three non-zero Wilson lines lead to a $\mathbb{Z}_3^2 \times \mathbb{Z}_2^3$ flavor symmetry.

This exercise can be performed for phenomenologically promising heterotic orbifolds. The use of the `orbifolder` [23] leads to large sets of models with *a priori* defined properties. Particularly, one can obtain models with the following features: 1) SM gauge group with non-anomalous and correctly normalized hypercharge, 2) three SM generations, 3) at least a couple of Higgses H_u and H_d, and 4) no chiral exotics. These models will be considered phenomenologically viable. Then, identifying their flavor symmetries can help envisage the phenomenological potential of these constructions.

We have performed non-exhaustive scans of all orbifolds listed in tables 2, with the geometries specified in the second column of each table in the notation of [27]. That is, we have used all possible geometries (ignoring roto-translations) of $\mathbb{Z}_N$ orbifolds and only the simplest ones for $\mathbb{Z}_N\times\mathbb{Z}_M$ orbifolds. Our results, that shall further detailed elsewhere, are statistically equivalent

Table 2: Statistically preferred flavor symmetries in phenomenologically viable $\mathbb{Z}_N$ and $\mathbb{Z}_N \times \mathbb{Z}_M$ heterotic orbifolds. The first two columns of each table correspond to the label of Abelian orbifolds, following the notation of [27]. The third column displays the most common flavor symmetry appearing in promising models; the symmetries in squared brackets are non-Abelian flavor symmetries that appear less frequently in these models. Additional Abelian factors are not displayed. The last column counts the number of independent non-trivial Wilson lines needed to build most frequent promising models over the maximal number of Wilson lines allowed by the geometry.

$\mathbb{Z}_N$		flavor symmetry	# WL
$\mathbb{Z}_3$	$(1,1)$	no viable model	–
$\mathbb{Z}_4$	$(1,1)$	no viable model	–
	$(2,1)$	$\mathbb{Z}_4$	3/3
	$(3,1)$	$\mathbb{Z}_4$	2/2
$\mathbb{Z}_6$-I	$(1,1)$	$\mathbb{Z}_6$	1/1
	$(2,1)$	$\mathbb{Z}_6$	1/1
$\mathbb{Z}_6$-II	$(1,1)$	D_4	2/3
	$(2,1)$	D_4	2/3
	$(3,1)$	D_4	2/3
	$(4,1)$	D_4	1/2
$\mathbb{Z}_7$	$(1,1)$	$\mathbb{Z}_7$	1/1
$\mathbb{Z}_8$-I	$(1,1)$	$\mathbb{Z}_8$ $[(D_4 \times \mathbb{Z}_8)/\mathbb{Z}_2]$	2/2
	$(2,1)$	$\mathbb{Z}_8$ $[(D_4 \times \mathbb{Z}_8)/\mathbb{Z}_2]$	2/2
	$(3,1)$	$\mathbb{Z}_8$	1/1
$\mathbb{Z}_8$-II	$(1,1)$	D_4 $[(D_4 \times \mathbb{Z}_8)/\mathbb{Z}_2]$	2/3
	$(2,1)$	D_4	1/2
$\mathbb{Z}_{12}$-I	$(1,1)$	$\mathbb{Z}_{12}$	1/1
	$(2,1)$	$\mathbb{Z}_{12}$	1/1
$\mathbb{Z}_{12}$-II	$(1,1)$	$\mathbb{Z}_{12}$ $[D_4]$	2/2

$\mathbb{Z}_N \times \mathbb{Z}_M$		flavor symmetry	# WL
$\mathbb{Z}_2 \times \mathbb{Z}_2$	$(1,1)$	$(D_4 \times D_4)/\mathbb{Z}_2$ $[D_4]$	4/6
$\mathbb{Z}_2 \times \mathbb{Z}_4$	$(1,1)$	$D_4 \times (D_4 \times \mathbb{Z}_4)/\mathbb{Z}_2$	2/4
		$[(D_4 \times D_4)/\mathbb{Z}_2]$	2/4
		$[D_4$ or $(D_4 \times \mathbb{Z}_4)/\mathbb{Z}_2]$	3/4
$\mathbb{Z}_2 \times \mathbb{Z}_6$-I	$(1,1)$	$\mathbb{Z}_2 \times \mathbb{Z}_6$ $[D_4]$	1/2
$\mathbb{Z}_2 \times \mathbb{Z}_6$-II	$(1,1)$	no viable model	–
$\mathbb{Z}_3 \times \mathbb{Z}_3$	$(1,1)$	$\Delta(54)$	2/3
		$[(\Delta(54) \times \Delta(54))/\mathbb{Z}_2]$	1/3
$\mathbb{Z}_3 \times \mathbb{Z}_6$	$(1,1)$	$\mathbb{Z}_3 \times \mathbb{Z}_6$ $[\Delta(54)]$	1/2
$\mathbb{Z}_4 \times \mathbb{Z}_4$	$(1,1)$	$((D_4 \times \mathbb{Z}_4)/\mathbb{Z}_2)^2/\mathbb{Z}_2$	1/3
		$[(D_4 \times \mathbb{Z}_4)/\mathbb{Z}_2]$	2/3
$\mathbb{Z}_6 \times \mathbb{Z}_6$	$(1,1)$	$\mathbb{Z}_6 \times \mathbb{Z}_6$	–

to those of [24, tab A.1], except in the $\mathbb{Z}_7$ orbifold where we do find one promising model with one Wilson line and in the $\mathbb{Z}_2 \times \mathbb{Z}_4$ orbifold, where we find that most of the models are endowed with two non-vanishing Wilson lines, instead of three.

Out of nearly 8,000 models, about 30% of them arise from $\mathbb{Z}_2 \times \mathbb{Z}_4$ models and around 10+10% from $\mathbb{Z}_4 \times \mathbb{Z}_4$ and $\mathbb{Z}_6 \times \mathbb{Z}_6$ orbifolds. About 35% of the models saturate the number of non-vanishing Wilson lines. In these cases, the flavor symmetries are purely Abelian, rendering the models less promising. These appear mostly in $\mathbb{Z}_{12}$-I and $\mathbb{Z}_6 \times \mathbb{Z}_6$ orbifolds, which may suggest that some efforts in the literature about $\mathbb{Z}_{12}$-I orbifolds should be redirected.

We have examined the most common flavor symmetries in each orbifold geometry. Our results are listed in table 2. There, we display only the non-Abelian flavor symmetries identified, omitting smaller Abelian factors, for the sake of simplicity. In the cases where only Abelian flavor symmetries appear, we present only the largest Abelian group. The symmetries in squared brackets correspond to less favored non-Abelian flavor symmetries that arise in our models.

In many cases where most of the models exhibit Abelian flavor symmetries, there are also some viable models with non-Abelian flavor symmetries. For example, in $\mathbb{Z}_8$-I with torus geometry $(2,1)$, besides the flavor group $\mathbb{Z}_8$, there are about 25% of the models endowed with a $(D_4 \times \mathbb{Z}_8)/\mathbb{Z}_2$ flavor symmetry.

Globally, we also find that about 30% of the models yield a D_4 symmetry multiplied by purely Abelian factors. Around 21% of the models have non-trivial combinations of D_4 with itself and

IOP Conf. Series: Journal of Physics: Conf. Series **912** (2017) 012011 doi:10.1088/1742-6596/912/1/012011

other Abelian factors or quotients, and close to 5% of the models are furnished with a $\Delta(54)$ flavor symmetry. That is, considering the models that exhibit these flavor symmetries together with those endowed with purely Abelian flavor groups, we obtain around 90% of all our models.

4. Phenomenological consequences of flavor symmetries in string models

Our results of the previous section show that most promising heterotic orbifolds enjoy a D_4 non-Abelian flavor symmetry. The phenomenological potential of this symmetry in this context has been studied in detail in [6, 30], where admissible Yukawa textures for quarks and (charged and neutral) leptons as well as promising supersymmetric soft terms (that help avoid FCNC in supersymmetric models) are achieved. It is interesting that the same flavor symmetry appears frequently in appealing D-brane models [11].

We have also found that $\Delta(54)$ is the second most favored non-Abelian flavor symmetry appearing in viable heterotic orbifolds and it has been largely ignored even from a bottom-up approach. Thus, it is necessary to study its phenomenological consequences. To do so, we recover here the main properties of the $\mathbb{Z}_3\times\mathbb{Z}_3$ orbifold model studied in [7]. In that work, one of the about 800 promising models was chosen due to its simplicity. The model is defined by the shift vectors

$$3V = \left(-\tfrac{1}{2},-\tfrac{1}{2},-\tfrac{1}{2},-\tfrac{1}{2},\tfrac{1}{2},\tfrac{1}{2},\tfrac{1}{2},\tfrac{1}{2};-2,0,0,1,1,1,1,4\right)\,, \tag{9a}$$

$$3W = \left(0,1,1,4,0,0,1,1;1,-1,4,-4,-1,0,0,1\right)\,, \tag{9b}$$

and the Wilson lines

$$3A_1 = 3A_2 = \left(-\tfrac{7}{2},-\tfrac{3}{2},\tfrac{9}{2},\tfrac{7}{2},-\tfrac{7}{2},-\tfrac{3}{2},\tfrac{5}{2},\tfrac{7}{2};-3,0,-2,0,-2,-4,3,-2\right)\,, \tag{10a}$$

$$3A_3 = 3A_4 = \left(3,3,-3,-2,-1,2,4,-4;-3,1,-1,-4,1,1,4,1\right)\,. \tag{10b}$$

These parameters yield the unbroken gauge group $\mathrm{SU}(3)_C\times\mathrm{SU}(2)_L\times\mathrm{U}(1)_Y\times[\mathrm{SU}(2)\times\mathrm{U}(1)^{11}]$, where the additional SU(2) factor is considered hidden because no SM–field carries a charge under that group. All fields in the spectrum are charged under the additional U(1) factors, but these shall be broken by flavon VEVs. The flavor symmetry is $\Delta(54)\times\mathbb{Z}_3^3$.

The relevant matter spectrum is displayed in table 3, where the first eight columns refer to SM matter states, including the Higgs fields and right-handed neutrinos, and the last five columns display the properties of flavon fields. In this model, SM fermion fields transform as triplets of the $\Delta(54)$ flavor symmetry while the Higgs fields do not transform because they are bulk fields.

Since the SM matter fields are charged under flavor symmetries, only the presence of the properly charged s and ϕ flavon fields allows for Yukawa couplings in the (non–renormalizable)

	Q_i	$\bar{d}_i^c$	$\bar{u}_i^c$	L_i	$\bar{e}_i^c$	$\bar{\nu}_i$	H_u	H_d	ϕ_i^u	$\phi_i^{(d,e)}$	$\bar{\phi}_i^\nu$	s^u	$s^{(d,e)}$
$\Delta(54)$	$\mathbf{3}_{11}$	$\mathbf{3}_{11}$	$\mathbf{3}_{11}$	$\mathbf{3}_{11}$	$\mathbf{3}_{11}$	$\mathbf{3}_{12}$	$\mathbf{1}_0$	$\mathbf{1}_0$	$\mathbf{3}_{11}$	$\mathbf{3}_{11}$	$\mathbf{3}_{12}$	$\mathbf{1}_0$	$\mathbf{1}_0$
$\mathbb{Z}_3^{(1)}$	ρ	1	ρ	1	ρ	1	1	1	1	1	1	ρ	ρ^2
$\mathbb{Z}_3^{(2)}$	1	ρ^2	1	ρ^2	1	ρ	1	1	1	ρ	ρ	1	1
$\mathbb{Z}_3^{(3)}$	ρ	1	ρ	1	ρ	1	1	1	1	1	1	ρ	ρ^2

Table 3: Flavor representations for the SM matter and flavon fields in a $\mathbb{Z}_3\times\mathbb{Z}_3$ sample model. The $\mathbb{Z}_3$ charges are defined by the field localizations with $\rho = e^{2\pi\mathrm{i}/3}$.

superpotential, given by

$$\begin{aligned} W_Y &= y^u_{ijk} Q_i H_u \bar{u}_j \phi^u_k s_u + y^d_{ijk} Q_i H_d \bar{d}_j \phi^{(d,e)}_k s^{(d,e)} + y^e_{ijk} L_i H_d \bar{e}_j \phi^{(d,e)}_k s^{(d,e)} \\ &+ y^\nu_{ijkl} L_i H_u \bar{\nu}_j + \lambda_{ijk} \bar{\nu}_i \bar{\nu}_j \bar{\phi}^\nu_k , \qquad i,j,k = 1,2,3, \end{aligned} \tag{11}$$

where the summation over repeated indices must follow the rules of the product of $\Delta(54)$ representations that lead to invariant singlets. We see from this superpotential, that quarks and charged leptons acquire masses through dimension–6 operators, and the Dirac neutrino masses as well as the right-handed Majorana neutrino masses are generated at renormalizable level. This is optimal because guarantees that the largest masses are those of right-handed neutrinos.

Assuming that most properties are conserved even after supersymmetry breakdown, we observe that choosing some *ad hoc* hierarchical flavon–VEV alignments results in the following flavor phenomenology features:

- correct masses for quarks and charged leptons;
- proper Gatto-Sartori-Tonin relation in the quark sector (although the other two mixing angles are very small);
- a mass relation between the down–quark sector and the charged leptonic sector

$$\frac{m_s - m_d}{m_b} \overset{!}{=} \frac{m_\mu - m_e}{m_\tau}; \tag{12}$$

- compatibility (only) with normal hierarchy of neutrino masses;
- smallest neutrino mass of order $6-7$ meV;
- total neutrino masses of order $65-70$ meV; and
- PMNS matrix compatible with current constraints (atmospheric and reactor mixing angles are in the 3σ region of the global best fit), with the atmospheric mixing angle greater than 45 degrees.

Interestingly, an inverted hierarchy being disfavored as well as the atmospheric mixing angle lying in the second octant, are features compatible with recent findings of the T2K collaboration [31]. These results render the neutrino sector of $\Delta(54)$ heterotic orbifolds much more promising than the other sectors and let us assert that $\mathbb{Z}_3\times\mathbb{Z}_3$ heterotic orbifolds and $\Delta(54)$ as a flavor symmetry provide a fertile playground for useful phenomenology.

5. Concluding remarks

In this paper, we firstly reviewed the geometric origin of flavor symmetries. They appear as a result of the effective properties of matter fields due to their localization in the compact space. The selection rules that define the non-zero couplings among fields are the key string ingredient in the construction of flavor symmetries. Non-Abelian symmetries turn out to be semi-direct products of a permutation symmetry, arising from the degeneracy of localized states in the absence of non-trivial Wilson lines, multiplied with some Abelian discrete symmetries inherited from the selection rules.

We have then performed a large scan of promising heterotic orbifolds and identified the statistically favored flavor symmetries in those models. We find that in about 50% of the explored models the D_4 flavor symmetry is preferred. The second most frequent non-Abelian flavor symmetry in phenomenologically viable models is $\Delta(54)$. It is somewhat surprising that about 35% of the models carry purely Abelian groups as flavor symmetries.

Since D_4 has been explored in the past, here we reviewed the main features of a $\mathbb{Z}_3\times\mathbb{Z}_3$ orbifold model giving rise to the SM particle spectrum that forms $\Delta(54)$ flavor representations.

Exploiting the couplings structure of the model and choosing some *ad hoc* flavon VEVs, the model reproduces all particle masses, but exhibits issues to yield the right quark and charged-lepton mixing angles. Yet the neutrino sector, that allows for see-saw neutrino masses, can meet the experimental constraints for all mixing angles, favors a normal hierarchy in the neutrino sector, and provides promising values for the left-handed neutrino masses. One is thus encouraged to explore other models with this symmetry to exhaust the phenomenological potential of $\Delta(54)$.

Two reasonable questions arise from this study. Given the most favored symmetries, can we provide a generic recipe for the flavor phenomenology arising from string models? Is it possible to falsify some of these promising heterotic orbifolds on the basis of their flavor symmetries? In clarifying these general questions, it would also be convenient to perform a more exhaustive scan, including all possible toroidal geometries of $\mathbb{Z}_N \times \mathbb{Z}_M$ orbifolds, roto-translations, and the enhancement(s) of these symmetries at special modular values. These open challenges shall be the aim of future works.

Acknowledgments
I thank B. Carballo-Pérez, E. Peinado and Y. Olguín-Trejo for fruitful collaborations on this subject. This work was partly supported by DGAPA-PAPIIT grant IN100217 and CONACyT grant F-252167. I would like to thank the ICTP for the kind hospitality and support received through its Junior Associateship Scheme during the realization of this work.

References

[1] Ishimori H, Kobayashi T, Ohki H, Shimizu Y, Okada H and Tanimoto M 2010 *Prog. Theor. Phys. Suppl.* **183** 1–163 (*Preprint* 1003.3552)
[2] King S F and Luhn C 2013 *Rept. Prog. Phys.* **76** 056201 (*Preprint* 1301.1340)
[3] Kobayashi T, Nilles H P, Plöger F, Raby S and Ratz M 2007 *Nucl. Phys.* **B768** 135–156 (*Preprint* hep-ph/0611020)
[4] Nilles H P, Ratz M and Vaudrevange P K S 2013 *Fortsch. Phys.* **61** 493–506 (*Preprint* 1204.2206)
[5] Kobayashi T, Raby S and Zhang R J 2004 *Phys. Lett.* **B593** 262–270 (*Preprint* hep-ph/0403065)
[6] Ko P, Kobayashi T, Park J h and Raby S 2007 *Phys. Rev.* **D76** 035005 [Erratum: Phys. Rev.D76,059901(2007)] (*Preprint* 0704.2807)
[7] Carballo-Pérez B, Peinado E and Ramos-Sánchez S 2016 *JHEP* **12** 131 (*Preprint* 1607.06812)
[8] Beye F, Kobayashi T and Kuwakino S 2014 *Phys. Lett.* **B736** 433–437 (*Preprint* 1406.4660)
[9] Abe H, Kobayashi T, Otsuka H, Takano Y and Tatsuishi T H 2016 *Phys. Rev.* **D94** 126020 (*Preprint* 1605.00898)
[10] Berasaluce-González M, Camara P G, Marchesano F, Regalado D and Uranga A M 2012 *JHEP* **09** 059 (*Preprint* 1206.2383)
[11] Marchesano F, Regalado D and Vázquez-Mercado L 2013 *JHEP* **09** 028 (*Preprint* 1306.1284)
[12] Burwick T T, Kaiser R K and Müller H F 1991 *Nucl. Phys.* **B355** 689–711
[13] Erler J, Jungnickel D, Spalinski M and Stieberger S 1993 *Nucl. Phys.* **B397** 379–416 (*Preprint* hep-th/9207049)
[14] Choi K S and Kobayashi T 2008 *Nucl. Phys.* **B797** 295–321 (*Preprint* 0711.4894)
[15] Dixon L J, Kaplunovsky V and Louis J 1991 *Nucl. Phys.* **B355** 649–688
[16] Hamidi S and Vafa C 1987 *Nucl. Phys.* **B279** 465–513
[17] Dixon L J, Friedan D, Martinec E J and Shenker S H 1987 *Nucl. Phys.* **B282** 13–73
[18] Casas J A, Gómez F and Muñoz C 1993 *Int. J. Mod. Phys.* **A8** 455–506 (*Preprint* hep-th/9110060)
[19] Kobayashi T and Ohtsubo N 1994 *Int. J. Mod. Phys.* **A9** 87–126
[20] Kobayashi T, Parameswaran S L, Ramos-Sánchez S and Zavala I 2012 *JHEP* **05** 008 [Erratum: JHEP12,049(2012)] (*Preprint* 1107.2137)
[21] Nilles H P, Ramos-Sánchez S, Ratz M and Vaudrevange P K S 2013 *Phys. Lett.* **B726** 876–881 (*Preprint* 1308.3435)
[22] Cabo Bizet N G, Kobayashi T, Mayorga Peña D K, Parameswaran S L, Schmitz M and Zavala I 2013 *JHEP* **05** 076 (*Preprint* 1301.2322)
[23] Nilles H P, Ramos-Sánchez S, Vaudrevange P K S and Wingerter A 2011 (*Preprint* 1110.5229)
[24] Nilles H P and Vaudrevange P K S 2015 *Mod. Phys. Lett.* **A30** 1530008 (*Preprint* 1403.1597)

[25] Parameswaran S L, Ramos-Sánchez S and Zavala I 2011 *JHEP* **01** 071 (*Preprint* 1009.3931)
[26] Dundee B, Raby S and Westphal A 2010 *Phys. Rev.* **D82** 126002 (*Preprint* 1002.1081)
[27] Fischer M, Ratz M, Torrado J and Vaudrevange P K S 2013 *JHEP* **01** 084 (*Preprint* 1209.3906)
[28] Plöger F, Ramos-Sánchez S, Ratz M, and Vaudrevange P K S 2007 *JHEP* **04** 063 (*Preprint* hep-th/0702176)
[29] Ishimori H, Kobayashi T, Ohki H, Okada H, Shimizu Y and Tanimoto M 2012 *Lect. Notes Phys.* **858** 1–227
[30] Lebedev O, Nilles H P, Raby S, Ramos-Sánchez S, Ratz M, Vaudrevange P K S and Wingerter A 2008 *Phys. Rev.* **D77** 046013 (*Preprint* 0708.2691)
[31] Abe K *et al.* (T2K) 2017 *Phys. Rev. Lett.* **118** 151801 (*Preprint* 1701.00432)

IOP Conf. Series: Journal of Physics: Conf. Series **912** (2017) 012012 doi:10.1088/1742-6596/912/1/012012

Helicity Amplitudes for production of massive gravitino/goldstino

Bryan O. Larios

Facultad de Ciencias Físico - Matemáticas, BUAP
Apdo. Postal 1364, C.P. 72000, Puebla, Pue. México

E-mail: bryanlarios@gmail.com

Abstract. The spinor helicity formalism (SHF) has been efficiently applied to compute perturbative amplitudes in plenty of processes and reactions in gauge theories (including gravity), mostly in the massless case. Some work has been done in order to extend the SHF to the massive case. We have used these powerful tools to evaluate amplitudes in a local supersymmetric model where the gravitino is the lightest supersymmetric particle (LSP). Two decays have been evaluated in order to show the capability of the SHF in the massive extension, namely the two body neutralino decays $\tilde{\chi}_0 \to V\bar{\Psi}^\mu$ with $V = \gamma, Z$. The comparisons of amplitudes with spin-3/2 gravitino and spin-1/2 goldstino are also presented.

1. Introduction

The traditional Feynman approach for perturbation theory has been developed for almost sixty years, and it has been undoubtedly a magnificent tool to evaluate amplitudes in gauge theories. Even with all the success of the Feynman rules, several computations for scattering processes especially in QCD suffer the lack of accuracy in order to be tested experimentally. Furthermore, processes with several external particles are extremely complicated to calculate with the traditional perturbative approach [1]. Recent developments for tree-level scattering amplitudes in the massless case using the helicity methods are now widely used [2, 3, 4, 5, 6, 7], allowing to efficiently evaluate amplitudes for processes with n external legs in pure Yang-Mills theories. A huge theoretical progress has been done during the last two decades in this active and growing field, making manifest certain mathematical structures that underlie quantum field theories [8, 9, 10, 11, 12, 13].

We shall try to present a pragmatic approach in this work, starting still `*a la Feynman* building the scattering amplitudes from the Feynman diagrams but expressing the usual spinors (in any representation) appearing in the amplitudes as momentum twistors, this step shall hugely facilitate the calculations, as we shall see later. This is one of the goals of the SHF. Addressing the massive case requires the implementation of the so-called light cone decomposition (LCD) for the massive momenta [14, 15, 16, 17]. This technique allow us to express a massive momentum, namely $p^2 = -m^2$ as a lineal combination of two massless momentum $p_i = r_i + \alpha q_i$ where $r_i^2 = q_i^2 = 0$ and $\alpha = -\frac{m^2}{2q_i \cdot r_i}$. Several processes of the electroweak Standard Model have been evaluated using the massive SHF in Ref. [18], reproducing the well known results. More recently we have implemented this method for $\mathcal{N} = 1$ supergravity with LSP gravitinos [19], showing that the formalism is well suitable for amplitudes involving massive gravitinos in the final state.

Considering scenarios where the gravitino is the LSP and hence a good dark matter candidate, the nature of the next-to-lightest supersymmetric particle (NLSP) determines its phenomenology [20, 21, 22, 23, 24]. Potential candidates for NLSP include the lightest neutralino [25, 26], the chargino [27], and the lightest charged slepton [28]. The weakness of the gravitational interactions suggest that the NLSP could have a long lifetime leading to scenarios with a metastable charged sparticles that could have striking signatures at colliders [29, 30] and it could also affect the Big Bang nucleosynthesis [31, 32, 33]. Here we shall consider several decay modes for the neutralino as the NLSP. We also have considered the case when the gravitino can be approximated by the goldstino state, this is for the region where the gravitino mass is small compared to the neutralino mass $\tilde{m} \ll m_{\tilde{\chi}_0}$. This proceeding contribution is structured as follows: Section 2 contains the solutions of the Rarita-Schwhinger equation for the gravitino. The four states of the gravitino are expressed in terms of the momentum twistors after LCD is applied. We shall see that this is the key to obtain compact expression for the decay amplitudes. In Section 3 the helicity amplitudes (HAs) for the two decays ($\tilde{\chi}_0 \to V\tilde{\Psi}^\mu$ with $V = \gamma, Z$) are shown in terms of very compact mathematical expressions, in addition the HAs with the golsdtino approximation are also shown. Finally in Section 4 we present some final remarks and comments.

2. Gravitino wave functions

2.1. Rarita-Schwinger equations

The wave function for massive gravitino is the solution of the Rarita-Schwhinger equation [34], this equation of motion (EOM) results from applying the Euler-Lagrange to the following lagrangian [35]

$$\mathcal{L} = -\frac{1}{2}\epsilon^{\mu\nu\rho\sigma}\Psi_\mu^T C^\dagger \gamma_5\gamma_\nu\partial_\rho\Psi_\sigma + \frac{1}{4}\tilde{m}\Psi_\mu^T C^\dagger[\gamma^\mu, \gamma^\nu]\Psi_\nu, \tag{1}$$

the EOM for the gravitino are equivalent to the following equations [36]

$$\gamma^\mu\Psi_\mu = 0, \tag{2}$$

$$\partial^\mu\Psi_\mu = 0, \tag{3}$$

$$(i\gamma^\mu\partial_\mu - \tilde{m})\Psi_\mu = 0, \tag{4}$$

where $\tilde{m}$ is the gravitino mass. Through all this work we shall use the following convention for the Minkowski metric $\eta^{\mu\nu} = \text{diag}(-1, 1, 1, 1)$, besides we have used the Dirac representation for the gamma matrices where they take the following form

$$\gamma^\mu = \begin{pmatrix} 0 & \sigma^\mu \\ \bar{\sigma}^\mu & 0 \end{pmatrix}, \tag{5}$$

with $\sigma^\mu = (1, \vec{\sigma})$ and $\bar{\sigma}^\mu = (1, -\vec{\sigma})$. Returning to the EOM Eqs. (2)-(4), these admit the following solution [36]

$$\tilde{\Psi}_\mu(\vec{p}, \lambda) = \sum_{s,m} \left\langle \left(\frac{1}{2}, \frac{s}{2}\right)(1, m) \middle| \left(\frac{3}{2}, \lambda\right) \right\rangle u(\vec{p}, s)\epsilon_\mu(\vec{p}, m), \tag{6}$$

this is a Clebsch-Gordon expansion with coefficient $\langle(\frac{1}{2},\frac{s}{2})\,(1,m)\big|\,(\frac{3}{2},\lambda)\rangle$, from Eq. (6), we know that there are four gravitino states, explicitly they are as follows

$$\tilde{\Psi}^{\mu}_{++}(p) = \epsilon^{\mu}_{+}(p)u_{+}(p), \tag{7}$$
$$\tilde{\Psi}^{\mu}_{--}(p) = \epsilon^{\mu}_{-}(p)u_{-}(p), \tag{8}$$
$$\tilde{\Psi}^{\mu}_{+}(p) = \sqrt{\frac{2}{3}}\epsilon^{\mu}_{0}(p)u_{+}(p) + \frac{1}{\sqrt{3}}\epsilon^{\mu}_{+}(p)u_{-}(p), \tag{9}$$
$$\tilde{\Psi}^{\mu}_{-}(p) = \sqrt{\frac{2}{3}}\epsilon^{\mu}_{0}(p)u_{-}(p) + \frac{1}{\sqrt{3}}\epsilon^{\mu}_{-}(p)u_{+}(p). \tag{10}$$

Replacing the momentum twistors in the four gravitino states Eqs. (7)-(10) is straightforward (See Ref. [19] for more details). Dealing with HAs involving gravitinos in the final state expressed now in these new variables shall avoid the large and messy expressions that appear when the trace technology is used (most of the time handled with `Mathematica`), we have to remember that the completeness relation for the gravitino takes the form

$$D_{\mu\nu}(p) = \sum_{\tilde{\lambda}=1}^{3}\tilde{\Psi}_{\mu}(\vec{p},\tilde{\lambda})\overline{\tilde{\Psi}}_{\nu}(\vec{p},\tilde{\lambda}) = -(\not{p}+\tilde{m})\times\left[\left(g_{\mu\nu}-\frac{p_\mu p_\nu}{\tilde{m}^2}\right)\right.$$
$$\left. -\frac{1}{3}\left(g_{\mu\sigma}-\frac{p_\mu p_\sigma}{\tilde{m}^2}\right)\left(g_{\nu\lambda}-\frac{p_\nu p_\lambda}{\tilde{m}^2}\right)\gamma^\sigma\gamma^\lambda\right]. \tag{11}$$

Fortunately in the SHF the fundamental building block are the HAs (without square modulus) and we shall not need Eq. (11), neither another completeness relation in general.

2.2. High energy equivalence theorem

We briefly discuss in this section the gravitino approximation to goldstino. When the gravitino mass ($\tilde{m}$) is small compared to the energy of the process, it is possible to applicate the equivalence theorem which roughly speaking allow us to replace the longitudinal components of the gravitino by the derivative of the goldstino field [37, 38, 39, 40, 41, 42]. In practice the equivalence theorem tell us that the four gravitino states Eqs. (7)-(10), go as follows; $\tilde{\Psi}^{\mu}_{++}(p)\approx 0$, $\tilde{\Psi}^{\mu}_{--}(p)\approx 0$, $\tilde{\Psi}^{\mu}_{-}(p)\approx\sqrt{\frac{2}{3}}\left(\frac{p^\mu}{\tilde{m}}\right)u_{-}(p)$ and $\tilde{\Psi}^{\mu}_{+}(p)\approx\sqrt{\frac{2}{3}}\left(\frac{p^\mu}{\tilde{m}}\right)u_{+}(p)$. At the end just two gravitino states shall survive in this approximation, one advantage is that they just depend on spinors and not anymore on the polarization vectors that makes calculations worse.

3. Decay Amplitudes with LSP gravitino/goldstino

In this section the idea is to present how the SHF works, with the amplitudes at hands the next step is to write the spinors and polarization vectors (for the gravitino) as momentum twistors, then the SHF makes its magic. We do not need to worry about addressing the massive case, because a massive momentum twistor is expressed in terms of two massless one [15, 19]. In this proceeding contribution two decay widths are worked out ($\tilde{\chi}_0 \to V\tilde{\psi}^\mu$, $V=\gamma, Z$), both within the local supersymmetric extension of the Standard Model with gravitino LSP in the final state and with the neutralino as the NLSP. These calculations have been already evaluated in References [43, 44, 45] using the Feynman approach with the trace technology.

3.1. Two-body Neutralino decay $\tilde{\chi}_0 \to \tilde{\Psi}^\mu\gamma$

We build out the amplitudes for the two-body neutralino decays applying the Feynman rules of Ref. [35] for the spin-3/2 gravitino interactions. We shall start with the decay $\tilde{\chi}_0 \to \tilde{\Psi}^\mu\gamma$

corresponding to the diagram of Figure 1. The $\tilde{\chi}_0(p_1)$ denotes the neutralino with momentum p_1 and $\tilde{\Psi}^\mu(p_2)$ represents the spin-3/2 gravitino (in the goldstino approximation will be $\tilde{G}(p_2)$), and $\gamma(p_3)$ denotes the photon with momentum p_3. The amplitude reads as follows

$$\mathcal{M}_{\lambda_1,\lambda_2,\lambda_3} = \frac{C_{\chi\gamma}}{4M}\bar{\Psi}_{\mu\lambda_2}(p_2)(p_{3\nu}[\gamma^\nu,\gamma^\sigma]\gamma^\mu)\epsilon_{\sigma\lambda_3}(p_3)u_{\lambda_1}(p_1). \tag{12}$$

where $C_{\chi\gamma} = U_{i1}\cos\theta_W + U_{i2}\sin\theta_W$, U_{ij} are the mixing matrices that diagonalize the neutralino factor, M is the Plank mass, then λ_1, λ_2 and λ_3 are the helicity labels corresponding to the neutralino, gravitino and photon. After applying the SHF to the amplitude of Eq. (12), one notice that there are in principle 16 HAs, but 14 of them vanish. The SHF helps to identify the spurious quantities from the beginning, making the process to find physical observables expeditious. The nonzero HAs are shown in the Table 1.

$\lambda_1, \lambda_2, \lambda_3$	$\mathcal{M}^{3/2}_{\lambda_1,\lambda_2,\lambda_3}$
$-,++,+$	$\frac{C_{\chi\gamma}[r_2q_2]^2}{M\langle r_1r_2\rangle}m_{\tilde{\chi}_0}$
$-,-,-$	$\frac{C_{\chi\gamma}s_{r_2q_2}}{\sqrt{3}M\tilde{m}[r_2q_2]}\langle r_2q_2\rangle[r_2r_1]$

Table 1. Helicity Amplitudes for the two-body neutralino decay $\tilde{\chi}_0 \to \tilde{\Psi}^\mu\gamma$ with LSP gravitino in the final state. We are using $s_{ij} = -(p_i+p_j)^2$ for the Mandesltam variable. There are two more HAs $(+,--,+$ and $-,-,-)$, but they are the complex conjugate of the two shown in this table, the same criterium shall apply to the next tables.

The squared and averaged amplitude takes the form

$$\langle|\mathcal{M}|^2\rangle = \frac{C^2_{\chi\gamma}}{2M^2}(|\mathcal{M}_{-,++,+}|^2 + |\mathcal{M}_{+,--,-}|^2 + |\mathcal{M}_{-,-,-}|^2 + \mathcal{M}_{+,+,+}|^2) \tag{13}$$

$$= \frac{C^2_{\chi\gamma}}{2M^2}\left(2\frac{s^2_{r_2q_2}m^2_{\tilde{\chi}_0}}{s_{r_1r_2}} + 2\frac{s^2_{r_2q_2}}{3\tilde{m}^2}s_{r_1r_2}\right) \tag{14}$$

$$= \frac{C^2_{\chi\gamma}}{M^2}\left(\frac{(m^2_{\tilde{\chi}_0}-\tilde{m}^2)^2}{3\tilde{m}^2}(3\tilde{m}^2+m^2_{\tilde{\chi}_0})\right) \tag{15}$$

$$= \frac{C^2_{\chi\gamma}m^6_{\tilde{\chi}_0}}{M^2}\left(1-\frac{\tilde{m}^2}{m^2_{\tilde{\chi}_0}}\right)^2\left(\frac{1}{3}+\frac{\tilde{m}^2}{m^2_{\tilde{\chi}_0}}\right). \tag{16}$$

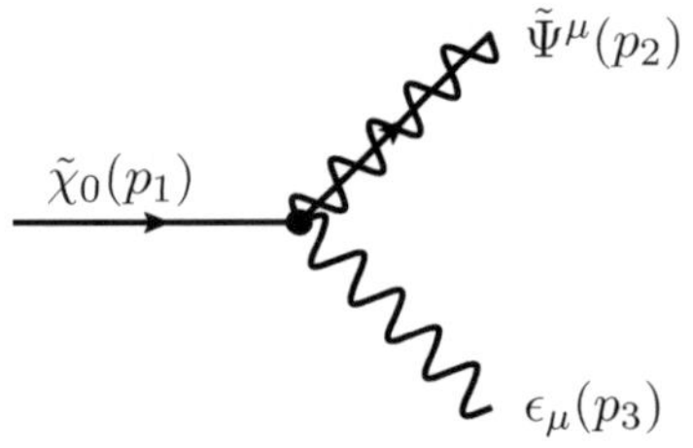

Figure 1. Feynman diagram for the 2-body neutralino decay.

The result of Eq. (16) conduces to the right decay width found in Ref. [44, 45], but our calculations result considerably simpler applying the SHF (massive).
We shall considerate now the approximation of the spin-3/2 gravitino to spin-1/2 goldstino, this

is due to the high energy equivalence theorem, physically the approximation is valid when the split mass between fermions and bosons is larger than the gravitino mass. Some work has been done in order to compare observables (lifetime τ) with gravitino and goldstino in the final state [24], basically comparing the squared amplitudes. It shall be interesting to exploit the power of the SHF for comparing the HAs with gravitino and goldstino for a given process or reaction.

The amplitude for the decay $\tilde{\chi}_0 \to \tilde{G}\gamma$ with goldstino in the final state is the following

$$\mathcal{M}_{\lambda_1,\lambda_2,\lambda_3} = \frac{C_{\chi\gamma} m_{\tilde{\chi}_0}}{2\sqrt{6}M\tilde{m}} \bar{\Psi}_{\lambda_2}(p_2)(p_{3\nu}[\gamma^\nu, \gamma^\sigma])\epsilon_{\sigma\lambda_3}(p_3)u_{\lambda_1}(p_1), \tag{17}$$

the kinematics of the process remains the same in the approximation to goldstino, but now the transversal degrees of freedom vanish. The nonzero HAs are shown in the Table 2

$\lambda_1, \lambda_2, \lambda_3$	$\mathcal{M}^{1/2}_{\lambda_1,\lambda_2,\lambda_3}$
$-,-,-$	$\frac{C_{\chi\gamma} s_{r_2 q_2}}{\sqrt{3}M\tilde{m}[r_2 q_2]}\langle r_2 q_2\rangle[r_2 r_1]$

Table 2. Helicity Amplitudes for the two-body neutralino decay $\tilde{\chi}_0 \to \tilde{G}\gamma$ with LSP goldstino in the final state.

In Table 1 we show two HAs, one of them includes the transversal d.o.f. $(-,++,+)$ corresponding to the full gravitino contribution, but Table 2 shows just one HA, $\mathcal{M}^{1/2}_{-,-,-}$, that is in fact identically to the HA $\mathcal{M}^{3/2}_{-,-,-}$.

The squared and averaged amplitude takes the following form

$$\langle|\mathcal{M}|^2\rangle = \frac{C^2_{\chi\gamma}}{2M^2}(|\mathcal{M}_{-,-,-}|^2 + |\mathcal{M}_{+,+,+}|^2) \tag{18}$$

$$= \frac{C^2_{\chi\gamma} s^2_{r_2 q_2}}{3M^2\tilde{m}^2} s_{r_2 r_1} \tag{19}$$

$$= \frac{C^2_{\chi\gamma}(m^2_{\tilde{\chi}_0} - \tilde{m}^2)^2 m^2_{\tilde{\chi}_0}}{3M^2\tilde{m}^2}, \tag{20}$$

this result can be obtained directly from Eq. (16) in the limit that the gravitino mass ($\tilde{m}$) is small compared to the neutralino mass ($m_{\tilde{\chi}_0}$).

3.2. 2-body neutralino decay $\tilde{\chi}_0 \to \tilde{\Psi}^\mu Z$

In the case that the vector boson is massive (but neutral) the decay becomes $\tilde{\chi}_0(p_1) \to \tilde{\Psi}^\mu(p_2)\, Z(p_3)$. Some technical complications appear from having the Z boson in the final state instead of the photon. The amplitude for this process with the spin-3/2 gravitino in the final state is as follows

$$\mathcal{M}_{\lambda_1,\lambda_2,\lambda_3} = \frac{C_{\chi Z}}{4M} \bar{\Psi}_{\mu\lambda_2}(p_2)(p_3^\rho[\gamma_\rho, \gamma_\sigma]\gamma^\mu)\epsilon^\sigma_{\lambda_3}(p_3)u_{\lambda_1}(p_1), \tag{21}$$

where $C_{\chi Z} = -U_{i1}\sin\theta_W + U_{i2}\cos\theta_W$, and everything else remains equal to the decay with a photon in final state ($\tilde{\chi}_0 \to \tilde{\Psi}^\mu\gamma$), but now the helicity label corresponding to the Z boson (λ_3) could be $+,-,0$.

From the 24 possible HAs resulting of Eq. (21), the nonzero are shown in the Table 3.

$\lambda_1,\lambda_2,\lambda_3$	$\mathcal{M}^{3/2}_{\lambda_1,\lambda_2,\lambda_3}$
$+,-,+$	$\frac{C_{\chi Z}}{\sqrt{3}M\tilde{m}s_{r_2q_2}[q_2r_2][r_1q_2]}\left(\tilde{m}^3M_Z^2s_{q_2r_1}+s^3_{r_2q_2}m_{\tilde{\chi}_0}\right)$
$-,-,0$	$\frac{\sqrt{2}C_{\chi Z}M_Z\langle r_2q_2\rangle}{\sqrt{3}Ms_{r_2q_2}\langle r_1q_2\rangle}\left(\tilde{m}s_{q_2r_1}+s_{r_2q_2}m_{\tilde{\chi}_0}\right)$
$-,--,+$	$-\frac{C_{\chi Z}\langle r_2q_2\rangle}{M\langle r_1q_2\rangle[r_2q_2]^2}(\tilde{m}M_Z^2m_{\tilde{\chi}_0}+s_{r_2q_2}s_{q_2r_1})$

Table 3. Helicity Amplitudes for the two-body Neutralino decay $\tilde{\chi}_0\to\tilde{\Psi}^\mu Z$ with LSP gravitino in the final state.

The squared and averaged amplitude is as follows

$$\langle|\mathcal{M}|^2\rangle=\frac{C^2_{\chi Z}}{2M^2}(|\mathcal{M}_{+,-,+}|^2+|\mathcal{M}_{-,+,-}|^2+|\mathcal{M}_{-,-,0}|^2+|\mathcal{M}_{+,+,0}|^2+|\mathcal{M}_{-,--,+}|^2+|\mathcal{M}_{+,++,-}|^2) \tag{22}$$

$$=\frac{C^2_{\chi Z}}{M^2}\left[\frac{1}{3\tilde{m}^2s^3_{r_2q_2}s_{r_1q_2}}\left(\tilde{m}^3M_Z^2s_{q_2r_1}+s^3_{r_2q_2}m_{\tilde{\chi}_0}\right)^2+\frac{2M_Z^2}{3s_{r_2q_2}s_{q_2r_1}}(s_{r_2q_2}m_{\tilde{\chi}_0}+\tilde{m}s_{q_2r_1})^2\right.$$
$$\left.+\frac{1}{s_{q_2r_1}s_{r_2q_2}}(\tilde{m}m_{\tilde{\chi}_0}M_Z^2+s_{q_2r_1}s_{r_2q_2})^2\right] \tag{23}$$

$$=\frac{C^2_{\chi Z}m^6_{\tilde{\chi}_0}}{M^2\tilde{m}^2}\left[\left(1-\frac{\tilde{m}^2}{m^2_{\tilde{\chi}_0}}\right)^2\left(\frac{1}{3}+\frac{\tilde{m}^2}{m^2_{\tilde{\chi}_0}}\right)-\frac{M_Z^2}{m^2_{\tilde{\chi}_0}}\left(1-\frac{M_Z^2}{m^2_{\tilde{\chi}_0}}\left(1-\frac{\tilde{m}^2}{3m^2_{\tilde{\chi}_0}}\right)-\frac{\tilde{m}^3}{m^3_{\tilde{\chi}_0}}\left(4-\frac{\tilde{m}}{3m_{\tilde{\chi}_0}}\right)\right.\right.$$
$$\left.\left.+\frac{M_Z^4}{3m^4_{\tilde{\chi}_0}}\right)\right], \tag{24}$$

this result conduces to the known decay width found in [44, 45].
We repeat the process of the first example and approximate the gravitino to goldstino, then we compute the HAs.

$$\mathcal{M}_{\lambda_1,\lambda_2,\lambda_3}=\frac{C_{\chi Z}m_{\tilde{\chi}_0}}{2\sqrt{6}M\tilde{m}}\bar{\Psi}_{\lambda_2}(p_2)(p_3^\rho[\gamma_\rho,\gamma_\sigma])\epsilon^\sigma_{\lambda_3}(p_3)u_{\lambda_1}(p_1), \tag{25}$$

the nonzero HA are shown in the Table 4

$\lambda_1,\lambda_2,\lambda_3$	$\mathcal{M}^{1/2}_{\lambda_1,\lambda_2,\lambda_3}$
$+,-,+$	$\frac{C_{\chi Z}m_{\tilde{\chi}_0}}{\sqrt{3}M\tilde{m}[q_2r_2][r_1q_2]}\left(\tilde{m}m_{\tilde{\chi}_0}M_Z^2+s_{r_2q_2}s_{q_2r_1}\right)$
$-,-,0$	$\frac{C_{\chi Z}M_Zm_{\tilde{\chi}_0}\langle r_2q_2\rangle}{\sqrt{3}M\tilde{m}s_{r_2q_2}\langle r_1q_2\rangle}\left(\tilde{m}s_{q_2r_1}+s_{r_2q_2}m_{\tilde{\chi}_0}\right)$

Table 4. Helicity Amplitudes for the two-body Neutralino decay $\chi_0\to Z\tilde{G}$ with goldstino in the final state.

We can notice in Table 3 that for the helicity labels $\lambda_1=-$, $\lambda_2=-$ and $\lambda_3=0$ the HA for the gravitino is up to a factor of $\sqrt{2}$ the same HA with goldstino, this is $\mathcal{M}^{3/2}_{-,-,0}=\sqrt{2}\mathcal{M}^{1/2}_{-,-,0}$.

Squaring and averaging the HAs of Table 4, these take the following form

$$\langle|\mathcal{M}|^2\rangle = (|\mathcal{M}_{+,-,+}|^2 + |\mathcal{M}_{-,+,+}|^2 + |\mathcal{M}_{-,-,0}|^2 + |\mathcal{M}_{+,+,0}|^2) \tag{26}$$

$$= \frac{C_{\chi Z}^2 m_{\tilde{\chi}_0}^2}{M^2\tilde{m}^2 s_{r_2q_2}s_{q_2r_1}}\left(M_Z^2(m_G s_{q_2r_1} + s_{r_2q_2}m_{\tilde{\chi}_0})^2 + 2(\tilde{m}m_{\tilde{\chi}_0}M_Z^2 + s_{r_2q_2}s_{q_2r_1})^2\right) \tag{27}$$

$$= \frac{2C_{\chi Z}^2 m_{\tilde{\chi}_0}^6}{M^2\tilde{m}^2}\left[\left(1-\frac{\tilde{m}^2}{m_{\tilde{\chi}_0}^2}\right)^2 - \frac{M_Z^2}{m_{\tilde{\chi}_0}^2}\left(\frac{1}{2}\left(1-6\frac{\tilde{m}}{m_{\tilde{\chi}_0}}+\frac{\tilde{m}^2}{m_{\tilde{\chi}_0}^2}\right)+\frac{M_Z^2}{2m_{\tilde{\chi}_0}^2}\right)\right] \tag{28}$$

In the traditional Feynman approach with trace technology, the calculations of the squared and averaged amplitudes with goldstinos are considerably simple, unlike amplitudes involving gravitinos. However, when the SHF is implemented the calculations with gravitino are still simple, and the helicity method is able to compute the decay widths without any approximation, as we have shown.

4. Conclusions

In this proceeding, we have presented how the spinor helicity formalism is well suitable to evaluate decay amplitudes, even in the massive case. Two examples have been recalculated with the SHF ($\tilde{\chi}_0 \to V\tilde{\Psi}^\mu$ with $V = \gamma, Z$). Furthermore, we briefly discussed the high energy equivalence theorem between gravitinos and goldstinos. It is possible in some cases to identify how the helicity amplitudes with goldstinos appear as a helicity amplitudes with gravitinos (for the longitudinal d.o.f.), this characteristic is dificult to identify in the traditional approach. We shall study and discuss more relationships between gravitino and goldstino amplitudes for another NLSP candidates in a future work to appear soon in a peer review journal.

5. Acknowledgements

The author wish to thank Lorenzo Diaz-Cruz for useful discussion, Roberto Garcia and Jose Zelaya for reading the manuscript. This work has been supported by CONACYT and was partly supported by DPC-SMF.

References

[1] L. Dixon, arXiv:hep-ph/9601359.
[2] S.J. Parke and T. Taylor, Phys. Lett. 157 B:81 (1985); Z. Kunszt, Nucl. Phys. B 271:333 (1986); M. Mangano and S. Parke, Phys. Rep. 200:301 (1991).
[3] M. Srednicki, *Quantum Field Theory*, Cambridge University Press, 2007.
[4] M. Peskin, http://arxiv.org/pdf/1101.2414.pdf.
[5] M. D. Schwarz, *Quantum Field Theory and the Standard Model*, Cambridge University Press, 2014.
[6] J. M. Henn, J. C. Plefka, *Scattering Amplitudes in Gauge Theories*, in: Lecture Notes in Physics, vol. 883, Springer, 2014.
[7] H. Elvang and Y. Huang, *Scattering Amplitudes in Gauge Theory and Gravity* (Cambridge University Press, 2015), arXiv:1308.1697v2 [hep-th].
[8] E. Witten, Commun. Math. Phys. 252 189-258 (2004), arXiv:hep-th/0312171.
[9] R. Britto, F. Cachazo, B. Feng, E. Witten, Phys. Rev. Lett. 94 (2005) 181602. hep-th/0501052.
[10] L.J. Mason, D. Skinner, Phys. Lett. B 636 (2006) 60. hep-th/0510262.
[11] Z. Bern, J. J. M. Carrasco, H. Johansson, Phys. Rev. D 78 (2008) 085011. arXiv:0805.3993.
[12] N. Arkani-Hamed, J. Trnka, J. High Energy Phys. 10 (2014) 030. arXiv:1312.2007.
[13] F. Cachazo, S. He, E.Y. Yuan, Phys. Rev. D 90 (2014) 065001. arXiv:1306.6575.
[14] K. Ozeren and W. Stirling, Eur. Phys. J. C 48 159 (2006), arXiv:hep-ph/0603071.
[15] J. Kuczmarski, arXiv:1406.5612 [hep-ph].
[16] D. A. Kosower, Phys. Rev. D 71, 045007 (2005), [arXiv:hep-th/0406175].
[17] C. Schwinn and S. Weinzierl, JHEP 0704 (2007) 072, arXiv:hep-ph/0703021 [HEP-PH].
[18] J. Lorenzo Diaz-Cruz, Bryan O. Larios, O. Meza-Aldama, J. Phys. Conf. Ser. 761 (2016) no.1, 012012, arXiv:1608.04129 [hep-ph].

[19] J. Lorenzo Diaz-Cruz, Bryan O. Larios, arXiv:1612.04331 [hep-ph].
[20] J. L. Feng, A. Rajaraman and F. Takayama, Phys. Rev. Lett. 91 (2003) 011302, arXiv:hep-ph/0302215; Phys. Rev. D 68 (2003) 063504, arXiv:hep-ph/0306024.
[21] J. L. Feng, S. Su and F. Takayama, Phys. Rev. D 70 (2004) 075019, arXiv:hep-ph/0404231.
[22] J. R. Ellis, K. A. Olive, Y. Santoso and V. C. Spanos, Phys. Lett. B 588 (2004) 7, arXiv:hep-ph/0312262.
[23] J. L. Diaz-Cruz, John Ellis, Keith A. Olive, Yudi Santoso, JHEP 0705:003, 2007, arXiv:hep-ph/0701229v1.
[24] J. Lorenzo Díaz-Cruz, Bryan O. Larios, Eur. Phys. J. C76 (2016) no.3, 157, arXiv:1510.01447v2 [hep-ph].
[25] F. D. Steffen, arXiv:hep-ph/0711.1240.
[26] M. Johansen, J. Edsj, S. Hellman, J. Milstead , JHEP 1008 1-27 (2010).
[27] G. D. Kribs, A. Martin, and T. S. Roy, JHEP 0901 (2009) 023, arXiv:hep-ph/0807.4936.
[28] J. Heisig, J. Heising, JCAP 04 (2014) 023, arXiv:1310.6352.
[29] J. R. Ellis, A. R. Raklev and O. K. Oye, JHEP 0610, 061 (2006), arXiv:hep-ph/0607261.
[30] K. Hamaguchi, Y. Kuno, T. Nakaya and M. M. Nojiri, Phys. Rev. D 70 (2004) 115007, arXiv:hep-ph/0409248.
[31] R. H. Cyburt, J. R. Ellis, B. D. Fields and K. A. Olive, Phys. Rev. D 67 (2003) 103521, arXiv:astro-ph/0211258; J. R. Ellis, K. A. Olive and E. Vangioni, Phys. Lett. B 619 (2005) 30, arXiv:astro-ph/0503023.
[32] M. Kawasaki, K. Kohri and T. Moroi, Phys. Lett. B 625 (2005) 7, arXiv:astro-ph/0402490; Phys. Rev. D 71 (2005) 083502, arXiv:astro-ph/0408426.
[33] K. Kohri and Y. Santoso, Phys. Rev. D 79, 043514 (2009), arXiv:0811.1119 [hep-ph].
[34] W. Rarita and J. Schwinger, Phys. Rev. 60, 61 (1941).
[35] Takeo Moroi, arXiv:hep-ph/9503210v1.
[36] P.R. Auvil and J.J. Brehm, Phys. Rev. 145 (1966) 1152.
[37] R. Casalbuoni, S. De Curtis, D. Dominici, F. Feruglio, and R. Gatto, Physics Letters B 215, 313-316, 1988.
[38] P. Fayet, Phys.Lett. 70B (1977) 461.
[39] P. Fayet, Phys. Lett. B. 175 (1986) 471.
[40] P. Fayet, Phys. Lett. B. 84 (1979) 421.
[41] P. Fayet, Phys. Lett. B. 86 (1979) 272.
[42] P. Fayet, Conference Proc. LPTENS-81-9 (1981) 347.
[43] Laura Covi, Jasper Hasenkamp, Stefan Pokorski, Jonathan Roberts, JHEP 0911:003, 2009, arXiv:0908.3399v1 [hep-ph].
[44] John Ellis, Keith A. Olive, Yudi Santoso, Vassilis Spanos, Phys. Lett. B588:7-16, 2004, arXiv:hep-ph/0312262v4.
[45] Jonathan L. Feng, Shufang Su, Fumihiro Takayama, Phys. Rev. D70:075019, 2004, arXiv:hep-ph/0404231v2.

QED and QCD self-energy corrections through the loop-tree duality

N. Selomit Ramírez-Uribe [1,2]∗, Roger J. Hernández-Pinto [1] and Germán Rodrigo [3]

[1] Facultad de Ciencias Físico-Matemáticas, Universidad Autónoma de Sinaloa, Avenida de las Américas y Boulevard Universitarios, Ciudad Universitaria, CP 80000, Culiacán, Sinaloa, México

[2] Facultad de Ciencias de la Tierra y el Espacio, Universidad Autónoma de Sinaloa, Avenida de las Américas y Boulevard Universitarios, Ciudad Universitaria, CP 80000, Culiacán, Sinaloa, México

[3] Instituto de Física Corpuscular, Universitat de València-Consejo Superior de Investigaciones Científicas, Parc Científic, E-46980 Paterna, Valencia, Spain

E-mail: selomitru@uas.edu.mx, roger@uas.edu.mx, german.rodrigo@csic.es

Abstract. The loop-tree duality (LTD) theorem establishes that loop contributions to scattering amplitudes can be computed through dual integrals, which are build from single cuts of the virtual diagrams. In order to build a complete LTD representation of a cross section and to achieve a local cancellation of singularities, it is crucial to include the renormalized self-energy corrections in an unintegrated form. In this document, we calculate the scalar functions related to the self-energy corrections in QED and QCD in the LTD formalism and extract explicitly their UV behaviour.

1. Introduction

The standard approach to perform perturbative calculations in QCD relies in the application of the subtraction formalism. There are diverse alternatives of the subtraction method at NLO and beyond [1, 2, 3, 4, 5, 6, 7, 8, 9, 10, 11, 12], that involve the treatment of real and virtual contributions separately. From a computational point of view, handling real and virtual contributions separately might not be efficient enough for multi-leg and multi-loop processes. This is due to the fact that the final-state phase-space (PS) of the different contributions implies different number of external particles and loop momenta. For instance, at NLO, virtual corrections with Born kinematics have to be combined with real contributions with one additional final-state particle. The infrared (IR) counter-terms in the subtraction formalism have to be local in the real PS, and analytically integrable over the extra-radiation in order to properly cancel the divergent structure present in the virtual corrections. Building these counter-terms represents a challenge and introduces a potential bottleneck to efficiently carry out the IR subtraction for multi-leg multi-loop processes.

With the purpose of building renormalized quantities, we work the renormalization constants with the application of the LTD [13, 14, 15, 16, 17, 18, 19, 20, 21, 22]. The LTD theorem establishes a direct connection among loop and phase-space integrals. Then, loop scattering amplitudes can be expressed as a sum of PS integrals, called dual integrals. Dual integrals and real-radiation contributions exhibit a similar structure and in consequence, the

∗ Speaker

divergent behaviour of both contributions is cancelled at integrand level, thus, rendering integrals finite. A remarkable implication of this formalism, is the possibility of carrying out purely four-dimensional implementations for any observable at NLO and higher orders.

In this document we compute the scalar functions for the photon, electron, quark and gluon self-energies, within the LTD framework. Furthermore, we focus on the singular behaviour of the dual-integrals and the location of the singularities in the loop momentum space.

2. Review of the loop-tree duality

In this section, we review the main ideas behind the LTD method. Using the LTD [23], loop contributions of scattering amplitudes in any relativistic, local and unitary quantum field theory can be computed through dual integrals, which are build from single cuts of the virtual diagrams. Let's consider a generic N-particle scalar one-loop integral, i.e.

$$L^{(1)}(p_1,\ldots,p_N) = \int_\ell \prod_{i\in\alpha_1} G_F(q_i) \,, \tag{1}$$

over Feynman propagators $G_F(q_i) = (q_i^2 - m_i^2 + \imath 0)^{-1}$, whose most general topology is shown in Fig. 1. The corresponding dual representation of Eq. (1) consists in the sum of N dual integrals:

$$L^{(1)}(p_1,\ldots,p_N) = -\sum_{i\in\alpha_1} \int_\ell \tilde{\delta}(q_i) \prod_{j\in\alpha_1,\, j\neq i} G_D(q_i;q_j) \,, \tag{2}$$

where

$$G_D(q_i;q_j) = \frac{1}{q_j^2 - m_j^2 - \imath 0\, \eta\cdot k_{ji}} \,, \tag{3}$$

are dual propagators, $k_{ji} = q_j - q_i$, and $i,j \in \alpha_1 = \{1,2,\ldots N\}$ label the available internal lines. In Eq. (1) and Eq. (2), the masses and momenta of the internal lines are denoted m_i and $q_{i,\mu} = (q_{i,0}, \mathbf{q}_i)$, respectively, where $q_{i,0}$ is the energy and $\mathbf{q}_i$ are the spatial components. In terms of the loop momentum ℓ and the outgoing momenta of the external particles p_i, the internal momenta are defined as

$$q_i = \ell + k_i \quad , \quad k_i = \sum_{j=1}^{i} p_j \,, \tag{4}$$

together with the constraint $k_N = 0$ imposed by momentum conservation.

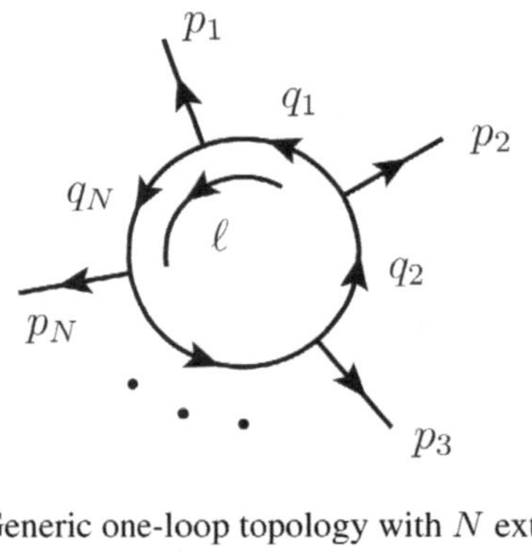

Figure 1. Generic one-loop topology with N external legs.

IOP Conf. Series: Journal of Physics: Conf. Series **912** (2017) 012013 doi:10.1088/1742-6596/912/1/012013

On the other hand, the d-dimensional loop measure is given by

$$\int_\ell \bullet = -\imath\mu^{4-d} \int \frac{d^d\ell}{(2\pi)^d} \bullet \ , \tag{5}$$

and

$$\tilde{\delta}\,(q_i) \equiv 2\pi\,\imath\,\theta(q_{i,0})\,\delta(q_i^2 - m_i^2) \tag{6}$$

used in Eq. (2) set the internal lines on-shell. It is worth mentioning that, the presence of the Heaviside function restricts the integration domain to the positive energy region (i.e. $q_{i,0} > 0$). Since LTD is derived through the application of the Cauchy's residue theorem, the remaining $d-1$ dimensional integration is performed over the forward on-shell hyperboloids defined by the solution of $G_F(q_i)^{-1} = 0$ with $q_{i,0} > 0$. Notice that these on-shell hyperboloids degenerate to light-cones when internal particles are massless.

The dual representation shown in Eq. (2) is built by adding all possible single-cuts from the original loop diagram. In this procedure, the propagator associated with the on-shell line is replaced by Eq. (6) where the remaining uncut Feynman propagators are promoted to dual ones. The introduction of dual propagators modifies the $\imath 0$-prescription since it depends on the sign of $\eta \cdot k_{ji}$, with η a *future-like* vector, $\eta^2 \geq 0$, with positive definite energy $\eta_0 > 0$, and k_{ji}, which is independent of the loop momentum ℓ at one-loop. According to the derivation shown in Ref. [23], η is arbitrary so we can chose $\eta_\mu = (1, \mathbf{0})$ to simplify the implementation.

The difference between LTD and the Feynman Tree Theorem (FTT) [24, 25], where the loop integral is obtained after summing over all possible m-cuts, is codified in the dual prescription, where correlations coming from multiple cuts in the FTT are recovered in the LTD by considering only single-cuts with the modified $\imath 0$-prescription. In other words, having different prescriptions for each cut is a necessary condition for the consistency of the method. As discussed in Ref. [26], the integrand in Eq. (2) becomes singular at the intersection of forward on-shell hyperboloids (FF case), and forward with backward ($q_{j,0} < 0$) on-shell hyperboloids (FB intersections). On one hand, the FF singularities cancel each other among different dual contributions; the change of sign in the modified prescription is crucial to enable this behaviour. On the other hand, singularities associated with FB intersections remain constrained to a compact region of the loop three-momentum space and are easily reinterpreted in terms of causality.

From a physical point of view, FB singularities take place when the on-shell virtual particle interacts with another on-shell virtual particle after the emission of outgoing on-shell radiation. The direction of the internal momentum flow establishes a natural causal ordering, and this interpretation is consistent with the Cutkosky rule. In fact, the total energy of the emitted particles, which is equal to $q_{i,0} - |q_{j,0}|$, has to be positive. Together with the positive energy constraint imposed by the delta distribution in Eq. (6), it restricts the possible kinematical configurations compatible with a sequential decay of on-shell physical particles.

Besides the mathematical understanding of the origin of the singularities by using the LTD, one of the most promising applications is the possible computation of scattering processes at NLO directly in four space-time dimensions without Dimensional Regularization (DREG). It has been also shown in Ref.[27] that one of the ingredients in order to build the 4D integrals are the renormalization constants. In fact, even if some integrals vanish at integral level, they are important for the local cancellation of singularities and the definition of renormalizad quantities in the LTD formalism. Therefore, in the next sections we compute the QED and QCD self-energy corrections by using the LTD theorem.

3. Photon and electron self-energies in QED

In the framework of QED, we need to compute the renormalization constants of the photon and the electron self-energies. In this handwritten, we will work in the massless approximation, labelling the internal lines as $q_1 = \ell$, $q_2 = \ell + p$ and considering $p^2 \neq 0$.

3.1. Photon self-energy

The one-loop photon self-energy is given by

$$\imath\,\Pi_{\mu\nu}(p^2) = -\int_\ell \text{Tr}\left[\imath e_0\gamma_\mu\left(\imath\frac{\not{q}_1+m_0}{q_1^2-m_0^2+\imath 0}\right)\imath e_0\gamma_\nu\left(\imath\frac{\not{q}_2+m_0}{q_2^2-m_0^2+\imath 0}\right)\right], \tag{7}$$

where due to gauge invariance

$$\Pi_{\mu\nu}(p^2) = (p^2\,g_{\mu\nu}-p_\mu p_\nu)\Pi(p^2)\,. \tag{8}$$

From Eq. (7) and Eq. (8) and taking the massless approximation,

$$\imath(p^2\,g_{\mu\nu}-p_\mu p_\nu)\Pi(p^2) = -e_0^2\int_\ell\frac{\text{Tr}\left[\gamma_\mu\,\not{q}_1\,\gamma_\nu\,\not{q}_2\right]}{(q_1^2+\imath 0)\,(q_2^2+\imath 0)}\,. \tag{9}$$

Contracting Eq. (10) with the metric tensor in d-dimensions we can extract the scalar function as,

$$\Pi(p^2) = \frac{8\,(1-\epsilon)\,e_0^2}{(3-2\epsilon)\,p^2}\int_\ell\frac{q_1\cdot q_2}{(q_1^2+\imath 0)\,(q_2^2+\imath 0)}\,. \tag{10}$$

Applying the LTD to the integral in Eq. (10) ,

$$\Pi(p^2) \;=\; -\frac{8\,(1-\epsilon)\,e_0^2}{(3-2\epsilon)\,p^2}\left[\int_\ell\tilde{\delta}(q_1)\frac{p\cdot q_1}{2p\cdot q_1+p^2-\imath 0}-\int_\ell\tilde{\delta}(q_2)\frac{p\cdot q_2}{-2p\cdot q_2+p^2+\imath 0}\right], \tag{11}$$

and taking the parametrization

$$\begin{aligned} p^\mu &= (p_0,\mathbf{0})\,,\\ q_i^\mu &= p_0\xi_{i,0}\left(1,2\sqrt{v_i(1-v_i)}\mathbf{e}_{i,\perp},1-2v_i\right)\,, \end{aligned} \tag{12}$$

$$\Pi(p^2) = -\frac{2^5\,(1-\epsilon)\,e_0^2}{(3-2\epsilon)}\left[\int d[\xi_{1,0}]d[v_1]\frac{\xi_{1,0}^2}{1+2\xi_{1,0}-\imath 0}-\int d[\xi_{2,0}]d[v_2]\frac{\xi_{2,0}^2}{1-2\xi_{2,0}+\imath 0}\right], \tag{13}$$

where

$$d[\xi_{i,0}] = \frac{(4\pi)^{\epsilon-2}}{\Gamma(1-\epsilon)}\left(\frac{4p^2}{\mu^2}\right)^{-\epsilon}\xi_{i,0}^{-2\epsilon}d\xi_{i,0}\quad\text{and}\quad d[v_i]=(v_i(1-v_i))^{-\epsilon}dv_i\,. \tag{14}$$

The parameters $\xi_{i,0}$ and v_i describe the energy and polar angle of the loop momentum respectively, defined in the regions $\xi_{i,0}\in[0,\infty)$ and $v_i\in[0,1]$. In order to explore the singular behaviour of $\Pi(p^2)$, we integrate Eq. (13) over the regions $i)$ $\xi_i\in[0,\omega]$ and $ii)$ $\xi_i\in[\omega,\infty)$. After integration, we found:

$$\Pi(p^2;\xi_i\le\omega) = -\frac{4\,e_0^2\,S_\epsilon}{3}\left[4\,\omega^2+\log(1+2\,\omega-\imath 0)+\log(1-2\,\omega+\imath 0)\right]+O(\epsilon) \tag{15}$$

and

$$\Pi(p^2;\xi_i\ge\omega) = -\frac{4\,e_0^2\,S_\epsilon}{3}\left[\frac{1}{\epsilon}-4\,\omega^2-\log(1+2\,\omega-\imath 0)-\log(1-2\,\omega+\imath 0)+\frac{5}{3}\right]+O(\epsilon)\,, \tag{16}$$

IOP Conf. Series: Journal of Physics: Conf. Series **912** (2017) 012013 doi:10.1088/1742-6596/912/1/012013

where $S_\epsilon = \frac{(4\pi)^{-2+\epsilon}}{\Gamma(1-\epsilon)\mu^{-2\epsilon}}(-p^2 - \imath 0)^{-\epsilon}$. From these results, it is clear that IR divergences are absent. Besides, $\Pi(p^2;\xi_i \leq \omega)$ is finite while the ultraviolet (UV) divergence is located fully in $\Pi(p^2;\xi_i \geq \omega)$. Since $\Pi(p^2;\xi_i \leq \omega)$ is IR finite, it admits a four dimensional integrand realization given by,

$$\Pi(p^2;\xi_i \leq \omega) = -\frac{e_0^2}{6\pi^2}\int_0^1 dv_1 \int_0^\omega \frac{d\xi_{1,0}\, 4\xi_{1,0}^2}{1+2\xi_{1,0}-\imath 0}\left[1+\log\left(\xi_{1,0}^2\right)\left(v_1\delta\left(v_1\right)+\left(1-v_1\right)\delta\left(1-v_1\right)\right)\right]$$
$$+\frac{e_0^2}{6\pi^2}\int_0^1 dv_2 \int_0^\omega \frac{d\xi_{2,0}\, 4\xi_{2,0}^2}{1-2\xi_{2,0}+\imath 0}\left[1+\log\left(\xi_{2,0}^2\right)\left(v_2\delta\left(v_2\right)+\left(1-v_2\right)\delta\left(1-v_2\right)\right)\right], \tag{17}$$

where the presence of the logarithmic terms are originated from the fact that we are using different coordinate systems for each dual integral.

The previous results shows one of the central implications of the use of the LTD theorem, resides in the location of singularities in physical regions. Hence, segmenting the integration domain, it is possible to find regions where the ϵ-parameter is no longer needed. It is also worth to highlight that in Eq. (17) is showing explicitly the cut dependency, this fact will be relevant when a physical process with photons is calculated at NLO accuracy.

3.2. Electron self-energy

The one-loop electron self-energy is given by

$$-\imath\Sigma(p) = \int_\ell \imath e_0\gamma^\mu\left(\imath\frac{\not{q}_2+m}{q_2^2-m_0^2+\imath 0}\right)\imath e_0\gamma^\nu\frac{-\imath}{q_1^2+\imath 0}g_{\mu\nu}\,, \tag{18}$$

where

$$\Sigma(p) = \not{p}\Sigma_V(p^2) - m\Sigma_S(p^2)\,. \tag{19}$$

Considering Eq. (18), Eq. (19) and the massless case, we get,

$$-\imath\not{p}\,\Sigma_V(p^2) = -e_0^2\int_\ell \frac{\gamma^\mu \not{q}_2\gamma_\mu}{(q_1^2+\imath 0)(q_2^2+\imath 0)}\,. \tag{20}$$

Extracting the scalar function $\Sigma_V(p^2)$, we find,

$$\Sigma_V(p^2) \quad = \quad -2\,(1-\epsilon)\,e_0^2\int_\ell \frac{1}{(q_1^2+\imath 0)\,(q_2^2+\imath 0)}\left(1+\frac{p\cdot q_2}{p^2}\right)\,. \tag{21}$$

Then, applying the LTD,

$$\Sigma_V(p^2) \quad = \quad 2(1-\epsilon)\,e_0^2\left(\int_\ell \frac{\tilde{\delta}(q_1)}{2p\cdot q_1+p^2-\imath 0}\left(1+\frac{p\cdot q_1}{p^2}\right)+\frac{1}{p^2}\int_\ell \frac{\tilde{\delta}(q_2)\,p\cdot q_2}{-2p\cdot q_2+p^2+\imath 0}\right), \tag{22}$$

working with the parametrization defined in Eq. (12), $\Sigma_V(p^2)$ becomes,

$$\Sigma_V(p^2) \quad = \quad 2(1-\epsilon)\,e_0^2\left(\int d[\xi_{1,0}]d[v_1]\frac{4\xi_{1,0}\,(1+\xi_{1,0})}{1+2\xi_{1,0}-\imath 0}+\int d[\xi_{2,0}]d[v_2]\frac{4\xi_{2,0}^2}{1-2\xi_{2,0}+\imath 0}\right). \tag{23}$$

To explore the behaviour of singularities from $\Sigma_V(p^2)$, we segment the integration region similar to the photon propagator case. Computing the integrals in both regions, we find,

$$\Sigma_V(p^2;\xi_i \leq \omega) = e_0^2\, S_\epsilon\left[-\log(1+2\omega-\imath 0)-\log(1-2\omega+\imath 0)\right]+O\left(\epsilon\right)\,, \tag{24}$$

$$\Sigma_V(p^2;\xi_i \geq \omega) = e_0^2\, S_\epsilon \left[-\frac{1}{\epsilon} + \log(1+2\omega - \imath 0) + \log(1-2\omega+\imath 0) - 1\right] + O(\epsilon)\,. \tag{25}$$

Based on these we find that $\Sigma_V(p^2)$ has not IR divergence and similar conclusions to the photon propagator case can be inferred. Since $\Sigma_V(p^2;\xi_i \leq \omega)$ is finite, we found that,

$$\begin{aligned}\Sigma_V(p^2;\xi_i \leq \omega) = {} & \frac{e_0^2}{8\pi^2}\int_0^1 dv_1 \int_0^\omega d\xi_{1,0} \frac{4\xi_{1,0}\,(1+\xi_{1,0})}{1+2\xi_{1,0}-\imath 0}\left[1+\log\left(\xi_{1,0}^2\right)\left[v_1\delta\left(v_1\right) + \left(1-v_1\right)\delta\left(1-v_1\right)\right]\right] \\ & + \frac{e_0^2}{8\pi^2}\int_0^1 dv_2 \int_0^\omega d\xi_{2,0} \frac{4\xi_{2,0}^2}{1-2\xi_{2,0}+\imath 0}\left[1+\log\left(\xi_{2,0}^2\right)\left[v_2\delta\left(v_2\right) + \left(1-v_2\right)\delta\left(1-v_2\right)\right]\right]\,, \end{aligned} \tag{26}$$

is a four dimensional representation of it. It is worth mentioning that the massive case has been studied in Ref.[28], but in this work we are only focused in the one-loop self-energies corrections to massless propagators.

4. Quark and gluon self-energies in QCD

In this section we explicitly compute the quark and gluon self-energy corrections at one-loop in QCD.

4.1. Quark self-energy

One loop correction to the quark self-energy has a similar structure to the electron propagator in QED. The one-loop quark self-energy differs from the one-loop electron self-energy, only by the replacement of e_0 by g_0 and the inclusion of the colour factor C_F.

Then, we have that the scalar function $\Sigma_V(p^2)$ is finite in the region $[0,\omega)$ and has a pole related to the presence of UV divergence at $[\omega,\infty]$. As in the electron case, there is a four dimensional realization given by

$$\begin{aligned}\Sigma_V(p^2;\xi_i \leq \omega) &= \frac{C_F\, g_0^2}{8\pi^2}\int_0^1 dv_1 \int_0^\omega d\xi_{1,0} \frac{4\xi_{1,0}\,(1+\xi_{1,0})}{1+2\xi_{1,0}-\imath 0}\left[1+\log\left(\xi_{1,0}^2\right)\left[v_1\delta\left(v_1\right) + \left(1-v_1\right)\delta\left(1-v_1\right)\right]\right] \\ &+ \frac{C_F\, g_0^2}{8\pi^2}\int_0^1 dv_2 \int_0^\omega d\xi_{2,0} \frac{4\xi_{2,0}^2}{1-2\xi_{2,0}+\imath 0}\left[1+\log\left(\xi_{2,0}^2\right)\left[v_2\delta\left(v_2\right) + \left(1-v_2\right)\delta\left(1-v_2\right)\right]\right]\,. \end{aligned} \tag{27}$$

4.2. Gluon self-energy

We analyse now the one-loop QCD correction to the gluon self-energy (in the Feynman gauge), which receives three different contributions, the quark, the gluon and the ghost-loop. For the gluon self-energy, the one-loop scalar function can be written as:

$$\Pi(p^2) = \Pi_q(p^2) + \Pi_1(p^2) + \Pi_2(p^2)\,, \tag{28}$$

where $\Pi_q(p^2)$, $\Pi_1(p^2)$ and $\Pi_2(p^2)$ are the corresponding scalar functions belonging to the quark, the gluon and the ghost-loop, respectively.

The quark-loop contribution differs from the one-loop photon self-energy only by the replacement of e_0 by g_0 and the inclusion of the factor n_f, representing the number of fermions in the loop, therefore;

$$\Pi_q(p^2) = \frac{8\, n_f\,(1-\epsilon)\, g_0^2}{(3-2\epsilon)\, p^2}\int_\ell \frac{q_1\cdot q_2}{(q_1^2+\imath 0)\,(q_2^2+\imath 0)}\,. \tag{29}$$

Now, the contribution coming from the gluon-loop is

$$\Pi_1(p^2) = -\frac{\imath C_A g_0^2}{2(3-2\epsilon)\,p^2}\int_\ell \frac{N}{(q_1^2+\imath 0)\,(q_2^2+\imath 0)}, \tag{30}$$

where the numerator is

$$\begin{aligned} N &= d((q_1+q_2)^2+(2q_1-q_2)^2+(2q_2-q_1)^2) \\ &- 2(q_1+q_2)\cdot(2q_1-q_2)-2q_2\cdot(2q_2-q_1)+2(2q_2-q_1)\cdot(2q_1-q_2)\,. \end{aligned} \tag{31}$$

Applying the LTD to the integral in Eq. (30) , we find,

$$\Pi_1(p^2) = -3\,C_A\,g_0^2\left(\int_\ell \frac{\tilde\delta(q_1)}{p^2+2p\cdot q_1-\imath 0}\left(1+\frac{p\cdot q_1}{p^2}\right)+\int_\ell \frac{\tilde\delta(q_2)}{p^2-2p\cdot q_2+\imath 0}\left(1-\frac{p\cdot q_1}{p^2}\right)\right). \tag{32}$$

Using the same parametrization, we find,

$$\Pi_1(p^2) = -\frac{12\,C_A\,g_0^2}{\Gamma(2-2\epsilon)}\left(\int d[v_1]\,d[\xi_{1,0}]\frac{\xi_{1,0}\,(1+\xi_{1,0})}{1+2\xi_{1,0}-\imath 0}+\int d[v_2]\,d[\xi_{2,0}]\frac{\xi_{2,0}\,(1-\xi_{2,0})}{1-\xi_{2,0}+\imath 0}\right). \tag{33}$$

Similar to the QED results, we find after integration

$$\Pi_1(p^2;\xi_i\le\omega) = \frac{C_A\,g_0^2\,S_\epsilon}{3}\left[4\omega^2-\log\left(1+2\omega-\imath 0\right)-\log\left(1-2\omega+\imath 0\right)\right]+O(\epsilon)\,, \tag{34}$$

$$\Pi_1(p^2;\xi_i\ge\omega) = \frac{C_A\,g_0^2\,S_\epsilon}{3}\left[-\frac{1}{\epsilon}-4\omega^2+\log\left(1+2\omega-\imath 0\right)+\log\left(1-2\omega+\imath 0\right)-\frac{4}{3}\right]+O\left(\epsilon\right). \tag{35}$$

Therefore, the four dimensional representation of $\Pi_1(p^2;\xi_i\le\omega)$ is given by,

$$\begin{aligned} \Pi_1(p^2;\xi_i\le\omega) &= -\frac{9C_A g_0^2}{4\pi^2}\int_0^1 dv_1\int_0^\omega \frac{d\xi_{1,0}\xi_{1,0}\,(1+\xi_{1,0})}{1+2\xi_{1,0}-\imath 0}[1+\log\left(\xi_{1,0}^2\right)\left[v_1\delta\left(v_1\right)+\left(1-v_1\right)\delta\left(1-v_1\right)\right]] \\ &- \frac{9C_A g_0^2}{4\pi^2}\int_0^1 dv_2\int_0^\omega \frac{d\xi_{2,0}\xi_{2,0}\,(1-\xi_{2,0})}{1-2\xi_{2,0}+\imath 0}[1+\log\left(\xi_{2,0}^2\right)\left[v_2\delta\left(v_2\right)+\left(1-v_2\right)\delta\left(1-v_2\right)\right]]\,. \end{aligned} \tag{36}$$

The last remaining piece is due to the ghost-loop, the interested integral is given by,

$$\Pi_2(p^2) = \frac{\imath\,C_A\,g_0^2}{(3-2\epsilon)p^2}\int_\ell \frac{q_1\cdot q_2}{(q_1^2+\imath 0)\,(q_2^2+\imath 0)} \tag{37}$$

applying LTD to the previous expression,

$$\Pi_2(p^2) = \frac{C_A\,g_0^2}{(3-2\epsilon)p^2}\left(\int_\ell \frac{\tilde\delta(q_1)q_1\cdot p}{p^2+2p\cdot q_1-\imath 0}-\int_\ell \frac{\tilde\delta(q_2)p\cdot q_2}{p^2-2p\cdot q_2+\imath 0}\right), \tag{38}$$

Following the precedent ideas, we find that the integral is decomposed as

$$\Pi_2(p^2;\xi_i\le\omega) = \frac{C_A g_0^2\,S_\epsilon}{3}\left[4\omega^2+\log\left(1+2\omega-\imath 0\right)+\log\left(1-2\omega+\imath 0\right)\right]+O\left(\epsilon\right), \tag{39}$$

$$\Pi_2(p^2;\xi_i \geq \omega) = \frac{C_A g_0^2 S_\epsilon}{3}\left[\frac{1}{\epsilon} - 4\omega^2 - \log\left(1+2\omega-\imath 0\right) - \log\left(1-2\omega+\imath 0\right) + 2\right] + O\left(\epsilon\right) . \tag{40}$$

Finally, we find that the four dimensional representation of $\Pi_2(p^2;\xi_i \leq \omega)$ is given by

$$\begin{aligned}\Pi_2(p^2;\xi_i \leq \omega) &= \frac{C_A g_0^2}{48\pi^2}\int_0^1 dv_1 \int_0^\omega \frac{d\xi_{1,0}4\xi_{1,0}^2}{1+2\xi_{1,0}-\imath 0}\left[1+\log\left(\xi_{1,0}^2\right)\left[v_1\delta\left(v_1\right)+\left(1-v_1\right)\delta\left(1-v_1\right)\right]\right] \\ &+ \frac{C_A g_0^2}{48\pi^2}\int_0^1 dv_2 \int_0^\omega \frac{d\xi_{2,0}4\xi_{2,0}^2}{1-2\xi_{2,0}+\imath 0}\left[1+\log\left(\xi_{2,0}^2\right)\left[v_2\delta\left(v_2\right)+\left(1-v_2\right)\delta\left(1-v_2\right)\right]\right]\end{aligned} \tag{41}$$

Gathering the four dimensional representations of the three different contributions to the gluon self-energy, the quark, the gluon and the ghost loop, we have that the four dimensional realization of $\Pi(p^2;\xi_i \leq \omega)$ is given by,

$$\Pi(p^2;\xi_i \leq \omega) = \Pi_q(p^2;\xi_i \leq \omega) + \Pi_1(p^2;\xi_i \leq \omega) + \Pi_2(p^2;\xi_i \leq \omega) . \tag{42}$$

5. Conclusions

In this paper we have shown that working with the LTD method the scalars functions from the one-loop self energies present only simple poles related to their UV-singular behaviour. Therefore a purely four dimensional representation could be always found for a region belonging to the ω-cut in the energy component. The possibility to obtain a four dimensional realization implies a major improvement in the computation of higher-order corrections in QFT, besides, allows a better understanding of the mathematical structures behind scattering amplitudes.

6. Acknowledgment

This work is supported by CONACyT, Mexico, PROFAPI 2015 grant number 121 and DPyC, by the Spanish Government and ERDF funds from European Commission (Grants No. FPA2014-53631-C2-1-P and SEV-2014-0398) and by Consejo Superior de Investigaciones Científicas (Grant No. PIE-201750E021)

7. References

[1] Z. Kunszt and D. E. Soper, "Calculation of jet cross-sections in hadron collisions at order alpha-s**3," Phys. Rev. D **46** (1992) 192.
[2] S. Frixione, Z. Kunszt and A. Signer, "Three jet cross-sections to next-to-leading order," Nucl. Phys. B **467** (1996) 399 [hep-ph/9512328].
[3] S. Catani and M. H. Seymour, "The Dipole formalism for the calculation of QCD jet cross-sections at next-to-leading order," Phys. Lett. B **378** (1996) 287 [hep-ph/9602277].
[4] S. Catani and M. H. Seymour, "A General algorithm for calculating jet cross-sections in NLO QCD," Nucl. Phys. B **485** (1997) 291 [Nucl. Phys. B **510** (1998) 503] [hep-ph/9605323].
[5] A. Gehrmann-De Ridder, T. Gehrmann and E. W. N. Glover, "Antenna subtraction at NNLO," JHEP **0509** (2005) 056 [hep-ph/0505111].
[6] S. Catani and M. Grazzini, "An NNLO subtraction formalism in hadron collisions and its application to Higgs boson production at the LHC," Phys. Rev. Lett. **98** (2007) 222002 [hep-ph/0703012].
[7] M. Czakon, "A novel subtraction scheme for double-real radiation at NNLO," Phys. Lett. B **693** (2010) 259 [arXiv:1005.0274 [hep-ph]].
[8] P. Bolzoni, G. Somogyi and Z. Trocsanyi, "A subtraction scheme for computing QCD jet cross sections at NNLO: integrating the iterated singly-unresolved subtraction terms," JHEP **1101** (2011) 059 [arXiv:1011.1909 [hep-ph]].
[9] V. Del Duca, C. Duhr, G. Somogyi, F. Tramontano and Z. Trócsányi, "Higgs boson decay into b-quarks at NNLO accuracy," JHEP **1504** (2015) 036 [arXiv:1501.07226 [hep-ph]].
[10] R. Boughezal, C. Focke, X. Liu and F. Petriello, "W-boson production in association with a jet at next-to-next-to-leading order in perturbative QCD," Phys. Rev. Lett. **115** (2015) 062002 [arXiv:1504.02131 [hep-ph]].
[11] J. Gaunt, M. Stahlhofen, F. J. Tackmann and J. R. Walsh, "N-jettiness Subtractions for NNLO QCD Calculations," JHEP **1509** (2015) 058 [arXiv:1505.04794 [hep-ph]].

IOP Conf. Series: Journal of Physics: Conf. Series **912** (2017) 012013 doi:10.1088/1742-6596/912/1/012013

[12] R. A. Fazio, P. Mastrolia, E. Mirabella, W. J. Torres-Bobadilla, "On the four-dimensional formulation of dimensionally regulated amplitudes," Eur. Phys. J.C. **74** (2014) 3197
[13] G. Rodrigo, S. Catani, T. Gleisberg, F. Krauss and J. C. Winter, "From multileg loops to trees (by-passing Feynman's Tree Theorem)," Nucl. Phys. Proc. Suppl. **183** (2008) 262 [arXiv:0807.0531 [hep-th]].
[14] I. Bierenbaum, S. Catani, P. Draggiotis and G. Rodrigo, "A Tree-Loop Duality Relation at Two Loops and Beyond," JHEP **1010** (2010) 073 [arXiv:1007.0194 [hep-ph]].
[15] I. Bierenbaum, S. Buchta, P. Draggiotis, I. Malamos and G. Rodrigo, "Tree-Loop Duality Relation beyond simple poles," JHEP **1303** (2013) 025 [arXiv:1211.5048 [hep-ph]].
[16] I. Bierenbaum, P. Draggiotis, S. Buchta, G. Chachamis, I. Malamos and G. Rodrigo, "News on the loop–tree Duality," Acta Phys. Polon. B **44** (2013) 2207.
[17] S. Buchta, "Theoretical foundations and applications of the Loop-Tree Duality in Quantum Field Theories," PhD thesis, Universitat de València, 2015, arXiv:1509.07167 [hep-ph].
[18] S. Buchta, G. Chachamis, P. Draggiotis, I. Malamos and G. Rodrigo, "Towards a Numerical Implementation of the Loop-Tree Duality Method," Nucl. Part. Phys. Proc. **258-259** (2015) 33 [arXiv:1509.07386 [hep-ph]].
[19] S. Buchta, "First Numerical Implementation of the Loop-Tree Duality Method," PoS EPS-HEP **2015** (2015) 430 [arXiv:1510.04105 [hep-ph].
[20] S. Buchta, G. Chachamis, P. Draggiotis and G. Rodrigo, "Numerical implementation of the Loop-Tree Duality method," arXiv:1510.00187 [hep-ph].
[21] R. J. Hernández-Pinto, G. F. R. Sborlini and G. Rodrigo, "Towards gauge theories in four dimensions," JHEP **1602** (2016) 044 [arXiv:1506.04617 [hep-ph]].
[22] G. F. R. Sborlini, "Loop-tree duality and quantum field theory in four dimensions," PoS RADCOR **2015** (2015) 082 [arXiv:1601.04634 [hep-ph]].
[23] S. Catani, T. Gleisberg, F. Krauss, G. Rodrigo and J. C. Winter, "From loops to trees by-passing Feynman's theorem," JHEP **0809** (2008) 065 [arXiv:0804.3170 [hep-ph]].
[24] R. P. Feynman, "Quantum theory of gravitation," Acta Phys. Polon. **24** (1963) 697.
[25] R. P. Feynman, "Closed Loop And Tree Diagrams. (talk)," In *Brown, L.M. (ed.): Selected papers of Richard Feynman* 867-887
[26] S. Buchta, G. Chachamis, P. Draggiotis, I. Malamos and G. Rodrigo, "On the singular behaviour of scattering amplitudes in quantum field theory," JHEP **1411** (2014) 014 [arXiv:1405.7850 [hep-ph]].
[27] G. F. R. Sborlini, R. Hernández-Pinto and G. Rodrigo, "From dimensional regularization to NLO computations in four dimensions," PoS EPS-HEP **2015** (2015) 479 [arXiv:1510.01079 [hep-ph]].
[28] G. F. R. Sborlini, Felix Driencourt-Mangin, German Rodrigo "Four-dimensional unsubtraction with massive particles," JHEP **1610** (2016) 162 [arXiv:1608.01584 [hep-ph]].

IOP Conf. Series: Journal of Physics: Conf. Series **912** (2017) 012014 doi:10.1088/1742-6596/912/1/012014

Feynman's path integral seen as a Henstock integral

M. Guadalupe Morales[1], Ricardo Gaitán[2]
Departamento de Física. FES Cuautitlán, Universidad Nacional Autónoma de México
Estado de México, 54770, México.
E-mail: [1]`mgumm@fesc.cuautitlan2.unam.mx`, [2]`rgaitan@unam.mx`

Abstract. The motivation of this paper is to give mathematical formalism to the Feynman path integral using the Henstock integral. This integral saves some of Feynman's integral difficulties and justifies Feynman's intuition: to interpret the state function as "a sum of complex contributions, one from each path in the region".

1. Introduction.

The interpretation of Richard Feynman, in order to solve the Schrödinger equation, (for non relativistic particle) involves taking limits of the sum over a space of paths of a functional of the action for each path. He did not provide proofs, see [1]. R. Feynman formulated quantum mechanics in terms of integrals over spaces of paths (Feynman path integrals). But the absolute value of Feynman's integrand is not integrable. A non-absolute method of integration is required in order to allow the cancellation effects described by Feynman to come into play. So, instead of Lebesgue integration, we use the non-absolute integral of Henstock. The main objective in this work is to realize a mathematical fundament where the Feynman path integral exists as an integral over the space of all paths.

It is well known that the Schrödinger equation describes the evolution of a state function for a particle of a constant mass m, moving in Euclidean space $\mathbb{R}^d$ in the presence of a potential $V(x)$. The state ψ at time $t = 0$ gives the initial condition for the equation and allows one to uniquely determine the state ψ function at all subsequent times:

$$\frac{\partial \psi}{\partial t} = i\left[\frac{1}{2m}\Delta - V\right]\psi$$

$$\psi|_{t=0} = \varphi,$$

where, $\hbar = 1$, Δ is the Laplacian operator in $\mathbb{R}^d$ and $V : \mathbb{R}^d \to \mathbb{R}$ is a measurable function. Suppose that we measure successive positions of a particle in one-dimensional space separated by a small time-interval ϵ, denote them by $x_1, x_2, x_3...$ Letting the intervals between measurements ϵ get smaller and smaller, we would expect the sequence $x_1, x_2, x_3...$ converges to a path of the particle, represented by a function of time $x(t)$. Then the state function can be represented as

$$\psi(x,t) \quad = \quad \lim_{\epsilon \to 0} \int_R e^{i \sum S(x_{i+1}, x_i)} ... \frac{dx_i}{A}\frac{dx_{i+1}}{A} ... \qquad (1)$$

IOP Conf. Series: Journal of Physics: Conf. Series **912** (2017) 012014 doi:10.1088/1742-6596/912/1/012014

where $\frac{1}{A}$ is the normalization factor and $S(x(t))$ is the classical action, that is the time integral of the Lagrangian along the path of the particle $S(x(t)) = \int L(\dot{x}(t), x(t))dt$. In the case when the Lagrangian has a simple form of a quadratic function of the velocity $L(\dot{x}(t), x(t)) = \frac{1}{2}m(\dot{x}(t))^2 - V(x(t))$, Feynman was able to define a state function. By Trotter's Theorem is obtained that, under certain conditions, the state function exists as limit of product of operators, see [2], [3] and [4], among others. Thus,

$$\psi(x,t) = \lim_{n\to\infty} \left(\frac{mn}{2\pi it}\right)^{nd/2} \underbrace{\int_{\mathbb{R}^d} \dots \int_{\mathbb{R}^d}}_{n\ times} e^{i\sum[\frac{m}{2}\frac{(x_j-x_{j-1})^2}{(t/n)^2} - V(x_j)]^{t/n}} \varphi(x_n)dx_1...dx_n$$

According to Feynman's fundamental postulate a resulting path of a particle is a sum over all paths and it is written in the form of an integral over the space of all paths, see [5]. Thus, this integral is written in the following way

$$Const \int_{\Omega_x} e^{iS(\omega,t)} \varphi(\omega(x)) D\omega, \tag{2}$$

where Ω_x is the set of all paths that start at x. We might interpret the limit (1) as a sum over all paths. Then it would be natural to represent it as an integral with respect to some measure in the path-space. However, there exist some difficulties, for example: *i*) the normalization "constant" has a meaning for every finite "n", but it becomes infinite as n goes to infinity; *ii*) and $\lim_{n\to\infty} \prod_{j=1}^{n} dx_j$ corresponds to some measure on a space of all paths, but this product is infinite, thus defined this way, the measure has no firm mathematical meaning.

These difficulties are saved by interpreting the Feynman path integral as Henstock integral over the space of all paths, $\mathbb{R}^T$, and it is justified the Feynman's intuition, to interpret the state function as "a sum of complex contributions, one from each path in the region". These problems are discussed in detail in [6], [7], [8], [9], [10] and [11]. It is well known that other methods exist to give solution to the Schrödinger equation, however the proposal of Feynman in addition to giving solution has implications in other areas as Quantum Mechanics, Financial Mathematics, among other, see for example, [12], [13], [14].

The objective of this paper is to present the main idea of this integral and its implications in Quantum Mechanics to give mathematical formalism.

2. Formalism. Finite-dimensional integration case.

We follow the notation from [9]. In order to present the Henstock integral theory over the space of all paths, we need to introduce basic definitions of the Henstock integral in the finite-dimensional case, $\mathbb{R}^n$ with $n \geq 1$. To see more about this integral consult [15], [16] and [17] among others.

Definition 2.1. A cell I in $\mathbb{R}^n$ consists in the product $I = I(N) = I_1 \times ... \times I_n$, where $N = \{1, ..., n\}$ and each I_j can have the form

$$(-\infty, a), \quad [u, v], \quad (b, \infty], \quad \text{or} \quad (-\infty, \infty).$$

The collection of all cells in $\mathbb{R}^n$ is denoted by $I(\mathbb{R}^n) = \{I(N)\}$. This definition is given by Riemman sums, thus each cell must be related or associated to a point $x \in \overline{\mathbb{R}}^n$, $x = (x_1, ..., x_n)$.

Definition 2.2. Let I be a cell in $\mathbb{R}^n$. The cell is associated to $x \in \overline{\mathbb{R}}^n$ if for each $j = 1, ..., n$ *i*) $x_j = -\infty$, if $I_j = (-\infty, a]$, *ii*) $x_j = u$ or $x_j = v$, if $I_j = [u, v]$, *iii*) $x_j = \infty$, if $I_j = (b, \infty)$, *iv*) $x_j = -\infty$ or $x_j = \infty$, if $I_j = (-\infty, \infty)$.

IOP Conf. Series: Journal of Physics: Conf. Series **912** (2017) 012014 doi:10.1088/1742-6596/912/1/012014

The association condition means that the point x should be in the interior or on the boundary of I.

Definition 2.3. A gauge in $\mathbb{R}^n$ is a positive function δ defined for $x \in \overline{\mathbb{R}}^n$. An associated point-cell pair (x, I) of $\mathbb{R}^n$ is $\delta - fine$ if, for each j, the pair (x_j, I_j) is $\delta - fine$ in $\mathbb{R}$; that is,

$$a < \frac{-1}{\delta(x)}; \quad u - v < \delta(x); \quad \text{or} \quad b > \frac{1}{\delta(x)}, \quad \text{respectively.}$$

A **partition** of $\mathbb{R}^n$ is a finite collection $\mathcal{P}$ of disjoint cells whose union is $\mathbb{R}^n$. A **division** $\mathcal{D}$ of $\mathbb{R}^n$ is a finite collection of associated point-cell pairs (x, I) whose cells form a partition of $\mathbb{R}^n$. Given a gauge $\delta : \overline{\mathbb{R}}^n \to \mathbb{R}^+$, a **division** $\mathcal{D}$ **is** $\delta - fine$ if each $(x, I) \in \mathcal{D}$ is $\delta - fine$, it is denoted by $\mathcal{D}_\delta$. The class of all the cells in $\mathbb{R}^n$ is denoted by $I(\mathbb{R}^n)$.

In general, an integrand in $\mathbb{R}^n$ is a point-cell function $h(x, I)$ defined in the product $\overline{\mathbb{R}}^n \times I(\mathbb{R}^n)$ to $\mathbb{C}$ (or $\mathbb{R}$); in particular it can be a product $f(x)h(I)$.

Definition 2.4. A function $h(x, I)$ is integrable in $\mathbb{R}^n$ with integral $\alpha = \int_{\mathbb{R}^n} h(x, I)$, given $\epsilon > 0$ there exists a gauge function δ in $\overline{\mathbb{R}}^n$ such that, for each $\delta - fine$ division $\mathcal{D}_\delta$ of $\mathbb{R}^n$, the corresponding Riemann sums satisfies

$$\left|\alpha - \mathcal{D}_\delta \sum h(x, I)\right| < \epsilon,$$

where $\mathcal{D}_\delta \sum h(x, I) = \sum_{(x,I) \in \mathcal{D}_\delta} h(x, I)$.

Example 2.5. The function $\cos(x^2)$ is not Lebesgue integrable on $(0, \infty)$. The function $\cos(x^2)|I|$, defined on $(0, \infty) \times I([0, \infty))$, is Henstock integrable.

First, we will show that $\cos(x^2)$ is not Lebesgue integrable, which means that it is not absolutely integrable on $(0, \infty)$. The function $\cos(x^2)|I|$ has zeros at

$$a_n := \sqrt{\frac{(2n+1)\pi}{2}},$$

where $n = 0, 1, 2, ...$, and let us define $b_n := \int_{a_n}^{a_{n+1}} \cos(x^2)dx$. Note that $|b_n| \to 0$ since $a_{n+1} - a_n \to 0$ as $n \to \infty$. Therefore, $\sum_{n=0}^{\infty} b_n$ converges non-absolutely. The Lebesgue integrability of $\cos(x^2)$ on $(0, \infty)$ is equivalent to the convergence of $\sum |b_n|$. However,

$$|b_n| > \frac{1}{2}(a_{n+1} - a_n) = \frac{1}{2}\sqrt{\frac{\pi}{2}}\left(\sqrt{2n+3} - \sqrt{2n+1}\right).$$

Since $\sum(\sqrt{2n+3} - \sqrt{2n+1})$ diverges, it is clear that $\cos(x^2)$ is not Lebesgue integrable on $(0, \infty)$. Now, we want to show that there exists a number α such that

$$\int_0^\infty \cos(x^2)|I| = \alpha \quad \text{where} \quad \alpha = \int_0^{a_0} \cos(x^2)dx + \sum_{n=0}^{\infty} b_n.$$

We have that

$$\int_0^{a_0} \cos(x^2)|I| = \int_0^{a_0} \cos(x^2)dx,$$

first being a Henstock integral and the second a Riemann integral. We denote $\alpha' = \sum_{n=0}^{\infty} b_n$, it is sufficient to prove that

$$\int_{a_0}^\infty \cos(x^2)|I| = \alpha'.$$

If $\epsilon > 0$ is given, n' can be chosen sufficiently large so that $\left|\sum_{n>n'} b_n\right| < \epsilon/2$, and then $\delta(\infty)$ can be defined as $a_{n'} > 1/\delta(\infty)$. Designating by n'' the smallest integer n for which $a_n > \delta(\infty)^{-1}$, the (infinite) group of terms $\sum_{n\geq n''} b_n$ though not zero, has value less than $\epsilon 2^{-1}$.

The next step is to compare each of the remaining (finite) number of terms b_n of the series $\sum b_n$ with the corresponding groups of terms of the Riemann sum, in order to use this comparison to define a positive function $\delta(x)$ when $(x \neq \infty)$. Accordingly, for $a_n \leq x \leq a_{n+1}$ and $n = 0, l, 2, ...,$ we choose $\delta_n(x)$ to satisfy:

$$\left|b_n - (D_{\delta_n}) \sum \cos(x^2)|I|\right| < \epsilon 2^{-(n+1)},$$

for all $\delta_n - fine$ division of $(a_n, a_{n+1}]$, and define $\delta(x)$ as follows:

- if $a_n < x < a_{n+1}$, then $\delta(x) < min\{x - a_n, a_{n+1} - x\}$;
- if $x = a_n$, $\delta(x) \leq min\{\delta_{n-1}(x), \delta_n(x)\}$;
- if $a_n < x < a_{n+1}$, $\delta(x) \leq \delta_n(x)$.

This ensures conformance, meaning that if $x \neq a_n$ for any n, and if (x, I) is $\delta - fine$, then I is a subset of some $(a_n, a_{n+1}]$, and cannot intersect with two of them. Therefore, any $\delta - fine$ division $D = (x, I)$ must include division points $x = a_0, a_1, ..., a_m$. Any $\delta - fine$ division then gives a Riemann sum of the form

$$(D_\delta) \sum \cos(x^2)|I| = \sum_{n=0}^{n''} \left((D_\delta) \sum \cos(x^2)|I|\right) = \sum_{n=0}^{n''} c_n.$$

Since

$$\left|\sum_{n=0}^{\infty} b_n - (D_\delta) \sum \cos(x^2)|I|\right| \leq \left|\sum_{n=0}^{n''-1} b_n - \sum_{n=0}^{n''-1} c_n\right| + \left|\sum_{n=n''}^{\infty} b_n\right|,$$

thus,

$$\left|\sum_{n=0}^{\infty} b_n - (D_\delta) \sum \cos(x^2)|I|\right| < \sum_{n=0}^{n''-1} \frac{\epsilon}{2^{n+1}} + \frac{\epsilon}{2} < \epsilon.$$

Therefore the function $\cos(x^2)|I|$ is Henstock integrable on $(0, \infty)$.

3. Infinite-dimensional integration case.

Let T be an interval in $(0, \infty)$. $\mathbb{R}^T$ represents the set $(x_t)_{t \in T}$, the set of real-valued functions defined on T. We start by considering cells in $\mathbb{R}^T$. $\mathcal{N} = \mathcal{N}(T)$ denotes the class of finite subsets N of T.

Definition 3.1. Let N be a finite set in $\mathcal{N}$, such that $t_1 < t_2 < ... < t_n$. A cell in $\mathbb{R}^T$ is

$$I[N] = I(N) \times \mathbb{R}^{T \setminus N}.$$

It is helpful to emphasize the "restricted" dimensions N of the cylindrical interval $I \subset \mathbb{R}^T$, written as $I[N]$, while with round brackets, is a finite-dimensional interval. The collection of all cells in $\mathbb{R}^T$ is denoted by $I(\mathbb{R}^T) = \{I[N] : N \in \mathcal{N}\}$. Note that $\mathbb{R}^{T \setminus N}$ denotes the set of real-valued functions defined on $T \setminus N$.

Definition 3.2. A partition of $\mathbb{R}^T$ is a finite collection $\mathcal{P}$ of disjoint cells $I[N]$ whose union is $\mathbb{R}^T$.

Definition 3.3. It is said that $(x, N, I[N])$ is associated in $\mathbb{R}^T$ to the corresponding finite-dimensional pair $(x(N), I(N))$, then it is associated in the finite-dimensional space $\mathbb{R}^N$.

Thus, an integrand in $\mathbb{R}^T$ may be expressed as $h(x, N, I[N])$, where the triple $(x, N, I[N])$ is associated if the pair $(x(N), I(N))$ is associated in $\mathbb{R}^n$. Note that $x(N) = (x(t_1), ..., x(t_n))$, $x_i = x(t_i)$ and $I_j = I_{t_j}$ for each $i = 1, .., n$.

In addition to a condition δ, on the lengths of the restricted edges I_j, a further condition is imposed: the dimension sets N (or sets of restricted dimensions) of partitioning cells $I[N]$ should, for each associated x, include some minimal set of dimensions $L(x)$. That is, we require $L(x) \subset N$, where $L(x)$ can be made successively larger, just as $\delta(x)$ is made successively smaller in forming Riemann sums. Due to this, a gauge in $\mathbb{R}^T$ is considered as a pair of mappings (δ, L).

Definition 3.4. A gauge γ in $\mathbb{R}^T$ is a pair of mappings (δ, L) such that

- $L : \overline{\mathbb{R}}^T \to \mathcal{N}$, it means, $x \to L(x) \in \mathcal{N}$
- $\delta : \overline{\mathbb{R}}^T \to (0, \infty)$, it means, $(x, N) \to \delta(x, N)$.

Definition 3.5. With $N = \{t_1, ..., t_n\} \in \mathcal{N}(T)$, an associated triple $(x, N, I[N])$ is $\gamma - fine$ if, $L(x) \subset N$ and (x_i, I_i) is $\delta - fine$ for $t_j \in N$, $1 \leq j \leq N$.

Definition 3.6. A division of $\mathbb{R}^T$ is a finite collection $\mathcal{D}$ of point-cell pairs $(x, I[N])$ such that the corresponding $(x, N, I[N])$ are associated, and the cells $I[N]$ form a partition $\mathcal{P}$ of $\mathbb{R}^T$.

Definition 3.7. Let be $\gamma = (L, \delta)$ a gauge function. A division $\mathcal{D}$ is a $\gamma - fine$ if each $(x, I[N]) \in \mathcal{D}$ is $\gamma - fine$. In that case, we can denote the $\gamma - fine$ division $\mathcal{D}$ by $\mathcal{D}_\gamma$.

Suppose $h = (x, N, I)$ is a real- or complex-valued function of associated elements $(x, N, I[N])$ in the infinite-dimensional domain $\mathbb{R}^T$ and $\mathcal{D} = \{(x, I[N]\}$ is a division of $\mathbb{R}^T$, then the corresponding Riemann sum for h is

$$(\mathcal{D})\sum h := \sum_{(x,I[N])\in\mathcal{D}} h(x, N, I[N]) = \sum\{h(x, N, I[N]) : (x, N, I[N]) \in \mathcal{D}\}.$$

Definition 3.8. A function h of associated triples $(x, N, I[N])$ is integrable on $\mathbb{R}^T$, with integral $\alpha = \int_{\mathbb{R}^T} h(x, N, I[N])$, if given $\epsilon > 0$, there exists a gauge γ so that, for each $\gamma - fine$ division $\mathcal{D}_\gamma$ of $\mathbb{R}^T$, the corresponding Riemann sum satisfies

$$|\alpha - (\mathcal{D}_\gamma)\sum h| < \epsilon.$$

We will examine some of the integrands encountered in $\mathbb{R}^T$.

Example 3.9. Consider $f(x) = \prod_{j=1}^{n} 2\pi(t_j - t_{j-1}))^{-1/2} e^{\left(-\frac{(x_j - x_{j-1})^2}{2(t_j - t_{j-1})}\right)}$, with $x_j = x(t_j)$ for $t_j \in N \in \mathcal{N}(T)$. In this case f depends not just on the variable coordinates x_j for variable x, but also on the variable $N = \{t_1, ..., t_n\}$, and this is something new which does not arise whenever T is a finite set. Since f depends explicitly on the variable elements $t_j \in T$ as well as on the variables $x_j = x(t_j)$, it is appropriate to write f as $f(x, N)$. Our theory of integration will enable us to consider the integrability of $h(x, I, N) = f(x, N)\mu(I)$, with $\mu(I) = |I[N]|$ and

$$\mu(I[N]) = |I[N]| = |I(N)| = \begin{cases} \prod_{j=1}^{n}(v_j - u_j) & \text{if} \quad I_j = (u_j - v_j] \\ 0 & \text{Otherwise} \end{cases}. \tag{3}$$

In the next section we will explain more about this type of integrand. The class of integrands that we are interested in consists of real- or complex-valued functions h defined on associated triples $(x, N, I[N])$:

$$\begin{array}{rcl} (\mathbb{R}^T, \mathcal{N}, I[N]) & \xrightarrow{h} & \mathbb{R} \ or \ \mathbb{C} \\ (x, N, I[N]) & \to & h(x, N, I[N]) \\ (x, N, I[N]) & \in & (\mathbb{R}^T, \mathcal{N}, I[N]) \\ x & \in & I[N] \end{array}$$

If T is a finite set, then integrands $h(x, N, I[N])$ reduce to functions of the form $h(x, I)$ where x and I are familiar finite-dimensional objects.

4. Fresnel Integral.

An integrand in $\mathbb{R}^T$ might then take the form $h(x, N, I[N]) = f(x)|I[N]|$ for some point function f, where $|I[N]|$ is given by expression (3). Because $\mathbb{R}^T$ is unbounded in each dimension it is easy to see that constant functions $f(x)$ are not integrable on $\mathbb{R}^T$ with respect to $|I[N]|$ unless, for instance, $f(x)$ is equal to zero for all x. That is why this approach will be based on an application of the Henstock integration technique to the Fresnel integrands. First, we will introduce *one-dimensional Henstock Fresnel integral* in order to present the infinite-dimensional case.

In [17], it was shown how the Feynman path integral leads to the Wiener kernel, with the purely imaginary diffusion coefficient and how this makes the corresponding measure not countably additive. The main obstacle consists in the exponentials with pure imaginary exponents. The improper Riemann integrals of such expressions are called the Fresnel integrals:

$$\int_{-\infty}^{\infty} e^{\frac{i}{2}x^2}\,dx. \tag{4}$$

It can be proved that the integral (4) exists as a Henstock integral, even more, is equal to $\sqrt{\frac{2\pi}{-i}}$, see [17]. This result can be generalized to any complex number $c = a + ib$, where $a \leq 0$, $b \geq 0$ and $c \neq 0$, see [9].

Let us define a complex-valued function $\varphi(x)$ for $x = (x_1, 1_2, ..., x_n) \in \mathbb{R}^n$ as:

$$\varphi(x) = \varphi_{n,\frac{i}{2}}(x) = e^{\frac{i}{2}(x_1^2+x_2^2+...+x_n^2)}.$$

Now we consider a "volume" function μ defined on a set $I(\mathbb{R}^n)$ of finite-dimensional cells $I(N) = I_1 \times ... \times I_n$ as the expression (3).

Consider the function $h(x, I(N)) = \varphi(x)\mu(I(N))$; we will refer to its integral as the Henstock Fresnel integral, and

$$\int_{\mathbb{R}^n} \varphi(x)\mu(I(N)) = \int_{\mathbb{R}^n} \prod_{j=1}^{n} e^{\frac{i}{2}x_j^2}|I_j| = \left(\sqrt{\frac{2\pi}{-i}}\right)^n,$$

see [17].

We would like to redefine or normalize the Fresnel integrand so that the new integral can be considered as a probability distribution function. If a cell $I \in I(\mathbb{R}^n)$ has an associated point $x \in \overline{\mathbb{R}}^n \setminus \mathbb{R}^n$, that is, $x = (x_1, ..., x_n)$ has some component $x_j = \pm\infty$, then a value zero was assigned to "volume" $\mu(I(N)) = |I|$, by convention. But since we need to construct a probability distribution function, nonzero values are assigned in all cases. Thus, we define the following function,

$$g_n(x)|I| = \begin{cases} \left(\sqrt{\frac{-i}{2\pi}}\right)^n e^{\frac{i}{2}x_1^2+...+x_n^2}|I_1|...|I_2| & \text{if} \quad x \in \mathbb{R}^n \\ \left(\sqrt{\frac{-i}{2\pi}}\right)^n \prod_{j=1}^n \int_{I_j} e^{\frac{i}{2}x_j^2}dx_j & \text{if} \quad x \in \overline{\mathbb{R}}^n \setminus \mathbb{R}^n \end{cases}. \tag{5}$$

For example, if $n = 1$ and $J = (u, \infty)$, $u > 0$, then

$$\begin{aligned} g_1(x)|J| &= \sqrt{\frac{-i}{2\pi}}\left[\frac{1}{2}\int_{-\infty}^{\infty} e^{\frac{i}{2}y^2}dy - \int_0^u e^{\frac{i}{2}y^2}dy\right] \\ &= \frac{1}{2} - \sqrt{\frac{-i}{2\pi}} g_1(x)|(0,u)|. \end{aligned}$$

The function $g_n(x)|I|$ is integrable in $\mathbb{R}^n$ in Henstock sense and $\int_{\mathbb{R}^n} g_n(x)|I| = 1$. For more details see [9]. If we consider $g_n(x)|I|$ as an n-dimensional Henstock-Fresnel density function, then we define its associated n-dimensional probability distribution function on cells,

$$G_n(I) = \begin{cases} \left(\sqrt{\frac{-i}{2\pi}}\right)^n \int_I e^{\frac{i}{2}x_1^2+\ldots+x_n^2}|I| & \text{if} \quad x \in \mathbb{R}^n \\ \left(\sqrt{\frac{-i}{2\pi}}\right)^n \prod_{j=1}^n \int_{I_j} e^{\frac{i}{2}x_j^2}dx_j & \text{if} \quad x \in \overline{\mathbb{R}}^n \setminus \mathbb{R}^n \end{cases}.$$

A figure in $\mathbb{R}^n$ is the union of a finite number of cells, denoted by E. The collection of all figures in $\mathbb{R}^n$ is denoted as $E(\mathbb{R}^n)$.

In fact, G_n is defined over the collection of figures $E(\mathbb{R}^n)$, and this is finitely additive on disjoint figures and $\int_{\mathbb{R}^n} G_n(I) = 1$; it means that is a probability distribution function, see [9, Theorem 147].

In order to show that $\int_{\mathbb{R}^n} G_n(I) = 1$ is easy to prove that if $\mathcal{D}$ is a division of $\mathbb{R}^n$, then

$$(\mathcal{D})\sum G_n(I) = \left(\frac{-i}{2\pi}\right)^n \left(\frac{2\pi}{-i}\right)^n = 1.$$

Additivity follows from the definition.

4.1. Fresnel Integral in $\mathbb{R}^T$.

The second step is to define a probability distribution function over the space of all paths. In addition, it should be noted that there are increments in the integrand, see expression (1). The increments in the variables are just translations that don't change significantly any results stated before. Thus from now on, we will use the following definitions for the density and probability distribution functions, $g^T(x(N))|I[N]|$ and $G^T([N])$, respectively

$$g^T(x(N))|I[N]| = \begin{cases} \left[\prod_{j=1}^n \left(\sqrt{\frac{-i}{2\pi t_j - t_{j-1}}}\right) e^{\frac{i}{2}\frac{(x_j-x_{j-1})^2}{(t_j-t_{j-1})}}\right] |I[N]| & \text{if} \quad x \in \mathbb{R}^T \\ \prod_{j=1}^n \left(\sqrt{\frac{-i}{2\pi t_j - t_{j-1}}}\right) \int_{I_j} e^{\frac{i}{2}\frac{(x_j-x_{j-1})^2}{(t_j-t_{j-1})}} dx_j & \text{if} \quad x \in \overline{\mathbb{R}}^T \setminus \mathbb{R}^T \end{cases}.$$

$$G^T(I[N]) = \prod_{j=1}^n \left(\sqrt{\frac{-i}{2\pi t_j - t_{j-1}}}\right) \int_{I_j} e^{\frac{i}{2}\frac{(x_j-x_{j-1})^2}{(t_j-t_{j-1})}} dx_j, \tag{6}$$

where $T \subset \mathbb{R}^+$ and $N = \{t_1, t_2, ..., t_n\} \in \mathcal{N}(T)$ with $0 = t_0 < t_1 < t_2 <, ..., t_n$. Here the differences $x_j - x_{j-1} = x(t_j) - x(t_{j-1})$ are the increments or transitions of the incremental or transitional Fresnel integrand.

By [9, Theorem 168] the function G^T in the expression (6) is integrable on $\mathbb{R}^T$ and defines a distribution function. Note that in the classical sense, this function would not be considered as a probability distribution function, because it can take negative or complex values. However, in the formulation of Quantum Mechanics, complex-valued functions play the role of probability distribution functions. Thus, in this context, we can refer $G^T(I)$ as a probability distribution function; it means that,

$$\int_{\mathbb{R}^T} G^T(I[N]) = 1, \tag{7}$$

and G^T is finitely additive on disjoint sets $E \in E(\mathbb{R}^T)$. In order to show that the expression (7) holds, it is enough to prove that:

$$(\mathcal{D})\sum G^T = \prod_{j=1}^{p}\left(\int_{\mathbb{R}} e^{\frac{i}{2}(x_j^2)}dx_j\right) = 1. \tag{8}$$

Let $\mathcal{D}$ be a division of $\mathbb{R}^T$, choose any $(x, I[N]) \in \mathcal{D}$, $N = \{t_1, ..., t_n\}$, so $I = I[N]$ is restricted in n dimensions. $\mathcal{D}$ can be composed of cells $I[N]$ with variable N, the partition $\mathcal{P}$ consists of cells $J[M]$, each having the same M. In other words, $\mathcal{D}$ gives a partition $\mathcal{P}$ of $\mathbb{R}^T$ such that each $I[N]$ of $\mathcal{D}$ is the union of a finite number of cells $J[M] \in \mathcal{P}$, and each cell J of $\mathcal{P}$ is restricted in the same dimensions $M = \{\tau_1, ..., \tau_p\}$, see [9, Partially Regular Partitions]. By additivity of the finite-dimensional Riemann and extended Riemann integrals involved,

$$(\mathcal{D})\sum G^T(I[N]) = (\mathcal{P})\sum G^T(J[M]).$$

The latter Riemann sum is easy to compute, because M is the same for each $J \in \mathcal{P}$ and we get the expression (8).

On the other hand, let us note that

$$\int_{\mathbb{R}^T} G^T(I[N]) = G^T(\mathbb{R}^T) = 1.$$

As the indefinite integral of an integrable function is an additive function on figures, see [9, Theorem 16], we get the result.

Example 4.1. Let us consider the function $g_0(x, N)$ as,

$$g_0(x, N) = e\left(\frac{i}{2}\sum_{j=1}^{n}\frac{(x_j - x_{j-1})^2}{t_j - t_{j-1}}\right)\prod_{j=1}^{n}(2\pi i(t_j - t_{j-1}))^{-1/2},$$

when $x \in \mathbb{R}^T$.

Note that $g_0(x, N) = g^T(x(N))$ if $x \in \mathbb{R}^T$. We redefine g^T as g_0 because we emphasize the free particle case, it means $V = 0$. In [8, Proposition 68] it is shown that g_0 is *Henstock Fresnel* integrable. The sketch of the proof is as follows. Let D a division of $\mathbb{R}^T$ and M the dimension set of I, $(x, I) \in \mathcal{D}$. Let $M = \{s_1, ..., s_{m-1}\}$, $s_0 = \tau'$ and $s_m = \tau$. For $y \in \mathbb{R}^m$, let

$$g(y) = e^{\frac{i}{2}\sum_{j=1}^{m}\frac{(y_j - y_{j-1})^2}{s_j - s_{j-1}}}\prod_{j=1}^{m}\left(\frac{-2\pi}{i}(s_j - s_{j-1})\right)^{-1/2}, \quad y_0 = \xi' \;\; y_m = \xi.$$

Note that

$$\frac{(y_{j+1} - y_j)^2}{s_{j+1} - s_j} + \frac{(y_j - y_{j-1})^2}{s_j - s_{j-1}} = \tag{9}$$

$$\frac{y_{j+1}^2}{s_{j+1} - s_j} + \frac{y_{j-1}^2}{s_j - s_{j-1}} + \frac{y_j^2 - 2y_j\left(\frac{y_{j+1}(s_j - s_{j-1})}{s_{j+1} - s_{j-1}} + \frac{y_{j-1}(s_{j+1} - s_j)}{s_{j+1} - s_{j-1}}\right)}{\frac{(s_{j-1} - s_j)(s_j - s_{j-1})}{s_{j+1} - s_{j-1}}} \tag{10}$$

Since $\int_0^\infty e^{\frac{i}{2}y^2}dy = \frac{1}{2}\sqrt{\frac{-2\pi}{i}}$, completing squares and by a change of variable we have that

$$\int_{-\infty}^{\infty}(9)dy_j = \int_{-\infty}^{\infty}(10)dy_j = \left(\frac{-2\pi}{i}\frac{(s_{j+1}-s_j)(s_j-s_{j-1})}{s_{j+1}-s_{j-1}}\right)^{1/2} e^{\frac{i}{2}\frac{(y_{j+1}-y_{j-1})^2}{s_{j+1}-s_{j-1}}}.$$

Thus,

$$\int_{\mathbb{R}^{m-1}} g(y)dy = \left(\frac{-2\pi}{i}(\tau-\tau')\right)^{-1/2} e^{\frac{i}{2}\frac{(\xi-\xi')^2}{\tau-\tau'}}.$$

It is possible to prove that for any figure E in $\mathbb{R}^T$, $\int_E g(y)dy$ exists and $\left|\int_{\mathbb{R}^{m-1}\setminus E(M)} g(y)dy\right| < \epsilon$, where $E(M)$ corresponds to those $(x, I) \in \mathcal{D}$ for which $y(M)$ has not infinite component. Therefore, we have that

$$\int_{\mathbb{R}^T} g(x, N, I[N]) = \left(\frac{-2\pi}{i}(\tau-\tau')\right)^{-1/2} e^{\frac{i}{2}\frac{(\xi-\xi')^2}{\tau-\tau'}}.$$

Observe that $T = (\tau', \tau)$. To prove that g_0 is not absolutely integrable it is enough to observe that the function e^{iu^2} is not Lebesgue integrable, according to the proof of Proposition 68 from [8]. Let us write

$$e^{iu^2} = \cos(u^2) + i\sin(u^2).$$

As

$$\int_{-\infty}^{\infty} e^{iy^2}dy = \sqrt{i\pi},$$

we have

$$\int_{[0,\infty)} \cos(u^2)du = \int_{[0,\infty)} \sin(u^2)du = \frac{1}{2}\sqrt{\frac{\pi}{2}}.$$

These integrals exist as extended Riemann integrals, but not as Lebesgue integrals. The graphs of $cos(u^2)$, $sin(u^2)$ oscillate periodically with constant amplitude 2 but with period decreasing to zero as $u \to \infty$.

As the complex number $c = \frac{i}{2}$ in the function g_0, we get the one-dimensional free particle propagator function of Quantum Mechanics. If c were $-\frac{i}{2}$, then

$$\int_{\mathbb{R}^T} g_0(x, N, I[N]) = \left(\frac{2\pi}{i}(\tau-\tau')\right)^{-1/2} e^{-\frac{i}{2}\frac{(\xi-\xi')^2}{\tau-\tau'}},$$

and gives us the probability density function or diffusion function of Brownian motion.

Now we turn to the general case when V is not the function zero. Let V be a real-valued function of a single real variable.

$$\begin{aligned}\phi_V(\xi,\tau) &= \int_{\mathbb{R}^T} e\left(\frac{i}{2}\sum_{j=1}^{n}\frac{(x_j-x_{j-1})^2}{t_j-t_{j-1}} V(x_{j-1})(t_j-t_{j-1})\right)\prod_{j=1}^{n}\left(\frac{2\pi}{i}(t_j-t_{j-1})\right)^{-1/2}|I[N]| \\ &= \int_{\mathbb{R}^T} g_V(x, N)|I[N]|.\end{aligned}$$

Then ϕ_V exists if V is a continuous functions. Now let

$$G_V(J[N]) := \int_{J(N)} g_V(x, N)|I[N]|.$$

Note that $G_V(J)$ corresponds to Feynman's probability amplitude.

Thus, a solution to Schrödinger equation is given by Henstock integral. Let $T = (\tau', \tau) \subset \mathbb{R}^+$ and V is continuous at $y = \xi$, $\tau = t$. If

$$\psi_1(\xi, \tau) = \int_{\mathbb{R}^T} e^{(-\frac{i}{2}(\sum_{j=1}^{n} V(x(t_{j-1}), t_{j-1})(t_j - t_{j-1})))} \prod_{j=1}^{n} \frac{1}{(2\pi i(t_j - t_{j-1}))^{1/2}} e^{\frac{i}{2}\frac{(x_j - x_{j-1})^2}{t_j - t_{j-1}}} |I[N]|$$

exists for each (ξ'', τ'') in a neighborhood of (ξ, τ), then $\psi_1(\xi, \tau)$ satisfies the Schrödinger equation, see [9, Theorem 221], where this integral is in Henstock sense.

5. Conclusions

In this work it is observed that there exists an equivalence between the Schrödinger equation solutions according to the perturbation theory (Trotters Theorem) and Henstock Theory. This follows from the uniqueness of the solution to the Schrödinger equation. On the other hand, the Henstock integral solves the problem that the absolute value of the Feynman integrand is not integrable because this approximation does not require the measure concept, for which the absolute value of the integrand does not need to be integrable.

References

[1] Feynman R P 1948 Space-time approach to non-relativistic quantum mechanics. *Rev. Modern Physics* **20** 367–387.

[2] Gel'fand I M and Jaglom A M 1960 Integration in functional spaces and its applications in quantum physics *J. Mathematical Phys.* **1** 48–69.

[3] Kato T 1951 Fundamental properties of Hamiltonian operators of Schrödinger type. *Trans. Amer. Math. Soc.* **70**, 195–211.

[4] Nelson E 1954 Feynman integrals and the Schrödinger equation. *J. Mathematical Phys.* **5** 332–343.

[5] Feynman R P and Hibbs A R 2010 *Quantum mechanics and path integrals. Emended edition.* (New York) (Dover Publications Inc.)

[6] Cao S S 1992 The Henstock integral for Banach-valued functions. *Southeast Asian Bull. Math.* **16** no. 1, 35–40.

[7] Henstock R, Muldowney P and Skvortsov V A 2006 Partitioning infinite-dimensional spaces for generalized Riemann integration. *Bull. London Math. Soc.* **38** no. 5, 795–803.

[8] Muldowney P 1987 *A general theory of integration in function spaces, including Wiener and Feynman integration.* (New York) (Longman Scientific & Technical, Harlow; John Wiley & Sons, Inc.)

[9] Muldowney P 2012 *A modern theory of random variation. With applications in stochastic calculus, financial mathematics, and Feynman integration* (New York) (John Wiley & Sons, Inc., Hoboken)

[10] Muldowney P 2000 Feynman's path integrals and Henstock's non-absolute integration. *J. Appl. Anal.* **6** no. 1, 1–24.

[11] Pardy M 1973 The Generalised Feynman Integral *Internat. J. Theoret. Phys.* **8** no. 1, 31–37.

[12] Bonotto E M, Federson M and Muldowney, P 2010/11 A Feynman-Kac solution to a random impulsive equation of Schrödinger type *Real Anal. Exchange* **36**, no. 1, 107–148.

[13] Muldowney P 2000/01 The Henstock integral and the Black-Scholes theory of derivative asset pricing. *Real Anal. Exchange* **26** no. 1, 117-131.

[14] Muldowney P 2000 Financial valuation and the Henstock integral. *Seminário Brasileiro de Análise* **60**, 79–108.

[15] Henstock R 1946 On Interval Functions and their Integrals. *Journal of the London Mathematical Society* **21**. no 3, 204–209.

[16] Henstock R, Muldowney P 2015 Ralph Henstock's Lectures on the Theory of Integration. notes of lectures given by Ralph Henstock at the New University of Ulster in 1970-71, 132pp.

[17] Nathanson E S 2014 Path integration with non-positive distributions and applications to the Schrodinger equation. Thesis (Ph.D.)The University of Iowa. ProQuest LLC, Ann Arbor, MI, 113 pp.

Acknowledgments

This work is partially supported from CONACYT-SNI. M.G.M. and R.G.L acknowledge support by project PAPIIT-IN113916 and program DGAPA-UNAM of postdoctoral fellowship.

Operation and performance of the CMS Resistive Plate Chambers during LHC run II

Jan Eysermans[1], María Isabel Pedraza Morales[1] on behalf of the CMS Collaboration

[1]Facultad de Ciencias Físico-Matemáticas, Benemérita Universidad Autónoma de Puebla, Mexico

E-mail: `jan.eysermans@cern.ch`

Abstract.
The Resistive Plate Chambers (RPC) at the Compact Muon Solenoid (CMS) experiment at the CERN Large Hadron Collider (LHC) provide redundancy to the Drift Tubes in the barrel and Cathode Strip Chambers in the endcap regions. Consisting of 1056 double gap RPC chambers, the main detector parameters and environmental conditions are carefully monitored during the data taking period. At a center of mass energy of 13 TeV, the luminosity reached record levels which were challenging from the operational and performance point of view. In this work, the main operational parameters are discussed and the overall performance of the RPC system is reported for the LHC run II data taking period. With a low amount of inactive chambers, a good and stable detector performance was achieved with high efficiency.

1. Resistive Plate Chambers at CMS

The Resitive Plate Chambers (RPC) at the Compact Muon Solenoid (CMS) experiment[1] at the CERN Large Hadron Collider (LHC) provide redundancy to the muon trigger system which consists of Drift Tubes (DT) in the barrel and Cathode Strip Chambers (CSC) in the endcap regions. Besides trigger redundancy it also contributes to the muon reconstruction and identification. The CMS RPCs consist of 5 barrel stations having 480 chambers whereas the endcaps consist of 576 chambers distributed over 4 stations on each side. Currently, the RPC pseudorapidity η coverage is 1.9 but an extension up to 2.4 is foreseen near 2020.

A CMS RPC chamber consists of two gaps operated in avalanche mode to ensure reliable operation at high rates. Each gap consists of two 2 mm thick high resistive High-Pressure Laminates (HPL) separated by a 2 mm gas gap. A graphite coating at the outer surface of the HPL plates guaranties a uniform distribution of the charges to achieve a uniform electric field over the entire gap area. A non-flammable 3-component gas mixture of 95.2% freon (C2H2F4, known as R134a), 4.5% isobutane (i-C4H10), and 0.3% sulphur hexafluoride (SF6) is used with a relative humidity of 40-50%. The readout plane is located between both gaps and consists of strips aligned in the $|\eta|$ direction with a pitch width between 2.28 and 4.10 cm in the barrel and between 1.74 and 3.63 cm in the encaps. In total, the entire system contains 137,000 of such copper strips covering an area of about 4000 m^2. The strip signals are asynchronously sent to the Front End Boards (FEBs) which shapes the signal before being sent to the RPC linkboard system and the CMS data acquisition system. The FEBs are electronically controllable by means of the signal threshold to handle the noise of the detector.

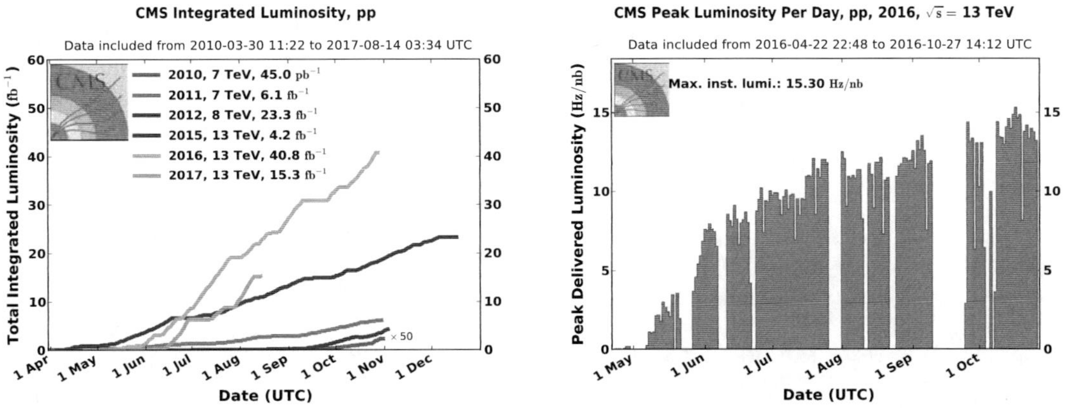

Figure 1. Delivered integrated luminosity over the years by the LHC (left) and instantaneous luminosity evolution during 2016 data taking (right)[2].

2. Run II in some numbers

After the Higgs boson discovery in 2012 during run I of the LHC data taking period, the accelerator and the experiments had their first Long Shutdown (LS1) of two years in 2013-2014. From the RPCs point of view, the fourth endcap station chambers were installed covering the full four disks in both endcaps as was initially foreseen. Besides this, a general maintenance was performed with the aim of recovering broken chambers and reparation of gas leaks.

After the LS1 in 2015, the LHC was again operational to deliver proton-proton collisions at an increased center-of-mass energy of $\sqrt{s} = 13$ TeV. Besides the energy increase, also the instantaneous luminosity was expected to increase throughout the years. 2015 was a warm up year delivering an integrated luminosity of 4.2 fb^{-1} (see Fig. 1) with a maximum luminosity achieved around 0.5×10^{34} cm^{-2}s^{-1}. During that year the newly installed RE4 stations were successfully commissioned providing good data to the CMS Muon system. In 2016, the LHC geared up rapidly, providing stable beam collisions with a record peak instantaneous luminosity of 1.53×10^{34} cm^{-2}s^{-1}. As a result, also the data collection reached a record level up to an integrated luminosity of 40.85 fb^{-1}.

During 2016, CMS collected 37.87 fb^{-1}, yielding a luminosity loss of 2.98 fb^{-1} which was assigned due to temporary failures of different subsystems. The RPC system experienced only very few hardware problems in 2016 which was due to channel readout problems and the partial failure of the high voltage power supply system. Besides these incidents, the RPC operation for both 2015 and 2016 was stable, with 98% active chambers.

3. Background rate

Besides detection of muons directly emerging from the proton-proton collissions, other particles such as gammas, alpha particles, electrons and neutrons are present due to nuclear interactions with the detector materials. The cumulative effect of these interacting particles is called the background rate and it can leave hits in the detector with a similar characteristic as muons. At high rates, this can obviously affect the muon reconstruction and therefore the behavior of the background rate versus instantaneous luminosity needs to be understood. In Fig. 2, the current through the RPC and hit rate are plotted versus instantaneous luminosity. In both cases, a natural linear relationship is observed, which was already known from run I, but in a lower instantaneous luminosity regime. Extrapolations towards higher luminosities up to 5×10^{34} cm^{-2}s^{-1} results in a good estimation of the currents and background rate expected for the High-Luminosity LHC (HL-LHC).

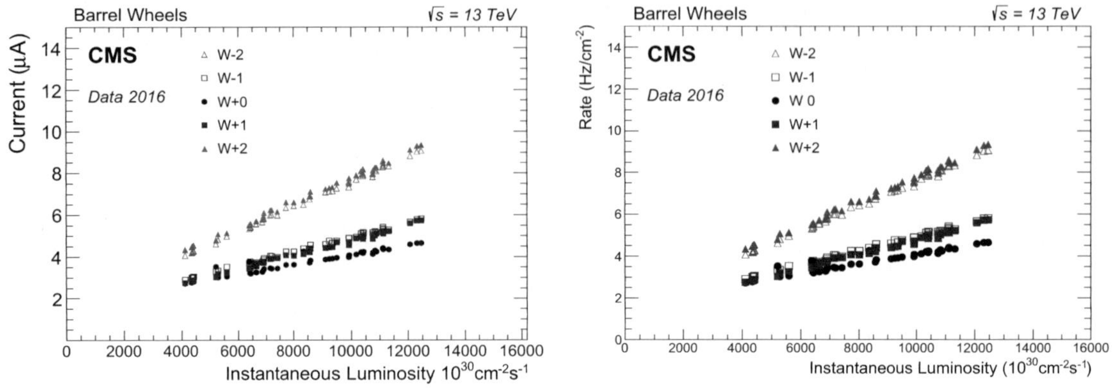

Figure 2. Current (left) and hit background rate (right) dependency of the instantaneous luminosity in the barrel region[3].

4. RPC Performance

The RPC system needs to fulfill several requirements in order to have a good performance and deliver qualitative physics to the CMS muon system. Several performance parameters are continuously monitored and the detector operational parameters needs to be adapted in order to fulfill these requirements and to optimize the physics performance. Furthermore, as the system gets older and aging effects might appear both in time or as a consequence of radiation, the hardware performance monitoring is also necessary and crucial. An overview of the three most important performance indicators are explained below and the results are given for the 2016 data taking period.

4.1. Working point calibration and efficiency

A high muon detection efficiency is the most basic and important requirement for the RPC system. A minimum of 95% is required over the entire system. The efficiency is directly related to the voltage applied and the threshold set on the electronics: the avalanches inside the gas gap have a higher probability to induce a detectable signal when the voltage is increased and the electronic threshold decreased. However, the range of both parameters are limited due to the detector noise and cluster size (see next sections respectively). The avalanche size also depends on the environmental conditions, in particular the pressure. Therefore, the working point HV_{WP} is corrected online as function of the pressure P according to:

$$HV_{corr} = HV_{WP}\left((1-\alpha) + \alpha\frac{P}{P_0}\right),$$

with $\alpha = 0.8$ and a reference pressure of $P_0 = 965$ mbar. The temperature variation is not (yet) taken into account.

The working point voltages of the RPCs are obtained by performing a high voltage scan during calibration runs: the efficiency is measured as function of the high voltage applied. The efficiency for each high voltage point can be obtained from two methods[4][5]:

- Segment extrapolation method: the muon tracks are extrapolated from the other muon detectors (DTs in the barrel and CSCs in the endcaps), and the corresponding muon hits are counted in the RPCs;
- Tag and Probe method: a new method using muon candidates from the tracker only.

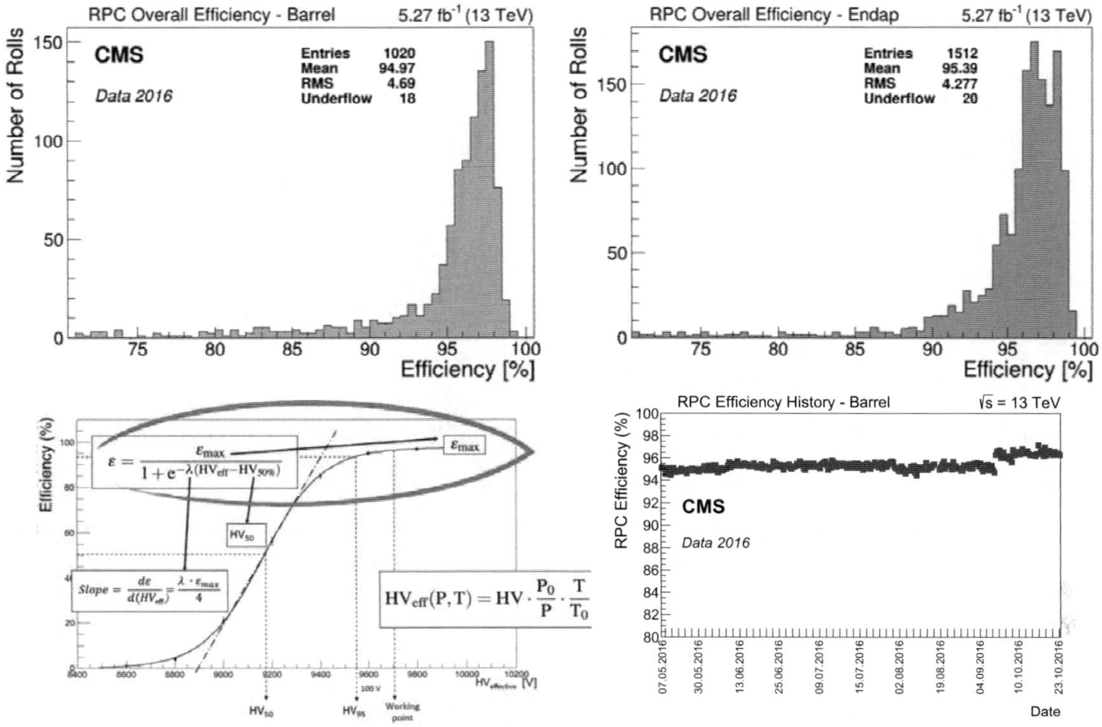

Figure 3. Top row: efficiency in barrel (left) and endcap (right) for the first period of 2016 data taking period. Bottom left: procedure to extract the working point from the high voltage scan. Bottom right: efficiency over time for the entire 2016 data taking period[3].

When plotting the efficiency versus high voltage, a sigmoid-like curve is obtained and the working point parameters are extracted from the fit, as shown in Fig. 3 (bottom-left). For the first data taking period in 2016, the results are shown in Fig. 3 (top row). An average of 95% is obtained for both barrel and endcap. Some RPCs with very low efficiency experience gas leaks or threshold control problems.

The efficiency at working point is continuously measured during data taking and is a good indicator of the stability of the RPC system. As shown in Fig. 3 (bottom right), the efficiency is very stable during the entire 2016 data taking period, resulting in a good performance. Also no detector degradation is observed as a direct effect of the high instantaneous luminosity or in-time aging. The efficiency increase at the end of the data taking period was after the recalibration of the working points and thresholds.

4.2. Intrinsic noise rate

The intrinsic noise rate is defined as the rate seen by the RPC chamber during cosmic data taking. In these circumstances, RPC high voltage is at its working point and only cosmic muons are detected. However, other contributions to the rate intrinsic to the detector can be detected due to noise coming from the high voltage system, tensions in wires, etc. Therefore it is crucial to keep this noise as low as possible, with a maximum of 5 Hz/cm^2 which is required by CMS.

The noise is measured between the LHC fills when the RPCs are at their working point or during calibration at the beginning of the year. The remote threshold control enables to control the noise per 8 strips in a chamber. Monitoring and adjusting during the data taking period is necessary to remove noisy channels which can affect the physics performance.

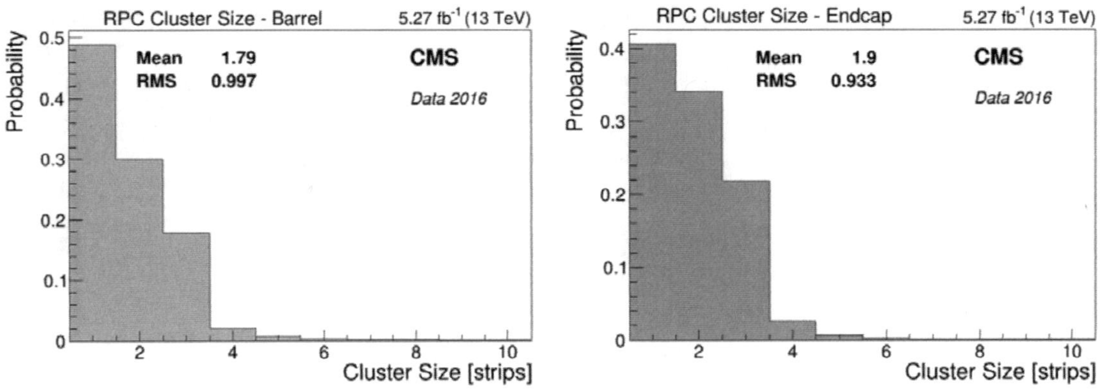

Figure 4. Cluster size distribution for barrel (left) and endcap (right) RPC chambers[3].

The intrinsic noise levels in 2015 and 2016 are continuously measured in both barrel and endcap and their values are not exceeding 0.15 Hz/cm^2, which is far below the limit. A slight dependency on the luminosity is observed which is the consequence of residual radiation after the beam dump.

4.3. Cluster Size

In ideal circumstances, only one strip is fired when a charged particle passes through both gas gaps, yielding the maximum power in position resolution of the muon reconstruction. However, fluctuations in the gas gain, high voltage, local environment and the bi-gap effect can alter the avalanche sizes, and therefore it is possible multiple strips are fired for the same muon. The cluster size is defined, on average basis, as the amount of strips fired per muon and can be maximal around 2-3 by CMS requirements.

In particular, the applied high voltage mainly determines the avalanche size and applying too high voltages can cause streamers with a large charge density in the gap covering several strips and hence the increase of the cluster size. On the other hand, the threshold applied is also sensitive to the cluster size. During operation, the cluster size is monitored by detailed analysis and the thresholds can be adapted accordingly if needed. In Fig. 4, the mean cluster size for barrel and endcap are shown for the 2016 data. The mean value of cluster size is lower than 2 and is stable during the data taking period.

5. Conclusions

During run II the CMS RPC system performed very well with a good and stable hardware operation. No major degradation has been observed concerning the hardware. Continuous and periodic detector calibrations assured good detector performance resulting in a high efficiency, low noise rate and low mean value of cluster size.

References

[1] CMS Collaboration, "The CMS experiment at the CERN LHC", JINST 3 (2008) S0800.
[2] https://twiki.cern.ch/twiki/bin/view/CMSPublic/LumiPublicResults
[3] https://twiki.cern.ch/twiki/bin/view/CMSPublic/RPCPlots
[4] M.Abbrescia et al., "Cosmic ray test of double-gap resistive plate chambers for the CMS experiment", Nucl. Instr. Meth. A550 (2005) 116.
[5] CMS Collaboration, "The performance of the CMS muon detector in proton-proton collisions at sqrt(s) = 7 TeV at the LHC", JINST 8 (2013) P11002.

MEXnICA, Mexican group in the MPD-NICA experiment at JINR

M. Rodríguez Cahuantzi for the MEXnICA group
Faculty of Physics and Mathematics, Autonomous University of Puebla
Av. San Claudio y 18 Sur, Edif. EMA3-231, Ciudad Universitaria 72570, Puebla, México
E-mail: mrodriguez@fcfm.buap.mx / mario.rodriguez@correo.buap.mx

Abstract. The Nuclotron Ion Collider fAcility (NICA) accelerator complex is currently under construction at the Joint Institute for Nuclear Research (JINR) laboratory located in the city of Dubna in the Russian Federation. The main goal of NICA is to collide heavy ion nuclei to study the properties of the phase diagram of strongly interacting matter at high baryon density. In this accelerator complex, two big particle detectors are planned to be installed: Spin Physics Detector (SPD) and Multi-Purpose Detector (MPD). At the design luminosity, the event rate in the MPD interaction region is about 6 kHz; the total charged particle multiplicity would exceeds 1000 in the most central Au+Au collisions at $\sqrt{s_{NN}}$= 11 GeV.

Since the middle of 2016 a group of researchers and students from Mexican institutions was formed (MEXnICA). The main goal of the MEXnICA group is to collaborate in the experimental efforts of MPD-NICA proposing a BEam-BEam counter detector which we called BEBE. In this written general aspects of MPD-NICA detector and BEBE are discussed. This material was shown in a contributed talk given at the XXXI Annual Meeting of the Mexican Division of Particles and Fields held in the Physics Department of CINVESTAV located in Mexico City during the last week of May 2017.

1. Introduction

Heavy-ion physics allow us to study the matter under extreme conditions of high temperature and energy density. The study of hot and dense baryonic matter should shed light on: in-medium properties of hadrons and the nuclear matter equation of state (EOS); the onset of deconfinement (OD) and/or chiral symmetry restoration (CSR); phase transition (PT), mixed phase and the critical end-point (CEP); possible local parity violation in strong interactions (LPV) [1].

Experimental evidences show that the deconfined phase of nuclear matter (Quark-Gluon Plasma, QGP) can be created in ultra-high-energy nuclear collisions [2]. In this type of collision, there are some signatures related with the production of QGP, i.e., collective flow, long-range angular correlations, suppression of hadrons with high transverse momentum (p_T) and enhancement of thermal photons and di-leptons. Theoretical calculations indicate that the deconfinement phase transition can be accomplished by partial restoration of the chiral symmetry in heavy-ion collisions leading to possible modifications of hadronic spectral functions in dense hadronic matter. RHIC experiments reported the observation of strong elliptic flow [3] and jet quenching (suppression of hadrons with large p_T) [4]. Recently, LHC experiments reported the observation of long-range, near-side angular correlations in Pb-Pb and p-Pb collisions [5]. In addition to this result, it was found that the hot dense matter created in heavy-ion collisions behaves like a fluid with almost zero friction [6], which undoubtedly shows an advance in the understanding of this state of matter. Unexpected phenomena like the enhanced production of multi-strange hadrons in high-multiplicity p-p collisions (proton collisions exhibits similar patterns to those observed in heavy-ion collisions) [7] have also been reported. Figure 1 shows the covered region of the diagram phase of the QCD by current and future experimental facilities like NICA at JINR [8].

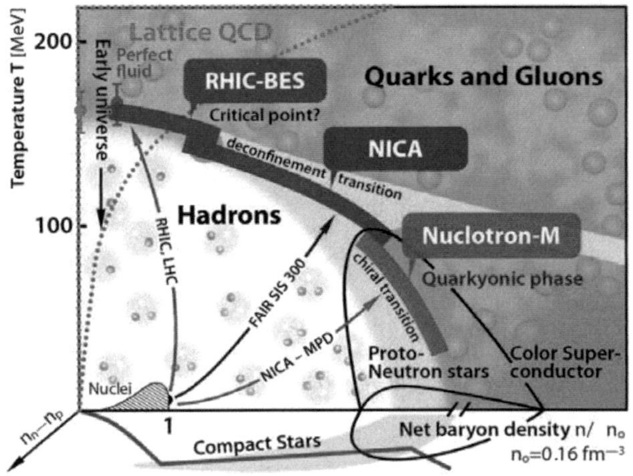

Figure 1. Diagram phase of the QCD. The energy range of NICA is an unexplored region to date. Experimental data on hadron production properties at SPS suggest that this transition occurs within the NICA energy range.

NICA is a new accelerator complex which will provide beams of ions over a wide range of atomic masses (from p to Au) at average luminosity of 10^{27} $cm^{-2}s^{-1}$ for gold-gold collisions and 10^{32} $cm^{-2}s^{-1}$ for proton-proton collisions. The center-of-mass energy will be between 4 to 11 GeV for Au+Au and up to 27 GeV for p+p. The energy range of NICA will allow to study environments with high net baryon density where the onset of deconfinement phase transition could occur.

The physics tasks of the NICA heavy-ion program are:

- event-by-event fluctuation in hadron productions
- femtoscopic correlation
- directed and elliptic flows for various hadrons
- multi-strange hyperon production (including hypernuclei): yield and spectra
- photon and electron probes
- charge asymmetry

To achieve a successfully study of matter under extreme conditions, the experimental program of NICA will be performed with the Multi-Purpose Detector (MPD).

2. The Multi-Purpose Detector (MPD) at NICA

The Multi-Purpose Detector (MPD) [9] is an experimental complex to be located at one of the interaction points of the NICA collider, see figure 2. The MPD detector is composed by central and forward detectors designed to detect charged hadrons, electrons and photons that will be produced in heavy-ion collisions at high luminosity in the energy range of the NICA collider. As the average transverse momentum of the particles produced in a collision at NICA energies is below 500 MeV/c, the detector design requires a very low material budget. The magnetic field produced by the MPD magnet will be of 0.6 Teslas. The general layout of the MPD apparatus is shown in figure 3.

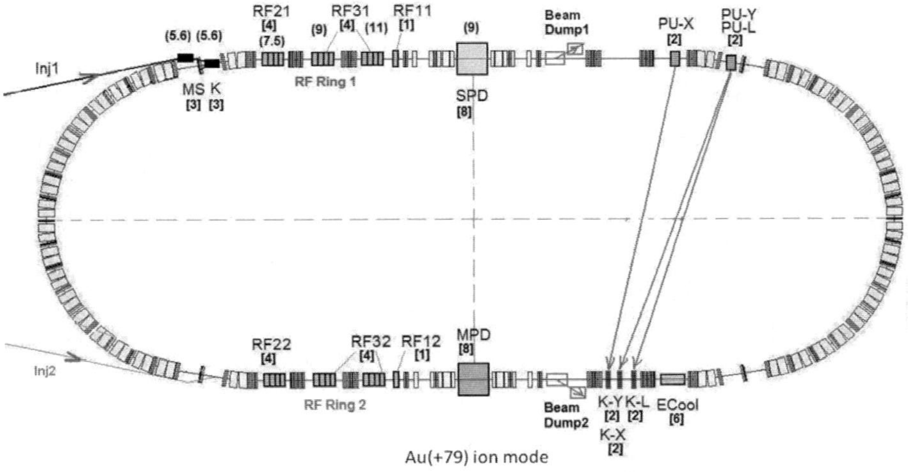

Figure 2. Schematic view of NICA facilities. MPD will be located at point 8. Taken from [10]

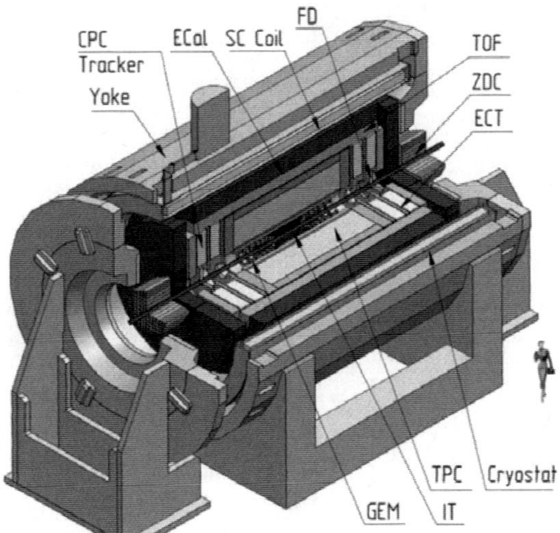

Figure 3. Multi-Purpose Detector at NICA. The detector consist of three major parts: CD-central detector, and (FS-A, FS-B) - two forward spectrometers (optional). The following subsystems are drawn: superconductor solenoid (SC Coil) and magnet yoke, inner detector (IT), straw-tube tracker (ECT), time-projection chamber (TPC), time-of-flight system (TOF), electromagnetic calorimeter (EMC), fast forward detectors (FFD), and zero degree calorimeter (ZDC) [11].

The MPD physics program will be carried out in two stages.

(i) participant detectors: TPC, TOF, ECAL, ZDC and FD. To study:
   - Particle yields and spectra (pions, Kaons, protons, etc)
   - Event-by-event fluctuations
   - Femtoscopy involving pions, Kaons and protons
   - Collective flow for identified hadrons
   - Electromagnetic probes (electrons, gammas)

(ii) participant detectors: TPC, TOF, ECAL, ZDC, FD + ITS. To study
   - total particles multiplicities
   - asymmetries studies (better reaction plane determination)
   - di-lepton production
   - charmonium production
   - production of soft photons and hyper-nuclei

Considering the fact that after the collision of two heavy-ion nuclei the matter behaves as a thermalized system and the particle production tracing a trajectory on phase diagram, maybe close to the CEP; a key study is the event-by-event fluctuations of the physical observables of particles in its final state, and thus, a particle detector capable to give a trigger signal for heavy-ion collisions is crucial. In this sense, since the middle of 2016 a group of researches and students from Mexican institutions was formed (MEXnICA). The main goal of the MEXnICA group is to collaborate in the experimental efforts of MPD-NICA proposing a BEam-BEam counter detector which we called BEBE.

3. MEXnICA group

The institutions involved in the MEXnICA group are:

- **BUAP** (Benemérita Universidad Autónoma de Puebla): Facultad de Ciencias Físico Matemáticas
- **CINVESTAV** (Centro de Invetigación y Estudios Avanzados del I.P.N.): Departamento de Física
- **UAS** (Universidad Autónoma de Sinaloa): Facultad de Ciencias Físico Matemáticas
- **UNISON** (Universidad de Sonora): Departamento de Física
- **UNAM** (Universidad Nacional Autónoma de México): Instituto de Ciencias Nucleares

The main goal of MEXnICA within MPD efforts is to contribute in the study of the QGP phase diagram from both, theoretical and experimental points of view in the search of the CEP:

(i) theoretical point of view
   - to study the mechanism responsible for the restoration of chiral symmetry
   - to study the QCD phase diagram at finite values of temperature and density

(ii) experimental point of view
   - to study signatures that will allow to locate the CEP
   - to study the inclusion of a detector that allows to MPD to increase its pseudo-rapidity coverage. With such detector, the MPD could increase its capabilities to determine the event plane resolution and trigger efficiency for general heavy-ion collisions and specific physical processes.

IOP Conf. Series: Journal of Physics: Conf. Series **912** (2017) 012016 doi:10.1088/1742-6596/912/1/012016

4. MEXnICA proposal

In collider experiments, the inclusion during commissioning or regular operations of a detector capable to monitor the beam activity is desirable. With the information provided with such kind of apparatus, it is possible to setup a trigger system to identify and to discriminate beam-beam minimum bias or centrality events from background and beam-gas interactions. In addition, these types of systems can be used for the reconstruction of physical observables of interest in heavy-ion collisions such as

- *multiplicity of charged particles*: key observable for the determination of the centrality of the collision events and the event plane resolution
- *luminosity measurement*: used to determine the absolute cross section of reaction processes

The technology and the type of materials for the development of beam monitors is diverse depending on their applications. In the specific case of particle detectors designed to collect information from colliders, it is often the use of materials and light sensors like diamond, silicon, quartz, plastic scintillator, MCTP, Si-PMT, APD and PMT.

Taking into consideration our previous experience on development of charge particle detectors [12, 13, 14], we propose a BEam-BEam counter detector (BEBE) for the MPD-NICA Project at JINR based on a granular plastic scintillator array and APD as light sensors.

As a first approach, the BEBE detector will consists of two detectors located at 2 meters away from the interaction point of the MPD-NICA, BEBE-1 and BEBE-2 respectively. Each detector is composed by an array of 80 cells made of plastic scintillator arranged in five rings forming a sixteen sectors disk of 1.52 meters of diameter (see figure 4). The pseudorapidity coverage of BEBE would be $1.69 < |\eta| < 4.36$.

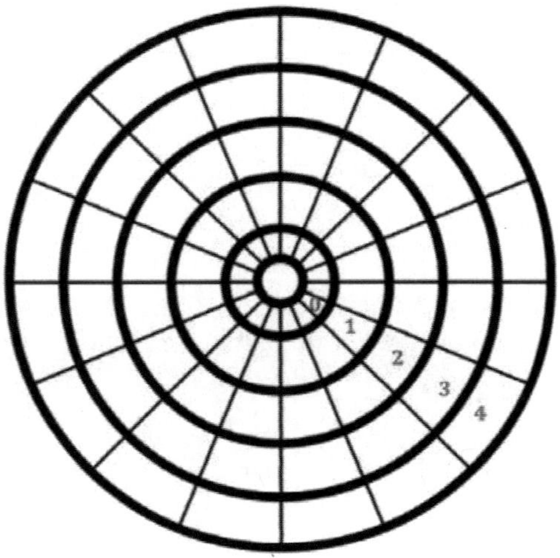

Figure 4. Schematic of BEBE detector. It will consists of an array of 80 cells distributed in 5 concentric rings.

To carry out a physics performance study of the BEBE detector within the experimental conditions of NICA, we have developed a dedicated module in the offline framework of the experiment: MPD-ROOT [15]. The results of the activity of BEBE during Au+Au collisions simulations at $\sqrt{s_{NN}} = 11\ GeV$ is shown in figure 5. In figure 6, a correlation between the BEBE mean multiplicity and the impact parameter of the simulated collisions is shown. As can be seen, BEBE is sensitive to the heavy-ion collisions in several impact parameter bins. This

IOP Conf. Series: Journal of Physics: Conf. Series **912** (2017) 012016 doi:10.1088/1742-6596/912/1/012016

type of studies are important because the centrality of the collisions is an important property that can not be accessed directly from the NICA beam online information. Thus, BEBE will play a crucial role in the identification of events per centrality range.

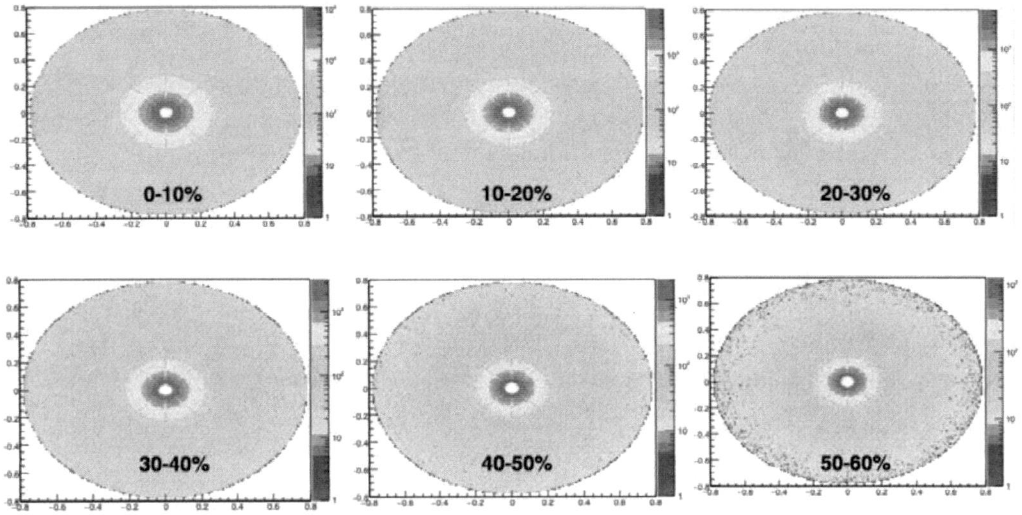

Figure 5. Monte Carlo hits of BEBE detector. The activity in each cell of the detector change with respect to the centrality of the collision, as expected.

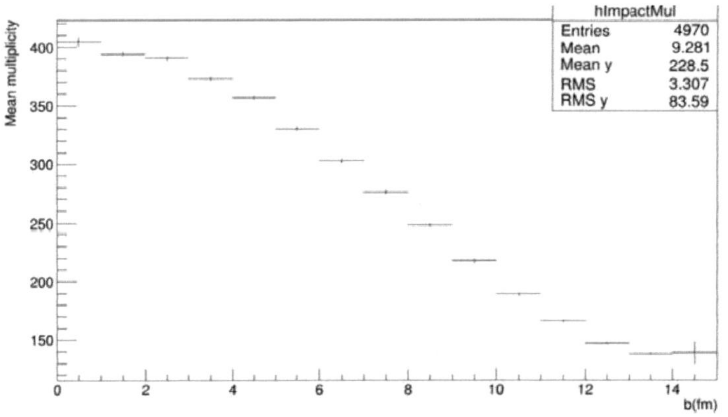

Figure 6. Mean multiplicity in BEBE versus impact parameter.

5. Conclusions

NICA has the potential for competitive research in the studies of dense baryonic matter. The MPD detector has many advantages and meets all the ambitious physics requirements for exploring the phase diagram of strongly interacting matter in a high track multiplicity environment. The MPD detector covers a large phase space; it is functional at high interaction rates; comprises high efficiency and excellent particle identification capabilities. MPD is based on recent detector developments and has comparatively reasonable cost.

MEXniCA group is currently working on the final design of BEBE. In parallel, several theoretical studies are being developed focused on MPD-NICA physics program. A niche of technological development will be explored by the MEXnICA group: construction of plastic scintillator and electronics acquisition system are activities that will be developed by MEXnICA group members once the BEBE geometry is defined.

MEXnICA group members

FISICA-UNISON: Dra. Maria Elena Tejeda Yeomans and Lic. Luis Valenzuela, **FCFM-UAS:** Dra. Isabel Domínguez Jiménez and Dr. Roger Hernández Pinto, **ICN-UNAM**: Dr. Wolfgang Bietenholz and Dr. José Alejando Ayala Mercado, **FISICA-CINVESTAV:** Dr. Luis Manuel Montaño Zetina and M.C. Heber Zepeda, **FCFM-BUAP:** C. Jorge Tolentino, Dr. Rodolfo Palomino Merino, Dr. Eduardo Moreno, Dr. Pedro Gonzalez Zamora, Dr. Guillermo Tejeda Muñoz and Dr. Mario Rodríguez Cahuantzi.

Acknowledgments

This work was partially support by VIEP-BUAP and CONACyT projects under grants 301 and 256494 respectively.

The author thank the inputs from all the members of MEXnICA group as well as the fruitful discussions with Dr. Slava Golovatyuk and Dr. Oleg Rogachevsky from JINR regarding the MEXnICA proposal. We also strongly thank to the Mexican workers who through their contributions make possible the development of basic and applied science in Mexico. Especially those who died, lost loved ones or became homeless as a result of devastating earthquakes in Mexico on September 8th and 19th, 2017. We hope that our modest efforts, written in this paper, will be reflected in the near future in benefit of all our community.

References

[1] V. Kekelidze, A. Kovalenko, R. Lednicky, V. Matveev, I. Meshkov, A. Sorin, G. Trubnikov, Status of the NICA project at JINR, In Nuclear and Particle Physics Proceedings, Volumes 273275, 2016, Pages 170-174, ISSN 2405-6014, https://doi.org/10.1016/j.nuclphysbps.2015.09.021.
[2] U. W. Heinz and M. Jacob,Evidence for a new state of matter: An Assessment of the results from the CERN lead beam program, nucl-th/0002042.
[3] STAR Collaboration, Phys. Rev. Lett. 86: 402-407, 2001
[4] PHENIX Collaboration, Phys. Rev. Lett. 88: 022301, 2002
[5] CMS Collaboration, Phys. Lett. B (718) 795-814, 2013
[6] ALICE Collaboration, PRL 105, 252302 (2015)
[7] ALICE Collaboration, Nature Physics 13, 535539 (2017)
[8] D. Blaschke, J. Aichelin, E. Bratkovskaya, V. Friese, M. Gazdzicki, J. Randrup, O. Rogachevsky, O. Teryaev and Viacheslav Toneev, Eur. Phys. J. A (2016) 52: 267
[9] V. Golovatyuk, V. Kekelidze, V. Kolesnikov, O. Rogachevsky and A. Sorin, Eur. Phys. J. A (2016) 52:212
[10] http://nica.jinr.ru/projects/collider.php
[11] http://nica.jinr.ru/projects/mpd.php
[12] ALICE Collaboration, 2013 JINST 8 P10016
[13] E. Cuautle, M. Rodríguez Cahuantzi, et al, Proceedings of the 30th ICRC, 2008, Vol. 5 (HE part 2), pages 12011204
[14] B. Alessandro, M. Rodríguez Cahuantzi, et al, Nuclear Instruments and Methods in Physics Research A 617 (2010) 5761
[15] http://mpd.jinr.ru/

Search for Dark Matter in the beam-dump of a proton beam with MiniBooNE

Alexis A. Aguilar-Arevalo for the MiniBooNE-DM Collaboration

Instituto de Ciencias Nucleares, Universidad Nacional Autónoma de México, CDMX, 04510, México

E-mail: alexis@nucleares.unam.mx

Abstract. A search for the production of sub-GeV dark matter via vector boson mediators was carried out using 8 GeV protons from the Fermilab Booster in a dedicated run where 1.86×10^{20} protons were delivered to the steel beam dump of the Booster Neutrino Beam (BNB). The MiniBooNE detector, 490 m downstream of the beam dump, is sensitive to the elastic scattering of dark matter particles off nucleons in the detector mineral oil, and neutrinos are an irreducible background. Here, a description of the detection principle and the analysis method of this novel and powerful technique to search for dark matter is given, and the results of a search in the context of a vector portal model of dark matter, as well as future prospects, are discussed.

1. Introduction

The strong evidence for dark matter (DM) from observations in a wide range of gravitational phenomena [1], and the lack of positive results from direct searches in the mass range from a few GeV to 10s of TeV [2] have motivated considering extensions of the Standard Model (SM) that include *Dark Sectors* (new fields with no charges under the symmetries of the SM) as an explanation for this major component of the Universe. Under appropriate assumptions, dark sector scenarios are able to provide DM candidates with sub-GeV masses. Models that connect the dark sector to the SM and are constrained by the SM symmetries are called *portals*, of which only three renormalizable versions exist: (*i*) the vector portal (the mediator is a vector), (*ii*) the Higgs portal (the mediator is a scalar), and (*iii*) the neutrino portal (the mediator is a fermion). Vector portal models are the most viable for thermal sub-GeV DM. Despite its sound theoretical motivation, sub-GeV DM has been significantly less explored experimentally, and is accessible, among others, with accelerator beam dump experiments [3].

The MiniBooNE (MB) experiment at Fermilab completed its original physics program on neutrino oscillation and cross section measurements in 2012. A special run was carried out in 2013-2014 where neutrino production was suppressed, enhancing the sensitivity to sub-GeV DM. The recently published results of this search [4] demonstrate the unique and powerful ability to search for DM with a neutrino experiment.

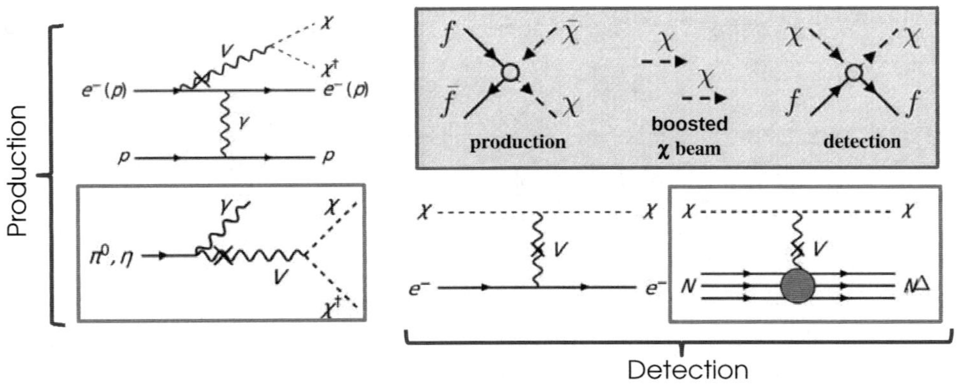

Figure 1. Production and detection of dark matter in a proton beam dump experiment. Production is dominated by proton bremsstrahlung and decay of secondary π^0's and η's; detection is expected through scattering off nucleons or electrons. The reactions in red boxes are the most relevant for this search.

2. The minimal kinetically mixed dark photon (MKMDP) model

A minimal extension to the SM featuring a single $U(1)_D$ (D for *dark*) gauge boson, dubbed the *dark photon*, and a dark matter candidate χ, has the effective Lagrangian [3]

$$\mathcal{L}_{V,\chi} = |D_\mu \chi|^2 - m_\chi^2 |\chi^2| - \frac{1}{4} V_{\mu\nu}^2 + \frac{1}{2} m_V^2 V_\mu^2 + \epsilon V_{\mu\nu} F_{\mu\nu} + \dots \ ,$$

$$D_\mu = \partial_\mu - i g_D V_\mu, \quad g_D = \sqrt{4\pi\alpha_D} \ ,$$

where $F^{\mu\nu}$ is the electromagnetic field-strength tensor of the SM, and $V_{\mu\nu}$ is the corresponding object associated with the dark photon field V_μ. Only four parameters control the physics: the DM particle mass m_χ, the mediator particle mass m_V, the kinetic mixing parameter ϵ, and the dark photon interaction coupling strength α_D. A $U(1)_D$ mediator can increase the DM annihilation cross section to give the observed relic density. Furthermore, a dark photon with a mass of $\mathcal{O} \sim 10 - 10^3$ MeV could explain the $(g-2)_\mu$ anomaly [5].

3. Dark matter beam and detection

The DM production can be achieved in high-energy collisions of SM particles (*eg.* protons), and the detection by elastic scattering of the DM particles in a sensitive detector. Figure 1 shows possible mechanisms. In the MKMDP model the event rate scales as $\epsilon^4 \alpha_D$, when $m_V > 2m_\chi$ (*invisible decay* of the V) which will be considered in what follows.

Neutrino interactions in the detector constitute an irreducible background to the DM search, hence neutrino production must be suppressed. Unlike a standard neutrino production target, a beam dump is a large piece of high-Z material where charged mesons will be either absorbed or stopped before decay, reducing the neutrino flux significantly.

4. The BNB and the MiniBooNE Detector

In its standard $\nu-$mode configuration, the BNB delivers 8 GeV protons to a 1.75 interaction length Be target in pulses with (3-5)$\times 10^{12}$ protons and 1.6 μs in duration, creating a large flux of charged mesons (predominantly π's). A focusing horn produces a pulsed toroidal magnetic field of 1.5 T, which focuses the charged mesons along a 1 m-diameter, 50 m-long decay pipe that terminates in a steel beam dump. This creates a large neutrino flux at the detector location, 541 m from the target. The polarity of the horn can be switched in order to create a beam of neutrinos or antineutrinos.

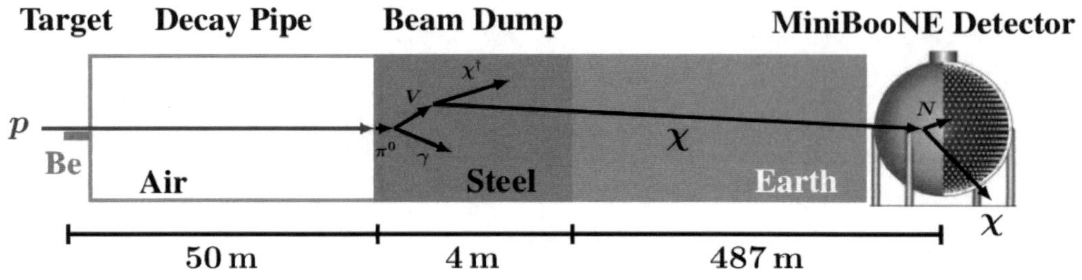

Figure 2. Schematic of the experimental setup in off-target mode. Taken from [4].

The MB detector is a 12.2 m-diameter spherical tank filled with 818 tons of mineral oil (CH_2). An inner wall at 5.5 m radius is lined with 1280 photomultiplier tubes (PMTs) that view the main tank volume. An optically separate 35 cm-thick outer shell viewed by 240 PMTs arranged in pairs comprises the veto region. PMT pulses with signal > 0.1 photoelectron are digitized in a 19.2 μs time window around the 1.6 μs proton pulses of the BNB. MB is primarily a Cherenkov detector, but is sensitive to sub-Cherenkov particles via a small amount of scintillation light emitted as these traverse the mineral oil. The detector is very well understood; it ran for more than 10 years in neutrino ($\nu-$) and antineutrino ($\bar{\nu}-$) modes producing 27 publications on neutrino oscillations, cross section measurements and other physics topics. Most relevant for this search are the measurements of the Charged Current Quasi-Elastic (CCQE) [6] and Neutral Current Elastic (NCE) [7] scattering, on which the DM search analysis relies.

5. Dark Matter search in beam dump mode

For this DM search, the beam line was configured in "off-target" mode, with the protons steered off the Be production target, keeping the magnetic horn powered off, and going directly into the steel beam dump, see Figure 2. This reduces the neutrino flux at the detector, 490 m from the beam dump, by a factor of ~ 30 compared to $\nu-$mode, while the event rate for CCQE events is ~ 50 times smaller. A total of 1.86×10^{20} POT (protons on target) were collected over a running period of 9 months between November 2013 and September 2014.

DM candidate events in the off-target data were selected as in the MB $\bar{\nu}-$NCE analysis [7], adding a reconstructed nucleon kinetic energy cut of $35 < T_n < 600$ MeV. This requires one time-cluster of hits in coincidence with the beam window having a time and spatial distribution consistent with a single nucleon and no pions. The selection criteria applied to the off-target data yields 1465 ± 38 events. In order to constrain the off-target neutrino flux, which gives rise to an irreducible background of neutrino-induced NCE events, a sample of CCQE events was selected from the off-target data using the cuts developed for the CCQE cross-section measurement [6]. In addition, and because NCE and CCQE cross sections are not known independently of MB data, two other large samples with the same NCE and CCQE cuts as for the off-target data were selected from previously collected $\nu-$mode data. The three "constraint" samples provide an improved estimate of beam-related backgrounds.

Prior to any constraints, a total background of 1579 ± 529 events (34% uncertainty) was expected. This is composed of 697 events from beam-unrelated backgrounds (measured with an out-of-beam random trigger), 775 events from detector beam-related backgrounds (dominated by NCE neutrino interactions in the detector), and 107 events from beam-related "dirt" backgrounds (neutrino-induced neutrons created outside the detector entering the main volume). The error estimates considered correlations between the different energy bins and samples. A combined background-only fit to the three constraint samples and the off-target sample yields

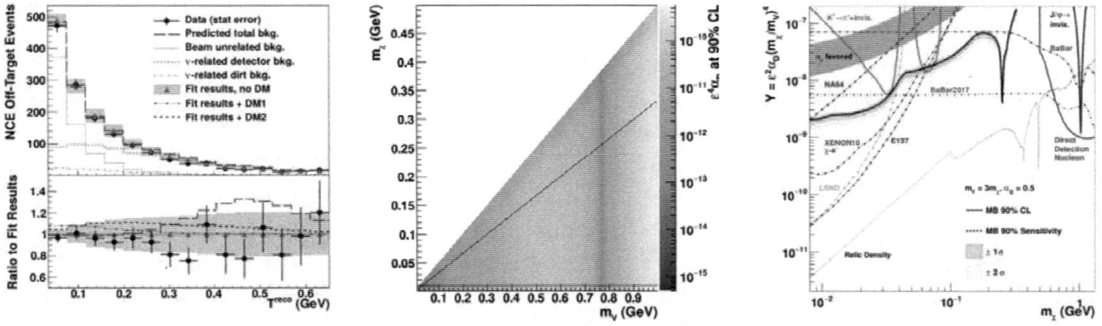

Figure 3. Left: Reconstructed nucleon kinetic energy distribution of DM candidates. Center: The 90% C.L. limit on $\epsilon^2\alpha_D$ for each point in the m_χ *vs* m_V plane obtained with this search. Right: The limit on the annihilation parameter $Y = \epsilon^2\alpha_D(m_\chi/m_V)^4$ that results from assuming the relation $m_V = 3m_\chi$ and $\alpha_D = 0.5$, compared to other results. The forthcoming results from BABAR [9] are expected to be stronger, especially for smaller values of α_D.

a constrained background prediction of 1548 ± 198 events (13% uncertainty).

The BdNMC DM simulation [8] was used to generate the energy distribution of the nucleons in χN scattering events for different values of $\epsilon^4\alpha_D$, m_V, and m_χ, which was passed on to the MB detector simulation. Corrections for bound nucleons were included using an effective efficiency calculated from the MB simulation.

6. Results

The signal prediction was included in a combined fit to the Q^2 distributions of the four mentioned samples, and subjected to a frequentist method to determine the 90% C.L. limit on $\epsilon^4\alpha_D$ for given m_V and m_χ. The results are shown in Figure 3. Fixing other DM parameters allows comparisons of experiments employing different methods. In the right-hand-side plot of the figure values of the annihilation cross-section parameter $Y = \epsilon^2\alpha_D(m_\chi/m_V)^4$ with $m_V = 3m_\chi$ and $\alpha_D = 0.5$ are plotted for this search and compared to other experimental results. Within the vector portal DM model and the chosen parameter constraints, the MB result excludes a vector mediator particle solution to the $(g-2)_\mu$ anomaly and sets stringent limits on DM with sub-GeV masses two orders of magnitude lower than attainable in nucleon direct detection experiments.

7. Lessons and future perspectives

Several lessons learned from this search are already proving useful for the planning of future measurements. Access to a decade-worth of data from a well characterized detector allowed for a good understanding of the backgrounds affecting the DM search. Having a large detector with a veto to tag incoming particles, reduced the events originating in the surrounding dirt. Being able to accurately measure the beam-unrelated backgrounds was crucial. Modeling the nuclear effects on the DM interaction with nucleons is an important characteristic that cannot be avoided in a realistic estimate of the experiment's sensitivity. Finally, the use of several constraint samples which may be correlated helps to reduce systematic uncertainties to a manageable level.

The collaboration is currently developing two analyses that will increase the sensitivity to vector portal DM with the off-target data, using new interaction channels: the "e-DM" analysis, where the DM particles elastically scatter off electrons, and the "*inelastic* π^0" analysis, where it will search for resonant production of π^0's in DM-nucleon interactions. An additional improvement will be the use of the timing structure of the Booster proton beam to look for characteristic intra-bunch (out of time) events in the detector associated subluminal massive DM particles, which is expected to increase the sensitivity to DM masses > 70 MeV.

Simulations have shown that replacing the BNB target assembly with a dedicated Fe or W beam dump, hence removing 50 m of air along the protons flight path, would reduce the neutrino flux by more than one order of magnitude compared to off-target mode. This would represent an non-trivial upgrade to the BNB and could entail an enhancement of the goals of the Fermilab Short Baseline Neutrino (SBN) Program to include a sub-GeV DM search. In this scenario, the SBN near detector (SBND), at 110 m from such beam dump, could achieve an order of magnitude better sensitivity in the e-DM and *inelastic* π^0 channels, with a short run collecting 2×10^{20} POT. An SBND with an improved beam dump would be most sensitive to leptophobic models, where the DM does not couple to electrons. These ideas have been put forward in presentations to the community [10, 11] and a recent LOI submitted to the Fermilab PAC.

8. Conclusions

The successful MB beam dump run is an example of the kind of low-cost sub-GeV DM searches that can be done with proton accelerators by re-tooling existing neutrino experiments. The MB result clearly establishes that such searches are achievable and lessons learned can be used to improve future measurements. A highly motivated and timely search for sub-GeV DM can be performed at Fermilab over a short 6-12 months run collecting data with the SBN near detector. This will require a modest investment for an improved beam dump, with an increased return in science, leveraging on the existing current neutrino program of the laboratory.

Acknowledgements

The collaboration thanks Fermilab Accelerator Divison for reconfiguring, operating, and understanding the off-target beam. We acknowledge Fermilab, the Department of Energy, the National Science Foundation, and Los Alamos National Laboratory for supporting the experiment. Fermilab is operated by Fermi Research Alliance, LLC under Contract No. DE-AC02-07CH11359 with the United States Department of Energy, The author acknowledges the support of DGAPA-UNAM, PAPIIT grant No. IN108917, and of CONACYT.

References

[1] C. Patrignani *et al.* (Particle Data Group), "The Review of Particle Physics", Chin. Phys. C **40**, 100001 (2016)
[2] J. Liu, X. Chen, X. Ji, "Current status of direct dark matter detection experiments", Nat. Phys. 13, 3 (2017)
[3] B. Batell, M. Pospelov, A. Ritz, "Exploring Portals to a Hidden Sector Through Fixed Targets", Phys. Rev. **D80**, 095024 (2009). P. deNiverville, D. McKeen, A. Ritz, "Signatures of sub-GeV dark matter beams at neutrino experiments", Phys. Rev. D **86**, 035022 (2012).
[4] A. Aguilar-Arevalo *et al.* [MiniBooNE-DM collaboration] "Dark Matter Search in a Proton Beam Dump with MiniBooNE". Phys. Rev. Lett. **118**, 22, 221803 (2017)
[5] P. Fayet, "U-boson production in e+ e- annihilations, psi and Upsilon decays, and Light Dark Matter", Phys. Rev. **D75**, 115017 (2007). M. Pospelov, "Secluded U(1) below the weak scale", Phys. Rev. D **80**, 095002 (2009).
[6] A. Aguilar-Arevalo *et al.* [MiniBooNE collaboration] "First Measurement of the Muon Neutrino Charged Current Quasielastic Double Differential Cross Section ", Phys. Rev. D **81**, 092005 (2010)
[7] A. Aguilar-Arevalo *et al.* [MiniBooNE collaboration] "Measurement of the Antineutrino Neutral-Current Elastic Differential Cross Section". Phys. Rev. D **91**, 012004 (2015)
[8] P. deNiverville, C.-Y. Chen, M. Pospelov, and A. Ritz, "Light dark matter in neutrino beams: production modelling and scattering signatures at MiniBooNE, T2K and SHiP", Phys. Rev. D **95**, 035006 (2017)
[9] J. P. Lees *et al.* [BABAr collaboration], "Search for invisible decays of a dark photon produced in e^+e collisions at BABAR", arXiv:1702.03327 [hep-ex] (2017)
[10] J. Alexander *et al.*, "Dark Sectors 2016 Workshop: Community Report", arXiv:1608.08632 [hep-ph] (2016)
[11] "Future Searches with Proton Fixed Target Experiments", talk by R. Van de Water at the workshop "U.S. Cosmic Visions: New Ideas in Dark Matter", Maryland, 23-25 March, 2017.

IOP Conf. Series: Journal of Physics: Conf. Series **912** (2017) 012018 doi:10.1088/1742-6596/912/1/012018

The detection of gravitational waves

Juan Carlos Degollado
Instituto de Ciencias Físicas, Universidad Nacional Autónoma de México, Apartado Postal 48-3, 62251, Cuernavaca, Morelos, México.
E-mail: `jcdegollado@ciencias.unam.mx`

Abstract. On February 11 2016 the scientific community was surprised by the announcement of the first observation of gravitational waves made by the LIGO and Virgo collaborations. The signal was named GW150914 indicating that the astrophysical event that produced it occurred the 14 of September of 2015. The detected waveform is consistent with the predictions of general relativity for the spiral and merger of a binary black hole system and subsequent relaxation period known as *ring down* of the resulting black hole. That detection was the first direct evidence of the existence of gravitational waves and the first observation of the merger of a binary black hole system [1].

1. Introduction

The first direct detection of a gravitational wave signal made by LIGO (Laser Interferometer Gravitational-Wave Observatory) represent a successful confirmation of a prediction by Einstein's theory of general relativity made more than one hundred years ago. The signal was detected by the two LIGO detectors located in Hanford, Washington and Livingston, Louisiana. The gravitational waves were produced by the collision of two black holes of approximately 36 and 29 solar masses [1, 2]. They were orbiting each other approximately 350km apart and subsequently merged to form another black hole. In order to extract information about the source that produced those gravitational waves many analytical and computational methods were required. Numerical Relativity templates and advanced data analysis algorithms were developed to pursuit that end [3].

In this respect, GW150914 allows to test General Relativity in the strong field. The signal was clear enough to allow us to perform some consistency tests with the predictions of General Relativity. The main results came from comparing the observed signal's amplitude and phase with numerical relativity predictions. These comparisons allowed to estimate the parameters of the binary, such as the mass of the individual black holes, their distance from the Earth and the spin of the final black hole [4].

Here, we present some of the basis of gravitational wave theory that will help us to understand the basic principles of one of the most important discoveries in physics.

2. Gravitational Waves. Foundations

Gravitational waves are produced by accelerating masses and they propagate at the speed of light. Detecting them has been very hard because they have almost no interaction with matter.

[1] Since then, there have been two more detections named GW151226 and GW170104 which are also consistent with the merger of two black holes [5, 6].

Even the strongest gravitational waves from extreme astrophysical events such as the merger of two black holes or neutron stars are only expected to produce relative length variations of order of 10^{-21}. As they travel, gravitational waves squash and stretch space-time in the plane perpendicular to their direction of propagation, hence are transversal waves.

General Relativity is a nonlinear theory. Therefore in general there is no clear distinction between waves and the rest of the metric. One can, however, use the notion of a wave in certain limiting situations: in a linearized theory, as small perturbations of a smooth background metric or in post-Newtonian theory. Perhaps the more straightforward way to introduce gravitational waves is through linearized theory around the space-time of Minkowski.

If the gravitational field strength is weak, then it is possible to find a Cartesian-like coordinate system such that the components of the metric tensor $g_{\mu\nu}$, will be small perturbations $h_{\mu\nu}$, of the flat Minkowski metric components $\eta_{\mu\nu}$, thus:

$$g_{\mu\nu} = \eta_{\mu\nu} + h_{\mu\nu} \;, \qquad |h_{\mu\nu}| << 1. \tag{1}$$

We will assume that $h_{\mu\nu}$ is a second rank tensor in the background metric and we will raise and lower the indices with $\eta^{\mu\nu}$ and $\eta_{\mu\nu}$. With this, we can linearize Einstein's field equations with respect to $h_{\mu\nu}$.

First, the linearized the Ricci tensor is

$$R_{\mu\nu} = \frac{1}{2}\left(\partial_\alpha\partial_\mu h^\alpha{}_\nu + \partial_\alpha\partial_\nu h^\alpha{}_\mu - \partial_\mu\partial_\nu h^\alpha{}_\alpha - \Box h_{\mu\nu}\right) + O(h^2) \;, \tag{2}$$

where the d'Alambertian operator in the flat Minkowski space-time is $\Box := \eta^{\mu\nu}\partial_\mu\partial_\nu$, the linearized Ricci scalar becomes

$$R = \eta^{\mu\nu}R_{\mu\nu} + O(h^2) = \partial_\mu\partial_\nu h^{\mu\nu} - \Box h_{\mu\nu} + O(h^2) \;. \tag{3}$$

The Einstein tensor, $G_{\mu\nu} := R_{\mu\nu} - \frac{1}{2}Rg_{\mu\nu}$ is written as

$$G_{\mu\nu} = \frac{1}{2}\left(\partial_\alpha\partial_\mu h^\alpha{}_\nu + \partial_\alpha\partial_\nu h^\alpha{}_\mu - \partial_\mu\partial_\nu h^\alpha{}_\alpha - \Box h_{\mu\nu} + \eta_{\mu\nu}(\Box h - \partial_\alpha\partial_\beta h^{\alpha\beta})\right) + O(h^2) \;. \tag{4}$$

It is possible to simplify the linearized Einstein tensor by introducing the trace reverse metric perturbation as

$$\bar{h}_{\mu\nu} := h_{\mu\nu} - \frac{1}{2}\eta_{\mu\nu}h \;, \tag{5}$$

and get

$$G_{\mu\nu} = \frac{1}{2}\left(\partial_\alpha\partial_\mu \bar{h}^\alpha{}_\nu + \partial_\alpha\partial_\nu \bar{h}^\alpha{}_\mu - \Box \bar{h}_{\mu\nu} - \eta_{\mu\nu}\partial_\alpha\partial_\beta \bar{h}^{\alpha\beta}\right) + O(h^2) \;. \tag{6}$$

Using a infinitesimal coordinate transformation –known as gauge transformation– of the form

$$x'^\alpha = x^\alpha + \xi^\alpha(x^\beta) \tag{7}$$

one can choose coordinates for which the following additional (harmonic gauge) conditions are fulfilled:

$$\partial_\beta \bar{h}^{\beta\alpha} = 0 \;. \tag{8}$$

If these conditions are satisfied, the linearized Einstein tensor simplifies considerably:

$$G_{\mu\nu} = -\frac{1}{2}\Box \bar{h}_{\mu\nu} + O(h^2) \;. \tag{9}$$

IOP Conf. Series: Journal of Physics: Conf. Series **912** (2017) 012018 doi:10.1088/1742-6596/912/1/012018

The linearized Einstein's field equations are:

$$\Box\bar{h}_{\mu\nu} = -\frac{16\pi G}{c^4}T_{\mu\nu} \ . \tag{10}$$

As a result, time-dependent solutions of these equations can be interpreted as weak gravitational waves. The simplest solution of Eqs. (10) in vacuum ($T_{\mu\nu} = 0$) is a monochromatic plane wave of the form

$$\bar{h}_{\mu\nu} = A_{\mu\nu}\cos(k_\alpha x^\alpha) \ . \tag{11}$$

$A_{\mu\nu}$ is the constant amplitude and we have chosen the initial phase of the wave as zero for simplicity. It is straightforward to show that (11) are solutions of (10) only if

$$\eta^{\alpha\beta}k_\alpha k_\beta = 0 \ , \tag{12}$$

i.e. k^α are the components of a null four vector. Consequently gravitational waves travel at the speed of light.

Einstein's field equations take the simple form (10) only if the harmonic gauge conditions (8) are satisfied. This leads to the requirement that the plane wave solution $\bar{h}_{\mu\nu}k^\nu$ is orthogonal to k^ν:

$$\bar{h}_{\mu\nu}k^\nu = 0 \ . \tag{13}$$

Yielding to the conclusion that gravitational waves are transversal. If we additionally introduce the angular frequency as $\omega = ck^0$. The frequency f of the wave, measured in hertz, is related with ω by $\omega = 2\pi f$. The 3-spatial components of k^μ form the wave vector, which points to the direction in which the wave propagates. Its Euclidean length is related to the wavelength λ as $\lambda = 2\pi/|k^i|$. The actual degrees of freedom of the wave field $h_{\mu\nu}$, are best represented choosing a suitable coordinate system as follows.

At each event in the space-time region covered by some almost Lorentzian and harmonic coordinates x^α let us choose time like unit vector u^μ and consider a gauge transformation generated by the functions ξ^α of the form:

$$\xi^\alpha = B^\alpha\cos(k_\beta x^\beta). \tag{14}$$

It is possible to choose the quantities B^α in such a way that in the new harmonic coordinates $x'^\alpha = x^\alpha + \xi^\alpha$ the following conditions are fulfilled [8]:

$$\bar{h}'_{\mu\nu}u'^\nu \ , \qquad \bar{h}'^{\ \mu}_{\mu} = 0 \ . \tag{15}$$

The transformations (14) preserve the conditions

$$\bar{h}'_{\mu\nu}k'^\nu = 0 \ , \tag{16}$$

and

$$\bar{h}'_{\mu\nu} = \bar{h}_{\mu\nu} \ . \tag{17}$$

Equations (15) define the *transverse and traceless* (TT) coordinate system related to the 4-vector u^μ. These equations comprise eight independent constraints on the components of the plane-wave solution $\bar{h}'_{\mu\nu}$: any plane monochromatic gravitational wave possesses two independent degrees of freedom, often also called waves polarizations. The simplest way to describe these two polarizations is by choosing u^μ of the form $u^\mu = (1, 0, 0, 0)$. Then the wave field satisfies

$$\bar{h}_{\mu 0} = 0 \ . \tag{18}$$

From now on, we drop the tilde in $h_{\mu\nu}$. If we assume that the gravitational wave propagates in the $+z$ direction. Then $k^\mu = (\omega/c, 0, 0, \omega/c)$ and

$$\bar{h}_{\mu 3} = 0 \ . \tag{19}$$

The traceless condition reduces to

$$\bar{h}_{11} + \bar{h}_{22} = 0 \ . \tag{20}$$

It is common to denote the independent components of the wave field as

$$h_+ = \bar{h}_{11} = -\bar{h}_{22} \ , \qquad h_\times = \bar{h}_{12} = \bar{h}_{21} \ . \tag{21}$$

The functions h_+ and $h_\times$ are called *plus* and *cross* polarizations of the wave. With this, the waveform in the TT gauge has the following matrix form:

$$h^{TT}_{\mu\nu}(t, \vec{x}) = \begin{pmatrix} 0 & 0 & 0 & 0 \\ 0 & h_+(t - z/c) & h_\times(t - z/c) & 0 \\ 0 & h_\times(t - z/c) & -h_+(t - z/c) & 0 \\ 0 & 0 & 0 & 0 \end{pmatrix} ,$$

where, for plane waves propagating in the $+z$ direction,

$$h_+ = A_+ \cos\left(\omega(t - \frac{z}{c})\right) \ , \qquad h_\times = A_\times \cos\left(\omega(t - \frac{z}{c})\right) \ . \tag{22}$$

where A_+ and $A_\times$ are constant amplitudes.

3. Response of test particles to gravitational waves

Gravitational waves cannot be detected locally due to their asymptotic character. According to the principle of equivalence the effects of gravity may be transformed away and the effect of a gravitational wave on a single point test particle has no measurable consequences. The coordinates of the particle remain unchanged when a gravitational wave passes [9]. Gravitational waves can be detected only by their influence on two or more test particle at different locations. Interferometer gravitational wave detectors are devices built to measure induced length changes [10]. The properties of the waves such as frequency and the amplitude are related to the motion of the test masses involved.

Gravitational waves represent propagating curvature perturbation and their effect on test particles is of tidal nature. In order to see the previous statement in detail, let us consider two nearby geodesics both parametrized by the same parameter λ, let ξ^μ be a vector connecting points on the two curves with the same value of λ. The connecting vector is said to measure the geodesic deviation of the curves. The equation that tell us how the geodesic deviation varies along the curves is given by

$$\frac{d^2\xi^\mu}{d\lambda^2} + R^\mu_{\alpha\nu\beta} u^\alpha \xi^\nu u^\beta = 0 \ , \tag{23}$$

where u^μ is the vector tangent to the geodesics [11]. The physical effect of gravitational waves is best explain with this equation, that, in the linearized limit for the comoving geodesic normal coordinates becomes:

$$\frac{d^2\xi^k}{dt^2} = -R^k_{0j0}\xi^j \ . \tag{24}$$

Substituting the components of the Riemann tensor in favor of the wave field in the TT gauge one gets:

$$\xi^k{}_{,00} = \frac{1}{2}\xi^k h^{\rm TT}_{ij,00} \ . \tag{25}$$

Inserting the components of h_{ij} and considering a wave propagating in the $+z$ direction one gets a set of equations of the form:

$$\xi^x_{,tt} = \frac{1}{2}\xi^x h_{xx,tt} + \frac{1}{2}\xi^y h_{xy,tt} \,, \tag{26}$$

$$\xi^y_{,tt} = \frac{1}{2}\xi^x h_{xy,tt} - \frac{1}{2}\xi^y h_{xy,tt} \,, \tag{27}$$

$$\xi^z_{,tt} = 0 \,. \tag{28}$$

These equations show that only the x and y components of ξ^i of the nearby geodesics will be disturbed by the gravitational wave traveling in the $+z$ direction. Hence test particles are influenced in the directions perpendicular to the wave propagation as previously stated. We can use the above equations to describe the behavior of a ring of free stationary free particles in the $x-y$ plane when a gravitational wave passes in the $+z$ direction. Let us consider two particles separated in the x direction. At lowest order, we can neglect the terms with ξ^y to get:

$$\xi^x_{,tt} = \frac{1}{2}\xi^x h_{xx,tt} \,, \qquad \xi^y_{,tt} = \frac{1}{2}\xi^x h_{xy,tt} \,. \tag{29}$$

Similarly, for two nearby particles separated in the y direction

$$\xi^y_{,tt} = \frac{1}{2}\xi^x h_{xy,tt} \,, \qquad \xi^y_{,tt} = -\frac{1}{2}\xi^x h_{xx,tt} \,. \tag{30}$$

Let us assume that a wave with $h_{xy} = 0$, $h_{xx} \neq 0$ hits the particles. First the particles along the x direction will come towards each other and they move away as h_{xx} changes sign. This is the h_+ polarization. If on the other hand, $h_{xy} \neq 0$, $h_{xx} = 0$ the movement of the ring will be the same but rotated 45 degrees. This is the $h_\times$ polarization.

Gravitational waves carry energy. This energy is released by accelerating masses and is taken away by gravitational waves. In the following we describe how to estimate the amount of energy radiated.

4. Energy and angular momentum carried by gravitational waves

The stress-energy carried by GWs cannot be localized within a wavelength [12]. Instead, one can say that a certain amount of stress energy is contained in a region of the space which extends over several wavelengths.

The energy-momentum tensor for gravitational waves $t_{\mu\nu}$, is obtained by averaging the squared gradient of the wave field over several wavelengths. In the TT gauge it has components:

$$t_{\mu\nu} = \frac{c^4}{32\pi G}\langle \partial_\mu h^{\rm TT}_{ij} \partial_\nu h^{ij}_{\rm TT}\rangle \,. \tag{31}$$

If one assumes that $h^{\rm TT}_{0\mu} = 0$ then

$$t_{\mu\nu} = \frac{c^4}{32\pi G}\sum_{i,j=1}^{3}\langle \partial_\mu h^{\rm TT}_{ij} \partial_\nu h^{\rm TT}_{ij}\rangle \,. \tag{32}$$

For a plane wave propagating in the z direction, the nonzero components of the tensor $t_{\mu\nu}$ are simply

$$t_{00} = -t_{0z} = -t_{z0} = t_{zz} = \frac{c^2}{16\pi G}\langle (\partial_t h_+)^2 + (\partial_t h_\times)^2\rangle \,. \tag{33}$$

A very instructive result arises if we consider the monochromatic wave (11) and substitute in the previous equation

$$t_{00} = \frac{c^2\omega^2}{16\pi G}\left(A_+^2\langle(\sin^2(\omega(t-z/c))\rangle + A_\times^2\langle(\sin^2(\omega(t-z/c))\rangle\right) . \tag{34}$$

The average of the sine squared over one period gives 1/2 and simplifying one obtains, in terms of the frequency measured in hertz:

$$t_{00} = \frac{\pi c^2 f^2}{8G}\left(A_+^2 + A_\times^2\right) . \tag{35}$$

This expression is very useful to give some orders of magnitude. For example, the energy flux in gravitational waves from the collapse of a supernova core to form a 10 solar masses black hole at a distance of 15 Mpc, will have frequencies about 10^3 Hz and amplitudes of the order of $A_+ \sim A_\times \sim 10^{-22}$. The computed flux is about 3 ergs/cm^2 sec. This is about ten orders of magnitude larger than the observed electromagnetic energy flux [8].

5. The quadrupole formalism

Perhaps the simplest technique for computing the gravitational wave field in the presence of matter is the quadrupole formalism. Einstein himself derived the quadrupole formula for gravitational radiation by solving the linearized equations (10). See for instance [13].

This formalism is especially important because it is very accurate for many astrophysical sources of gravitational waves and is easy to calculate. It does not require for high accuracy in the computation on the strength of the source's internal gravity, but requires that internal motions inside the source are slow compared to the speed of light. This requirement implies that the wavelength of the gravitational waves emitted is much larger than the source's characteristic size, $\lambda >> L_s$.

The lowest allowed multipole for gravitational radiation is the quadrupole. The monopole is forbidden as a result of mass conservation. In a similar way, dipole radiation is absent as a result of momentum conservation.

In order to gain some insight about the quadrupole formalism, let us introduce some coordinate system $(t, \vec{x'})$ centered on the gravitational wave source. If we assume that the source has weak internal gravity and small internal stresses, Newtonian gravity is a good approximation to general relativity inside and near the source. Let us assume that an observer at rest (with respect to the source) measures the gravitational wave field $h^{\mathrm{TT}}_{\mu\nu}$.

The quadrupole formalism allow us to get an expression for the gravitational wave field:

$$h_{ij}(t, x'^i) = \frac{2}{R}\frac{G}{c^4}\ddot{Q}^{\mathrm{TT}}_{ij}\left(t - \frac{R}{c}\right) , \tag{36}$$

where R is the euclidean distance from the point x'^i to the center of the source and t is the proper time measured by the observer. The quantity Q_{ij} is the symmetric part of the second moment of the sources mass density ρ computed in a Cartesian coordinate system centered on the source evaluated at the retarded time $t_{ret} = t - \frac{R}{c}$

$$Q^{\mathrm{TT}}_{ij} = \int d^3x\rho\left(x_i x_j - \frac{1}{3}\delta_{ij}r^2\right) , \tag{37}$$

the superscript TT means that one needs to keep only the part that is transverse to the direction of propagation of the wave and is traceless. Furthermore Q_{ij}, is the coefficient of the $1/r^5$ term in the multipolar expansion of the source's Newtonian gravitational potential Φ [14].

$$\Phi = -\frac{GM}{r} - \frac{3G}{2}\frac{Q_{ij}x^i x^j}{r^5} - \frac{5G}{2}\frac{Q_{ijk}x^i x^j x^k}{r^7} + \ldots . \tag{38}$$

From the quadrupole formula (37) and Eq. (32) for the energy-momentum tensor, one can compute the fluxes of energy and angular momentum carried by the wave.

The rates of emission of energy $\mathscr{F}_E$ and angular momentum $\mathscr{F}_{J_i}$ in the quadrupole approximation are [8].

$$\mathscr{F}_E = \frac{G}{5c^5} \sum_{i,j=1}^{3} \langle(\dddot{Q}_{ij})^2\rangle \ , \tag{39}$$

and

$$\mathscr{F}_{J_i} = \frac{2G}{5c^5} \sum_{j,k,l=1}^{3} \epsilon_{ijk} \langle \ddot{Q}_{jl} \dddot{Q}_{kl}\rangle \ . \tag{40}$$

Because of the conservation of energy and angular momentum the source's energy and angular momentum should decrease at rates equal to minus rates given by Eqs. (39) and (40), thus one arrives to the energy and angular momentum loss equations:

$$\frac{dE^{\rm source}}{dt} = -\frac{G}{5c^5} \sum_{i,j=1}^{3} \langle(\dddot{Q}_{ij})^2\rangle \ , \tag{41}$$

and

$$\frac{dJ_i^{\rm source}}{dt} = -\frac{2G}{5c^5} \sum_{j,k,l=1}^{3} \epsilon_{ijk} \langle \ddot{Q}_{jl} \dddot{Q}_{kl}\rangle \ . \tag{42}$$

The estimation of the energy and angular momentum radiated using (41) and (42) is very accurate for all sources as long as the wavelength is much longer than the source size L_s. In the following section we give an example of how to compute the energy radiated by a slow moving system of compact objects.

6. The two-body problem

One of the most important applications of the quadrupole formalism as an approximation to general relativity is the two-body problem, i.e. the problem of finding the motion and gravitational radiation of self gravitating relativistic systems of two extended bodies. This calculation is based on a Newtonian description of the dynamics of the two bodies. The Newtonian dynamics will depart from the relativistic description when the relative velocity of the binary is comparable with the speed of light or when the gravitational energy becomes large compared to the rest mass energy of the system.

Relativistic binary systems exist in the universe, examples of this are neutron stars or black holes binaries. These systems emit gravitational waves and are the most prominent sources of gravitational radiation that can be detected by LIGO. Surprisingly one can learn a lot from the collision of two bodies using the quadrupole formalism presented above.

A Newtonian binary system is illustrated in Fig. 1. For simplicity let us assume that the system follow circular orbits and the masses are equal. Thus

$$M_1 = M_2 = M \ , \qquad \text{and} \qquad R_1 = R_2 = R \ . \tag{43}$$

Introducing cylindrical coordinates

$$x(t) = R\cos\Omega t \ , \qquad y(t) = R\cos\Omega t \ , \qquad z(t) = 0 \ . \tag{44}$$

The nonzero components of the quadrupole moment (37) are

$$Q_{xx} = MR^2\left(\frac{1}{3} + \cos 2\Omega t\right) \ , \qquad Q_{yy} = MR^2\left(\frac{1}{3} - \cos 2\Omega t\right) \ , \qquad Q_{zz} = -\frac{2}{3}MR^2 \ . \tag{45}$$

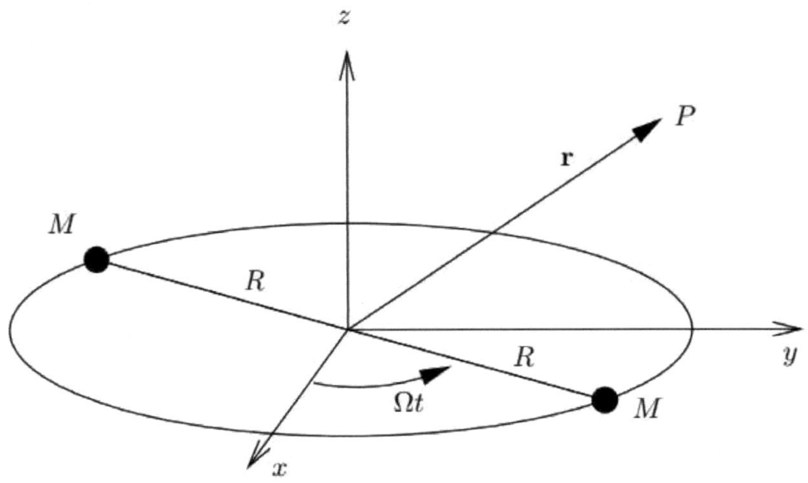

Figure 1. Two bodies with equal masses orbiting around their center of mass in circular orbits.

Taking the third time derivative and replacing into the expression to compute the gravitational wave luminosity as the energy lost by the source $L_{gw} = -\frac{dE^{\text{source}}}{dt}$

$$L_{gw} = \frac{G}{5c^5}\langle \dddot{Q}^{ij}\dddot{Q}_{ij}\rangle \tag{46}$$

$$= \frac{G}{5c^5}\langle(\dddot{Q}_{xx})^2 + 2(\dddot{Q}_{xy})^2 + (\dddot{Q}_{yy})^2 + (\dddot{Q}_{zz})^2\rangle \tag{47}$$

$$= \frac{128G}{5c^5}\Omega^6 M^2 R^4 \,. \tag{48}$$

The frequency Ω and the period T are related by $\Omega = 2\pi/T$. Furthermore, using Kepler's third law: $R^3 = \frac{GMT}{16\pi^2}$ the gravitational wave luminosity can be expressed in terms of T

$$L_{gw} = \frac{128}{5}2^{2/3}\frac{c^5}{G}\left(\frac{\pi GM}{c^3 T}\right)^{10/3} . \tag{49}$$

(50)

This may be an enormous luminosity, For example, the estimated luminosity for the event GW150914 is $L \sim 0.2 \times 10^{-3} L_0$, where the Planck luminosity is $L_0 := c^5/G \sim 10^{59}$ erg/s. By comparison, the luminosity of the sun is $L_\odot = 3.839 \times 10^{33}$ erg/s.

The Newtonian binding energy of the binary is $E = -\frac{1}{2}\frac{GmM}{R}$, and taking the derivative

$$\frac{dE}{dt} = \frac{GmM}{2R^2}\frac{dR}{dt} \,, \tag{51}$$

thus as the gravitating system losses energy by emitting radiation, the distance between the two bodies decreases at a rate

$$\frac{dR}{dt} = -\frac{64}{5}\frac{G^3}{c^5}\frac{M^3}{R^3} \,. \tag{52}$$

The size of the binary orbit decreases and the components move faster leading to emission of gravitational waves with increasing amplitude and frequency. This last stage is known as the *chirp* signal.

The orbital frequency increases accordingly to .

$$\frac{1}{T}\frac{T}{dt} = \frac{3}{2}\frac{1}{R}\frac{dR}{dt} \ , \tag{53}$$

and the system will merge after a time $t_{\rm merger}$

$$t_{\rm merger} = \frac{5}{256}\frac{c^5}{G^3}\frac{R_0^4}{M^5}, \tag{54}$$

where R_0 is the initial separation.

The previous analysis can be generalized to consider elliptic orbits, but it is possible to shown that gravitational wave emission circularize the orbits faster than the coalescence timescale, making the study of circular orbits quite useful [8]. For instance, applying this formalism to the binary Pulsar of Hulse and Taylor [2]: with parameters $R_0 = 5$Kpc, $M_1 \sim M_2 \sim 1.4M_\odot$, $dT/dt = 7$h 45min. one gets:

$$T \sim 2.4 \times 10^{-12}{\rm sec/sec}, \qquad f_{\rm gw} \sim 7 \times 10^{-5}{\rm Hz}. \qquad t_{\rm merger} \sim 3.5 \times 10^{8}{\rm yr}. \tag{55}$$

which is in excellent agreement with the observed values for $dT/dt = (2.30 \pm 0.22) \times 10^{-12}$sec/sec [16].

7. Outlook

Gravitational-wave detection will allow, in a near future, new and more precise measurements of astrophysical events. We may be able to determine the formation mechanism of two merging black holes by measuring accurately their spins. Although LIGO was not able to measure the magnitude of the spins of the colliding black holes very accurately, better measurements might be possible with more sensitive detectors and better data analysis techniques.

The implications of gravitational wave astronomy for astrophysics are impressive. Multiple detections may allow us to determine the rate at which black holes merge and to test the models used to describe the formation of binary systems. Furthermore, electromagnetic counterparts of gravitational waves may be used to have a better understanding of astrophysical phenomena [17].

With the detection of LIGO, we are entering in a new age of gravitational-wave astronomy. With this new window we will be able to detect phenomenon that was not possible to detect using the standard tools. Surprisingly the first detection of LIGO came from the merger of two black holes, the objects we cannot see with electromagnetic radiation and represent one of the most precious jewels of General Relativity. Certainly, we are in the verge of dazzling discoveries.

Acknowledgments

The author thanks the organizers of the XXXI Anual meeting of the Division of Particles and Fields of the Mexican Physical Society for their hospitality. This work is partially supported by the Network Project 280908 Agujeros Negros y Ondas Gravitatorias" of Consejo Nacional de Ciencia y Tecnologia (CONACYT) Mexico.

References

[1] B. P. Abbott *et al.* [LIGO Scientific and Virgo Collaborations], Phys. Rev. Lett. **116** (2016) no.6, 061102.
[2] B. P. Abbott *et al.* [LIGO Scientific and Virgo Collaborations], Phys. Rev. Lett. **116** (2016) no.22, 221101.

[2] The Hulse and Taylor binary neutron star system provided the first indirect astrophysical evidence of the existence of gravitational waves [15].

[3] Baumgarte, T. W., and Shapiro, S. L., "Numerical Relativity: Solving Einstein's Equations on the Computer", Cambridge University Press, (2010). New York, NY, USA.
[4] B. P. Abbott *et al.* [LIGO Scientific and Virgo Collaborations], Phys. Rev. D **93** (2016) no.12, 122003.
[5] B. P. Abbott *et al.* [LIGO Scientific and Virgo Collaborations], Phys. Rev. Lett. **116** (2016) no.24, 241103.
[6] B. P. Abbott *et al.* [LIGO Scientific and VIRGO Collaborations], Phys. Rev. Lett. **118** (2017) no.22, 221101.
[7] B. P. Abbott *et al.* [LIGO Scientific and Virgo Collaborations], "The basic physics of the binary black hole merger GW150914," Annalen Phys. (2016) [Annalen Phys. **529** (2017) 0209].
[8] M. Maggiore, *Gravitational waves. Volume 1: Theory and Experiments*, Oxford, New York (2008).
[9] C.W. Misner, K.S. Thorne and J.A. Wheeler, *Gravitation*, W.H. Freeman and Company, San Francisco, (1973).
[10] M. Pitkin, S. Reid, S. Rowan and J. Hough, "Gravitational Wave Detection by Interferometry (Ground and Space)," Living Rev. Rel. **14**, 5 (2011).
[11] R. M. Wald, *General Relativity*, Chicago Univ. Press (1984).
[12] B. S. Sathyaprakash and B. F. Schutz, "Physics, Astrophysics and Cosmology with Gravitational Waves," Living Rev. Rel. **12**, 2 (2009).
[13] S.L. Shapiro y S.A. Teukolsky *Black Holes, White Dwarfs and Neutron Stars: The Physics of Compact Objects*, John Wiley & Sons, New York, (1983).
[14] James B. Hartle, *Gravity: An Introduction to Einstein's General Relativity, Addison Wesley*, 1st Edition (2003) ISBN 0-8053-8662-9.
[15] J.H. Taylor, *Pulsar timing and relativistic gravity, Class. Quantum. Grav.* **10** (1993) S167.
[16] C.M. Will, "The Confrontation between General Relativity and Experiment," Living. Rev. Rel. **9** (2006).
[17] E. Berti *et al.*, Class. Quant. Grav. **32**, 243001 (2015)

IOP Conf. Series: Journal of Physics: Conf. Series **912** (2017) 012019 doi:10.1088/1742-6596/912/1/012019

New sources of CP violation

A. Moyotl

Ingeniería en Mecatrónica, UPP, Tercer Carril del Ejido "Serrano" s/n San Mateo Cuanalá. Juan C. Bonilla, Puebla, Pue., México.

E-mail: agustin.moyotl@uppuebla.edu.mx

Abstract. We present a short review of CP-violating effects induced by radiative corrections in a framework of extensions of the Standard Model: (EM, Weak, Chromo) electric dipole moments of heavy fermions, trilinear neutral gauge boson couplings and decays of the Higgs boson. We show that in order to induce CP-violating effects, non-diagonal couplings with complex coupling constant are required and the respective CP-odd term is proportional to the imaginary part of the product of coupling constants involved in the process, which is mathematically consistent with the respective CP-odd Lagrangian.

1. Introduction

In the universe observed there exists an evident matter-antimatter asymmetry, which is also known as baryon asymmetry in particle physics. But, the Standard Model (SM) fails to provide an explanation to the baryon asymmetry in the universe. However, it is assumed that the matter and antimatter were created in equal amounts after the Big-Bang. Then, to explain this asymmetry in a way consistent with the CPT theorem, the baryogenesis mechanism was proposed. According to Sakharov's criteria, there are three conditions required for the baryogenesis:

- At least one B-number violating process.
- Charge conjugation (C) and Charge-Parity (CP) violation.
- Interactions outside of thermal equilibrium.

The non-conservation of CP symmetry is thus a necessary requirement for the baryon asymmetry to occur. CP violation could be included in the SM in three different ways; through the strong interactions that is known as the Strong CP problem, if neutrinos are massive and finally by the complex phase of the CKM matrix. Small amounts of CP violation in neutral mesons decays have been measured, but unfortunately, it is not enough to explain the baryon asymmetry, so it is important the study of new sources of CP violation beyond the SM.

Theoretically speaking, we show that in order to induce an electric dipole moment, the CP-odd part of the trilinear neutral gauge boson couplings and Higgs to gauge boson decays, are necessary non-diagonal couplings and complex coupling constants in some SM extensions. Additionally, we show that the respective CP source is proportional to the imaginary part of the product of the coupling constants, which is mathematically consistent with the non-conservation of the respective discrete symmetries.

The paper is organized as follows: in Sec. 2, we define the mathematical properties of the electric dipole moments, and we mention some examples in some SM extensions. In Sec. 3, we show the general mathematical structure of the trilinear neutral gauge boson couplings, and we mention some contributions to the CP-odd part of the $ZZ^*\gamma$ and ZZZ^* couplings. In Sec. 4, we present a study of the CP violation in Higgs boson decays to gauge bosons, we examined their experimental potential in LHC. Finally, we present our conclusions in the Sec. 5.

2. Electric Dipole Moments

In general, the magnetic (MDM) and electric (EDM) dipole moments induced by a neutral gauge boson $V = \gamma, Z, G$ can be described through the following interaction Lagrangian:

$$\begin{aligned}\mathcal{L}^{\text{spin}-1/2} &= -\frac{i}{2}\bar{f}\sigma_{\mu\nu}\gamma_5 f(ed_f F^{\mu\nu} + gd_f^W Z^{\mu\nu} + g_s d_q^G G^{\mu\nu}) \\ &+\frac{1}{4m_f}\bar{f}\sigma_{\mu\nu}f(ea_f F^{\mu\nu} + ga_f^W Z^{\mu\nu} + g_s a_q^G G^{\mu\nu}), \end{aligned} \tag{1}$$

where $F^{\mu\nu}$, $Z^{\mu\nu}$ and $G^{\mu\nu}$ are the strength tensors associated to a neutral gauge boson V, while $g_V = e, g, g_s$ is the respective charge induced by the V boson. The Lorentz structure associated with d^V violate parity and time-reversal symmetry, and due to the CPT theorem is equivalent to CP violation. The MDM and EDM arise at the loop level and can be extracted from the matrix element $ig_V\bar{\text{u}}(p')\Gamma_V^\mu\text{u}(p)$, where Γ_V^μ is given by:

$$\begin{aligned}\Gamma_\mu^V(q^2) &= F_A^V(q^2)(\gamma_\mu\gamma_5 q^2 - 2m_f\gamma_5 q_\mu) + F_1^V(q^2)\gamma_\mu \\ &+i\frac{F_2^V(q^2)}{2m}\sigma_{\mu\nu}q^\nu + \frac{F_3^V(q^2)}{2m}\sigma_{\mu\nu}\gamma_5 q^\nu, \end{aligned} \tag{2}$$

with $q = p' - p$ is the four-momentum of the gauge boson V, m is the mass of the fermion f and $F_k^V(q^2)$ are an one-loop form factors. Thus, the MDM and the EDM are defined as:

$$a_f^V = -F_2^V(q^2 = m_V^2) \tag{3}$$

$$d_f^V = -eF_3^V(q^2 = m_V^2) \tag{4}$$

respectively, where m_V is the rest mass of the neutral gauge boson. The MDMs are induced at one-loop level in the SM, but the EDMs are induced at multi-loop level and therefore are very suppressed in the SM framework. However, in general the MDM and EDM are proportional to the mass of the fermion in question [1]; therefore the respective values for the tau lepton, bottom and top quarks are expected to be greater than the other fermions. However, the lifetimes of tau and top quark are very short to measure their interaction with an electromagnetic field. Then, the current bound have been measured indirectly, this is, through deviations from the SM cross section of decay and production process associated with those fermions [2, 3, 4, 5, 6]. Nonetheless, the electromagnetic EDM of the tau lepton also can be proved using radiative $\tau \to \ell\gamma\nu\bar{\nu}$ decay at high-luminosity B factories [7]. But in general, the tau lepton decays provide opportunities to search effects of new physics [8]. In the next subsections we go to analyze qualitatively some results of EDMs in several SM extensions.

2.1. Electromagnetic electric dipole moment

The electromagnetic case is the well known EDM and the associate gauge boson is the photon; for a charged lepton we have experimental bounds given in the Table 1.

Theoretical prediction in SM for the tau lepton is $d_\tau^{\text{SM}} < 10^{-34}e$ cm [12], but the EDMs for the other charged leptons are more suppressed. However, in some SM extensions is possible to obtain predictions competitive with the experimental bound, and less suppressed than the SM

lepton	d_ℓ
e [9]	$d_e < (1.6 \pm 7.4) \times 10^{-28} e$ cm
μ[10]	$d_\mu < (-0.1 \pm 0.9) \times 10^{-19} e$ cm
τ [11]	$-0.22 < \mathrm{Re}(d_\tau)/10^{-16} e \text{ cm} < 0.45$
	$-0.25 < \mathrm{Im}(d_\tau)/10^{-16} e \text{ cm} < 0.008$

Table 1. Electric dipole moment bounds for charged leptons, at 90% C.L. for the electron and 95% C.L. for muon and tau leptons.

predictions. An example is the called *Unparticle* framework [13], which has been explored by CMS at the LHC[14, 15, 16]. In this model the non-diagonal coupling with fermions is induced at tree level and it is an important ingredient to induce CP violations effects. For the case of spin-1 unparticle $\mathcal{O}^\mu_\mathcal{U}$, the effective lagrangian is given by:

$$\mathcal{L}_{\mathcal{U}_V} = \frac{\lambda^{ij}_V}{\Lambda^{d_\mathcal{U}-1}_\mathcal{U}} \bar{f}_i \gamma_\mu f_i \mathcal{O}^\mu_\mathcal{U} + \frac{\lambda^{ij}_A}{\Lambda^{d_\mathcal{U}-1}_\mathcal{U}} \bar{f}_i \gamma_\mu \gamma_5 f_i \mathcal{O}^\mu_\mathcal{U}, \tag{5}$$

and the respective propagator can be written as:

$$\Delta^{\mu\nu}_F(p^2) = \frac{A_{d_\mathcal{U}}}{2\sin(d_\mathcal{U}\pi)} (-p^2 - i\epsilon)^{d_\mathcal{U}-2} \Big[- g^{\mu\nu} + \frac{p^\mu p^\nu}{p^2} \Big], \tag{6}$$

where $\Lambda_\mathcal{U}$ and $d_\mathcal{U}$ are the scale energy and dimension of the unparticle, respectively, while p is the 4-momentum of the unparticle and $\lambda^{ij}_{A,V}$ is a complex coupling constant. A detailed discussion of the eqs. (5) and (6) can be found in [17]. Then, after resolving the one-loop integral, the authors of the reference [18] have obtained the following results for the EDM:

$$d^\mathcal{U}_i = \frac{eA_{d_\mathcal{U}}}{32\pi^2 \sin(\pi d_\mathcal{U}) m_i} \Big(\frac{m_i^2}{\Lambda^2_\mathcal{U}}\Big)^{d_\mathcal{U}-1} \sum_{j=e,\mu,\tau} \mathrm{Im}(\lambda^{ij}_V \lambda^{*ji}_A) g_{V,A}(\sqrt{r_i}, d_\mathcal{U}), \tag{7}$$

the explicit form of $g_{V,A}(\sqrt{r_i}, d_\mathcal{U})$ is given in [18], where $\sqrt{r_i} = m_j/m_i$. It is very interesting to note that this result is proportional to $m_i^{2(d_\mathcal{U}-1)}/m_i$ and to the imaginary part of the product of the vector and the axial coupling constants, such as we mentioned in the introduction. For selected values of the free parameters, the numerical result for the muon is $d^\mathcal{U}_\mu \simeq (10^{-17} - 10^{-22})\mathrm{Im}(\lambda^{j\mu}_V \lambda^{*\mu j}_A) e$ cm for the range of $1.1 < d_\mathcal{U} < 1.9$. Further, we note that there is not EDM contribution if all the coupling constants are real numbers.

On the other hand, the contribution of spin-0 unparticles to EDM of the tau lepton is $d^\mathcal{U}_\tau \simeq \mathrm{Im}(\lambda^{*P}_{\tau\mu} \lambda^S_{\tau\mu})(0.69 + 3.17i) \times 10^{-20} e$ cm for $\Lambda_\mathcal{U} = 10$ TeV [19]. Other contributions to tau lepton have been reported in the MSSM with an additional vector-like multiplet [20], and renormalizable scalar leptoquark interactions [21]. But theoretically speaking they have common aspects, these contributions were induced by non-diagonal couplings and they are proportional to the imaginary part of the product of the coupling constant.

2.2. Weak electric dipole moment

The weak EDM is associated with the Z^0 boson, and it is the generalization of the electromagnetic case; mathematically speaking, with the appropriate change in the free parameters, we can obtain the usual EDM induced by the photon. On the other hand, due to the extremely short range of the weak interaction, the best possibility to measure a weak EDM is only by the deviation from the SM cross section for the production of heavy fermions, such as the tau lepton and top quark. Nevertheless, the weak EDM remain almost unexplored up to date, where the experimental current bounds were obtained in the $e^+e^- \to \tau^+\tau^-$ process by the ALEPH collaboration [4]. These bounds are the following:

$$\mathrm{Re}(d_\tau^W) \; < \; 0.5 \times 10^{-17} \; e \text{ cm}, \tag{8}$$

$$\mathrm{Im}(d_\tau^W) \; < \; 1.1 \times 10^{-17} \; e \text{ cm}, \tag{9}$$

and the SM prediction is $d_\tau^W < 8 \times 10^{-34} e$ cm [22]. Nonetheless, some SM extensions can give predictions much larger than the SM value. An example is the renormalizable *Leptoquark* interactions framework [23, 24]. In this context, the model with a scalar leptoquark doublet $R_2^T = (R_{\frac{1}{2}}, R_{-\frac{1}{2}})$ with quantum mumbers (3; 2; 7/6) under the $SU_c(3) \times SU_L(2) \times U_Y(1)$ gauge group has the following effective Lagrangian interaction [25, 26]:

$$\mathcal{L}_{F=0} = \bar{e}^i (\lambda_L^{ij} P_L + \lambda_R^{ij} P_R) u^j S_1^* + \bar{e}^i \eta_R^{ij} P_R d^j S_2^* + H.c., \tag{10}$$

where $S_1 = R_{\frac{1}{2}}$ and $S_2 = R_{-\frac{1}{2}}$ are scalar leptoquarks with electric charge $5/3e$ and $2/3e$, respectively. While, $\lambda_{L,R}^{ij}$ and η_R^{ij} are complex coupling constants again. Further, the leptoquark coupling with the photon and Z^0 boson are extracted from the respective leptoquark kinetic Lagrangian. Then, the contribution to the weak EDM of the fermion f_i can be written as [21]:

$$d_i^W = \frac{3e\sqrt{x_i x_j}}{32 m_i s_W c_W \pi^2} \mathrm{Im}(\lambda_L^{ij} \lambda_R^{ij*}) G(x_i, x_j, x_Z), \tag{11}$$

here $x_A = m_A^2/m_{S_k}^2$ with m_{S_k} the leptoquark mass and m_A the mass of any fermion or the Z^0 boson; furthermore, the explicit form of the $G(x_i, x_j, x_Z)$ function can be found in [21]. The importance of the equation (11) is that the weak EDM is proportional to the external fermion mass and to the imaginary part of the product of left and right handed coupling constants. Then, the coupling constants should be complex to obtain nonzero weak EDM, such as the EDM of the Eq. (7). Numerical results for the tau lepton are $\mathrm{Re}(d_\tau^W) \simeq 10^{-22} \sin\delta \; e$ cm and $\mathrm{Im}(d_\tau^W) \simeq 10^{-24} \sin\delta \; e$ cm, with δ the relative imaginary phase between left and right coupling constants. Although these results are much larger than the SM values, they are still far from the experimental limits. Other contributions to weak EDM have been reported in SM extensions as unparticle [19], MSSM [29] and THDM [27, 28], where to induce the $F_3^Z(m_Z^2)$ form factor was necessary to have non-diagonal interactions.

2.3. Chromo electric dipole moment

Since the gluon does not have rest mass, mathematically the chromo EDM is a variation of the electromagnetic EDM, with the difference that the chromo EDM is induced only in fermions with color charge. Additionally, because most quarks are confined in a compound state, the experimental measurement of these proprieties is done indirectly, and the best possibilities are associated with the heavy quarks. For the top quark the most recent bound to chromo EDM has been obtained with the Higgs boson production at the LHC, where the bound is $|d_t| < 4.8 \times 10^{-16} e$ cm [30]. The SM prediction is negligibly small, i.e $d_t < 10^{-30} e$ cm [31]. To induce a chromo EDM less suppressed, we can use the *Two Higgs Doublet Model* (THDM) with a sequential fourth family [32]. In this version of THDM, the fourth family interacts only with the heavy scalar bosons, and thus the theoretical prediction for Higgs boson production in the LHC remains unchanged. In this model we have also non-diagonal couplings given by the following Lagrangian:

$$\mathcal{L}_{\mathrm{FV}} = \frac{g}{2m_W} f_\beta^\phi \bar{f}_i \left[g_s^\phi + g_p^\phi \gamma_5 \right] f_j \phi^0, \tag{12}$$

here the scalar (g_s^ϕ) and pseudo-scalar (g_p^ϕ) couplings are mathematical structures that are composed by fermion masses and complex mixing matrices [1]. Then, the contribution to chromo EDM of the top quark can be written as [34]:

$$d_t^\phi = -\mathrm{Im}\big(G_s^\phi G_p^{\phi *}\big)\frac{m_t^2 g^2 f_\phi^2 r_{t'}}{2^4\pi^2 m_W^2}\int_0^1 \frac{x^2 dx}{x r_{t'}^2 - (x-1)(x-r_\phi^2)} \tag{13}$$

where $r_k = m_k/m_t$, $G_{s,p}^\phi = g_{s,p}^\phi/m_t$ and $\phi = H^0, A^0, H^\pm$. Furthermore, we can observe that this result is proportional to $\mathrm{Im}\big(G_s^\phi G_p^{\phi *}\big)$ and m_t^2, such as the electromagnetic an weak cases. For neutral heavy scalar bosons the numerical result is around of $10^{-24}e$ cm, where they set $m_{t'} = 600$ GeV and $m_\phi = 300$ GeV [34]. Another results for the chromo EDM can be found for Models with Vector Like Multiplets [35] and THDM with CP violation [36].

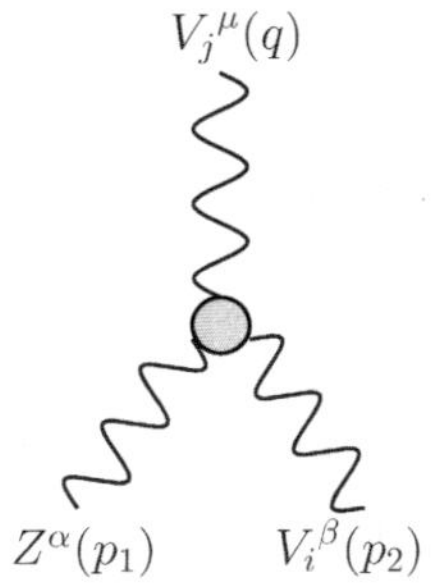

Figure 1. Diagram for the vertex function $ie\ \Gamma^{\alpha\beta\mu}_{ZV_iV_j}(p_1,p_2,q)$ for the ZV_iV_j ($V_{i,j} = Z,\ \gamma$) couplings, all momenta are outgoing.

3. Trilinear neutral gauge boson coupling

In SM the triple coupling VW^+W^- ($V = Z,\gamma$) is the only interaction between electroweak gauge bosons at tree level. On the other hand, by Landau-Yang's theorem the trilinear neutral gauge boson couplings (TNGBCs) VVV are induced only at one-loop level, with at least one off-shell neutral gauge boson. In this context, the different TNGBs are $Z\gamma\gamma$, $ZZ\gamma$ and ZZZ, while the $\gamma\gamma\gamma$ coupling is forbidden by Furry's theorem. Then for one off-shell neutral gauge boson there are four types of TNGBCs, which are described for the vertices $ZV^*\gamma$ and ZZV^*. The respective effective Lagrangians for each coupling are given for [37]:

$$\begin{aligned}
\mathcal{L}_{ZV^*\gamma} &= \frac{e}{m_Z^2}\Big\{-[h_1^\gamma(\partial^\alpha F_{\alpha\mu}) + h_1^Z(\partial^\alpha Z_{\alpha\mu})]Z_\beta F^{\mu\beta}\\
&\quad -\frac{1}{m_Z^2}\Big[h_2^\gamma(\partial_\alpha\partial_\beta\partial^\rho F_{\rho\mu}) + h_2^Z[\partial_\alpha\partial_\beta(\partial^2+m_Z^2)Z_\mu]\Big]Z^\alpha F^{\mu\beta}\\
&\quad -[h_3^\gamma(\partial_\beta F^{\beta\mu}) + h_3^Z(\partial_\beta Z^{\beta\mu})]Z^\alpha\tilde{F}_{\mu\alpha}\\
&\quad +\frac{1}{2m_Z^2}\Big[h_4^\gamma(\partial^2\partial^\beta F^{\mu\alpha}) + h_4^Z[(\partial^2+m_Z^2)\partial^\beta Z^{\mu\alpha}]\Big]Z_\beta\tilde{F}_{\mu\alpha}\Big\}, \qquad (14)\\
\mathcal{L}_{ZZV^*} &= \frac{e}{m_Z^2}\Big[-[f_4^\gamma(\partial_\mu F^{\mu\beta}) + f_4^Z(\partial_\mu Z^{\mu\beta})]Z_\alpha(\partial^\alpha Z_\beta)\\
&\quad +[f_5^\gamma(\partial^\alpha F_{\alpha\mu}) + f_5^Z(\partial^\alpha Z_{\alpha\mu})]\tilde{Z}^{\mu\beta}Z_\beta\Big]. \qquad (15)
\end{aligned}$$

[1] For more details see [33].

here h_i^V and f_j^V are one-loop form factors, while $\tilde{V}_{\mu\nu} = \epsilon_{\mu\nu\alpha\beta}V^{\alpha\beta}/2$ with $V_{\mu\nu} = \partial_\mu V_\nu - \partial_\nu V_\mu$ standing for the strength tensor of the neutral gauge boson. The operators associated with $h_{2,4}^V$ have dimension eight, and the remaining ones are of dimension six. Furthermore, the Lorentz structure associated with $h_{1,2}^V$ and f_4^V are CP-odd, $h_{4,5}^V$ and f_5^V are CP-even. Then, from the effective Lagrangians (14) and (15) we obtain the respective vertex functions and they respect Lorentz covariance, Bose symmetry and $U_{\rm em}(1)$ gauge invariance, which are given by [38, 39]:

$$\begin{aligned}\Gamma^{\alpha\beta\mu}_{ZV^*\gamma}(p_1,p_2,q) &= i\frac{(p_2^2-m_V^2)}{m_Z^2}\Big[h_1^V(q^\beta g^{\alpha\mu}-q^\alpha g^{\beta\mu})+\frac{h_2^V}{m_Z^2}p_2^\alpha\big[(q\cdot p_2)g^{\beta\mu}-q^\beta p_2^\mu\big]\\ &\quad -h_3^V\epsilon^{\beta\alpha\mu\rho}p_{2\rho}-\frac{h_4^V}{m_Z^2}p_2^\alpha\epsilon^{\beta\mu\rho\sigma}p_{2\rho}q_\sigma\Big], \qquad (16)\\ \Gamma^{\alpha\beta\mu}_{ZZV^*}(p_1,p_2,q) &= i\frac{(q^2-m_V^2)}{m_Z^2}\big[f_4^V(q^\alpha g^{\mu\beta}+q^\beta g^{\mu\alpha})-f_5^V\epsilon^{\mu\alpha\beta\rho}(p_1-p_2)_\rho\big]. \qquad (17)\end{aligned}$$

where the four-momenta of the gauge bosons are defined in Figure 1. Thus, the CP-odd form factor parameterizes the effect of the CP-violation in the TNGBs. At one-loop level in the SM only the CP-even part is induced via fermion triangle [39], but to induce the CP-odd parts other theoretical mechanisms are necessary. One possible theoretical framework is a renormalizable effective theory with several physical charged scalar bosons, which includes non-diagonal couplings. In this context, the Lagrangian describing the non-diagonal coupling between the Z^0 boson and the charged scalar bosons can be written as:

$$\mathcal{L}_{\rm FV} = i\sum_{i\neq j} g_{ij}^Z Z_\mu \Phi_i^+ \overleftrightarrow{\partial}^\mu \Phi_j^- + \text{H.c.} \qquad (18)$$

here g_{ij}^Z is a complex coupling constant. The diagonal coupling $\gamma\Phi_i^+\Phi_i^-$ has the usual gauge invariant interaction for a scalar sector of any renormalizable theory. Then, the CP-odd part of the TNGBs can be induced via the non-diagonal interaction (18). In the next sub-sections we show some results reported previously for the $ZZ^*\gamma$ and ZZZ^* couplings.

3.1. $ZZ^*\gamma$ coupling

To induce this vertex four diagrams are necessary, where two diagrams are required by Bose symmetry, whereas two diagrams involve the exchange of the virtual scalar bosons and they are necessary to cancel out ultraviolet divergences. Thus, after the transversality and mass-shell conditions for the gauge bosons are considered, the results are the following [40]:

$$\begin{aligned}h_1^Z &= \frac{m_Z^2 g_{ii}^\gamma \text{Im}(g_{ij}^Z g_{ji}^{Z*})}{12\pi^2(m_Z^2-p_2^2)^3}\Big[(m_i^2-m_j^2)(m_Z^2-p_2^2)-2(m_Z^2-p_2^2)\big[m_i^2B_{ii}(0)-m_j^2B_{jj}(0)\big] \qquad (19)\\ &-(m_i^2-m_j^2)\big[(m_Z^2-p_2^2)B_{ij}(0)+3(m_Z^2+p_2^2)B_{ij}(p_2^2)-6m_Z^2B_{ij}(m_Z^2)\big]\\ &+3(m_Z^2-p_2^2)\big[m_j^2(m_i^2-m_j^2-m_Z^2)C_{ijj}(p_2^2)+m_i^2(m_i^2-m_j^2+m_Z^2)C_{jii}(p_2^2)\big]\Big],\\ h_2^Z &= \frac{m_Z^4 g_{ii}^\gamma \text{Im}(g_{ij}^Z g_{ji}^{Z*})}{6\pi^2(m_Z^2-p_2^2)^4}\Big[2(m_i^2-m_j^2)(m_Z^2-p_2^2)-4(m_Z^2-p_2^2)\big[m_i^2B_{ii}(0)-m_j^2B_{jj}(0)\big] \qquad (20)\\ &-(m_i^2-m_j^2)\big[2(m_Z^2-p_2^2)B_{ij}(0)+3(m_Z^2+3p_2^2)B_{ij}(p_2^2)-3(3m_Z^2+p_2^2)B_{ij}(m_Z^2)\big]\\ &+3(m_Z^2-p_2^2)\big[m_j^2(2m_i^2-2m_j^2-m_Z^2-p_2^2)C_{ijj}(p_2^2)\\ &+m_i^2(2m_i^2-2m_j^2+m_Z^2+p_2^2)C_{jii}(p_2^2)\big]\Big].\end{aligned}$$

where p_2 is the four-momentum of the off-shell Z^0 boson, $m_{i,j}$ is the mass of the charged scalar boson and it was introduced the shorthand notation in terms of Passarino-Veltman scalar functions. We note that this result also is proportional to $\mathrm{Im}(g_{ij}^{Z}g_{ji}^{Z*})$, and it is evident that it vanishes when $i = j$. The numerical results for values above 300 GeV, for the charged scalar bosons are $|h_1^Z| \sim 10^{-5} - 10^{-4}$ and $|h_2^Z| \sim 10^{-7} - 10^{-6}$.

The CP-violating vertices $ZZ^*\gamma$ and $Z\gamma^*\gamma$ may induce CP-odd asymmetries in the radiative decay $Z \to \mu^+\mu^-\gamma$ that could be used to set bounds on these vertices [41].

3.2. ZZZ^* coupling

For this coupling twelve diagrams are necessary, and in order to satisfy the Bose symmetry $3! = 6$ diagrams are required, and it corresponds to the different permutations of the Z^0 bosons. Additionally, six diagrams involve the exchange of the virtual scalar bosons that are added to cancel out ultraviolet divergences. Then, after the Passarino-Veltman method is applied, the result is the following:

$$f_4^Z = \frac{m_Z^2 g_{ii}^Z \mathrm{Im}(g_{ij}^Z g_{ji}^{Z*})}{24\pi^2 s_W q^2(q^2 - m_Z^2)(q^2 - 4m_Z^2)^2}\Xi(q^2, m_i^2, m_j^2) \tag{21}$$

where now q is the four momentum of the off-shell Z^0 boson and the explicit form of $\Xi(q^2, m_i^2, m_j^2)$ can be found in [40]. We see the same properties for the $ZZ^*\gamma$ coupling; the form factor (21) is proportional to $\mathrm{Im}(g_{ij}^Z g_{ji}^{Z*})$ and the $\Xi(q^2, m_i^2, m_j^2)$ function vanishes when $i = j$. The numerical result is around $|f_4^Z| \sim 10^{-6} - 10^{-5}$ for the same values of the scalar bosons in the previous case.

To study the TNGBCs, the experimentalists focused their attention on pair production of neutral gauge bosons; but in the SM the non-resonant neutral diboson production proceeds mainly through quark-antiquark (electron-positron) $t-$ and $u-$ channel scattering diagrams, though also a contribution about of the 6% proceeds via gluon-gluon fusion into box quark loops in hadronic processes. On the other hand, the SM offers precise predictions of production rates associated with self-interactions of the neutral gauge bosons [42]. Then the production of neutral gauge bosons not only provides an important background for the studies of the Higgs boson, it also allows to test the electroweak sector of the SM. Consequently, with precise measurements of the cross sections of production of neutral diboson, any observed deviation from the SM prediction would be an indication of new SM physics, such as the TNGBCs. But up to now, the measured total cross section is statistically consistent with the respective SM prediction [43, 44, 45, 46, 47, 48, 49, 50, 51, 52, 53] and therefore, no evidence of NTGBCs has been found. Nevertheless, all results of the analysis have been used to set limits on NTGBCs at the 95% confidence level. Particularly, the most stringent limits to date on CP-violating form factors are shown in the table 2.

Experiment	limits
DELPHI [53]	$\lvert h_1^Z\rvert < 0.23$
L3 [52]	$-0.087 < h_2^Z < 0.079$
CMS[43]	$-0.00117 < f_4^Z < 0.00110$
CMS[43]	$-0.00133 < f_4^\gamma < 0.00132$

Table 2. The most stringent limits up to date on CP-violating NTGBCs.

4. Higgs to gauge boson decays

In the SM, the Higgs mechanism generates the masses of the electroweak gauge bosons, while the photon and gluon remain massless. Consequently, at tree level the Higgs boson interact only with $W^{\pm}$ and Z^0. The Lagrangian for this interaction is the following:

$$\mathcal{L}_{\text{tree}} = gm_W h W^{\dagger}_{\mu} W^{\mu} + \frac{1}{2}\frac{gm_Z}{c_W} h Z^{\dagger}_{\mu} Z^{\mu}, \tag{22}$$

these interactions have been observed since discovery of Higgs boson, and they are statistically consistent with the SM [54, 55, 56, 57]. However, the Higgs boson decay into a photon pair was also measured by ATLAS [54] and CMS [55] collaborations. Additionally, the Higgs boson production at the LHC is dominated by the gluon fusion process [58]. Although the coupling of the Higgs boson with photons and gluons is absent at tree level in the SM, these interactions are induced at one-loop level and they are also consistent with the SM [54, 55, 56, 57]. Nevertheless, at one-loop level new contributions may arise in some SM extension and they are called anomalous couplings. The Lagrangian for these interactions has four and six dimension operators and is given by:

$$\mathcal{L}^{hVV}_{\text{eff}} = gm_V a^V_1 h V_{\mu} V^{\mu} + a^V_2 h V_{\mu\nu} V^{\mu\nu} + a^V_3 h \tilde{V}_{\mu\nu} V^{\mu\nu} \tag{23}$$

here $V = \gamma, Z^0, W, G$ and a^V_k is an one-loop form factor, while $\tilde{V}_{\mu\nu} = \epsilon^{\mu\nu\alpha\beta} V_{\alpha\beta}$ and the operator $Z_{\mu\nu}(Z^{\mu}\partial^{\nu}h - Z^{\nu}\partial^{\mu}h)$ is related with a^V_1 and a^V_2 terms. On the other hand, the Lorentz structures associated to $a^V_{1,2}$ are CP-even and for a^V_3 is CP-odd. Therefore, a nonzero a^V_3 form factor parameterizes the effect of CP-violation in the interactions of the Higgs boson with gauge bosons. Then from (23), the vertex function for the coupling $hV^{\mu}(p_1)V^{\nu}(p_2)$ is the following:

$$\Gamma^{\mu\nu} = gm_V a^V_1 g^{\mu\nu} + a^V_2 (p_1 \cdot p_2 g^{\mu\nu} - p^{\mu}_2 p^{\nu}_1) + a^V_3 \epsilon^{\mu\nu\alpha\beta} p_{1\alpha} p_{2\beta} \tag{24}$$

where $a^{Z,W}_1$ is a correction of the respective tree-level coupling, while that $a^{\gamma,g}_2$ explain the gluon fusion and Higgs boson decay into a photon pair at LHC. Further, in the SM the $a^{Z,W}_1$ and a^V_2 can be induced through a loop involving top or bottom quarks, but to induce a^V_3 form factor is necessary some additional theoretical mechanisms.

A possibility consists in considering flavor changing neutral currents (FCNC). But the FCNC is induced only at one-loop in some SM extensions and since the photon and the gluon are massless, the FCNC associated with these gauge bosons are very suppressed [2]. Then the best opportunity resides in the Z^0 boson [3]. However, considering an effective lagrangian for the $Z^0 \bar{\ell}_i \ell_j$ coupling, that includes both vector and axial complex coupling constants, the results for a^Z_3 is extremely suppressed [62]. The importance of this analysis is that the result is proportional to $\mathrm{Im}(g^{ij}_V g^{ji*}_A)$, and vanishes when we consider diagonal couplings. From the experimental side, the CMS collaboration has been exploring the a^V_3 coupling in different Higgs boson decays to neutral diboson [63, 64, 65]. In those experimental analysis the effective fractional cross sections f_{a_i} and phases ϕ_{a_i} are defined as follows:

$$f_{a_i} = |a_i|^2 \Big/ \sum_j |a_j|^2 \sigma_j, \tag{25}$$

$$\phi_{a_i} = \arg(a_i/a_1). \tag{26}$$

Where σ_j is the cross section for the $h \to ZZ/Z\gamma^*/\gamma^*\gamma^* \to 2e2\mu$ process. Thus, from the combined Run 1 and Run 2 analysis, the current results are $f_{a_3} = 0.00^{+0.26}_{-0.09}$ with 68% C.L., and $-0.38 < \cos(\phi_{a_3}) < 0.46$ with 95% C.L. [63].

[2] BR($\tau \to \mu\gamma$)$< 4.4 \times 10^{-8}$, BR($\tau \to e\gamma$)$< 3.3 \times 10^{-8}$ [59] and BR($\mu \to e\gamma$)$< 4.2 \times 10^{-13}$ [60].
[3] BR($Z^0 \to e\mu$)$< 1.7 \times 10^{-6}$, BR($Z^0 \to e\tau$)$< 9.8 \times 10^{-6}$ and BR($Z^0 \to \mu\tau < 1.2 \times 10^{-5}$) [61].

5. Conclusions

In this report we have shown that the radiative corrections are a powerful tool to study new sources of CP-violation. In this context, we have explored the mathematical potential of the electric dipole moments, trilinear neutral gauge boson couplings and Higgs to gauge boson decays. These sources can be induced in some SM extension framework, and we showed that in order to induce the respective source of CP-violation non-diagonal couplings are necessary. Additionally, we showed that the respective CP-odd term is proportional to the imaginary part of the product of the coupling constants, which is necessary to obtain nonzero results and is consistent with the Lorentz structure of the respective Lagrangian.

Acknowledgement

I am grateful to the conference organizers for their kind invitation to present our research work and their kind hospitality. I also would like to thank discussions with Professor Miguel Angel Pérez Angón and I acknowledge support from CONACYT (Mexico).

References

[1] K. S. Babu, S. M. Barr, and I. Dorsner, *Phys. Rev.* **D64**, 053009 (2001).
[2] P. D. Acton *et al.* [OPAL Collaboration], (1992) *Phys. Lett.* **B281**, 405.
[3] M. Acciarri *et al.* [L3 Collaboration], (1998) *Phys. Lett.* **B426**, 207.
[4] A. Heister *et al.* [ALEPH Collaboration], (2003) *Eur. Phys. J.* **C30**, 291. [hep-ex/0209066].
[5] J. Abdallah *et al.* [DELPHI Collaboration], (2004) *Eur. Phys. J. C* **35**, 159.
[6] U. Baur, A. Juste, L. H. Orr and D. Rainwater, (2005) *Phys. Rev.* **D71**, 054013.
[7] S. Eidelman, D. Epifanov, M. Fael, L. Mercolli, and M. Passera, *JHEP* **03**, 140 (2016).
[8] D. A. Epifanov [Belle Collaboration], *Nucl. Part. Phys. Proc.* 287-288, 7 (2017).
[9] Regan, B. C. and Commins, E. D. and Schmidt, C. J. and DeMille, D., (2002) *Phys. Rev. Lett.* **88** 071805.
[10] Bennett, G. W. *et al.* [Muon (g-2) collaboration], (2009) *Phys. Rev.* **D80** 052008.
[11] Inami, K. *et al.* [Belle Collaboration], (2003) *Phys. Lett.* **B551**, 16-26.
[12] F. Hoogeveen, *Nucl. Phys.* **B341** 322 (1990).
[13] H. Georgi, (2007), *Phys. Rev. Lett.* **98**, 221601.
[14] Sirunyan, Albert M and *et al.* [CMS Collaboration], (2017) *JHEP* **03**, 061.
[15] Khachatryan, Vardan and *et al.* [CMS Collaboration], (2016) *Phys. Rev. D93* 052011.
[16] Khachatryan, Vardan and *et al.* [CMS Collaboration], (2015) *Eur. Phys. J. C75* no.5, 235.
[17] K. Cheung, W. -Y. Keung and T. -C. Yuan, (2007) *Phys. Rev.* **D76**, 055003.
[18] Moyotl, A. and Rosado, A. and Tavares-Velasco, G., (2011) *Phys. Rev.* **D84**, 073010.
[19] Moyotl, A. and Tavares-Velasco, G., (2012) *Phys. Rev.* **D86**, 013014.
[20] T. Ibrahim and P. Nath, (2010) *Phys. Rev.* **D81**, 033007.
[21] Bolaos, A. and Moyotl, A. and Tavares-Velasco, G., (2014) *Phys. Rev.* **D89** no.5, 055025.
[22] M. J. Booth, hep-ph/9301293.
[23] S. Davidson, D. C. Bailey, and B. A. Campbell, *Z. Phys.* **C61**, 613 (1994).
[24] J. L. Hewett and T. G. Rizzo, *Phys. Rev.* **D56**, 5709 (1997).
[25] W. Buchmuller, R. Ruckl, and D. Wyler, *Phys. Lett.* **B191**, 442 (1987).
[26] J. Mizukoshi, O. J. Eboli, and M. Gonzalez-Garcia, *Nucl. Phys.* **B443**, 20 (1995).
[27] J. Bernabeu, D. Comelli, L. Lavoura, and J. P. Silva, *Phys. Rev.* **D53**, 5222 (1996).
[28] D. Gomez-Dumm and G. Gonzalez-Sprinberg, *Eur. Phys. J.* **C11**, 293 (1999).
[29] W. Hollik, J. I. Illana, S. Rigolin, and D. Stockinger, Phys.Lett. B425, 322 (1998).
[30] Hayreter, Alper and Valencia, German, *Phys. Rev* **D.88** 034033 (2013).
[31] A. Soni and R. M. Xu, *Phys. Rev. Lett.* **69**, 33 (1992).
[32] S. Bar-Shalom, S. Nandi, and A. Soni, *Phys. Rev* **D84**, 053009, (2011).
[33] S. Chamorro-Solano, A. Moyotl, M.A. Pérez, arXiv:1707.00100.
[34] A. Hernández-Juárez, A. Moyotl and G. Tavares, work presented in this meeting.
[35] T. Ibrahim and P. Nath, *Phys. Rev.* **D84**, 015003 (2011).
[36] Gaitán R, Garcés E A, de Oca J H M and Martinez R, 2015 *Phys. Rev.* **D92**(9) 094025.
[37] G. Gounaris, J. Layssac, and F. Renard, *Phys. Rev.* **D61**, 073013 (2000).
[38] K. Hagiwara, R. Peccei, D. Zeppenfeld, and K. Hikasa, *Nucl. Phys.* **B282**, 253 (1987).
[39] G. Gounaris, J. Layssac, and F. Renard, *Phys. Rev.* **D62**, 073013 (2000).
[40] A. Moyotl, J.J. Toscano, G. Tavares-Velasco, *Phys. Rev.* **D91** 093005 (2015).

[41] M. A. Perez and F. Ramirez-Zavaleta, *Phys. Lett.* **B609**, 68 (2005).
[42] Campbell, John M. and Ellis, R. Keith,*Phys. Rev.* **D60**, 113006 (1999).
[43] CMS Collaboration, CMS-PAS-SMP-16-017 (2017).
[44] Aaboud, Morad and *et al.* [ATLAS Collaboration], *JHEP* **1701**, (2017) 099.
[45] G. Aad *et al.* [ATLAS Collaboration], *Phys. Rev. Lett.* **108**, 041804 (2012), 1110.5016.
[46] G. Aad *et al.* [ATLAS Collaboration], *JHEP* **1303**, 128 (2013), 1211.6096.
[47] S. Chatrchyan *et al.* [CMS Collaboration], *JHEP* **1301**, 063 (2013), 1211.4890.
[48] S. Chatrchyan *et al.* [CMS Collaboration], *JHEP* **1310**, 164 (2013), 1309.1117.
[49] V. Khachatryan *et al.* [CMS Collaboration], *Phys. Lett.* **B740** (2015) 250.
[50] Abazov *et al.* [D0 Collaboration], *Phys. Rev. Lett.* 100, 131801 (2008).
[51] V. M. Abazov *et al.* [D0 Collaboration], *Phys. Rev.* **D85**, 052001 (2012).
[52] P. Achard *et al.* [L3 Collaboration], *Phys. Lett.* **B597**, 119 (2004).
[53] J. Abdallah *et al.* [DELPHI Collaboratio]), *Eur. Phys. J.* **C51**, 525 (2007)
[54] G. Aad *et al.* [ATLAS Collaboration], *Phys. Lett.* **B716**, 1 (2012), 1207.7214.
[55] S. Chatrchyan *et al.* [CMS Collaboration], *Phys. Lett.* **B716**, 30 (2012), 1207.7235.
[56] G. Aad *et al.* [ATLAS Collaboration], *Phys. Rev.* **D90**, 052004 (2014), 1406.3827.
[57] G. Aad *et al.* [ATLAS, CMS Collaborations], *Phys. Rev. Lett.* **114**, 191803 (2015), 1503.07589.
[58] Q. Li, M. Spira, J. Gao, and C. S. Li, *Phys. Rev.* **D83**, 094018 (2011).
[59] B. Aubert *et al.* [BaBar Collaboration], *Phys. Rev. Lett.* **104**, 021802 (2010), 0908.2381.
[60] A. M. Baldini *et al.* [MEG Collaboration], *Eur. Phys. J.* **C76**, 434 (2016), 1605.05081.
[61] C. Patrignani *et al.* (Particle Data Group), *Chin. Phys. C*, **40**, 100001 (2016).
[62] A. Moyotl, work in progress (2017).
[63] A. M. Sirunyan *et al.* [CMS Collaboration], (2017), 1707.00541.
[64] C. Collaboration (CMS) CMS-PAS-HIG-17-011 (2017).
[65] V. Khachatryan *et al.* [CMS Collaboration], *Phys. Rev.* **D92**, 012004 (2015), 1411.3441.

G-parity breaking contributions in $\tau^- \to \eta^{(\prime)}\pi^-\nu_\tau$ decays induced at one loop level by the anomalous $\gamma\gamma\eta^{(\prime)}$ vertices

G. Hernández-Tomé

Departamento de Física, Centro de Investigación y de Estudios Avanzados del Instituto Politécnico Nacional, Apdo. Postal 14-740, 07000 México D.F., México

E-mail: ghernandez@fis.cinvestav.mx

Abstract. Second class current processes (SCC) are forbidden in the limit of exact isospin symmetry by G-parity. Since G-parity invariance is slightly broken by the $u-d$ mass difference and electromagnetic effects, SCC can appear but they are naturally suppressed. $\tau^- \to \eta^{(\prime)}\pi^-\nu_\tau$ decays represent a clean test of SCC and according to theoretical estimations and thanks to the increased luminosity contemplated at the Belle-II collaboration, their discovery should be finally possible. Consequently, a careful evaluation of G-parity breaking contributions results mandatory in order to isolate possible new physics contributions. In this work, we evaluate the G-parity breaking contributions in $\tau^- \to \eta^{(\prime)}\pi^-\nu_\tau$ decays induced at one loop level by the anomalous electromagnetic $\gamma\gamma\eta^{(\prime)}$ vertices.

1. Introduction

The study of the G-parity [1] invariance plays a fundamental role in considering the effects of strong interactions on weak processes, based on that Weinberg established a classification for non-strange weak (V-A) hadronic currents [1]. According to this classification, weak hadronic currents with isospin 0, 1 quantum numbers can be split into two classes. The first class includes currents with quantum numbers $J^{PG} = 0^{++}, 0^{--}, 1^{+-}, 1^{-+}$, whereas the second class currents (SCC) have opposite G-parity $J^{PG} = 0^{+-}, 0^{-+}, 1^{++}, 1^{--}$. SCC can not occur if G-parity were an exact symmetry, but since G-parity invariance is slightly broken by isospin non-conservation, electromagnetic effects and the mass difference of $u-d$ quarks can induce the hadronization of the standard model (SM) currents into states that mimic the effects of a genuine SCC [2].

Several channels for SCC have been proposed involving nuclear β decay processes, where the many form factors and small momentum transfer have driven to a complicated search [4]. On the other hand, a clean test for the existence of SCC would be provided by the observation of the semileptonic transitions $\tau^- \to \eta^{(\prime)}\pi^-\nu_\tau$ [6], since the G-parity of the system is -1, which is opposed to the vector current that drives the decay in the SM. Consequently, the S(P)-wave of the $\pi^-\eta^{(\prime)}$ system gives $J^{PG} = 0^{+-}(1^{--})$, which as mentioned before corresponds to

[1] G operator is defined as G$= Ce^{i\pi I_2}$, with C the charge conjugation operator and I_2 the second component of the isospin rotation operator.

[2] For genuine SCC we mean effects due to New Physics; for example, the ones induced by the exchange of charged Higgs or leptoquarks particles. A recent study of the effects of New Physics in $\tau^- \to \eta^{(\prime)}\pi^-\nu_\tau$ decays using the framework of a general effective field theory can be found in [5].

a SCC independently of possible intermediate resonant states. Regarding experimental search of $\tau^- \to \eta^{(\prime)}\pi^-\nu_\tau$ decays, CLEO and Belle collaborations reported previous limits on their branching fractions, but the most stringent bounds available are based on searches by the BaBar collaboration [7] corresponding to $BR(\tau^- \to \eta\pi^-\nu_\tau) < 9.9 \times 10^{-5}$ and $BR(\tau^- \to \eta'\pi^-\nu_\tau) < 7.2 \times 10^{-6}$ [8], which lie close to the estimates based on isospin symmetry breaking [9] for the $BR(\tau^- \to \eta^{(\prime)}\pi^-\nu_\tau)$ decays mainly induced by the $u-d$ quark mass difference [10, 11].

We are interested in the study of the G-parity breaking contribution in $\tau^- \to \eta^{(\prime)}\pi^-\nu_\tau$ decays induced at one loop level by the anomalous electromagnetic $\gamma\gamma\eta^{(\prime)}$ vertices. In spite of this kind of contribution is expected to give only a minor correction with respect to tree level ($(\pi^0-)\eta - \eta'$ mixing) contributions, it is crucial to have a reliable estimate of these effects in order to eliminate a possible source of background for a genuine SCC [3]. In order to compare the different contributions to $\tau^- \to \eta^{(\prime)}\pi^-\nu_\tau$ decays, we show in section 2 the relevant formulas for the tree level result [4], whereas a brief explanation of the approach used in order to calculate the one-loop contribution is given in section 3. Then, an evaluation of our results is presented in section 4. Finally, the conclusions are presented in section 5.

2. Tree level contribution (Hadronic matrix elements)

Into the SM, the semileptonic $\tau^- \to \eta^{(\prime)}\pi^-\nu_\tau$ decays are induced at tree level due to the charged weak interaction between the tau lepton and its corresponding neutrino and later hadronization of the current. Since parity conservation implies that only weak vector current contributes to the hadronic part, the amplitude of the decay $\tau^- \to \pi^-\eta^{(\prime)}\nu_\tau$ it can be written as

$$\mathcal{M} = \frac{G_F}{\sqrt{2}} V_{ud}\bar{u}(p_{\nu_\tau})\gamma_\mu(1-\gamma_5)u(p_\tau)\langle\pi^-\eta^{(\prime)}|\bar{d}\gamma^\mu u|0\rangle\,. \tag{1}$$

where the hadron matrix element is

$$\langle\pi^-\eta^{(\prime)}|\bar{d}\gamma^\mu u|0\rangle = \left[(p_{\eta^{(\prime)}} - p_\pi)^\mu + \frac{\Delta_{\pi^-\eta^{(\prime)}}}{s}q^\mu\right] c^V_{\pi^-\eta^{(\prime)}} F_+^{\pi\eta^{(\prime)}}(s) + \frac{\Delta^{QCD}_{K^0K^+}}{s} q^\mu c^S_{\pi^-\eta^{(\prime)}} F_0^{\pi^-\eta^{(\prime)}}(s) \tag{2}$$

and the following definitions have been used $s = q^2 = (p_\pi + p_{\eta^{(\prime)}})^2$, $c^V_{\pi^-\eta^{(\prime)}} = \sqrt{2}$, $c^S_{\pi^-\eta} = \sqrt{2/3}$, $c^S_{\pi^-\eta'} = 2/\sqrt{3}$, $\Delta_{\pi^-\eta^{(\prime)}} = m^2_{\pi^-} - m^2_{\eta^{(\prime)}}$ and the superscript QCD indicates that the K^0K^+ electromagnetic mass difference does not contribute to eq. (2).

The differential partial decay width, as a function of the $\pi^-\eta^{(\prime)}$ invariant mass, is

$$\begin{aligned}\frac{d\Gamma\left(\tau^- \to \pi^-\eta^{(\prime)}\nu_\tau\right)}{d\sqrt{s}} &= \frac{G_F^2 M_\tau^3}{24\pi^3 s} S_{EW}\left|V_{ud}F_+^{\pi^-\eta^{(\prime)}}(0)\right|^2\left(1-\frac{s}{M_\tau^2}\right)^2 \\ &\times\left\{\left(1+\frac{2s}{M_\tau^2}\right)q^3_{\pi^-\eta^{(\prime)}}(s)|\widetilde{F}_+^{\pi^-\eta^{(\prime)}}(s)|^2\right. \\ &\left.+\frac{3\Delta^2_{\pi^-\eta^{(\prime)}}}{4s}q_{\pi^-\eta^{(\prime)}}(s)|\widetilde{F}_0^{\pi^-\eta^{(\prime)}}(s)|^2\right\},\end{aligned} \tag{3}$$

[3] Another possible source of background to $\tau^- \to \eta^{(\prime)}\pi^-\nu_\tau$ decays is given by the $\tau^- \to \eta^{(\prime)}\pi^-\nu_\tau\gamma$ channels since they are not suppressed by G-parity considerations, however, this kind of background can be eliminated considering appropriate cuts in the energy of the photons [12].

[4] A complete explanation of the theoretical construction of the participant vector and scalar form factors in eqs. (2) and (3) can be found in Ref. [13], where the authors employ the state-of-the-art analysis of meson-meson scattering within unitarized Chiral Perturbation Theory [15] in order to obtain the scalar form factors.

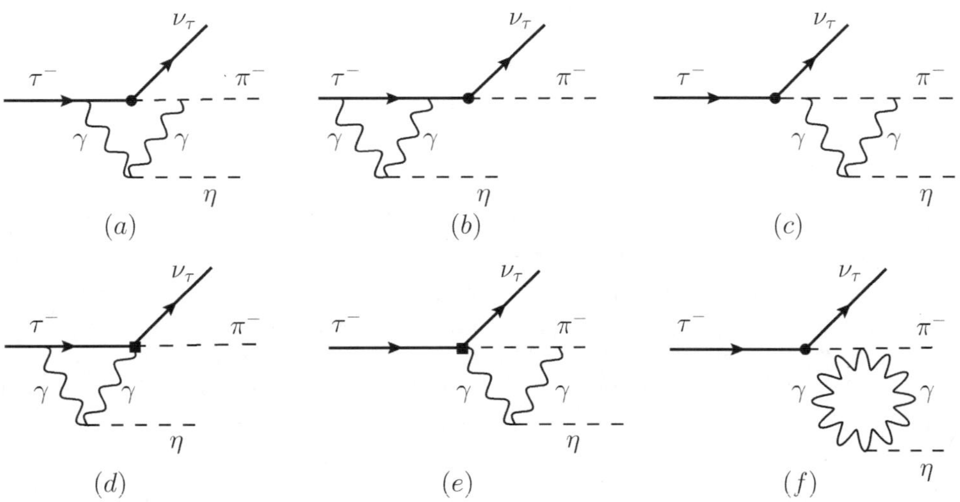

Figure 1. Feynman diagrams for the $\tau^- \to \eta^{(\prime)}\pi^-\nu_\tau$ decays at one loop level induced by $\gamma\gamma\eta^{(\prime)}$ vertices. It is easy to check that the contributions from diagrams (c), (e) and (f) are identically zero (see text).

where $\widetilde{F}_{+,0}^{\pi^-\eta^{(\prime)}}(s) = \frac{F_{+,0}^{\pi^-\eta^{(\prime)}}(s)}{F_{+,0}^{\pi^-\eta^{(\prime)}}(0)}$ stand for the two form factors normalized to unity at the origin. In eq. (3) the short-distance electroweak correction factor, $S_{EW} = 1.0201$ [26], has been introduced and

$$q_{PQ}(s) = \frac{\sqrt{s^2 - 2s\Sigma_{PQ} + \Delta_{PQ}^2}}{2\sqrt{s}}, \quad \Sigma_{PQ} = m_P^2 + m_Q^2 . \tag{4}$$

3. $\tau^- \to \eta^{(\prime)}\pi^-\nu_\tau$ decays induced at one loop level

Besides the contribution given by the eq. (1), electromagnetic interactions also break isospin symmetry and will contribute to $\tau^- \to \eta^{(\prime)}\pi^-\nu_\tau$ decays. This can occur at the one-loop level, *via* the emission of a pair of photons from $\tau^- \to \pi^-\nu_\tau$ decays and their later conversion into an $\eta^{(\prime)}$ meson through the anomalous $\gamma\gamma\eta^{(\prime)}$ vertices as shown in Figure 1 [5]. In order to calculate this kind of contribution, we have considered, in the low-energy limit, a point-like interaction for the $\tau^- \to \nu_\tau\pi^-$ vertex, which can be described by the Lagrangian density

$$\mathcal{L} = G_F V_{ud} f_\pi \bar{\nu}_\ell \gamma_\mu (1 - \gamma_5) \ell \partial^\mu \pi^+ + \text{h.c.}, \tag{5}$$

where $f_\pi = F_\pi/\sqrt{2} = 92.2$ MeV. A detailed explanation of the low- and high-energy limits of QCD for the structure of the pion electromagnetic coupling and the two-photon coupling of the neutral meson is discussed in appendix A of [21], where we also explain the cancellation of ultraviolet divergences in the loop integrals.

We have verified that the contributions from diagrams (c) and (f) vanish due to the conservation of P and CP by strong and electromagnetic interactions, whereas the contribution from diagram (e) vanishes when considering the loop integration because it is odd in the

[5] Obviously, this type of vertex would also contribute to the $\tau^- \to \pi^-\pi^0\nu_\tau$ decays. However, as we will check, it is negligible in this case because the tree level contribution is not suppressed.

integration variable. On the other hand, after performing the loop integration for diagrams in Fig. 1 (a), (b) and (d) and employing the Chisholm identity [6] for the Levi-Civita tensor contracted with a gamma matrix, we get the following generic form for each of the non-vanishing amplitudes

$$\mathcal{M}^k = \frac{e^4 G_F V_{ud} f_\pi}{16\pi^2} g_{\gamma\gamma\eta^{(\prime)}} \bar{u}(p_\nu) \left[F_0^k P_R + F_1^k \not{p}_\pi P_L \right] u(p_\tau), \tag{6}$$

where the superindex $k = a, b, d$ labels the non-vanishing Feynman diagrams in Figure 1.

The form factors $F_{0,1}^k$ are generated by the loop integration, they have been obtained using the Mathematica packages FeynCalc [22] and are given in terms of the invariant Passarino-Veltman (PaVe) scalar functions [20]. Further, they can be understood as corrections to the form factors appearing in eq. (2), the relations between both will be presented in the next section. The factor $g_{\gamma\gamma\eta^{(\prime)}}$ stands for the value of the $\gamma^{(*)}\gamma^{(*)}\eta^{(\prime)}$ form factor for on-shell photons, which is a global dependence of the matrix element (6) and is given by

$$g_{\gamma\gamma\eta} = \left(\frac{5}{3} C_q - \frac{\sqrt{2}}{3} C_s \right) g_{\gamma\gamma\pi^0}, \qquad g_{\gamma\gamma\eta'} = g_{\gamma\gamma\eta} \left(C_q \to C_{q'}, C_s \to -C_{s'} \right), \tag{7}$$

where the input values for the mixing coefficients can be found in Ref. [17].

The square of the total decay amplitude ($\mathcal{M} = \sum_k \mathcal{M}^k$) is given by

$$\begin{aligned} |\mathcal{M}|^2 &= \frac{\left(e^4 |V_{ud}| G_F f_\pi g_{\gamma\gamma\eta^{(\prime)}}\right)^2}{128\pi^4} \left[p_\nu \cdot p_\tau \left(|F_0|^2 - |F_1|^2 m_\pi^2 \right) \right. \\ &\left. +2\, p_\nu \cdot p_\pi \left(|F_1|^2 p_\tau \cdot p_\pi + m_\tau \mathrm{Re}[F_0 F_1^*] \right) \right], \end{aligned} \tag{8}$$

where $F_0 = \sum_{k=a,b,d} F_0^k$ and $F_1 = \sum_{k=a,b,d} F_1^k$. The explicit expression for these form factors can be found as well in appendix B of [21], they are given in terms of the independent kinematical scalars $s_{12} = (p_\pi + p_\eta)^2 = (p_\tau - p_\nu)^2$ (the square of the invariant-mass of the hadronic system) and $s_{13} = (p_\pi + p_\nu)^2 = (p_\tau - p_\eta)^2$.

4. Evaluation

An evaluation of the individual contribution of eq. (8) can be done straightforwardly considering the well know kinematics for a three body decay. In order to distinguish this individual contribution, we will employ the notation $BR_{P^0}^{\gamma\gamma} \equiv BR(\tau^- \to P^0 \pi^- \nu_\tau)$ when P^0 is produced from a two-photon intermediate state ($P = \pi, \eta, \eta'$), we obtain

$$BR_{\pi^0}^{\gamma\gamma} = 5.3 \cdot 10^{-13}, \quad BR_\eta^{\gamma\gamma} = 5.2 \cdot 10^{-13}, \quad BR_{\eta'}^{\gamma\gamma} = 0.8 \cdot 10^{-16}. \tag{9}$$

Notice that in the case of $P = \pi^0$ the ratio between the number in eq. (9) and the corresponding measured branching fraction ($BR_{\pi^0} \sim 25\%$) is at the level 10^{-11}, which validates neglecting the one-loop contribution in the $\tau^- \to \pi^- \pi^0 \nu_\tau$ decays, as we anticipated. On the other hand, the ratio for the $\eta^{(\prime)}$ modes between the numbers in eq. (9) and their respective branching fractions predicted at the tree level in Ref. [13] ($BR_\eta^{\mathrm{tree}} \sim 1.7 \cdot 10^{-5}$ and $10^{-7} \leq BR_{\eta'}^{\mathrm{tree}} \leq 10^{-8}$) are of the order of 10^{-8}.

Now, we turn to the evaluate the branching ratio including the sum of the tree level [13] and the one-loop contribution. We note that there is an equivalence between the $F_{0,+}$ form factors appearing in eq. (2) and our $F_{0,1}$ form factors in Eq. (6). It turns out appropriate to use an upper index $\gamma\gamma$ for the $F_{0,1}$ form factors defined in eq. (6) in order to avoid confusion with the F_0

[6] Chisholm identity reads $\gamma_\mu \epsilon^{\mu\nu\rho\sigma} = i\left(\gamma^\nu \gamma^\rho \gamma^\sigma - g^{\nu\rho}\gamma^\sigma - g^{\rho\sigma}\gamma^\nu + g^{\nu\sigma}\gamma^\rho\right)\gamma^5$.

form factor appearing in eq. (2). In this way, we can include the electromagnetic contribution into the vector and scalar form factors by shifting $F_{+,0}^{\pi\eta^{(\prime)}}(s_{12}) \to F_{+,0}^{\pi\eta^{(\prime)}}(s_{12}) + F_{+,0}^{\gamma\gamma}(s_{12}, s_{13})$, where:

$$F_{+}^{\gamma\gamma} = -\frac{e^4 f_\pi g_{\gamma\gamma\eta^{(\prime)}} F_1}{64\pi^2}, \qquad F_0^{\gamma\gamma} = \frac{\frac{f_\pi}{16\sqrt{2}\pi^2} e^4 g_{\gamma\gamma\eta^{(\prime)}} \left[\frac{F_0}{m_\tau} + \frac{F_1}{2}\left(1 + \frac{\Delta_{\pi\eta^{(\prime)}}}{s_{12}}\right)\right]}{c_{\pi\eta}^S \frac{\Delta_{K^0K^+}^{QCD}}{s_{12}}}, \tag{10}$$

with $c_{\pi\eta}^S$ and Δ_{PQ} defined previously. Now, using the fortran version of LoopTools we have obtained

$$BR_\eta^{\text{tree}+\gamma\gamma} - BR_\eta^{\text{tree}} \in \left[-5\cdot 10^{-9}, 2\cdot 10^{-9}\right], \tag{11}$$

$$BR_{\eta'}^{\text{tree}+\gamma\gamma} - BR_{\eta'}^{\text{tree}} \in \left[-3\cdot 10^{-12}, 3\cdot 10^{-12}\right], \tag{12}$$

here the difference comes mainly from the interference or tree- and loop-level contributions and –as in Ref. [13]– the quoted errors only arise from the uncertainties in the overall normalization factor $F_+^{\pi\eta^{(\prime)}}(0)$.

5. Conclusions

$\tau^- \to \eta^{(\prime)}\pi^-\nu_\tau$ decays belong to the SCC processes, these modes are naturally suppressed by approximate G-parity symmetry in the SM. This suppression provides an ideal place to look for New Physics contributions, which can be of the similar size as the SM effects. Consequently, a precise evaluation of the possible G-parity breaking contributions is required in order to isolate the New Physics.

In this work, we have calculated the one-loop contribution to the $\tau^- \to \eta^{(\prime)}\pi^-\nu_\tau$ decays induced by the $\gamma\gamma\eta^{(\prime)}$ vertices; as far as we know, this is the first time in considering this contributions to these decays. In the case of final state with an η meson, the two-photon intermediate states contribute -at most- with corrections at the 10^{-4} level and in the case of a decay with an η' their maximum relative size can vary between $3\cdot 10^{-4}$ and $3\cdot 10^{-5}$ depending on the value of the tree level branching ratio [7]. At the same time, we have verified that the analogous contributions for the $\tau \to \pi^0\pi^-\nu_\tau$ channel are negligible, as was expected.

We can conclude that searches at forthcoming flavor factories will not be sensitive to the one loop effects in $\tau^- \to \eta^{(\prime)}\pi^-\nu_\tau$ decays considered in this work. Even if $\tau^- \to \eta^{(\prime)}\pi^-\nu_\tau$ decays are finally observed at Belle-II it will be very difficult to achieve a measurement with a few percent accuracy even with the complete Belle-II data sample.

Acknowledgements

This work has been supported by Conacyt Projects No. FOINS-296-2016 ('Fronteras de la Ciencia') and 236394 and 250628 ('Ciencia Básica'). It has been a pleasure collaborate with G. L. C and P. R. in the publication of this work.

References

[1] S. Weinberg, Phys. Rev. **112**, 1375 (1958).
[2] T. D. Lee and C. N. Yang, Nuovo Cim. **10**, 749 (1956).
[3] C. Patrignani *et al.* [Particle Data Group], Chin. Phys. C **40**, 100001 (2016).

[7] This is due to the fact that the tree level contribution to the η' channel has a big uncertainty [13] ($\sim 90\%$) which is dominated by the error on $F_+^{\pi\eta'}(0)$.

[4] L. Grenacs, Ann. Rev. Nucl. Part. Sci. **35**, 455 (1985). doi:10.1146/annurev.ns.35.120185.002323
[5] E. A. Garcs, M. H. Villanueva, G. L. Castro and P. Roig, arXiv:1708.07802 [hep-ph].
[6] C. Leroy and J. Pestieau, Phys. Lett. **72B**, 398 (1978).
[7] P. del Amo Sánchez *et al.* [BaBar Collaboration], Phys. Rev. D **83**, 032002 (2011).
[8] B. Aubert *et al.* [BaBar Collaboration], Phys. Rev. D **77**, 112002 (2008).
[9] A. Bramon, S. Narison and A. Pich, Phys. Lett. B **196**, 543 (1987); A. Pich, Phys. Lett. B **196**, 561 (1987); J. L. Díaz-Cruz and G. López Castro, Mod. Phys. Lett. A **6**, 1605 (1991); H. Neufeld and H. Rupertsberger, Z. Phys. C **68**, 91 (1995); S. Nussinov and A. Soffer, Phys. Rev. D **78**, 033006 (2008); Phys. Rev. D **80**, 033010 (2009); N. Paver and Riazuddin, Phys. Rev. D **82**, 057301 (2010); Phys. Rev. D **84**, 017302 (2011); M. K. Volkov and D. G. Kostunin, Phys. Rev. D **86**, 013005 (2012); S. Descotes-Genon and B. Moussallam, Eur. Phys. J. C **74**, 2946 (2014).
[10] S. Weinberg, Physica A **96**, 327 (1979); J. Gasser and H. Leutwyler, Annals Phys. **158**, 142 (1984); Nucl. Phys. B **250**, 465 (1985).
[11] G. Ecker, G. Müller, H. Neufeld and A. Pich, Phys. Lett. B **477**, 88 (2000).
[12] A. Guevara, G. López Castro and P. Roig, Phys. Rev. D **95**, 054015 (2017).
[13] R. Escribano, S. Gonzàlez-Solís and P. Roig, Phys. Rev. D **94**, 034008 (2016).
[14] M. Fujikawa *et al.* [Belle Collaboration], Phys. Rev. D **78**, 072006 (2008).
[15] Z. H. Guo and J. A. Oller, Phys. Rev. D **84**, 034005 (2011); Z. H. Guo, J. A. Oller and J. Ruiz de Elvira, Phys. Rev. D **86**, 054006 (2012).
[16] S. L. Adler, Phys. Rev. **177**, 2426 (1969); J. S. Bell and R. Jackiw, Nuovo Cim. A **60**, 47 (1969).
[17] P. Roig, A. Guevara and G. López Castro, Phys. Rev. D **89**, 073016 (2014).
[18] G. 't Hooft, Nucl. Phys. B **72**, 461 (1974); **75**, 461 (1974); E. Witten, Nucl. Phys. B **160**, 57 (1979).
[19] H. Leutwyler, Nucl. Phys. Proc. Suppl. **64**, 223 (1998); R. Kaiser and H. Leutwyler, In *Adelaide 1998, Nonperturbative methods in quantum field theory* 15-29, Eur. Phys. J. C **17**, 623 (2000).
[20] G. Passarino and M. J. G. Veltman, Nucl. Phys. B **160**, 151 (1979).
[21] G. Hernández-Tomé, G. López-Castro and P. Roig, "G-parity breaking in $\tau^- \to \eta^{(\prime)}\pi^-\nu_\tau$ decays induced by the $\eta^{(\prime)}$ transition form factor," first version of this paper, which can be found at arXiv:1707.03037 [hep-ph] (v1).
[22] R. Mertig, M. Bohm and A. Denner, Comput. Phys. Commun. **64**, 345 (1991).
[23] G. J. van Oldenborgh and J. A. M. Vermaseren, Z. Phys. C **46**, 425 (1990).
[24] T. Hahn and M. Pérez-Victoria, Comput. Phys. Commun. **118**, 153 (1999).
[25] G. Ecker, J. Gasser, A. Pich and E. de Rafael, Nucl. Phys. B **321**, 311 (1989); G. Ecker, J. Gasser, H. Leutwyler, A. Pich and E. de Rafael, Phys. Lett. B **223**, 425 (1989).
[26] A. Sirlin, Rev. Mod. Phys. **50**, 573 (1978); Nucl. Phys. **B71**, 29 (1974); W. J. Marciano and A. Sirlin, Phys. Rev. Lett. **61**, 1815 (1988), *ibid.* **71**, 3629 (1993); W. J. Marciano and A. Sirlin, Phys. Rev. Lett. **56**, 22 (1986); A. Sirlin, Nucl. Phys. **B196**, 83 (1982); J. Erler, Rev. Mex. Fis. **50**, 200 (2004). J. Erler, Rev. Mex. Fis. **50**, 200 (2004).
[27] H. Czyz, S. Ivashyn, A. Korchin and O. Shekhovtsova, Phys. Rev. D **85**, 094010 (2012).
[28] S. J. Brodsky and G. R. Farrar, Phys. Rev. Lett. **31**, 1153 (1973); G. P. Lepage and S. J. Brodsky, Phys. Rev. D **22**, 2157 (1980).

IOP Conf. Series: Journal of Physics: Conf. Series **912** (2017) 012021 doi:10.1088/1742-6596/912/1/012021

Lepton flavor changing Higgs boson decays in some extensions of the Standard Model

S. Chamorro-Solano[1], A. Moyotl[2] and M. A. Perez[2].

[1]Departamento de Ciencias Naturales y Exactas, Universidad de la Costa, Calle 58 num. 55-66, Barranquilla, Colombia.

E-mail: schamorr1@cuc.edu.co

[2]Departamento de Física, Centro de Investigación y de Estudios Avanzados del Instituto Politecnico Nacional, Apdo. Postal 14-740, 07000 México D.F., México.

E-mail: mchamorro@fis.cinvestav.mx
amoyotl@fis.cinvestav.mx
mperez@fis.cinvestav.mx

Abstract. We present our results for the one-loop contributions to the Higgs boson flavor-changing decays $h \to \mu\tau$ in two extensions of the Standard Model: the Little Higgs Model with T-parity and the Two Higgs Doublet Model with a fourth generation of fermions. In both cases we find that the respective branching ratio BR($h \to \mu\tau$) is of order $10^{-4} - 10^{-6}$. In the case of the 4G2HDM, the one-loop radiative correction is of the same order of magnitude as the tree level branching ratio. We find that in both models the branching ratios for the decay modes $h \to e\tau$, $e\mu$ are even more suppressed.

1. Introduction

In the Standard Model (SM) the flavor-changing Higgs decays $h \to \mu\tau$, $e\tau$ and $e\mu$ are forbidden at tree level due to the Glashow-Iliopoulos-Maiani mechanism [1]. The one-loop contributions to these decay channels in the SM with massive neutrinos are highly suppressed, of order 10^{-29} [2]. Due to this expectation, the excess reported by the CMS [3] and the ATLAS [4] Collaborations for the branching ratio BR($h \to \mu\tau$) stimulated great interest among theoreticians [7–11]. In particular, in some extensions of the SM this branching ratio could be at the level of few percent, close to the upper limits included in the first reports of CMS and ATLAS [9–11]. Recently, CMS have published new results on this decay mode; however, their search does not throw a clear conclusion [5,6] and the respective branching ratio now has an upper limit of about 10^{-3}.

In general the amplitude of the $h \to \ell_i \ell_j$ decay is given by:

$$\mathcal{M} = i\bar{u}(p_1, m_i)(A + iB\gamma_5)u(p_2, m_j), \quad (1)$$

where A and B are one-loop form factors, while m_k is the mass of the lepton ℓ_k. In the approximation of massless external leptons the decay width can be written as:

$$\Gamma(h \to \tau\mu) = \frac{m_h}{8\pi}(|A|^2 + |B|^2), \tag{2}$$

in the next sections we mention the potential some results of two extensions of the SM; the Littlest Higgs Model with T-parity (LHM+T), and the Two Higgs doublet model with fourth family (4GTHDM).

The present report is organized as follow. In section we mention some general aspect of the LHM with T-parity and its respective contribution to $h \to \tau\mu$. In section 3 we analyzed the fermionic fourth family in the context of the Two Higgs doublet model, as well as the results for $h \to \ell_i \ell_j$ decays. Finally, our conclusions and outlook are presented in the section 3.

2. The Littlest Higgs Model with T-Parity

The LHM+T comes from a Littlest Higgs model type proposed by Cheng and Low [12], which finally protects the mass of the Higgs boson and gives a more elegant solution to the problem of the precision electroweak that had the LHM initially. This model is based of nonlinear sigma model with a global $SU(5)$ symmetry, which is broken down to its subgroup $SO(5)$ via a vacuum expectation value of order $f \sim \mathcal{O}(TeV)$. At the same time a subgroup $[SU(2) \otimes U(1)]^2$ is also broken in the SM gauge group. By introducing an additional symmetry, the T-parity can relax the constraint of the EW precision data, leading to a scale f as 500 GeV. The T-parity exchange the two $SU(2) \otimes U(1)$ factors, thus the number of parameters are reduced. To implement the T-parity the introduction of mirror fermions is required in an extra multiplet, which are necessary in order to eliminate possible quadratic divergences in the Higgs mass. Now the SM-gauge light bosons are T-even while the new heavy bosons are T-odd; also fermions are T-even.

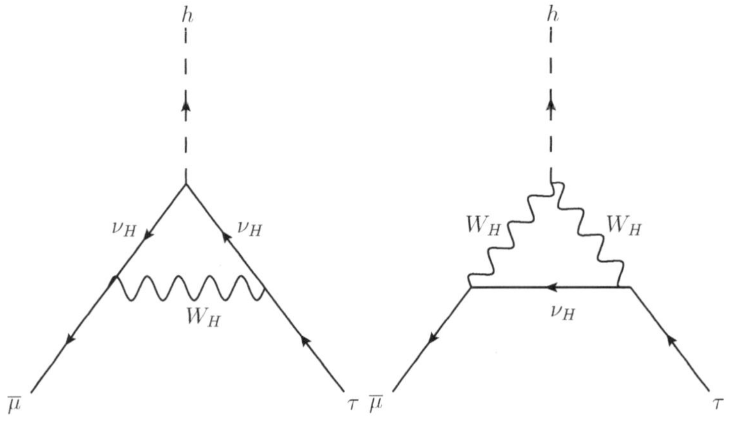

Figure 1. Feynman diagrams inducing the $h \to \tau\mu$ decay in the LHM with T-parity. We show only the contribution induced by the heavy charged gauge boson W_H and the heavy mirror neutrino ν_H. Also bubble diagrams are necessaries to remove the divergence.

The new particles of the gauge boson sector of this model includes W_H, Z_H and A_H with masses of order f. There are also heavy charged mirror leptons (ℓ_H) and neutrinos (ν_H) with mass of the same order of magnitude as the heavy gauge bosons. Due to the T-parity symmetry, the heavy charged mirror leptons do not couple directly with the SM Higgs boson [13, 14].

The T-parity forbids any contribution at tree level that includes heavy gauge bosons, then in the LHM+T the decay mode $h \to \mu\tau$ is induced at one loop-level by the heavy gauge bosons W_H and Z_H with its respective mirror leptons. We considered a symmetry breaking scale f between 500 and 2000 GeV. Our results indicate that main contribution comes of the heavy charged gauge boson $W_H^{\mp}$, where the respective branching ratio is given by $(10^{-3} - 10^{-1})|V_{H\mu}V_{H\tau}|^2$, where the mixing matrices $V_{H\ell}$ correspond to the mixing of the heavy gauge bosons of the model and they satisfy the relation $V_{H\nu}V_{H\ell} = V_{PMNS}$, i.e., the Pontecorvo-Maki-Nakagata-Saki matrix. However, this result decreases with reasonable scenarios for these mixing matrices [15].

3. Two Higgs Doublet Model with a fourth Generation

In this model the flavor changing transitions involving the SM model Higgs boson are induced at tree level and, of course, also at one-loop level. We choose the 4G2HDM version, where the heavy scalar bosons couple only to the fourth generation fermions while the SM Higgs boson couples only to the light fermions. This guarantees that the 4G2HDM is in concordance with the known data coming from the LHC and the electroweak precision observables [16,17].

The fourth generation can be implemented in three different scenarios, we considered only the scenario where one of the two Higgs doublets (ϕ_1) gives masses only to the 4th family fermions and the other one (ϕ_2) generates the masses of the rest of the known fermions. However, there is a Flavor Changing Neutral Couplings (FCNC) induced at tree level between scalar bosons and fermions. The Lagrangians that describe the change of flavor through neutral and charged currents, are respectively

$$\mathcal{L}_{\rm NC} = \frac{g}{4m_W} g_\phi \bar{\ell}_i \left[\left(g_s^\phi\right)_{ij} + \left(g_p^\phi\right)_{ij} \gamma_5 \right] \ell_j \phi, \tag{3}$$

$$\mathcal{L}_{\rm CC} = \frac{g}{2\sqrt{2}m_W} g_\phi \bar{\nu}_i \left[\left(g_s^\phi\right)_{ij} + \left(g_p^\phi\right)_{ij} \gamma_5 \right] \ell_j H^+, \tag{4}$$

where ϕ is any neutral scalar boson and i, j run for for each generation of fermions, g_s and g_p are the scalar and pseudoscalar coupling constants. While g_ϕ is a constant that takes different values for each scalar bosons, the explicit form of these constants can be found in [17].

The 4G2HDM has several free parameters such as the masses of the new scalar bosons and the heavy fermions, as well as tan$\beta = v_1/v_2$, the ratio of the vacuum expectation values of the two Higgs doublets, and additional mixing matrices among the known fermions and the new 4th generation fermions. In our calculation we considered the mass range for the 4th family lepton between 100 and 350 GeV and for the heavy neutral Higgs boson between 200 and 700 GeV. In the 4G2HDM scenario considered $\tan\beta$ should be of order unit [17].

Our results indicate that the dominant contribution to the one-loop radiative correction to the BR($h \to \mu\tau$) in this model is associated with the virtual exchange of the heavy charged Higgs boson $H^{\pm}$ in a wide range of the free parameters involved in this model [16, 17]. We also calculated the one-loop contribution coming from the virtual exchange of heavy neutral bosons

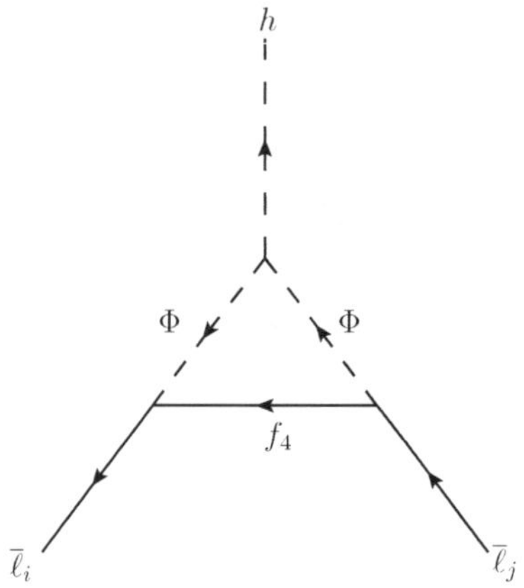

Figure 2. Feynman diagram inducing the $h \to \ell_i \ell_j$ decay in the 4GTHDM, where for $\Phi = h^0$, H^0, A^0 we have $f_4 = \ell_4$ and for $\Phi = H^\pm$ we have $f_4 = \nu_4$.

but they are smaller than the heavy charged Higgs boson contribution.

It is interesting to note that the 4G2HDM allows a tree level FCNC coupling that induces a branching ratio for the decay mode $h \to \mu\tau$ of the same order of magnitude than the charged Higgs boson one-loop contribution. However, the tree level results depends on the χ parameter that involves the mixing angles involved in the model and thus the respective predictions can be highly suppressed [16].

4. The decay modes $h \to e\tau$, $e\mu$

It is possible to use or results for the decay mode $h \to \mu\tau$ in order to obtain the respective branching ratios expected in both models for the other two FCNC decay modes $h \to e\tau$, $e\mu$. Since our $h \to e\tau$ results for the one-loop contributions to the case $h \to \mu\tau$ involve exact analytical expressions, the respective results for the other two FCNC decay modes are easily obtained by just substituting the new masses and coupling constants. Our results indicate that these two decay modes are even more highly suppressed: BR($h \to e\mu$)$\sim 10^{-11}$ [16].

5. Acknowledgements

We appreciate discussions with our colleagues H. Castilla Valdez and P. Roig. We thank support from CONACyT (Mexico).

References

[1] S. L. Glashow, J. Iliopoulos, and L. Maiani, *Phys. Rev.* D, 2:1285-1292, Oct 1970.
[2] L. Diaz-Cruz and J. J. Toscano, *Phys. Rev.*, D62:116005, 2000.
[3] V. Khachatryan et al [CMS Collaboration], *Phys. Lett.*, **B749**:337-362, 2015.
[4] G. Aad et al [ATLAS Collaboration], *JHEP*, 03:076, 2013.
[5] CMS Collaboration, CMS-PAS-HIG-16-005.
[6] CMS Collaboration, CMS-PAS-HIG-17-001.
[7] C. Alvarado et al., *Phys. Rev.*, D94(7):075010, 2016.

[8] A. Lami and P. Roig, *Phys. Rev.*, D94(5):056001, 2016.
[9] J. Herrero-Garca, T. Ohlsson, S. Riad, and J. Wirn, *JHEP*, 04:13, 2017.
[10] D. Aristizabal Sierra and A. Vicente, *Phys. Rev.*, **D90**(11):115004, 2014.
[11] F. del Aguila et al., *JHEP* 1708 (2017) 028.
[12] H. C. Cheng, I. Low, and L. T. Wang, *Phys. Rev.* **D74**, 055001 (2006).
[13] N. Arkani-Hamed, A. G. Cohen, E. Katz, and A. E. Nelson, *JHEP*, 07:034, 2002.
[14] J. Hubisz, Seung J. Lee, and G. Paz, *JHEP*, 06:041, 2006.
[15] S. Chamorro-Solano, A. Moyotl, and M. A. Pérez, *J. Phys. Conf. Ser.*, 761(1):012051, 2016.
[16] A. Moyotl, S. Chamorro-Solano, and M.A. Pérez, *Nuclear and Particle Physics Proceedings*, 287288:205-207, 2017.
[17] Chamorro-Solano, S. and Moyotl, A. and Pérez, M. A., arXiv:1707.00100 (2017).

Unitarity and stability conditions in a 4-Higgs doublet model with an S_3-family symmetry.

C Espinoza[1], E A Garcés[2], M Mondragón[3] and H Reyes-González[3].

[1]Cátedras CONACyT - Instituto de Física, Universidad Nacional Autónoma de México, Apdo. Postal 20-364, Cd. México 01000, México.

[2]Departamento de Física, Centro de Investigación y de Estudios Avanzados del IPN, Apdo. Postal 14-740 07000 Cd. México, México.

[3]Instituto de Física, Universidad Nacional Autónoma de México, Apdo. Postal 20-364, Cd. México 01000, México.

E-mail: `myriam@fisica.unam.mx`

Abstract. In order to obtain a dark matter candidate we propose an extension of the S_3 symmetric 3-Higgs doublet model, adding a new scalar doublet, occupying all the irreducible representations of the discrete symmetry. To ensure stability of the dark matter particle we impose an extra Z_2 symmetry. We find the analytical masses of the scalar particles and constrain their values using stability and unitarity conditions.

1. Introduction.

Finding what dark matter (DM) is made of is one of the main challenges in particle physics and cosmology. It has been proposed in [1] that DM is conformed by neutral scalar particles, a very interesting hypothesis as these particles have the characteristics expected for DM: Neutral, cold and weakly interacting (see for example [2, 3, 4, 5, 6]).

Among the numerous proposals to extend the scalar sector of the standard model, the 3-Higgs Doublet Model with an S_3-family symmetry (3H-S_3) presents interesting phenomenology, such as the prediction of a non zero reactor neutrino mixing angle θ_{13} and of a CKM matrix in accordance with experimental results (see e.g. [7, 8, 9, 10, 11, 12, 13, 14])

A natural generalization embracing these two ideas suggests itself: enlarge the 3H-S_3 model with an additional doublet representing a dark matter candidate. In this letter we present an analysis of the S_3-symmetric 4-Higgs Doublet Model (4HDM) in which we occupy all irreducible representations of the S_3 symmetry: one symmetric singlet, one antisymmetric singlet and one doublet. The 3H-S_3 is constituted by the symmetric singlet and doublet representations, while for the fourth Higgs doublet (in the antisymmetric representation) we impose a Z_2 symmetry ensuring the stability of the potential dark matter candidates. In the following we present the analytical calculation of the masses of the scalar particles at tree level and constrain their values using unitarity and stability conditions. A complete analysis of the model with its dark matter phenomenology will be presented elsewhere [15].

IOP Conf. Series: Journal of Physics: Conf. Series **912** (2017) 012022 doi:10.1088/1742-6596/912/1/012022

2. The model.

The most general $SU(2)_L \times U(1)_Y$ invariant renormalizable scalar potential for a 4-Higgs doublet model with additional $S_3 \times Z_2$ symmetry is given by:

$$\begin{aligned}
V = {} & \mu_0^2 H_s^\dagger H_s + \mu_1^2 (H_1^\dagger H_1 + H_2^\dagger H_2) + \mu_2^2 H_a^\dagger H_a \\
& + \lambda_1 (H_1^\dagger H_1 + H_2^\dagger H_2)^2 + \lambda_2 (H_1^\dagger H_2 - H_2^\dagger H_1)^2 \\
& + \lambda_3 [(H_1^\dagger H_1 - H_2^\dagger H_2)^2 + (H_1^\dagger H_2 + H_2^\dagger H_1)^2] \\
& + \lambda_4 [(H_s^\dagger H_1)(H_1^\dagger H_2 + H_2^\dagger H_1) + (H_s^\dagger H_2)(H_1^\dagger H_1 - H_2^\dagger H_2) + \text{h.c.}] \\
& + \lambda_5 (H_s^\dagger H_s)(H_1^\dagger H_1 + H_2^\dagger H_2) \\
& + \lambda_6 [(H_s^\dagger H_1)(H_1^\dagger H_s) + (H_s^\dagger H_2)(H_2^\dagger H_s)] \\
& + \lambda_7 [(H_s^\dagger H_1)(H_s^\dagger H_1) + (H_s^\dagger H_2)(H_s^\dagger H_2) + \text{h.c.}] \\
& + \lambda_8 (H_s^\dagger H_s)^2 \\
& + \lambda_{10} (H_a^\dagger H_a)(H_1^\dagger H_1 + H_2^\dagger H_2) \\
& + \lambda_{11} [(H_a^\dagger H_1)(H_1^\dagger H_a) + (H_a^\dagger H_2)(H_2^\dagger H_a)] \\
& + \lambda_{12} [(H_a^\dagger H_1)(H_a^\dagger H_1) + (H_a^\dagger H_2)(H_a^\dagger H_2) + \text{h.c.}] \\
& + \lambda_{13} (H_a^\dagger H_a)^2 \\
& + \lambda_{14} (H_s^\dagger H_a H_a^\dagger H_s).
\end{aligned} \tag{1}$$

Here we have arranged the $SU(2)_L$ doublets H_1 and H_2 into a column vector transforming in the doublet representation of S_3, while H_s and H_a are required to transform in the symmetric and antisymmetric singlet representations of S_3 respectively. In the following we will assume all the couplings λ_i, $i = 1...14$ to be real, CP conserving, and $|\lambda_i| < 4\pi$ in order to fulfill perturbativity. Note that there are no terms with odd powers of H_a, the only field assumed odd under Z_2.

After electroweak symmetry breaking all Higgs doublets acquire a vacuum expectation value (vev) which we denote respectively by v_0, v_1, v_2 and v_a. Nevertheless, to avoid the breaking of the Z_2 symmetry we fix v_a identically to zero, and from the minimization conditions (a.k.a. tadpole equations) of the potential (1) the fourth equation $\partial V/\partial v_a = 0$ is therefore automatically satisfied. The three minimization equations left ($\partial V/\partial v_i = 0$, $i = 1, 2, 3$) reduce to those of the 3H-S_3, from which we reproduce the following conditions to the quadratic couplings:

$$\begin{aligned}
\mu_0^2 &= -(\lambda_5 + \lambda_6 + 2\lambda_7)(v_1^2 + v_2^2) - 2\lambda_8 v_0^2 + \frac{\lambda_4 (v_2^2 - 3v_1^2) v_2}{v_0} \\
\mu_1^2 &= -(\lambda_5 + \lambda_6 + 2\lambda_7) v_0^2 - 2(\lambda_1 + \lambda_3)(v_1^2 + v_2^2) - 6\lambda_4 v_2 v_0 \\
\mu_1^2 &= -(\lambda_5 + \lambda_6 + 2\lambda_7) v_0^2 - 2(\lambda_1 + \lambda_3)(v_1^2 + v_2^2) + 3\lambda_4 \frac{v_0 (v_2^2 - v_1^2)}{v_2}.
\end{aligned} \tag{2}$$

The self consistency of the above conditions require either $\lambda_4 = 0$ or the alignment of the vacuum expectation values v_1 and v_2:

$$v_1 = \sqrt{3} v_2. \tag{3}$$

Since the case $\lambda_4 = 0$ is phenomenologically unappealing in the context of the 3HDM, we will assume the alignment (3) in the rest of this work.

2.1. The masses.

We parametrize the Higgs doublets by

$$H_s = \begin{pmatrix} h_s'^{+} \\ v_s + h_s'^{n} + i h_s'^{p} \end{pmatrix} \tag{4}$$

and similarly for H_1, H_2 and H_a. Here the indexes n and p refer to *neutral (scalar)* and *(neutral) pseudoscalar* and we use primes to distinguish from the mass eigenstates denoted by the same letters. The masses of the scalar particles are found by diagonalizing the corresponding mass matrices, e.g. for the neutral scalar fields the following matrix:

$$(\mathcal{M}^2_{h'^n})_{ij} = \frac{1}{2} \frac{\partial^2 V}{\partial h_i'^n \partial h_j'^n}\bigg|_{\min}, \tag{5}$$

which is block diagonal such that the primed fields $h_s'^n$, $h_1'^n$ and $h_2'^n$ mix into the mass eigenstates h_s^n, h_1^n and h_2^n and the Z_2 odd field remains unmixed $h_a'^n = h_a^n$.

The expressions for the masses presented below are separated by neutral scalar, neutral pseudo scalar and charged particles.

- Masses of the neutral scalar particles:

$$\begin{aligned}
m^2_{h_s^n} &= -18\lambda_4 v_0 v_2 \\
m^2_{h_a^n} &= \mu_2^2 + \lambda_{14} v_0^2 + 4(\lambda_{10} + \lambda_{11} + 2\lambda_{12}) v_2^2 \\
m^2_{h_1^n} &= (\frac{1}{v_0})(2\lambda_8 v_0^3 + v_2(3\lambda_4 v_0^2 + 8(\lambda_1 + \lambda_3) v_0 v_2 - 4\lambda_4 v_2^2) + \\
&\quad ((4\lambda_8^2 v_0^6 - 12\lambda_4\lambda_8 v_0^5 v_2 + (9\lambda_4^2 + \\
&\quad 16((\lambda_5 + \lambda_6 + 2\lambda_7)^2 - 2(\lambda_1 + \lambda_3)\lambda 8)) v_0^4 v_2^2 + \\
&\quad 16\lambda_4(3(\lambda_1 + \lambda_3 + 2(\lambda_5 + \lambda_6 + 2\lambda_7)) - \lambda_8) v_0^3 v_2^3 + \\
&\quad 8(8(\lambda_1 + \lambda_3)^2 + 21\lambda_4^2) v_0^2 v_2^4 + 64(\lambda_1 + \lambda_3)\lambda_4 v_0 v_2^5 + \\
&\quad 16\lambda_4^2 v_2^6)^{1/2}) \\
m^2_{h_2^n} &= (\frac{1}{v_0})(2\lambda_8 v_0^3 + v_2(3\lambda_4 v_0^2 + 8(\lambda_1 + \lambda_3) v_0 v_2 - 4\lambda_4 v_2^2) \\
&\quad - (4\lambda_8^2 v_0^6 - 12\lambda_4\lambda_8 v_0^5 v_2 + (9\lambda_4^2 + \\
&\quad 16((\lambda_5 + \lambda_6 + 2\lambda_7)^2 - 2(\lambda_1 + \lambda_3)\lambda_8)) v_0^4 v_2^2 + \\
&\quad 16\lambda_4(3(\lambda_1 + \lambda_3 + 2(\lambda_5 + \lambda_6 + 2\lambda_7)) - \lambda_8) v_0^3 v_2^3 + \\
&\quad 8(8(\lambda_1 + \lambda_3)^2 + 21\lambda_4^2) v_0^2 v_2^4 + 64(\lambda_1 + \lambda_3)\lambda_4 v_0 v_2^5 + \\
&\quad 16\lambda_4^2 v_2^6)^{1/2}).
\end{aligned} \tag{6}$$

- Masses of the neutral pseudo scalar particles:

$$\begin{aligned}
m^2_{h_s^p} &= 0 \\
m^2_{h_a^p} &= \mu_2^2 + \lambda_{14} v_0^2 + 4(\lambda_{10} + \lambda_{11} - 2\lambda_{12}) v_2^2 \\
m^2_{h_1^p} &= -\frac{2(2\lambda_7 v_0^3 + 5\lambda_4 v_0^2 v_2 + 8\lambda_2 v_0 v_2^2 + 8\lambda_3 v_0 v_2^2)}{v_0} \\
m^2_{h_2^p} &= -\frac{2(2\lambda_7 v_0 + \lambda_4 v_2)(v_0^2 + 4v_2^2)}{v_0}.
\end{aligned} \tag{7}$$

- Masses of the charged particles:

$$\begin{aligned} m_{h_s^\pm} &= 0 \\ m_{h_a^\pm} &= \mu_2^2 + 4\lambda_{10} v_2^2 \\ m_{h_1^\pm} &= -(\lambda_6 + 2\lambda_7)v_0^2 - 10\lambda_4 v_0 v_2 - 16\lambda_3 v_2^2 \\ m_{h_2^\pm} &= -\frac{(\lambda_6 v_0 + 2\lambda_7 v_0 + 2\lambda_4 v_2)(v_0^2 + 4v_2^2)}{v_0}. \end{aligned} \tag{8}$$

2.2. Stability conditions.

The stability conditions, ensure that the vacuum is stable i. e. that the potential has a minimum. They were found for the potential (1), following the procedure in [14] for the 3H-S_3:

$$\begin{aligned} \lambda_8 &> 0 \\ \lambda_1 + \lambda_3 &> 0 \\ \lambda_5 &> -2\sqrt{(\lambda_1+\lambda_3)\lambda_8} \\ \lambda_5 + \lambda_6 - 2|\lambda_7| &> \sqrt{(\lambda_1+\lambda_3)\lambda_8} \\ \lambda_1 - \lambda_2 &> 0 \\ \lambda_1 + \lambda_3 + \mid 2\lambda_4 \mid + \lambda_5 + 2\lambda_7 + \lambda_8 &> 0 \\ \lambda_{13} &> 0 \\ \lambda_{10} &> -2\sqrt{(\lambda_1+\lambda_3)\lambda_{13}} \\ \lambda_{10} + \lambda_{11} - 2|\lambda_{12}| &> \sqrt{(\lambda_1+\lambda_3)\lambda_{13}} \\ \lambda_{14} &> -2\sqrt{\lambda_8\lambda_{13}}. \end{aligned} \tag{9}$$

2.3. Unitarity constraints.

Unitarity constraints over the quartic parameters can be obtained from the elegant LQT method [16]. On account of renormalizability, scattering amplitudes cannot exhibit unphysical growth in the limit of high energies. Thus, one is led to impose (tree level) unitarity on different sets of scattering processes, in particular those of two particle states. Defining the matrix $M_{ij} = \mathcal{M}_{i\to j}$ where the indices stand for all possible two particle processes, it suffices to consider processes involving Higgs scalars and longitudinal vector bosons. The eigenvalues of the matrix M_{ij}, denoted $a_i^\pm$ and b_i for charged and neutral two particle channels, will then be constrained according to $\mid a_i^\pm \mid, \mid b_i \mid < 16\pi$, reflecting the physical fact that the coefficient a_0 of the s-wave term in a partial wave expansion of the scattering amplitude is bounded (e.g. $|a_0| < 1/2$) in the limit of high energy exchange.

Of course, the eigenvalues corresponding to physical processes already present in the 3H-S_3 are the same in this 4H-S_3, so here we only report on the new constraints. Their expressions in

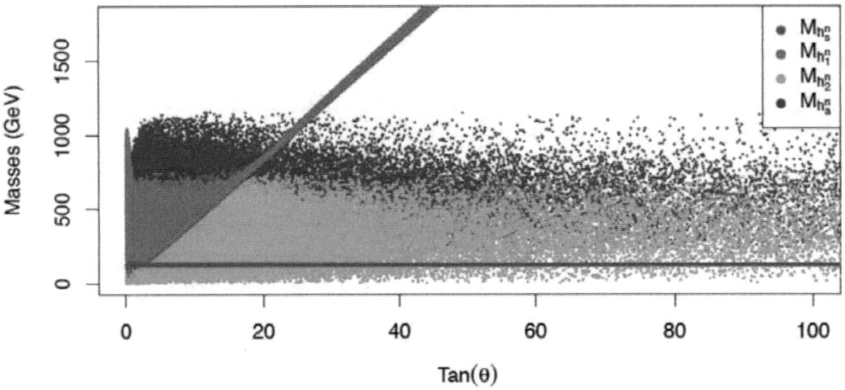

Figure 1. Mass scan for neutral scalar particles, where h_s^n is the Higgs boson of the SM.

terms of λ_i are:

$$\begin{aligned}
a_7^{\pm} &= \pm\sqrt{\lambda_{10}^2 + 2\lambda_{10}(\lambda_{11} - 4\lambda_{12}) + 2(\lambda_{11}^2 - 6\lambda_{11}\lambda_{12} + 10\lambda_{12}^2)} \\
a_8^{\pm} &= \pm\sqrt{(\lambda_{10} - 2\lambda_{12})(\lambda_{10} + 2\lambda_{11} - 6\lambda_{12})} \\
a_9^{\pm} &= \lambda_{10} \pm \lambda_{11} \\
a_{10}^{\pm} &= \lambda_{10} \pm 2\lambda_{12} \\
a_{11}^{\pm} &= \pm\lambda_{14} \\
a_{12}^{\pm} &= \pm\frac{\lambda_{14}}{2} \\
b_7 &= \lambda_{10} + 2\lambda_{11} + 6\lambda_{12}.
\end{aligned} \tag{10}$$

Eigenvalues $a_i^{\pm}$ and b_i where $i = 1, 2, ..., 6.$ were found by Das and Dey for the 3HDM in [12], whereas $a_i^{\pm}$ where $i = 7, 8, ..., 12$ and b_7 correspond to 4H-S_3 interactions decoupled from the 3H-S_3 matrices, e.g. $h_1^n h_a^n h_1^n h_a^n$. The dispersion matrices involving fields from 4 Higgs Doublet, e.g. $h_a^n h_a^n h_s^n h_s^n$ were solved numerically during the scan with pseudorandom values for the λ_i couplings.

3. Numerical analysis.

To perform the analysis is convenient to parametrize the vevs in spherical coordinates, so

$$\begin{aligned}
v_0 &= v\cos\theta \\
v_1 &= v\sin\theta\cos\phi \\
v_2 &= v\sin\theta\sin\phi,
\end{aligned} \tag{11}$$

IOP Conf. Series: Journal of Physics: Conf. Series **912** (2017) 012022 doi:10.1088/1742-6596/912/1/012022

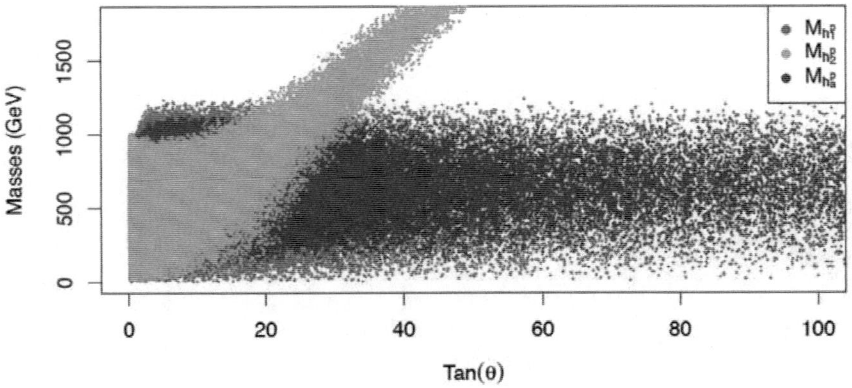

Figure 2. Mass scan for neutral pseudoscalar particles, where h_s^n is the Higgs boson of the SM.

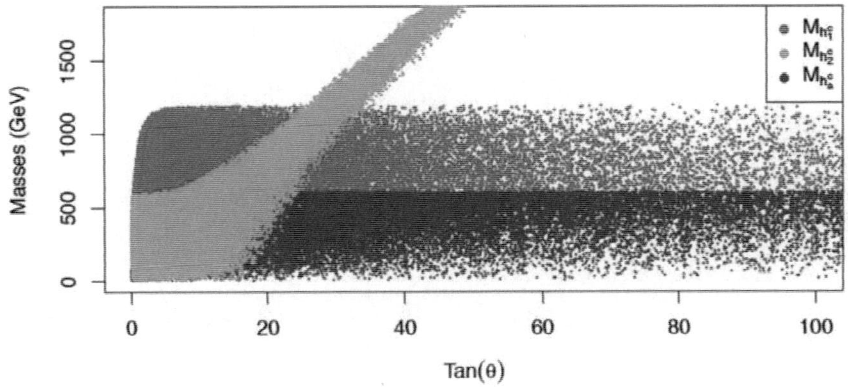

Figure 3. Mass scan for charged particles, where h_s^n is the Higgs boson of the SM.

where $\theta \in (0, \pi)$ and $\phi \in (0, 2\pi)$. With this parametrization and the relation found in Equation 3 we get

$$\tan^2 \phi = \frac{1}{3}. \tag{12}$$

IOP Conf. Series: Journal of Physics: Conf. Series **912** (2017) 012022 doi:10.1088/1742-6596/912/1/012022

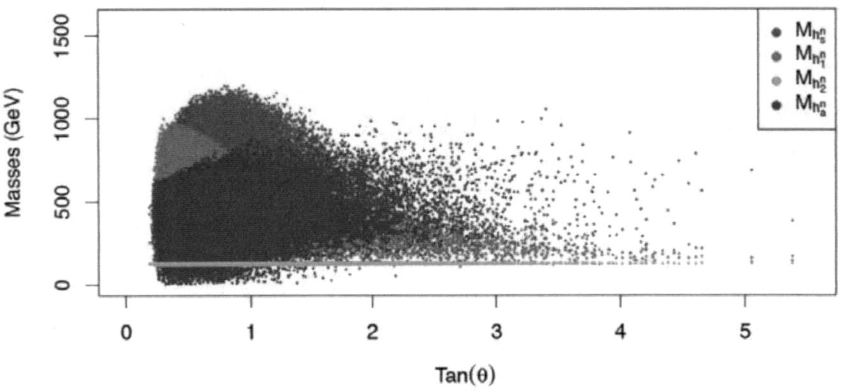

Figure 4. Mass scan for neutral scalar particles, where h_2^n is the Higgs boson of the SM.

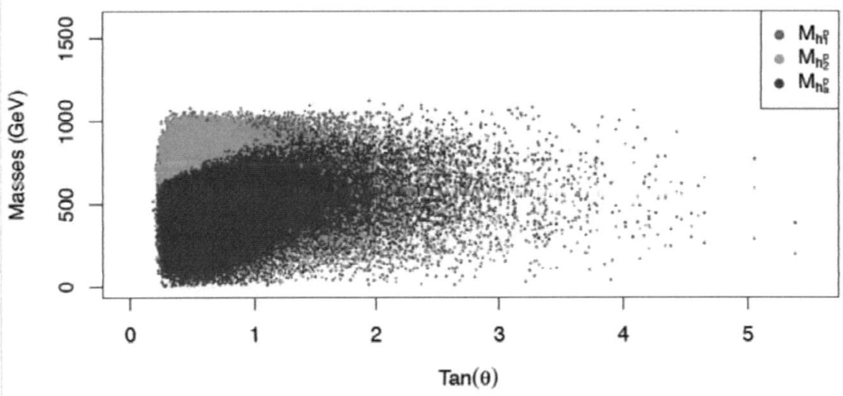

Figure 5. Mass scan for neutral pseudoscalar particles, where h_2^n is the Higgs boson of the SM.

Hence the vevs only depend on θ:

$$\begin{aligned} v_0 &= v\cos\theta \\ v_2 &= \frac{1}{2}v\sin\theta. \end{aligned} \tag{13}$$

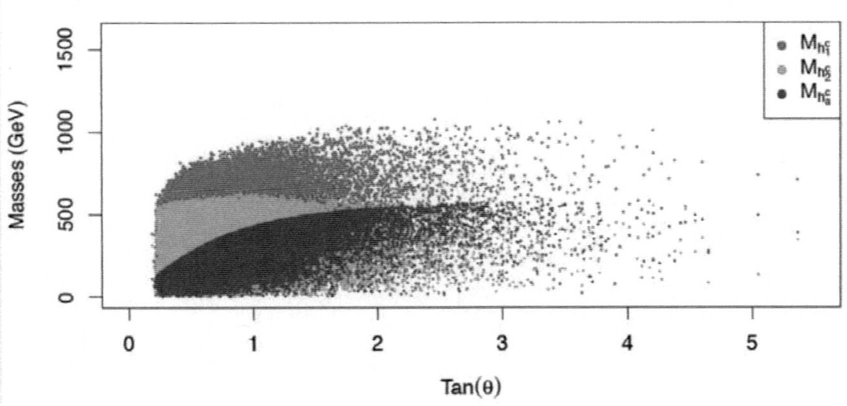

Figure 6. Mass scan for charged particles, where h_2^n is the Higgs boson of the SM.

We computed the dependence on $\tan\theta$ of the mass spectrum of the Higgs bosons. The figures 1, 2 and 3 correspond to the neutral scalar, neutral pseudoscalar and charged higgses where h_s^n corresponds to the SM Higgs boson, while the figures 4, 5 and 6 correspond to the neutral scalar, neutral pseudoscalar and charged higgses where h_2^n corresponds to the SM Higgs boson. All the figures above mentioned are scatter plots of mass vs. $\tan\theta$ and generated varying the λ_i parameters randomly, where we choose only the points that satisfy the unitarity and stability conditions previously calculated.

4. Conclusions.

The S3 flavor symmetry has been very successful accommodating fermion masses and mixings; in this work we use four $SU(2)$ doublets, using all irreducible S3 representations, which provides a rich phenomenology where one or several dark matter candidates can be feasible. We found analytical expression for the masses of the scalar particles and constrained their numerical values applying stability and unitarity conditions. The mass spectra was computed for two cases: 1) the SM Higgs boson corresponds to h_s^n and 2) the SM Higgs boson corresponds to h_2^n. In the plots 1, 2 and 3 corresponding to the first case, we found that the masses of h_1^n, h_2^p and h_2^c are heavier than the rest for large $\tan\theta$. While, in the plots 4, 5 and 6 all masses have nearly the same range, with an upper limit close to 1TeV, the exception being $m_{h_2^c}$ and $m_{h_a^c}$ with a 500GeV upper limit. As a perspective of these results we are currently working in the complete analysis of the model and on the dark matter phenomenology in order to constrain the parameter space of the model by relic density measurements.

Acknowledgements
This work was supported by PAPIIT IN111115 and CONACYT. C.E. acknowledges the support of CONACYT. E.A.G. thanks support by CONACYT Project No. FOINS-296-2016 (Fronteras de la Ciencia).

IOP Conf. Series: Journal of Physics: Conf. Series **912** (2017) 012022 doi:10.1088/1742-6596/912/1/012022

References.

[1] Silveira V and Zee A 1985 *Phys. Lett.* **B161** 136–140
[2] Tytgat M H G 2011 *PoS* **IDM2010** 126 (*Preprint* 1012.0576)
[3] McDonald J 1994 *Phys. Rev.* **D50** 3637–3649 (*Preprint* hep-ph/0702143)
[4] Bento M C, Bertolami O, Rosenfeld R and Teodoro L 2000 *Phys. Rev.* **D62** 041302 (*Preprint* astro-ph/0003350)
[5] Lopez Honorez L, Nezri E, Oliver J F and Tytgat M H G 2007 *JCAP* **0702** 028 (*Preprint* hep-ph/0612275)
[6] Ilnicka A, Krawczyk M and Robens T 2016 *Phys. Rev.* **D93** 055026 (*Preprint* 1508.01671)
[7] Mondragon A and Rodriguez-Jauregui E 2000 *AIP Conf. Proc.* **531** 310–314 [AIP Conf. Proc.490,393(1999)]
[8] Kubo J, Mondragon A, Mondragon M and Rodriguez-Jauregui E 2003 *Prog. Theor. Phys.* **109** 795–807 [Erratum: Prog. Theor. Phys.114,287(2005)] (*Preprint* hep-ph/0302196)
[9] Mondragon A, Mondragon M and Peinado E 2007 *Phys. Rev.* **D76** 076003 (*Preprint* 0706.0354)
[10] Gonzalez Canales F, Mondragon A and Mondragon M 2013 *Fortsch. Phys.* **61** 546–570 (*Preprint* 1205.4755)
[11] González Canales F, Mondragón A, Mondragón M, Saldaña Salazar U J and Velasco-Sevilla L 2013 *Phys. Rev.* **D88** 096004 (*Preprint* 1304.6644)
[12] Das D and Dey U K 2014 *Phys. Rev.* **D89** 095025 [Erratum: Phys. Rev.D91,no.3,039905(2015)] (*Preprint* 1404.2491)
[13] Barradas-Guevara E, Félix-Beltrán O and Rodríguez-Jáuregui E 2014 *Phys. Rev.* **D90** 095001 (*Preprint* 1402.2244)
[14] Emmanuel-Costa D, Ogreid O M, Osland P and Rebelo M N 2016 *JHEP* **02** 154 (*Preprint* 1601.04654)
[15] Espinoza C Garcés E A M M and H R G In preparation.
[16] Lee B W, Quigg C and Thacker H B 1977 *Phys. Rev.* **D16** 1519

IOP Conf. Series: Journal of Physics: Conf. Series **912** (2017) 012023 doi:10.1088/1742-6596/912/1/012023

The ALICE experiment upgrades for LHC Run 3 and beyond: contributions from mexican groups

L. Valencia Palomo

Facultad de Ciencias en Física y Matemáticas, Universidad Autónoma de Chiapas

E-mail: lizardo.valencia.palomo@cern.ch

Abstract. The ALICE experiment at the LHC was designed and built to study the Quark-Gluon Plasma, a state of matter where quarks and gluons are not confined to hadrons and that can be recreated using high-energy heavy-ion collisions. The plans for ALICE after 2020 include the collection of more than 10 nb^{-1} of heavy-ion collisions at luminosities up to $6x10^{27}$ cm^{-2} s^{-1}, corresponding to collision rates of 500 kHz. Such scenario imposes stringent constraints to the detector performances, forcing a major upgrade of the experiment. In this proceeding the main characteristics of the detector upgrades are presented, in particular the contributions from mexican groups are highlighted.

1. Introduction

Quantum Chromodynamics (QCD) is the correct gauge theory to describe the interactions among partons, but it is not enough, as it is also necessary to study the phases of the quark matter. Since equilibrium and phase transitions involve quarks and gluons interacting over a large distance scales, perturbative QCD cannot be applied. The solution to this problem is provided by lattice QCD, a non perturbative treatment of QCD formulated on a discrete lattice of space-time coordinates, where the extraction of the corresponding thermodynamic variables is possible.

Lattice QCD predicts that, at very high temperature and energy densities, ordinary nuclear matter undergoes a phase transition where quarks and gluons are no longer confined into hadrons. This new state of matter is called Quark-Gluon Plasma (QGP) and existed during the early Universe, a few microseconds after its formation. It is also believed to exist in the core of compact stars.

Theoretical calculations indicate that the transition between hadronic matter and the QGP at zero baryo-chemical potencial (u_B) and high temperature is not a sharp phase transition, but rather a smooth cross-over. Nowadays it has been established that the critical temperature where this transition occurs is $T_c \approx 150$ MeV [1]. By increasing u_B lattice QCD predicts the existence of a Critical End-Point, where the phase transition is expected to become of first order. This picture of strongly-interacting matter is called QCD phase diagram.

In the laboratory the only way to recreate the QGP is through high-energy heavy-ion collisions. Different regions of the QCD phase diagram can be accessed by changing the center of mass energy of the collision. Indeed, as the energy increases, the initial temperature of the system increases and its baryo-chemical potential decreases.

The LHC provides the optimal experimental conditions to study the QGP because [2]:

- The net baryon density in the central (mid-rapidity) region is very small, corresponding to the conditions of the early Universe.
- The initial temperature and energy density are the highest achievable in the laboratory.
- The large collision energy ensures an abundance of perturbatively calculable hard QCD processes.

The study of the strongly-interacting matter in the second generation of the LHC heavy-ion studies following the Long Shutdown 2 (2020-2021) will focus on rare probes and the study of their coupling with the medium and hadronization processes. Major highlights of the proposed program focus on the following physics questions [3]:

- Heavy flavour. Precise characterization of the quark mass dependence of in-medium parton energy loss; study of the transport and possible thermalization of heavy quarks in the medium; study of heavy quark hadronization mechanisms in a partonic environment. These studies require measurements of the production and azimuthal anisotropy of several charm and beauty hadron species, over a broad momentum range, as well as of b-tagged jets. ALICE will mainly focus on the low-momentum region, down to zero p_T and on a reconstruction of several heavy-flavour hadron species (including baryons). ATLAS and CMS will focus on b-tagged jets and D and B mesons at higher p_T. LHCb has a strong potential for all these measurements, although the detector performance in central Pb-Pb collisions is still not clear.
- Quarkonia. Study of quarkonium dissociation and possible regeneration as probes of deconfinement and of the medium temperature. ALICE will carry out precise measurements, starting form zero p_T, of J/ψ yields and azimuthal anisotropy, $\psi(2S)$ and Υ yields, at both central and forward rapidity. ATLAS and CMS will carry out precise multi-differential measurements of the Υ states to map the dependencies of their suppression pattern. They will also complement the charmonium measurements to high momentum. Also in this case LHCb has a strong potential but the detector performance in central Pb-Pb collisions is still not clear.
- Jets. Detailed characterization of the in-medium parton energy loss mechanism, that provides both a testing ground for the multi-particle aspects of QCD and a probe of the QGP density. The relevant observables are: jet structure and di-jet imbalance at TeV energies, b-tagged jets and jet correlations with photons and Z^0 bosons. These studies are crucial to address the flavour dependence of the parton energy loss and will be the main focus of ATLAS and CMS, which have unique high-p_T and triggering capabilities. ALICE will complement in the low-momentum region, and carry out measurements of the flavour dependence of medium-modified fragmentation functions using light flavour, strange and charm hadrons reconstructed within jets.
- Low-mass dileptons and thermal photons. These observables are sensitive to the initial temperature and the equation of state of the medium, as well as to the chiral nature of the phase transition. The study will be carried out by ALICE, which will strengthen its unique electron and muon reconstruction capabilities down to almost zero p_T.

2. The ALICE experiment upgrade strategy

ALICE is the only experiment at the LHC that was designed and built to study the QGP created in high-energy heavy-ion collisions. However the physics program also includes in-depth research on proton-proton and proton-ion collisions. The ALICE experiment and its detectors during LHC Run 2 (2015-2018) are shown in figure 1.

The plans for the ALICE experiment before Long Shutdown (LS) 2 include collecting 1 nb^{-1} of Pb-Pb collisions at a peak luminosity of $L = 10^{27}$ cm^{-2} s^{-1}, corresponding to a collision rate

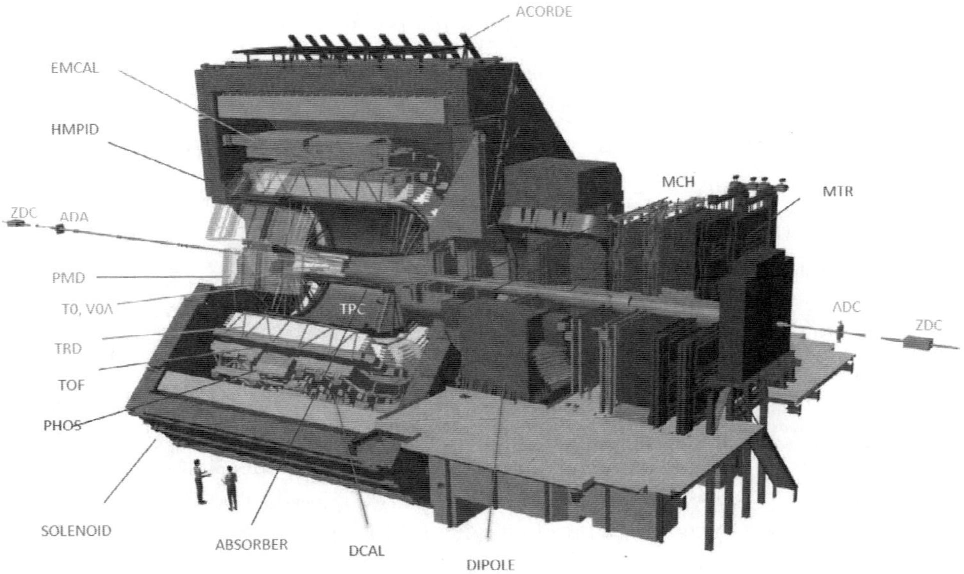

Figure 1. The ALICE experiment during LHC Run 2.

of 8 kHz [4]. During LHC Run 1 (2009-2013), the maximum readout rate was limited to 500 Hz of Pb-Pb events.

The ALICE upgrade strategy beyond LS2 is based on collecting more than 10 nb^{-1} of Pb-Pb collisions at luminosities up to $6x10^{27}$ cm^{-2} s^{-1}, corresponding to collision rates of 50 kHz. In this scenario each collision is shipped to the online systems, either upon a minimum bias trigger or in a self-triggered or continuous fashion. This will result in the collection and inspection of a data volume of heavy-ion events roughly 100 times greater than that of Run 1, implying an upgrade strategy also for the online and offline computing system.

The plan also considers the collection of 6 pb^{-1} of pPb collisions, both at a leveled collision rate of 200 kHz.

The operation of the Time Projection Chamber (TPC) is one of the limiting factors to achieve the desired readout rate. Actually, the TPC is based on a gated readout with Multi Wire Proportional Chambers (MWPC) [5]. The electron drift time of 100 μs from the central electrode to the readout chambers, together with the ion drift time of 180 μs from the sense wires to the gating grid, limits the readout rate to less than 3.5 kHz.

Operating the TPC in ungated mode, this is, leaving the gating grid continuously open, would result in a severe build up of space charge in the drift volume due to back drifting ions. At a minimum bias event rate of 50 kHz this would lead to a sustained ion accumulation from several thousand collision events piling up in the drift volume. Under such conditions, the space charge would distort the electron drift paths in such a way that a meaningful reconstruction of the particle trajectories would no longer be possible. In addition, the anticipated particle rates at the wire chambers would reach 100 kHz/cm^2, for which space charge effects in the amplification region result in a few percent gain drop, thus deteriorating the dE/dx performance of the detector.

In view of the challenging scenario for LHC Run 3 (2021-2023) and beyond, the ALICE Collaboration has scheduled an important upgrade of the experiment during LS2.

3. Upgrades

The upgrades are driven by the following requirements:

- Increase the readout capabilities in order to record all the Pb-Pb interactions (each collision will be shipped to the online systems).
- Improve the track reconstruction performance, in particular for low-$p_{\rm T}$ particles.
- Enhance particle identification (PID) capabilities.

Although not a detector itself, the beam pipe will also be replaced. Indeed, the reduction of the beam pipe diameter in the center of the experiment is one of the main ingredients in view of the improvement of the impact parameter resolution.

3.1. Inner Tracking System

The present Inner Tracking System (ITS) consists of six cylindrical layers of silicon detectors placed coaxially around the beam pipe [6]. The mains tasks of the ITS are to determine the interaction vertex, to track and identify particles at low-$p_{\rm T}$, to measure the distance of closest approach of charged particles to the interaction vertex and to the Particle IDentification (PID). The inner radius is the minimum allowed by the radius of the beam pipe and the outer one is determined by the necessity to match tracks with those from the TPC. The impossibility to access the present ITS detector for maintenance and repair interventions during the yearly shutdowns, together with the limited read-out rate of 1 kHz are key priorities in the design of the new ITS.

For the ITS upgrade, the first detection layer will be closer to the interaction point, as the reduction of the beampipe diameter in the center of ALICE is one of the main ingredients to improve the measurement of the impact parameter resolution [7]. The new detector will use Monolithic Active Pixel Sensors (MAPS) that will reduce the material budget and will allow the tracking performance and momentum resolution to be significantly improved. Also a new layer will be added, enhancing even more, the physics performance capabilities of the upgraded ITS. For example, simulations show that the pointing resolution provided by this detector in Pb-Pb will be improved, with respect to the current ITS, by a factor of 5 and 3 along the z axis and in the $r\phi$ plane, respectively (left plot of Figure 2). Another important point is that the new ITS is designed to be able to read the data related to each individual interaction up to a rate of 100 kHz for Pb-Pb and 400 kHz for pp collisions, a factor two higher than the ALICE upgrade requirements.

3.2. Muon Forward Tracker

The Muon Spectrometer is composed of tracking and trigger chambers, a dipole magnet, a rear absorber, an iron wall, a beam shield and a front absorber [8]. The latter is made out of different materials and its goal is to absorb the large flux of particles resulting from the collision. The distance from the interaction point is constrained by the ITS dimension along the z axis.

However, the current ALICE muon physics program suffers from various limitations, basically because of the multiple scattering induced on the muon tracks by the front absorber. In this way the details of the vertex region are completely smeared out, and as a consequence it is very difficult to reject muons coming from semi-muonic decays of pions and kaons, representing an important background (specially at low-$p_{\rm T}$). Besides this, the degradation of the kinematics, imposed by the presence of the hadron absorber, plays a crucial role in determining the mass resolution for the resonances (in particular in the region of the ϕ and ω mesons). Finally the lack of details in the vertex region prevents to differentiate between prompt and displaced J/ψ production, as well as to disentangle open charm and open beauty production without making assumption relying on physics model.

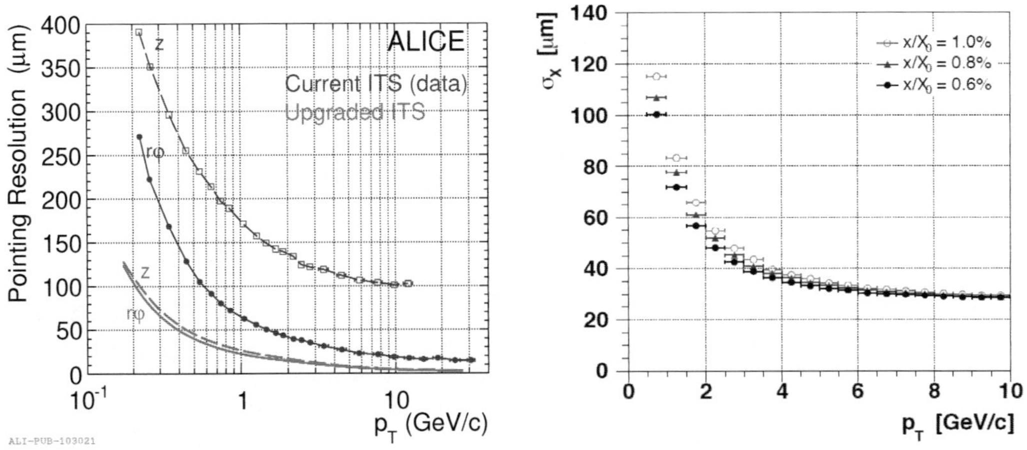

Figure 2. *Left:* Comparison between current and upgraded ITS pointing resolution in Pb-Pb data and simulations, respectively. *Right:* Pointing resolution of the MFT in the transverse plane using Pb-Pb simulations. Different markers/colors indicate the sensitivity of the pointing resolution to the material budget per layer.

The Muon Forward Tracker (MFT) will be used to overcome these limitations. This new element in ALICE will be a silicon pixel detector (identical to that of the new ITS) that will be located upstream the front absorber and in the acceptance of the Muon Spectrometer [9]. The idea is to match the extrapolated muon tracks, coming from the tracking chambers after the absorber, with the clusters measured in the MFT before the absorber. In this way the muon tracks will gain enough pointing accuracy to allow a reliable measurement of their pointing resolution with respect to the interaction vertex.

The MFT consists of two half cones, each one in turn is made of five half-disks positioned along the beam axis. A half-disk includes a disk spacer, a disk support, two printed circuit boards and sensors. Current simulations of Pb-Pb collisions indicate that at low-p_{T} the pointing resolution has an important dependence on the material budget (right plot of Figure 2).

3.3. Other detectors

The Muon Chamber System is presently limited to 1 kHz readout rate and will change the entire readout electronics using the SAMPA chip to digitize the detector signals. The Muon Trigger detector is also limited to the same readout rate, and as the upgrade strategy does not foresee a muon trigger, all events will be read upon the interaction trigger. Consequently the detector will be called Muon IDentifier (MID).

The Transition Radiation Detector (TRD) is presently limited to a few kHz readout rate. By reducing the data volume from the detector, using tracklets and increasing the data throughput of the off-detector electronics, a readout rate of 100 kHz for Pb-Pb and pp collisions can be achieved. However, since the front-end electronics does not support the use of multi event buffers, a 50 kHz trigger rate in Pb-Pb events corresponds to 75% of the collisions being read-out. Going beyond this threshold is not conceivable because a change of the on-detector electronics would be needed, which requires a removal and disassembly of the TRD modules.

The readout rate of the Time Of Flight detector is at present limited to 40 kHz by the throughput of the VME system located in the crates at the end of the detector modules. An upgrade of this element will increase, by a factor five, the readout rate capabilities.

The Zero Degree Calorimeter will change the readout electronics, it will also be able to provide trigger information that can be used to clean the interaction trigger.

The PHOton Spectrometer (PHOS) and ElectroMagnetic CALorimeter (EMCAL) use the same readout electronics and was already upgraded during Long Shutdown 1. These systems accomplish the desired performance.

3.4. Online-offline computing system

The computing upgrade concept consists on transferring all detector data to the computing system [10]. The data volume reduction will be performed by processing the data on the fly in parallel with the data collection and not by rejecting the complete events as the high-level triggers or event filter farms of most high-energy physics experiments do. The online-offline (O^2) system will perform a partial calibration and reconstruction online and replace the original raw data with compressed data. The online detector calibration and data reconstruction will, therefore, be instrumental in keeping the total data throughput within an envelope compatible with the available computing resources. The main role of the O^2 system will be to perform detector calibration and data reconstruction concurrently with the data taking.

Figure 3 shows how the new detectors inside ALICE will look like after LS2 when looking to the direction of the front absorber of the Muon Spectrometer.

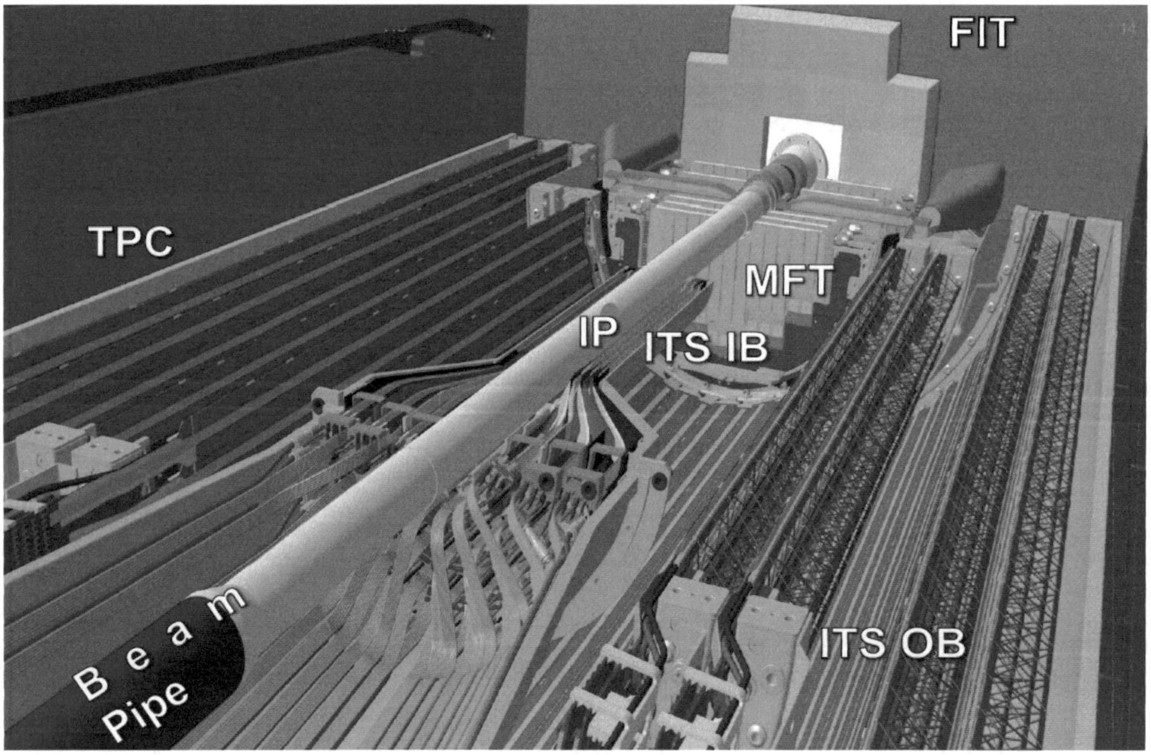

Figure 3. Schematic view of the of the new beam pipe, ITS, TPC, MFT and FIT detectors inside ALICE. IP stands for Interaction Point, IB for Inner Barrel and OB for Outer Barrel.

IOP Conf. Series: Journal of Physics: Conf. Series **912** (2017) 012023 doi:10.1088/1742-6596/912/1/012023

4. Contributions from Mexican groups

Mexican groups were involved in the design and/or construction of the following detectors: Alice COsmic Ray DEtector (ACORDE), Alice Diffraction (AD), TPC and V0A. So far there are no official plans for ACORDE and AD upgrades.

4.1. Time Projection Chamber

Limitations of the actual TPC have already been mentioned. The new TPC will use Gas Electron Multipliers (GEM) technology for the new readout chamber, since they feature intrinsic ion-blocking capabilities that avoid massive charge accumulation in the drift volume from the back-drifting ions, and prevent excessive space-charge distortions [11]. However GEMs do not feature the same opacity for ions as a gating grid. The requirement to keep the distortions at a tolerable level leads to an upper limit of 1% for the fractional ion backflow at a gas gain of 2000. The resulting space-charge distortions are less than 10 cm in most of the TPC drift volume and can be calibrated with sufficient precision. The replacement of the existing MWPC-based readout chambers by GEMs also implies also the necessity for new readout electronics that accommodate the negative signal polarity and enable continuous data readout.

In this sense the contribution of a Mexican group in this project is the development of the new high voltage mechanism that will be used by the TPC.

4.2. Fast Interaction Trigger

Currently ALICE employs two forward detectors (T0 and V0) in order to provide minimum bias trigger, multiplicity measurements, beam-gas rejection, event plane determination and collision time for TOF. However for LHC Run 3 the upgrades of these detectors (T0+ and V0+) will be merged into a single system: the Fast Interaction Trigger (FIT) [12]. This new system will discriminate beam-beam interactions with 99% efficiency and will act as a Level 0 trigger, this is, it will provide a start signal for the rest of the ALICE detectors with a time resolution better than 30 ps.

Each T0+ module consists of a 2 cm thick quartz radiator coupled to a modified Planacon MCP-PMT, the size of the radiator matches that of the photocathode. In order to improve the detector performance at high particle multiplicities, the granularity of the sensors has been increased. The best performance of the T0+ has been reached by grouping the anodes and dividing the radiator into four sectors.

The V0+ will consist of a 4 cm thick ring made of plastic scintillation with an inner diameter of 8 cm and an outer one of 148 cm, it will be located at 330 cm from the interaction point on the opposite to the MFT. The plastic scintillator will be divided into five concentric rings and eight 45° sectors in the azimuthal angle direction. The light collection will be performed by 1 mm diameter optical fibers, assembled on a two-dimensional grid with 5 mm pitch. As compared to the actual V0 detector, V0+ will have 25% more readout channels, 3.2 times larger active area and an improved time resolution. Besides this, it is expected to diminish aging problems and after-pulses observed in the current V0.

The physics performance of both detectors has been simulated within the AliRoot package, that is the official software of the ALICE experiment. A sample of simulated Pb-Pb collisions at $\sqrt{s_{\rm NN}} = 2.76$ TeV was used to compute the event plane resolution (Figure 4 left) and the centrality resolution (Figure 4 right).

Mexican groups have an important role in the FIT group, as they are in charge of the design and construction of the V0+.

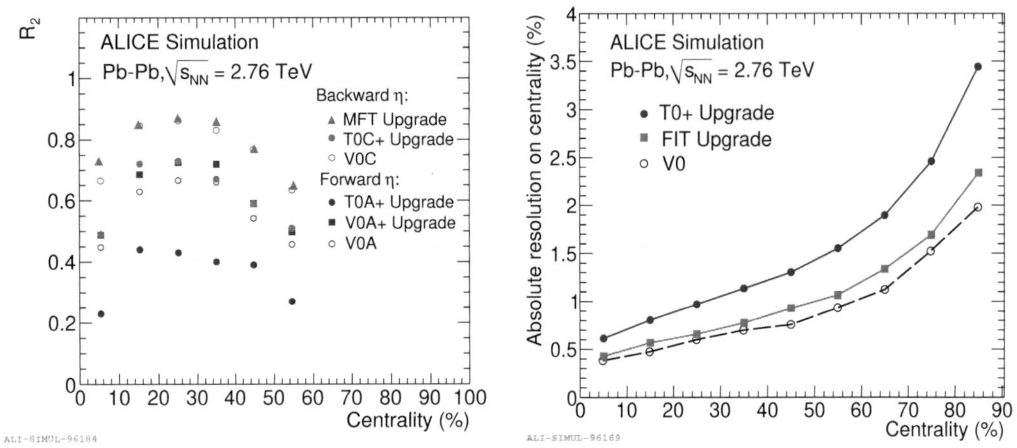

Figure 4. *Left:* Event plane resolution as a function of the centrality and for different detectors. Smaller values of the centrality indicate a more central collision. *Right:* Centrality resolution. Both plots are obtained from a Pb-Pb collisions event generator.

5. Summary

There are still many physics questions related to the QGP waiting for an answer, for this reason more data from high-energy heavy-ion collisions is needed. The ALICE experiment is looking forward to collect more than 10^{-1} nb of Pb-Pb collisions during LHC Run 3 and Run 4 at an interaction rate of 50 kHz, approximately 100 times the interaction rate during Run 1. In order to achieve this goal, the ALICE experiment is preparing a major upgrade of many detectors. Among these, mexican groups have an active contribution in this process, as they are deeply involved in the upgrades of the TPC and the V0+.

References

[1] S. Borsanyi *et al.*, "Full result for the QCD equation of state with 2+1 flavors", *Phys. Lett. B 730 (2014) 99.*
[2] ALICE Collaboration, "Upgrade of the ALICE Experiment: Letter of Intent", *Jour. of Phys. G: Nucl. Part. Phys. 41 (2014) 087001.*
[3] P. Antonioli *et al.*, "INFN What Next", *Frascati Phys. Ser. 62 (2016).*
[4] ALICE Collaboration, "Upgrade of the ALICE Readout and Trigger System", *CERN-LHCC-2013-019.*
[5] ALICE Collaboration, "ALICE Technical Design Report of the Time Projection Chamber", *CERN-LHCC-2000-001.*
[6] ALICE Collaboration, "ALICE Technical Design Report of the Inner Tracking System", *CERN-LHCC-1999-012.*
[7] ALICE Collaboration, "Technical Design Report for the Upgrade of the ALICE Inner Tracking System", *Jour. of Phys. G: Nucl. Part. Phys. 41 (2014) 087002.*
[8] ALICE Collaboration, "Addendum to the ALICE Technical Design Report of the Dimuon Forward Spectrometer", *CERN-LHCC-2000-046.*
[9] ALICE Collaboration, "Technical Design Report for the Muon Forward Tracker", *CERN-LHCC-2015-001.*
[10] ALICE Collaboration, "Technical Design Report for the Upgrade of the Online-Offline Computing System",
[11] ALICE Collaboration, "Upgrade of the ALICE Time Projection Chamber", *CERN-LHCC-2013-020. CERN-LHCC-2015-006.*
[12] W. H. Trzaska, "New Fast Interaction Trigger for ALICE", *Nuc. Inst. and Meth. in Phys. A 845 (2017) 463.*

IOP Conf. Series: Journal of Physics: Conf. Series **912** (2017) 012024 doi:10.1088/1742-6596/912/1/012024

Topology in the 2d Heisenberg Model under Gradient Flow

I O Sandoval[1], W Bietenholz[1], P de Forcrand[2,3], U Gerber[1,4] and H Mejía-Díaz[1]

[1] Instituto de Ciencias Nucleares, Universidad Nacional Autónoma de México, A.P. 70-543, C.P. 04510 Ciudad de México, Mexico

[2] Institut für Theoretische Physik, ETH Zürich, Wolfgang-Pauli-Str. 27, CH–8093 Zürich, Switzerland

[3] CERN, Theory Division, CH-1211 Genève 23, Switzerland

[4] Instituto de Física y Matemáticas, Universidad Michoacana de San Nicolás de Hidalgo, Edificio C-3, Apdo. Postal 2-82, C.P. 58040, Morelia, Michoacán, Mexico

E-mail: `ilya.orson@ciencias.unam.mx`

Abstract. The 2d Heisenberg model — or 2d O(3) model — is popular in condensed matter physics, and in particle physics as a toy model for QCD. Along with other analogies, it shares with 4d Yang-Mills theories, and with QCD, the property that the configurations are divided in topological sectors. In the lattice regularisation the topological charge Q can still be defined such that $Q \in \mathbb{Z}$. It has generally been observed, however, that the topological susceptibility $\chi_t = \langle Q^2 \rangle / V$ does not scale properly in the continuum limit, *i.e.* that the quantity $\chi_t \xi^2$ diverges for $\xi \to \infty$ (where ξ is the correlation length in lattice units). Here we address the question whether or not this divergence persists after the application of the Gradient Flow.

1. The 2d O(3) model on the lattice

We consider square lattices of volume $V = L \times L$, and we refer to lattice units, *i.e.* the spacing between lattice sites is set to 1. At each site x there is a 3-component classical spin variable of length 1, $\vec{e}_x \in S^2$. The standard lattice action of a configuration $[\vec{e}]$ is given by

$$S[\vec{e}] = \beta \sum_{\langle xy \rangle} (1 - \vec{e}_x \cdot \vec{e}_y) \ , \tag{1.1}$$

where the sum runs over the nearest neighbour lattice sites. We assume periodic boundary conditions and $\beta > 0$. Obviously, this model is symmetric under global O(3) spin rotations.

In solid state physics this represents a model for a ferromagnet. Its rôle as a toy model for QCD is based on asymptotic freedom [1], a dynamically generated mass gap (which was computed with the Bethe ansatz [2]), and the existence of topological sectors.

2. Monte Carlo simulation

Since the action is real positive for any configuration, $S[\vec{e}] \geq 0$, it can be employed to define a probability

$$p[\vec{e}] = \frac{1}{Z} e^{-S[\vec{e}]} \ , \quad Z = \int D\vec{e} \ e^{-S[\vec{e}]} \ . \tag{2.1}$$

It is normalised by the partition function Z, which is given by a functional integral over all configurations.

A Monte Carlo simulation generates a large set of random configurations with this probability distribution, which enable numerical measurements. To this end, we used the highly efficient cluster algorithm [3], both in its single-cluster and its multi-cluster version. It is far superior to local update algorithms, which suffer *e.g.* from a very long auto-correlation time with respect to the topological charge Q, in particular close to criticality.

3. Scale and parameters

As usual, the intrinsic scale of the system is given by its correlation length ξ. It describes the decay of the correlation function, which can be computed as the correlation between layer averages,

$$\langle \vec{s}_{x_2} \cdot \vec{s}_{y_2} \rangle \propto \cosh\left(- \frac{|x_2 - y_2| - L/2}{\xi} \right) , \quad \vec{s}_{x_2} = \frac{1}{L} \sum_{x_1} \vec{e}_x , \quad x = (x_1, x_2) . \tag{3.1}$$

This proportionality relation holds if the size L is large compared to ξ, and the numerator $|x_2 - y_2| - L/2$ is sufficiently small. We determined ξ by a fit in the interval $L/3 \leq |x_2 - y_2| \leq 2L/3$, as suggested in Ref. [4]. For a variety of parameters, our results for ξ are consistent with values given in the literature, for instance in Refs. [4–6].

The correlation length depends essentially on the parameter β, and to some extent also on the size L. As we vary L at fixed β, ξ is asymptotically stable in large volumes, $L \gg \xi$. Generally, the *finite-size effects* are suppressed by the ratio L/ξ. Our study was performed in boxes of constant size,

$$L \simeq 6 \, \xi , \tag{3.2}$$

which suppresses the finite-size effects quite well. This required a fine-tuning of β in each volume.

On the other hand, the *lattice artifacts* depend on the ratio of the correlation length and the lattice spacing. Since we are using lattice units, this ratio is simply given by ξ. Our values of β and ξ, in the range $L = 24 \dots 404$, are listed in Table A1, in the appendix.

Hence we study the convergence to the continuum limit by increasing L, and increasing β accordingly such that $L/\xi \simeq 6$ persists. This amounts to an amplification of ξ at small finite-size effects, which are kept of the same magnitude, *i.e.* we perform a controlled extrapolation towards the continuum.

4. Topological charge and susceptibility

The geometric definition of the topological charge Q [7] of a lattice configuration $[\vec{e}]$ has the virtue that it provides integer values, $Q[\vec{e}] \in \mathbb{Z}$ for all configurations (up to a subset of measure zero). We split each plaquette into two triangles, in an alternating order, as illustrated in Figure 1 on the left. The spins at the vertices of one triangle, say $(\vec{e}_x, \vec{e}_y, \vec{e}_z)$, span a spherical triangle on S^2. We refer to the spherical triangle with minimal area, and a fixed orientation (which determines the sign). This oriented area A_{xyz} defines the topological charge density $A_{xyz}/4\pi$, and therefore the winding number, or topological charge

$$Q[\vec{e}] = \frac{1}{4\pi} \sum_{\langle xyz \rangle} A_{xyz} \in \mathbb{Z} , \tag{4.1}$$

where the sum runs over all triangles (for obtaining integer Q-values, it is crucial to account for the periodic boundary conditions). The explicit formulae are given in Refs. [7–9].

IOP Conf. Series: Journal of Physics: Conf. Series **912** (2017) 012024 doi:10.1088/1742-6596/912/1/012024

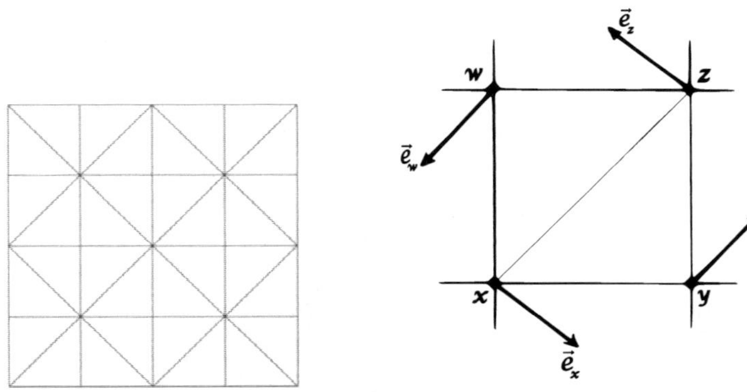

Figure 1. On the left: illustration of the division of the lattice plaquettes into triangles, in an alternating order. On the right: the spins at the vertices of each triangle — here $(\vec{e}_x, \vec{e}_y, \vec{e}_z)$ and $(\vec{e}_x, \vec{e}_z, \vec{e}_w)$ — span a spherical triangle, with the oriented minimal areas A_{xyz} and A_{xzw}, respectively. Their sum defines the topological charge density at the plaquette under consideration.

Parity symmetry implies $\langle Q \rangle = 0$, hence the topological susceptibility takes the form

$$\chi_{\rm t} = \frac{1}{V} \left(\langle Q^2 \rangle - \langle Q \rangle^2 \right) = \frac{\langle Q^2 \rangle}{V} \ . \tag{4.2}$$

Figure 2 shows an example for a histogram of the topological charge distribution, which tends to be approximately Gaussian.[1]

For the topological susceptibility to have a sound continuum limit, in a large volume, the (dimensionless) physical quantity $\chi_{\rm t}\xi^2$ should converge to a finite constant,

$$\lim_{\xi \to \infty} \chi_{\rm t}\xi^2 = \text{constant} \ . \tag{4.3}$$

The question whether this is actually the case has been debated since the 1980s. While it was controversial for a while — based on considerations of various lattice actions and definitions of the lattice topological charge — the consensus is now that this limit diverges, *i.e.* the topology of this model is not well-defined in the continuum limit. This appears as a conceptual disease of the 2d O(3) model.

After the first numerical evidence for this divergence [7], a semi-classical argument was elaborated in Ref. [11]: it considers very small topological windings of a lattice configuration ("dislocations"). For increasing β they are suppressed, but the semi-classical picture suggests that this suppression is not sufficient to compensate for the entropy growth due to the increase in ξ.

Later a sophisticated version of a (truncated) classically perfect lattice action was applied, which suppresses such dislocations by numerous additional couplings, beyond nearest neighbour sites [12]. However, the numerical results with this action (which were also obtained at $L \simeq 6\,\xi$) suggest that the term $\chi_{\rm t}\xi^2$ still does not converge to a finite value in the continuum limit. That study observed a logarithmic divergence of $\chi_{\rm t}\xi^2$ with ξ.

[1] Results for the kurtosis term $c_4 = \left(\langle Q^2 \rangle^2 - \langle Q^4 \rangle \right) / V$, as a measure for the deviation from a Gaussian distribution, are given in Ref. [10].

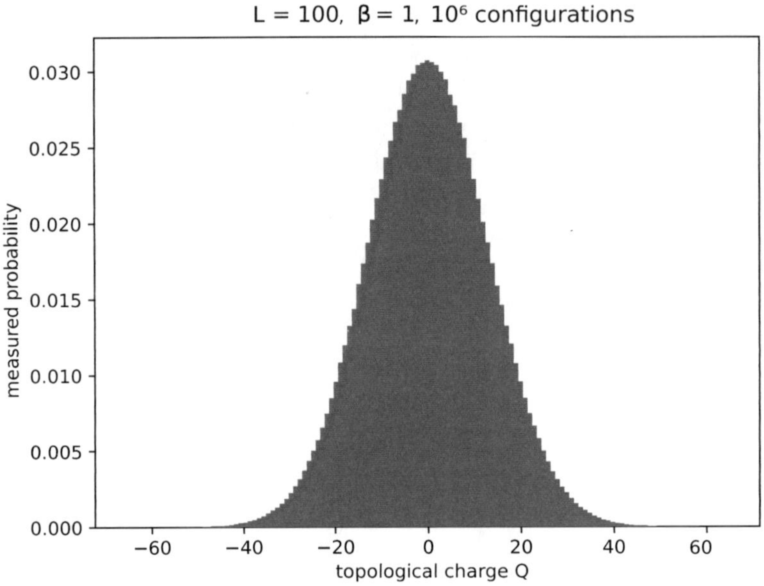

Figure 2. Histogram of the topological charges, obtained at $L = 100$, $\beta = 1$ with a statistics of 10^6 configurations. Its shape is approximately Gaussian, with a width of $\langle Q^2 \rangle = \chi_t V$.

5. Gradient Flow

In recent years, the Gradient Flow has attracted considerable attention in the lattice community. This interest was boosted in particular by Refs. [13]. Unlike previously popular methods, where lattice configurations were smoothened *ad hoc*, the Gradient Flow performs such a smoothing in a controlled manner, which corresponds to a renormalisation group flow. When applied to the 2d O(3) model, one could intuitively imagine that it removes the (small) dislocations, while preserving topological winding on a large scale — in the semi-classical simplification they are represented by instantons with large radii. Hence, the question arises if the application of the Gradient Flow leads to a finite continuum limit of $\chi_t \xi^2$.

The formula for the Gradient Flow in the 2d $O(N)$ models has been written down in Ref. [14]. Here we reproduce it for the reader's convenience. In the continuum, the spin components $e(x)^i$, $i = 1 \dots N$, are modified according to the differential equation

$$\partial_t \, e(t,x)^i = P^{ij}(t,x) \, \Delta \, e(t,x)^j \; , \quad P^{ij}(t,x) = \delta^{ij} - e(t,x)^i \, e(t,x)^j \; , \tag{5.1}$$

where t is the Gradient Flow time (which generically has the dimension $[\text{length}]^2$) starting at $t = 0$, and Δ is the Laplace operator. On the lattice we replace $\vec{e}(t,x)$ by the spin variable at one site, $\vec{e}(t)_x$, and we apply the standard discretisation of the Laplacian,

$$\Delta e(t,x)^j \longrightarrow e(t)^j_{x_1+1,x_2} + e(t)^j_{x_1,x_2+1} + e(t)^j_{x_1-1,x_2} + e(t)^j_{x_1,x_2-1} - 4e(t)^j_{x_1,x_2} \; . \tag{5.2}$$

For the corresponding spin rotations we apply the Runge-Kutta 4-point method. In practice we proceed as follows: for a given configuration, the gradients are computed for all spin variables, at the flow time instants which are needed for the Runge-Kutta scheme. This is done in a fixed configuration; then all the spins are simultaneously modified with a Gradient Flow time step of $dt = 10^{-4}$. (After each step, the normalisation of the modified spins is re-adjusted.)

This value of dt seems to be sufficiently small to avoid significant artifacts due to the flow time discretisation, whereas some discretisation effects were observed at $dt = 10^{-3}$. On the

IOP Conf. Series: Journal of Physics: Conf. Series **912** (2017) 012024 doi:10.1088/1742-6596/912/1/012024

other hand, for $dt = 10^{-4}$ we did not find any significant difference when we modified the spins one by one lexicographically.

In order to explore the effect of the Gradient Flow on the topology towards the continuum limit, we have to set a scale for the flow time. Thus the results at various L and β can be related. We follow the recipe of Refs. [13] by considering the energy density. In our case, it is calculated as

$$E_x = 4 - \vec{e}_x \cdot (\vec{e}_{x_1+1,x_2} + \vec{e}_{x_1,x_2+1} + \vec{e}_{x_1-1,x_2} + \vec{e}_{x_1,x_2-1}) \tag{5.3}$$

at some lattice site x (it vanishes for a uniform configuration, and grows the more the spin directions differ). Its expectation value is trivially related to the mean value of the action density, $\langle E \rangle = \langle S \rangle / (\beta V)$. In QCD, Lüscher suggested to choose the Gradient Flow time unit t_0 such that $\langle E \rangle t_0^2 = 0.3$ [13]. In a 2d theory the corresponding dimensionless product reads $\langle E \rangle t$, and we had to fix a lower reference value, which we chose as

$$\langle E \rangle t_0 = 0.08 \, . \tag{5.4}$$

As the flow time proceeds, the term $\langle E \rangle t$ rises from 0 to some maximum before gradually decreasing again. If the reference value is taken too large, it is not even attained for all parameter sets in our study (if one increases L and β more and more, this maximum decreases monotonously). The above reference value of 0.08 captures lattice sizes up to $L = 606$ [15], where β is always tuned such that relation (3.2) holds. Figure 3 shows the evolution of the term

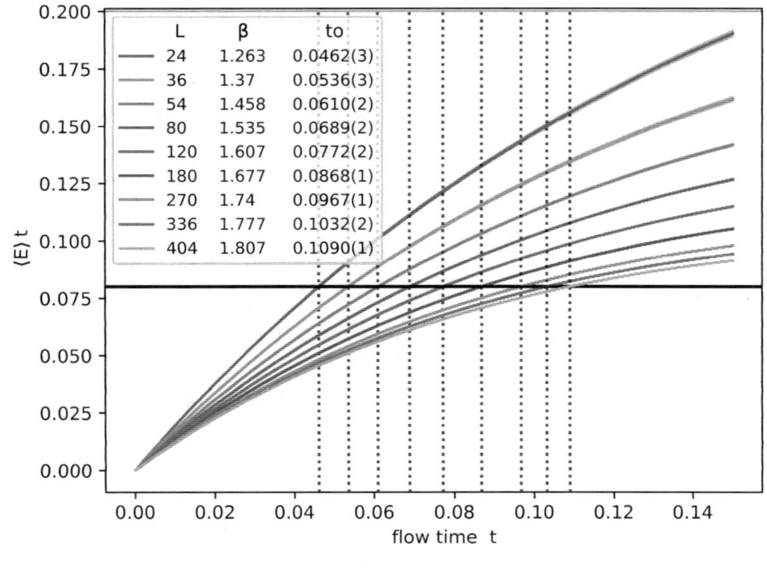

Figure 3. The Gradient Flow time evolution of the dimensionless term $\langle E \rangle t$, for a variety of lattice sizes $L = 24 \dots 404$, always with $\xi \simeq L/6$. We illustrate in particular the time t_0, where the reference value of 0.08 is attained for the first time, so it matches the condition (5.4).

$\langle E \rangle t$ under Gradient Flow for a variety of volumes (with the suitable β-value), and the time t_0 where it amounts to 0.08 for the first time (in the long-time evolution it decreases again down to this value and below). The resulting t_0-values are given in the plot of Figure 3 and in Table A1; they are rather small compared to typical values in QCD.

Next we consider the correlation function (3.1),

$$C(r) = \langle \vec{s}_{x_2} \cdot \vec{s}_{x_2+r} \rangle \ . \tag{5.5}$$

It coincides with the connected correlation function due to the O(3) rotation symmetry, which implies $\langle \vec{s}_{x_2} \rangle = \vec{0}$. Figure 4 shows an example for the behaviour of the correlation function under

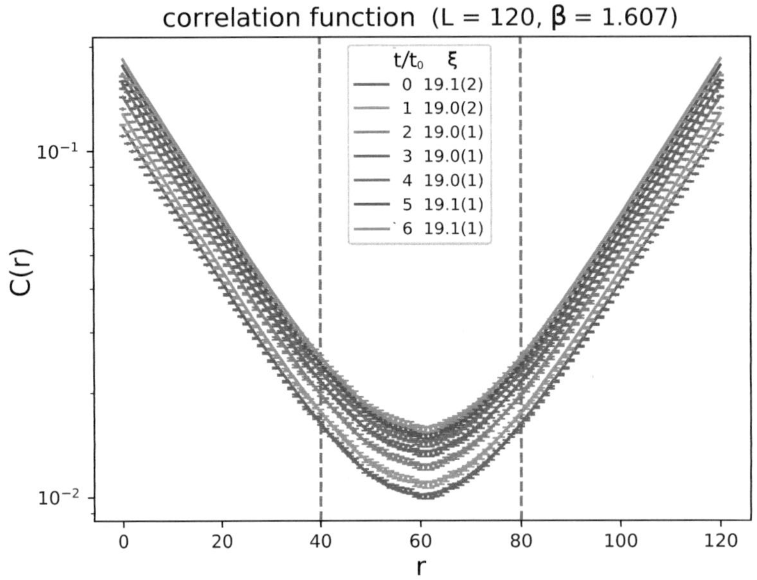

Figure 4. The correlation function $C(r)$ at $L = 120$, $\beta = 1.607$ (as an example). We see that the correlation at a fixed layer separation r increases due to the Gradient Flow, but the correlation length ξ remains almost unaltered, up to flow times as long as $6t_0$. This is the generic behaviour that we also observed in all other volumes under consideration. In this regime, the value of ξ is a long-range property, which is hardly affected by the Gradient Flow.

Gradient Flow. As the flow time t proceeds, the correlation at a fixed distance r is getting stronger, as one might expect. However, when we perform the fit to measure the correlation length ξ, according to the formula (3.1), we see that ξ hardly changes. This observation holds for all parameter sets (L, β) that we considered. Hence the intrinsic scale of the system is almost constant under the Gradient Flow, at least up to about $6\,t_0$, although flow times of this magnitude smoothen the configurations significantly over distances below ξ, as we will see in Figure 5.

Now we consider the quantity $\chi_t\,\xi^2$, which is supposed to be the scaling term towards the continuum limit, as we mentioned before. Figure 5 shows that this term is suppressed significantly for the flow times in our study — in our three largest volumes ($L \geq 270$), $t = 3t_0$ already reduces $\chi_t\,\xi^2$ below half of its initial value. Since ξ hardly changes, this reduction is due to the destruction of topological windings. This seems compatible with the picture of the elimination of dislocations. The final question is whether this effect is sufficient to entail a finite continuum limit of $\chi_t\,\xi^2$, after a fixed multiple of the flow time unit t_0.

6. Continuum limit

According to Lüscher, in QCD any finite amount of Gradient Flow removes the UV divergences of the original theory [13]. This motivates us to investigate whether the same effect takes place in the 2d O(3) model, such that the Gradient Flow cures its topological UV behaviour.

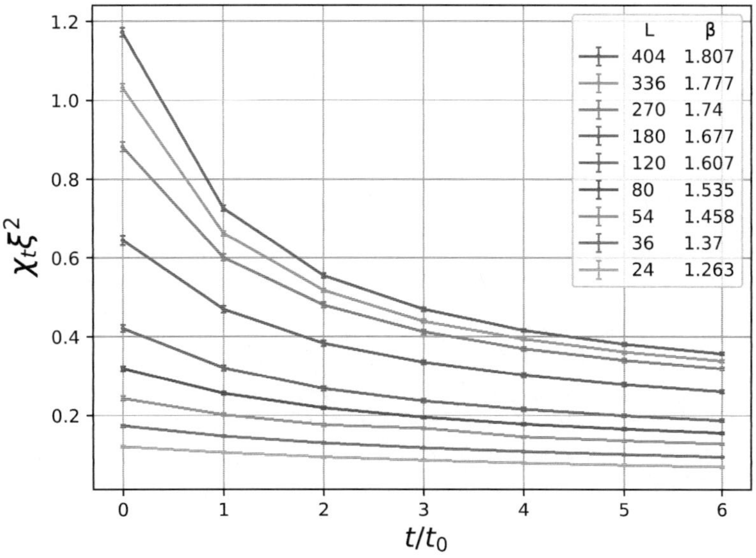

Figure 5. The evolution of the "scaling quantity" $\chi_t\,\xi^2$ under the Gradient Flow, for several volumes, up to $6t_0$. We see that this quantity is strongly reduced by the smoothing procedure. Since ξ remains almost constant, this demonstrates the destruction of a significant part of the topological windings. In this respect, the magnitude of our t_0-values is perfectly sensible.

Hence we finally arrive at the crucial question how $\chi_t\,\xi^2$ behaves when we approach the continuum limit. This behaviour is shown in Figure 6, based on 10^5 configurations in each volume, and part of the data are given in Table A1. Our study extends up to $\xi \simeq 67.7(3)$, which is close to the continuum limit indeed, but we cannot see any trend towards a convergence of $\chi_t\,\xi^2$ to a finite value, at any fixed ratio $t/t_0 = 1, 2 \dots 6$.

At $t = 0$ the data are very well compatible with a logarithmic divergence of the form $\chi_t\,\xi^2 = c_1 \ln(c_2\xi + c_3)$ (where c_i are constants), as it was observed before for a classically perfect action [12], and for two types of topological lattice actions [8]. After application of the Gradient Flow, the quality of the fits to this function decreases somewhat. Figure A1 shows the data along with the logarithmic fits at flow time $t = 0$, $2t_0$, $4t_0$ and $6t_0$. For comparison, we considered another 3-parameter fit to a power-law of the form $\chi_t\,\xi^2 = c_1\xi^{c_2} + c_3$. At $t = 0$ it is excellent too, but after the Gradient Flow it is a little worse than the logarithmic fits. Table A2 displays the χ^2/d.o.f-values for both fitting functions. In particular, at $t = 0$ a power-law cannot be ruled out by the present data, although a logarithmic divergence is expected. We hope for the extension of this study to even larger volumes [15] to be helpful also in this regard.

7. Conclusions

The outcome of our study is illustrated in Figure 6, and the most important data are given in Table A1. After applying the Gradient Flow, with a fixed ratio t/t_0 (where t_0 has be determined according to eq. (5.4)), the quantity $\chi_t\,\xi^2$ is reduced, which reveals the destruction of a significant part of the topological windings. However, as ξ increases, $\chi_t\,\xi^2$ still does not show any trend of a convergence towards a finite continuum value. Instead our data suggest a divergence in the continuum limit, as it was observed previously without Gradient Flow for the standard lattice action [7], a classically perfect action [12], and for topological lattice actions [8].

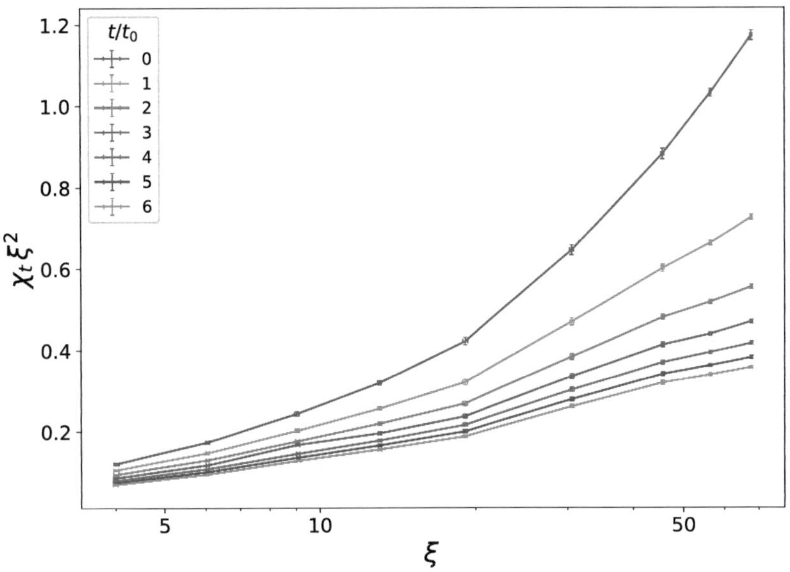

Figure 6. The "scaling quantity" $\chi_{\rm t}\,\xi^2$, at flow times $t/t_0 = 0\dots6$, and correlation length $\xi \simeq 4\dots67.3$ (at $L/\xi \simeq 6$). Thus our study advances up to very fine lattices, but $\chi_{\rm t}\,\xi^2$ still does not seem to scale towards a finite continuum limit at fixed t/t_0.

We add that prominent gauge theories with topological sectors, in particular SU(N) Yang-Mills theories ($N \geq 2$) and QCD, suffer — by default — from the same problem: if we write the lattice topological susceptibility as $\chi_{\rm t} = \sum_x \langle q_0 q_x \rangle$ (where q is the topological charge density, in a conventional formulation), one encounters a divergence, due to the point $x = 0$.[2] However, in those models the problem is overcome by the application of the Gradient Flow [13].[3]

In contrast, at this point we conclude that the topology of the 2d O(3) model seems to be ill-defined in the continuum limit, even after the application of the Gradient Flow. However, this study is going to be extended to even larger L and ξ, in order to further check this conclusion [15].

Acknowledgments
Our interest in the issue of this article originates from a remark by Martin Lüscher. We thank him, as well as the LOC of the XXXI Reunión Anual de la División de Partículas y Campos de la Sociedad Mexicana de Física, where this talk was presented by IOS. This work was supported by DGAPA-UNAM, grant IN107915, and by the *Consejo Nacional de Ciencia y Tecnología* (CONACYT) through project CB-2013/222812. The computations were performed on the cluster of ICN/UNAM; we thank Luciano Díaz and Eduardo Murrieta for technical assistance.

Appendix A. Numerical data and quality of the fits

Table A1 displays the most relevant numerical results in the lattice volumes under consideration, for the quantities t_0, ξ and $\chi_{\rm t}$. Next we refer to our $\chi_{\rm t}\,\xi^2$-values as a function of ξ, shown in Figure 6. Figure A1 illustrates the logarithmic fits to the data at flow time $t = 0$, $2t_0$, $4t_0$ and $6t_0$. Finally, Table A2 gives the quality of the fits to a logarithmic and a power-law function, as described in the last paragraph of Section 6.

[2] In this notation, the point $x = 0$ also causes the divergence of $\chi_{\rm t}\xi^2$ in the 2d O(3) model [8].
[3] For alternative solutions in those models, we refer to Refs. [16].

L	β	t_0	ξ		χ_t (in units of 10^{-3})			
			$t=0$	$6\,t_0$	$t=0$	t_0	$3\,t_0$	$6\,t_0$
24	1.263	0.0462(3)	4.01(5)	4.00(4)	7.51(4)	6.58(4)	5.38(3)	4.38(2)
36	1.37	0.0536(3)	6.05(5)	6.04(3)	4.74(3)	4.03(2)	3.22(2)	2.60(1)
54	1.458	0.0610(2)	9.0(1)	9.10(7)	2.99(2)	2.47(1)	2.04(1)	1.550(9)
80	1.535	0.0689(2)	13.1(1)	13.1(1)	1.86(1)	1.495(8)	1.139(6)	0.907(5)
120	1.607	0.0772(2)	19.1(2)	19.1(2)	1.154(5)	0.879(4)	0.649(3)	0.513(2)
180	1.677	0.0868(1)	30.6(3)	30.6(3)	0.691(3)	0.503(2)	0.358(2)	0.278(1)
270	1.74	0.0967(1)	45.6(3)	45.6(3)	0.424(2)	0.289(1)	0.1983(9)	0.1534(7)
336	1.777	0.1032(2)	56.4(2)	56.3(2)	0.324(1)	0.2084(9)	0.1381(6)	0.1066(5)
404	1.807	0.1090(1)	67.7(3)	67.7(3)	0.256(1)	0.1585(7)	0.1024(5)	0.0779(4)

Table A1. A summary of our numerical results in the nine volumes $V = L \times L$ that we investigated. In each volume, β was tuned such that $L/\xi \simeq 6$, and t_0 was determined by the condition $\langle E \rangle\, t_0 = 0.08$. We see that the correlation length ξ hardly changes under Gradient Flow, but the topological susceptibility is significantly reduced as we proceed up to $6t_0$. These results are based on a statistics of 10^5 configurations in each volume.

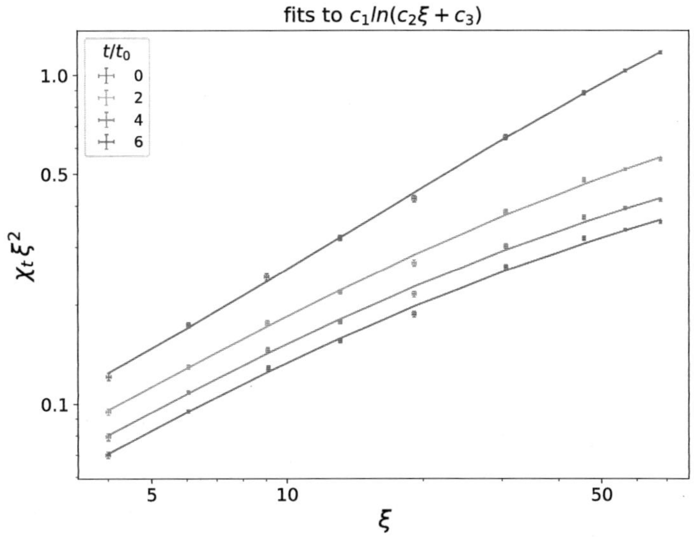

Figure A1. The fits of our data in Figure 6, at various Gradient Flow times, to the logarithmic ansatz $\chi_t\,\xi^2 = c_1 \ln(c_2\xi + c_3)$ (where c_i are constants). These fits work well, and they suggest the divergence of $\chi_t\,\xi^2$ in the continuum limit.

fitting function	$t=0$	t_0	$2t_0$	$3t_0$	$4t_0$	$5t_0$	$6t_0$
$c_1\,\ln(c_2\,\xi + c_3)$	1.07	1.34	1.63	2.20	1.82	1.89	1.91
$c_1\,\xi^{c_2} + c_3$	1.01	1.58	1.94	2.32	2.12	2.16	2.18

Table A2. The χ^2/d.o.f-values for two fits of our $\chi_t\,\xi^2$-results as a function of ξ. We consider two 3-parameter fitting functions, with a logarithmic and a power-law divergence in the continuum limit. At Gradient Flow time $t = 0$ both fits are excellent, but in the range $t = t_0 \ldots 6t_0$ they become somewhat worse, in particular for the power-law.

IOP Conf. Series: Journal of Physics: Conf. Series **912** (2017) 012024 doi:10.1088/1742-6596/912/1/012024

References

[1] Polyakov A M 1975 *Phys. Lett.* B **59** 79
[2] Hasenfratz P, Maggiore M and Niedermayer F 1990 *Phys. Lett.* B **245** 522
[3] Wolff U 1989 *Phys. Rev. Lett.* **62** 361
[4] Wolff U 1990 *Nucl. Phys.* B **334** 581
[5] Apostolakis J, Baillie C F and Fox G C 1991 *Phys. Rev.* D **43** 2687
[6] J-K Kim 1994 *Phys. Rev.* D **50** 4663
[7] Berg B and Lüscher M 1981 *Nucl. Phys.* B **190** 412
[8] Bietenholz W, Gerber U, Pepe M and Wiese U-J 2010 *JHEP* **1012** 020
[9] Bautista I *et al.* 2015 *Phys. Rev.* D **92** 114510
[10] Bietenholz W, Cichy K, de Forcrand P, Dromard A and Gerber U 2016 *PoS* **LATTICE2016** 321
[11] Lüscher M 1982 *Nucl. Phys.* B **200** 61
[12] Blatter M, Burkhalter R, Hasenfratz P and Niedermayer F 1996 *Phys. Rev.* D **53** 923
[13] Lüscher M 2010 *JHEP* **1008** 071, *PoS* **LATTICE2010** 015
[14] Makino H and Suzuki H 2015 *PTEP* **2015** 033B08
[15] Bietenholz W, de Forcrand P, Gerber U, Mejía-Díaz H and Sandoval I O, in preparation
[16] Giusti L, Rossi G C and Testa M 2004 *Phys. Lett.* B **587** 157
Lüscher M 2004 *Phys. Lett.* B **593** 296
Giusti L and Lüscher M 2009 *JHEP* **0903** 013
Lüscher M and Palombi F 2010 *JHEP* **1009** 110

IOP Conf. Series: Journal of Physics: Conf. Series **912** (2017) 012025 doi:10.1088/1742-6596/912/1/012025

Majorana versus Dirac neutrino in $\nu - e$ scattering process and its possible applications

J Barranco*, **D Delepine**†, **M Napsuciale**‡ **and A Yebra**§

Departamento de Física, División de Ciencias e Ingenierías, Campus León, Universidad de Guanajuato, León 37150, México

E-mail: *jbarranc@fisica.ugto.mx, †delepine@fisica.ugto.mx, ‡mauro@fisica.ugto.mx §azarael@fisica.ugto.mx,

Abstract. In previous works it was established that if neutrino helicity differs from minus one, then the Dirac and Majorana cross sections in neutrino-electron scattering are different, despite of the smallness of the neutrino mass. Considering this fact, it is possible to employ that difference in order to establish a method to derive an upper bound for neutrino magnetic moment.

1. Introduction

Determining the neutrino nature is an open, important and fundamental question in particle physics. It refers to establish if neutrinos are Dirac or Majorana particles. If neutrinos are Dirac particles, then they will be different from its own antiparticles. On the other hand, if they are Majorana particles, neutrino and anti-neutrino must be the same particle. In that case, certain kind of reactions where the lepton number is no longer conserved are possible. The experimental observation of those reactions would constitute a direct determination of the Majorana nature of the neutrino. One representative example from those reactions is the well known neutrinoless double beta decay [1].

Another way to try to settle this problem is to use the elastic neutrino-electron scattering process. Common wisdom assumes that differences between both cross sections are proportional to terms that depend on the neutrino mass, which is usually very small. As a result, it is very difficult to have a significant difference in the measurement of the scattering cross sections for Dirac or Majorana neutrinos. Nevertheless, if the neutrino polarization is taken into account, or in other words, if we consider that the neutrino longitudinal polarization, $s_{||}$, may differ from minus one ($s_{||} \neq -1$), then, a measurable difference between the Dirac or Majorana cross section is possible [2, 3, 4]. The changes in the neutrino helicity can be induced by the interaction of the neutrino magnetic moment μ_ν with an astrophysical external magnetic field. The changes of $s_{||}$ is determined by the Bargmann-Michel-Telegdi equation [5, 6].

Thus, the experimental verification of the neutrino nature will rely on the total number of events detected for astrophysical neutrinos that have changed their longitudinal polarization by the interaction of the neutrinos magnetic moment with an external magnetic field.

Solar neutrinos offer a possible source where such differences should appear. No appreciable differences will imply that the change in $s_{||}$ is very small and thus implies a limit on the neutrino magnetic moment μ_ν. For example, we will briefly discuss how Borexino [7], which has no seen

IOP Conf. Series: Journal of Physics: Conf. Series **912** (2017) 012025 doi:10.1088/1742-6596/912/1/012025

any appreciable difference of their expected number of events and the values measured imply that there is no appreciable difference between Dirac and Majorana neutrinos within its relative error. Therefore, we can set a new upper bound for the neutrino magnetic moment [4].

In the next section we will show the differences between the Dirac or Majorana neutrino-electron scattering cross section as a function of the neutrino longitudinal polarization $s_{||}$. Then, we will show how Borexino detection of ^{7}Be line solar neutrinos imply a determination within a 4% error in the measured cross section and thus a limit on the neutrino magnetic moment.

2. Dirac and Majorana neutrinos

As we mentioned above, the fact that neutrinos can be its own antiparticles or not is the central question about its nature. Moreover, one direct implication for Majorana neutrinos is that additional contributions from antineutrinos in the scatter on electrons must be included in the calculation of the amplitude, which is not necessary for Dirac neutrinos. Thus, changes on its cross sections are expected.

Let us star with the effective Lagrangian that describes the neutrino-electron scattering process at low energies. It is given by [3]:

$$\mathcal{L} = \frac{G_F}{\sqrt{2}} \left[\bar{u}^f_{\nu_\ell} \gamma^\mu \left(1 - \gamma^5\right) u^i_{\nu_\ell}\right] \left[\bar{u}^f_e \gamma_\mu \left(g^\ell_V - g^\ell_A\right) u^i_e\right], \tag{1}$$

where the effective coupling constants are usually defined as

$$g^\ell_V = -\frac{1}{2} + 2\sin^2\theta_W + \delta_{\ell e}, \quad g^\ell_A = -\frac{1}{2} + \delta_{\ell e}, \tag{2}$$

where θ_W is the weak mixing angle and note that $\delta_{\ell e}$ is added in order to include the charged current contribution.

From this Lagrangian we can obtain the usual amplitudes for neutrino-electron scattering process in Dirac and Majorana case, which are given by:

$$\mathcal{M}_D = -i\frac{G_F}{\sqrt{2}} \left[\bar{u}^f_{\nu_\ell} \gamma^\mu \left(1 - \gamma^5\right) u^i_{\nu_\ell}\right] \left[\bar{u}^f_e \gamma_\mu \left(g^\ell_V - g^\ell_A \gamma^5\right) u^i_e\right], \tag{3}$$

when the neutrino is a Dirac particle, and

$$\mathcal{M}_M = i\frac{2G_F}{\sqrt{2}} \left[\bar{u}^f_e \gamma^\mu \left(g^\ell_V - g^\ell_A \gamma^5\right) u^i_e\right] \left[\bar{u}^f_{\nu_\ell} \gamma_\mu \gamma^5 u^i_{\nu_\ell}\right], \tag{4}$$

for Majorana neutrinos. It is important to note that one crucial difference from equations (3) and (4) is given by the fact that there is a cancellation of the vector part in second term of Eq. (4) that it is not present in the Dirac case. Last two amplitudes look very different. At this point the computation of the cross sections will lead to different results. Nevertheless, the left-handedness of the neutrino implies an important role: since neutrinos are born with $s_{||} \simeq -1$ then an extra term $(1 - \gamma^5)/2$ is usually added in both eqs. (3) and (4) making them identical and thus, no appreciable difference arises. Addition of this factor $(1 - \gamma^5)/2$ in fact leads to the usual cross section, which in the massless limit is:

$$\frac{d^2\sigma}{dTdE} = \frac{m_e G_F^2}{2\pi} \left[(g_A + g_V)^2 + (g_V - g_A)^2 \left(1 - \frac{T}{E_\nu}\right)^2 - (g_V^2 - g_A^2)\frac{m_e T}{E_\nu}\right]. \tag{5}$$

That is the reason why it is very difficult to distinguish between Dirac and Majorana neutrinos in neutrino-electron scattering processes. In Eq. (5) the electron recoil energy is T and the energy of the incident neutrino is E_ν.

Nevertheless, let us consider that $s_\parallel \neq -1$ and then compute the corresponding cross sections with Eqs. (3) and (4). In previous works [2, 3] authors have demonstrated that when $s_\parallel \neq -1$ there is an additional contribution to the usual cross sections, proportional to $s_\parallel$, that can improve the difference between Dirac and Majorana neutrinos. In other words, the longitudinal neutrino polarization would be able to enhance the difference between Dirac and Majorana cross sections for ν-e elastic scattering process, but only if it differs from minus one. When the incident neutrino polarization if defined in its rest frame as $s_\nu = (0, s_\perp, 0, s_\parallel)$, with a straightforward calculation we can obtain [2, 3, 4] for Dirac case:

$$\begin{aligned}\frac{d^2\sigma^D}{dEdT} = \frac{m_e G_F^2}{4\pi P^2}\Bigg\{ & \left[(g_A - g_V)^2 (E_\nu - T)^2 + m_e T\left(g_A^2 - g_V^2\right)\right]\left(1 - \frac{E_\nu}{P}s_\parallel\right) \\ & + m_\nu^2\left[\left(g_A^2 - g_V^2\right)\left(1 - \frac{T}{P}s_\parallel\right) + (g_A - g_V)^2\left(E_\nu - T\left(1 + \frac{T}{m_e}\right)\frac{s_\parallel}{P}\right)\right] \\ & + (g_A + g_V)^2 E_\nu^2\left(1 - \frac{P}{E_\nu}s_\parallel\right)\Bigg\}, \end{aligned} \tag{6}$$

, and in the case where the neutrino is a Majorana particle:

$$\begin{aligned}\frac{d^2\sigma^M}{dEdT} = \frac{m_e G_F^2}{2\pi P^2}\Bigg\{ & \left(g_A^2 + g_V^2 - 2\frac{E_\nu}{P}g_A g_V s_\parallel\right)E_\nu^2 + \left(g_A^2 - g_V^2\right)m_e T \\ & + \frac{m_\nu^2}{m_e}T\left[2\frac{m_e}{T}\left(2g_A^2 - g_V^2\right) - 4g_A g_V s_\parallel\frac{E_\nu}{P}\left(1 - \frac{T}{2E_\nu}\right) + g_A^2 + g_V^2\right] \\ & + \left(g_A^2 + g_V^2 + 2\frac{E_\nu}{P}g_A g_V s_\parallel\right)(E_\nu - T)^2\Bigg\}, \end{aligned} \tag{7}$$

Equations (7) and (6) are the differential cross sections in the laboratory frame. Here, $P = \left|\vec{P}_\nu^i\right|$ is the magnitude of the momentum of the incident neutrino.

Note that in the case $s_\parallel$=-1, those two cross sections differ only by terms proportional to the neutrino mass, as common wisdom said. Nevertheless, even if $m_\nu \to 0$, IF $s_\parallel \neq -1$ those two cross sections are different regardless the neutrino energy and thus, measurable differences should be observable at current neutrino detectors.

3. Applications

According to last equations, the differential cross sections depend on the longitudinal polarization $s_\parallel$. The total cross section is the obtained by integrations over recoil and neutrino incident energies and it will depend on $s_\parallel$, i.e.

$$\sigma^{M,D}(s_\parallel) \equiv \int_{T_{min}}^{T_{max}} dT \int_0^\infty dE_\nu \lambda(E_\nu)\frac{d^2\sigma^{M,D}}{dE_\nu dT}(E_\nu, T, s_\parallel). \tag{8}$$

The ingredients that we need to know in order to obtain the total cross section given by Eq. (8) are i) *the expression for differential cross section in Dirac or Majorana case*, ii) *a spectrum* $\lambda(E_\nu)$ *of energy for neutrinos*, and finally iii) *an interval of integration over recoil energy* T. For example, in a detector with the features of Borexino, the electron recoil energy can be integrated over the interval $T \in [250, 750]$ keV while the neutrino spectrum is the well know ^{7}Be line of solar neutrino spectrum given by $\lambda(E_\nu) = \delta(E_\nu - 0.862 \text{ MeV})$.

Once we have calculated the total cross section the next step is to construct a function so as to be able to quantify the percentage of difference between Dirac and Majorana cross sections:

$$D(s_{||}) \equiv \frac{|\sigma^M(s_{||}) - \sigma^D(s_{||})|}{\sigma^D(s_{||})}. \tag{9}$$

This function depends only on $s_{||}$ and allows to evaluated how difference between Dirac and Majorana changes for different values of the polarization. The next question related to last equation is how can we change the neutrino helicity. We actually know that although neutrino magnetic moment is very small [10], it is not equal to zero. This fact recovers importance because it has been showed that particles with magnetic moment different from zero that are embedded in a external magnetic field can change its helicity according to the equation [5, 6, 9]

$$\frac{ds_{||}}{dr} = -2\mu_\nu B_\perp s_{||}\,, \tag{10}$$

Here, $B_\perp$ is the perpendicular component of that external magnetic field, with respect to the direction in which the particle is propagating and μ_ν is the neutrino magnetic moment. In the case of solar neutrinos, a magnetic profile of the Sun is needed. For definitiveness we will use the profile proposed in [11]. In that work, authors have obtained a magnetic profile through self-consistent and analytical solutions of the magneto-hydrodynamic equations inside the Sun (See reference [11] for more details). Solution of eq. 10 can be obtained at the surface of the Sun, $r = R_\odot$, for a given value of μ_ν and a maximum value of $B_\perp$ at the convective zone. Current limits fixes $B_\perp^{max} 10 - 20$ KG. Thus we can obtain the final neutrino polarization $s_{||}(r = R_\odot, \mu_\nu, B_\perp^{max} = 10\text{KG})$

At this point we can put all ingredients together. As we discussed above, function (9) gives a measure of the difference between Dirac and Majorana cross sections. On the other hand, if we take into account the number of events reported by Borexino collaboration, which is N=49 $\pm$ 1.5stat$^{+1.5}_{-1.6}$syst counts/day/100 ton [7, 8], and assuming that no error arises from the neutrino probability, this measurement gives a determination of the cross section within a error given by the error in the number of events. That is, the relative error in the determination of the number of events reported by Borexino is 4.4%.

In other words, if there is a difference between the Majorana and the Dirac cross section this should be less than the relative error reported by Borexino, thus $D(s_{||})<0.044$. Using Eqs. 6 and 7 we can solve $D(s_{||})<0.044$ for $s_{||}$ and this gives $s_{||}<-0.68$.

Finally, solving $s_{||}(r = R_\odot, \mu_\nu, B_\perp^{max} = 10\text{KG}) < -0.68$ for the neutrino magnetic moment, we obtain the new upper bound $\mu_\nu < 1.4 \times 10^{-12}\mu_B$. See [4] for more details.

4. Conclusion

To determine the neutrino nature through neutrino-electron scattering process is difficult because any possible difference will be suppressed by terms that are proportional to neutrino mass unless $s_{||} \neq -1$, [2, 3, 4]. In that case, measurable differences are possible. Solar neutrinos offer a possible source where such differences should appear since its magnetic field may induce a change in $s_{||}$ due to the interaction of the magnetic moment of the neutrino with the external magnetic field of the Sun. No appreciable differences will imply that the change in $s_{||}$ is very small and thus implies a limit on the neutrino magnetic moment μ_ν.

The number of events reported by Borexino collaboration of Solar neutrinos of the ^{7}Be line implies a determination of the cross section within a error given by the error in the number of events. which is of the order of the 4%. Then, if there is a difference between the Majorana and the Dirac cross section this should be less than the relative error reported by Borexino, which implies a limit on the neutrino magnetic moment given by $\mu_\nu < 1.4 \times 10^{-12}\mu_B$.

Acknowledgments
This work was partially supported by Conacyt project CB-259228, Conacyt SNI and DAIP-UG.

References
[1] Schechter J and Valle J W F 1982 *Phys. Rev.* D **25** 2951
[2] Kayser B 1982 *Phys. Lett.* B **112** 137-42
[3] Barranco J, Delepine D, Gonzalez-Macias V, Lujan-Peschard C and Napsuciale M 2014 *Phys. Lett.* B **739** 343-7
[4] Barranco J, Delepine D, Napsuciale M and Yebra A, arXiv:1704.01549 [hep-ph].
[5] V. Bargmann, L. Michel and V. L. Telegdi, Phys. Rev. Lett. **2**, 435 (1959). doi:10.1103/PhysRevLett.2.435
[6] V. Semikoz, Phys. Rev. D **48** (1993) 5264 [Erratum-ibid. D **49** (1994) 6246].
[7] Arpesella C et al. [Borexino Collaboration] 2008 *Phys. Rev. Lett.* **101** 091302
[8] Bellini G et al. 2011 *Phys. Rev. Lett.* **107** 141302
[9] Semikoz V 1993 *Phys. Rev.* D **49** 5264 [Erratum-ibid. D **49** (1994) 6246]
[10] K. Fujikawa and R. Shrock, Phys. Rev. Lett. **45** (1980) 963.
[11] Miranda O G, Pena-Garay C, Rashbe T I, Semikoz V and Valle J W F 2001 *Nucl. Phys.* B **595** 360

IOP Conf. Series: Journal of Physics: Conf. Series **912** (2017) 012026 doi:10.1088/1742-6596/912/1/012026

Two sum rules involving leading vector form factors in hyperon semileptonic decays

[1]Flores-Mendieta Rubén and [2]Padrón-Stevens Roberto

[1,2] Instituto de Física, Universidad Autónoma de San Luis Potosí, Álvaro Obregón 64, Zona Centro, San Luis Potosí, S.L.P. 78000, México

Abstract. When the weak and electromagnetic currents are considered to be in the same $SU(3)$ octet, it is possible to derive two sum rules involving leading vector form factors in hyperon semileptonic decays in the limit of exact flavor $SU(3)$ symmetry. According to the Ademollo-Gatto theorem, deviations of these sum rules from this limit are expected to occur due to second-order flavor $SU(3)$ symmetry-breaking effects, which are evaluated within the $1/N_c$ expansion of QCD. Accordingly, one sum rule acquires no deviations whereas the other one obtains contributions coming from the $\mathbf{10}+\overline{\mathbf{10}}$ representation. The sum rules are validated using results obtained in heavy baryon chiral perturbation theory. Up to order $\mathcal{O}(p^2)$ in the chiral expansion, the sum rules are fulfilled.

1. Introduction

Despite the enormous progress achieved in the understanding of the fundamental interactions with the Standard Model, first principles calculations of the properties of hadrons are not possible because the theory is strongly coupled at low energies. In order to resort this situation, several methods have been introduced to understand the low-energy QCD hadron dynamics and the $1/N_c$ expansion [1] has played a crucial role. This method promotes QCD to a $SU(N_c)$ non-Abelian gauge theory, where N_c is the number of colors.

In this tenor, hyperon semileptonic decays (HSD) offer some challenges until today. In a recent paper [2], two sum rules involving the leading vector form factors f_1 were derived by considering that the weak currents and the electromagnetic current are members of the same $SU(3)$ octet. According to the Ademollo-Gatto theorem [3], the leading vector form factors f_1 are protected against $SU(3)$ symmetry breaking (SB) corrections to lowest order in $m_s - \hat{m}$, where $\hat{m}$ denotes the mean mass of the up and down quarks. The purpose of the present paper is to provide some generalities about the method used to obtain those sum rules and to discuss some important implications.

The organization of this paper is as follows. In section II, the two sum rules involving leading vector form factors are derived by taking the matrix element of the vector current between eigenstates of the V^2 operator. The resulting expressions are valid in the exact flavor $SU(3)$ symmetry limit. Subsequently, the two sum rules are modified to account for SB. In section III, a brief review about how Ademollo and Gatto came up with their result on the nonrenormalization of the electromagnetic and vector currents at first-order in SB is provided, followed by a survey on the $1/N_c$ expansion in section IV. In section V, the expressions containing the operators that explicitly break symmetry at first- and second-order are introduced. Restrictions on the nonrenormalization of the electric charge serve as an important factor to understand which operator coefficients (since they are the ones responsible for SB effects) will be present in the expressions for f_1. Collecting all partial findings, the operators responsible for SB in the sum rules are properly identified. Next, results obtained in the frame of chiral perturbation theory are used to validate the sum rules. Up to $\mathcal{O}(p^2)$ in the chiral expansion, the sum rules are satisfied. A validation beyond this chiral order is not currently possible because there are some partial results available. Finally, a numerical evaluation of the sum rules is also performed through a least-squares fit to data. Results are encouraging. Some closing remarks are given in section VI.

2. Sum rules for baryon vector form factors

The Cabibbo model for weak hadronic currents based on $SU(3)$ symmetry has been a key approach to describe HSD [4]. Since flavor $SU(3)$ symmetry is broken, its exact symmetry limit has been scrutinized using several methods in order to find discrepancies between theory and experiment and the leading weak form factors f_1 in HSD are the usual probes.

The starting point of the present analysis is the observation made by Pham about the matrix elements of the $V=1$ V-spin multiplet [5]. He showed that these matrix elements can be related to each other in the limit of exact flavor $SU(3)$ symmetry. From the two I-spin relations,

$$\langle I=1, I_3=0|\,\overline{u}\Gamma^n d\,|I=1, I_3=1\rangle = -\langle I=1, I_3=-1|\,\overline{u}\Gamma^n d\,|I=1, I_3=0\rangle\,, \tag{1a}$$

and

$$\langle I=0, I_3=0|\,\overline{u}\Gamma^n d\,|I=1, I_3=1\rangle = \langle I=1, I_3=-1|\,\overline{u}\Gamma^n d\,|I=0, I_3=0\rangle\,, \tag{1b}$$

the rotated V-spin versions of the above relations become

$$\langle V=1, V_3=0|\,\overline{u}\Gamma^n s\,|V=1, V_3=1\rangle = -\langle V=1, V_3=-1|\,\overline{u}\Gamma^n s\,|V=1, V_3=0\rangle\,, \tag{2a}$$

and

$$\langle V=0, V_3=0|\,\overline{u}\Gamma^n s\,|V=1, V_3=1\rangle = \langle V=1, V_3=-1|\,\overline{u}\Gamma^n s\,|V=0, V_3=0\rangle\,, \tag{2b}$$

where the bilinear forms $\overline{q}_1\Gamma^n q_2$ are given in terms of quark fields q_i and the matrices $\Gamma^n_{\alpha\beta}$ are the usual ones that appear in applications of Dirac theory.

Decomposing the $SU(3)$ baryon octet into eigenfunctions of V^2 yields

$$|\Xi^-\rangle = |V=1, V_3=1\rangle\,, \tag{3}$$

$$\left|\frac{1}{2}\Sigma^0+\frac{\sqrt{3}}{2}\Lambda\right\rangle = |V=1, V_3=0\rangle\,, \tag{4}$$

$$|p\rangle = |V=1, V_3=-1\rangle\,, \tag{5}$$

$$\left|\frac{\sqrt{3}}{2}\Sigma^0-\frac{1}{2}\Lambda\right\rangle = |V=0, V_3=0\rangle\,. \tag{6}$$

Relations (2) are general enough to apply to matrix elements of any $SU(3)$ octet $\Delta S=1$ operator. The Gell-Mann–Okubo mass formula is obtained from (2a) by relating $\langle B_2|\,\overline{u}\mathcal{I}s\,|B_1\rangle$ to $\left[f_1^{SU(3)}\right]_{B_1B_2}(M_{B_2}-M_{B_1})$, where $f_1^{SU(3)}$ stands for the leading vector form factor at zero recoil in the limit of exact flavor $SU(3)$ symmetry and M_{B_i} is the mass of the B_i baryon. Applications to the axial-vector to vector form factor ratios g_1/f_1 are obtained when $\Gamma^n=\gamma_5\gamma_\mu$ [5].

The vector current can be straightforwardly analyzed for $\Gamma^n=\gamma_\mu$. Consequently, there are two nontrivial expressions that relate the matrix elements of the vector current between baryon octet states, namely,

$$\left\langle\frac{1}{2}\Sigma^0+\frac{\sqrt{3}}{2}\Lambda\middle|\overline{u}\gamma_\mu s\middle|\Xi^-\right\rangle = -\left\langle p\middle|\overline{u}\gamma_\mu s\middle|\frac{1}{2}\Sigma^0+\frac{\sqrt{3}}{2}\Lambda\right\rangle, \tag{7a}$$

$$\left\langle \frac{\sqrt{3}}{2}\Sigma^0 - \frac{1}{2}\Lambda \middle| \overline{u}\gamma_\mu s \middle| \Xi^- \right\rangle = \left\langle p \middle| \overline{u}\gamma_\mu s \middle| \frac{\sqrt{3}}{2}\Sigma^0 - \frac{1}{2}\Lambda \right\rangle. \tag{7b}$$

Hence, in the limit of exact flavor $SU(3)$ symmetry and neglecting isospin breaking, two relations between the leading vector form factors in HSD arise as a consequence of (7a) and (7b), namely,

$$\left[f_1^{SU(3)}\right]_{\Xi^-\Sigma^0} + \sqrt{3}\left[f_1^{SU(3)}\right]_{\Xi^-\Lambda} + \frac{1}{\sqrt{2}}\left[f_1^{SU(3)}\right]_{\Sigma^-n} + \sqrt{3}\left[f_1^{SU(3)}\right]_{\Lambda p} = 0, \tag{8a}$$

and

$$\sqrt{3}\left[f_1^{SU(3)}\right]_{\Xi^-\Sigma^0} - \left[f_1^{SU(3)}\right]_{\Xi^-\Lambda} - \sqrt{\frac{3}{2}}\left[f_1^{SU(3)}\right]_{\Sigma^-n} + \left[f_1^{SU(3)}\right]_{\Lambda p} = 0. \tag{8b}$$

When SB is present, relations (8) must be modified. After minor rearrangements, sum rules (8) thus read

$$\frac{1}{4}\left[\frac{f_1}{f_1^{SU(3)}}\right]_{\Xi^-\Sigma^0} + \frac{3}{4}\left[\frac{f_1}{f_1^{SU(3)}}\right]_{\Xi^-\Lambda} - \frac{1}{4}\left[\frac{f_1}{f_1^{SU(3)}}\right]_{\Sigma^-n} - \frac{3}{4}\left[\frac{f_1}{f_1^{SU(3)}}\right]_{\Lambda p} = \delta_1^{SB}, \tag{9a}$$

and

$$\frac{3}{4}\left[\frac{f_1}{f_1^{SU(3)}}\right]_{\Xi^-\Sigma^0} - \frac{3}{4}\left[\frac{f_1}{f_1^{SU(3)}}\right]_{\Xi^-\Lambda} + \frac{3}{4}\left[\frac{f_1}{f_1^{SU(3)}}\right]_{\Sigma^-n} - \frac{3}{4}\left[\frac{f_1}{f_1^{SU(3)}}\right]_{\Lambda p} = \delta_2^{SB}. \tag{9b}$$

A few remarks are in order here. The coefficients on the right-hand-side of Eqs. (9) have been chosen to keep some resemblance to their counterpart in K_{l3} decays [6], namely,

$$\frac{1}{4}\left|f_+^{K^+\pi^0}\right|^2 + \frac{3}{4}\left|f_+^{K^0\eta}\right|^2 = 1 + \mathcal{O}(\epsilon^2), \tag{10}$$

where ϵ will be referred to as a measure of SB hereafter. Similarly, δ_k^{SB}, which arise from SB, are formally of order $\mathcal{O}(\epsilon^2)$ according to the Ademollo-Gatto theorem. Obtaining δ_k^{SB} requires and extra effort and it will be performed within the $1/N_c$ expansion.

3. Some generalities about the Ademollo-Gatto theorem

Following the assumption that the vector and electromagnetic currents are members of the same unitary octet and that the breaking of the unitary symmetry behaves as the eighth component of an octet, M. Ademollo and R. Gatto set up an important theorem on the nonrenormalization of the strangeness-violating vector currents [3].

In the Ademollo and Gatto formalism, the ath component of the current $\mathcal{J}^a$ to first-order in SB is written as

$$\begin{aligned}
\mathcal{J}^a + \epsilon\delta\mathcal{J}^a \;=\;& a_0 \operatorname{Tr}\left(\overline{B}B\lambda^a\right) + b_0 \operatorname{Tr}\left(\overline{B}\lambda^a B\right) \\
& + \epsilon a\left[\operatorname{Tr}\left(\overline{B}B\left\{\lambda^a,\lambda^8\right\}\right) - \frac{1}{8}\delta^{a8}\operatorname{Tr}\left(\overline{B}B\left\{\lambda^c,\lambda^c\right\}\right)\right] \\
& + \epsilon b\left[\operatorname{Tr}\left(\overline{B}\left\{\lambda^a,\lambda^8\right\}B\right) - \frac{1}{8}\delta^{a8}\operatorname{Tr}\left(\overline{B}\left\{\lambda^c,\lambda^c\right\}B\right)\right] \\
& + \epsilon c\left[\operatorname{Tr}\left(\overline{B}\lambda^a B\lambda^8\right) - \operatorname{Tr}\left(\overline{B}\lambda^8 B\lambda^a\right)\right] + \epsilon g \operatorname{Tr}\left(\overline{B}B\right)\operatorname{Tr}\left(\lambda^a\lambda^8\right) \\
& + \epsilon h\left[\operatorname{Tr}\left(\overline{B}\lambda^a B\lambda^8\right) + \operatorname{Tr}\left(\overline{B}\lambda^8 B\lambda^a\right) - \frac{1}{8}\delta^{a8}\operatorname{Tr}\left(\overline{B}\lambda^c B\lambda^c\right)\right. \\
& \left. -\frac{6}{5}d^{a8c}d^{cgh}\operatorname{Tr}\left(\overline{B}\lambda^g B\lambda^h\right)\right],
\end{aligned} \tag{11}$$

IOP Conf. Series: Journal of Physics: Conf. Series **912** (2017) 012026 doi:10.1088/1742-6596/912/1/012026

where B represents the baryon matrix, λ^a stand for the Gell-Mann matrices and a_0, b_0, ..., h are coupling constants. The parameter ϵ is a reminder of the order SB is accounted for. At the end ϵ can be set to one.

The electromagnetic current is defined in the usual way as

$$\mathcal{J}_{\text{em}} = \mathcal{J}^Q \equiv \mathcal{J}^3 + \frac{1}{\sqrt{3}}\mathcal{J}^8, \tag{12}$$

so the baryon charges can be obtained from (12) without any difficulties. For instance, the neutron charge, including first-order SB is

$$Q_n + \epsilon\delta Q_n = -\frac{2}{3}a_0 - \frac{2}{3}b_0 + \frac{4}{3\sqrt{3}}\epsilon a - \frac{8}{3\sqrt{3}}\epsilon b + \frac{2}{\sqrt{3}}\epsilon c + \frac{2}{\sqrt{3}}\epsilon g + \frac{\sqrt{3}}{10}\epsilon h. \tag{13}$$

Similar relations can be obtained for the remaining baryons. Therefore, there are eight equations (one for each baryon charge) and seven parameters to be found. However, the isospin relation

$$\frac{1}{2}(Q_{\Sigma^+} + Q_{\Sigma^-}) = Q_{\Sigma^0}, \tag{14}$$

leads to seven linear independent equations. Solving the system for the seven unknowns yields

$$a_0 = -\frac{1}{2}, \; b_0 = \frac{1}{2}, \; a = b = c = g = h = 0. \tag{15}$$

The solution of the system shows that the electromagnetic charge acquires no corrections to first-order SB (as it is expected since the electric charge remains unrenormalized to all orders in perturbation theory). As a direct consequence, the vector current is also protected against first-order SB. This is the original conclusion reached by Ademollo and Gatto [3].

4. Generalities on the $1/N_c$ expansion

The formalism on the $1/N_c$ expansion for baryons can be found in Ref. [8], so a few salient facts will be repeated. In the large-N_c limit, the baryon sector has a contracted $SU(2N_f)$ spin-flavor symmetry, where N_f is the number of light quark flavors [7]. The $1/N_c$ expansion is given in terms of $1/N_c$-suppressed operators with well-defined spin-flavor transformation properties. For $N_f = 3$, the lowest-lying baryon states fall into a representation of the spin-flavor group $SU(6)$.

The $1/N_c$ expansion of any QCD operator transforming according to a given $SU(2)\otimes SU(N_f)$ spin-flavor representation is expressed in terms of n-body operators $\mathcal{O}_n$ as [8]

$$\mathcal{O}_{\text{QCD}} = \sum_n c_{(n)} \frac{1}{N_c^{n-1}} \mathcal{O}_n, \tag{16}$$

where the operator coefficients $c_{(n)}$ are undetermined and have power series expansions in $1/N_c$ beginning at order one. The operators $\mathcal{O}_n$ are polynomials in the spin-flavor generators J^k, T^c, and G^{kc}, which are 1-body operators acting on the baryon states. They read

$$J^k = \sum_\alpha^{N_c} q_\alpha^\dagger \left(\frac{\sigma^k}{2} \otimes \mathcal{I}\right) q_\alpha, \tag{17a}$$

$$T^c = \sum_\alpha^{N_c} q_\alpha^\dagger \left(\mathcal{I} \otimes \frac{\lambda^c}{2}\right) q_\alpha, \tag{17b}$$

$$G^{kc} = \sum_\alpha^{N_c} q_\alpha^\dagger \left(\frac{\sigma^k}{2} \otimes \frac{\lambda^c}{2}\right) q_\alpha. \tag{17c}$$

Operators $q_\alpha^\dagger$ and q_α create and annihilate states in the fundamental representation of $SU(6)$ and σ^k and λ^c are the Pauli spin and Gell-Mann flavor matrices, respectively.

IOP Conf. Series: Journal of Physics: Conf. Series **912** (2017) 012026 doi:10.1088/1742-6596/912/1/012026

5. The leading baryon vector form factor in the $1/N_c$ expansion

In this section a brief description of the baryon vector current will be provided. The operator is constructed [2] in such a way that its matrix elements give the actual values of the leading vector form factor as is defined in HSD.

To start with, let V^{0c} denote the flavor octet baryon charge [9]

$$V^{0c} = \left\langle B_2 \left| \left(\overline{q}\gamma^0 \frac{\lambda^c}{2} q \right)_{\text{QCD}} \right| B_1 \right\rangle, \tag{18}$$

where the subscript QCD indicates that the quark fields are QCD quark fields [9]. V^{0c} is a spin-0 and a flavor octet; it transforms as a $(0,\mathbf{8})$ under $SU(2)\otimes SU(N_f)$. The actual values of the leading vector form factors $f_1^{SU(3)}$ at zero recoil can be easily obtained by computing the appropriate matrix elements between baryon octet states for $c = 4 \pm i5$.

In QCD, flavor SB is due to the strange quark mass m_s and transforms as the eight component of an octet [8]. First-order SB corrections arise from the tensor product $(0,\mathbf{8})\otimes(0,\mathbf{8})$ so that the $SU(3)\otimes SU(3)$ representations to be included are $(0,\mathbf{1})$, $(0,\mathbf{8})$, $(0,\mathbf{10}+\overline{\mathbf{10}})$ and $(0,\mathbf{27})$ [10].

Now let $V^c + \epsilon\delta V^c$ be the most general $(0,\mathbf{8})$ operator containing first-order SB. Its $1/N_c$ expansion is then

$$\begin{aligned}
V^c + \epsilon\delta V^c \; = \; & c_{(1)}^{\mathbf{8}} T^c + c_{(2)}^{\mathbf{8}} \frac{1}{N_c}\{J^r, G^{rc}\} + \epsilon N_c a_{(0)}^{\mathbf{1}} \delta^{c8} + \epsilon N_c a_{(1)}^{\mathbf{8}} d^{c8e} T^e + \epsilon a_{(2)}^{\mathbf{8}} \frac{1}{N_c} d^{c8e}\{J^r, G^{re}\} \\
& + \epsilon a_{(3)}^{\mathbf{10}+\overline{\mathbf{10}}} \frac{1}{N_c^2} \left(\{T^c, \{J^r, G^{r8}\}\} - \{T^8, \{J^r, G^{rc}\}\} \right) \\
& + \epsilon a_{(2)}^{\mathbf{27}} \frac{1}{N_c} \Bigg[\{T^c, T^8\} - \frac{N_f - 2}{2N_f(N_f^2-1)} N_c(N_c + 2N_f)\delta^{c8} - \frac{2}{N_f^2-1}\delta^{c8} J^2 \\
& \qquad - \frac{N_f-4}{N_f^2-4}(N_c+N_f) d^{c8e} T^e - \frac{2N_f}{N_f^2-4} d^{c8e}\{J^r, G^{re}\} \Bigg] \\
& + \epsilon a_{(3)}^{\mathbf{27}} \frac{1}{N_c^2} \Bigg[\{T^c, \{J^r, G^{r8}\}\} + \{T^8, \{J^r, G^{rc}\}\} - \frac{4}{N_f(N_f+1)}(N_c+N_f)\delta^{c8} J^2 \\
& \qquad - \frac{2}{N_f+2}(N_c+N_f) d^{c8e}\{J^r, G^{re}\} - \frac{2}{N_f+2} d^{c8e}\{J^2, T^e\} \Bigg].
\end{aligned} \tag{19}$$

The expansion (19) has been truncated at the physical value $N_c = 3$, so up to three-body operators are retained. Also, singlet and octet representations are explicitly subtracted off from the two- and three-body operators of the $(0,\mathbf{27})$ representation to keep only the truly $(0,\mathbf{27})$ contributions. Finally, operator coefficients $c_{(n)}^{\mathbf{rep}}$ accompany operators that belong to the exact flavor $SU(3)$ limit whereas $a_{(n)}^{\mathbf{rep}}$ accompany n-body operators of the representation **rep** that explicitly breaks flavor symmetry. Higher-order operators are constructed by anticommuting J^2 with the already existing ones as $\mathcal{O}_{n+2} = \{J^2, \mathcal{O}_n\}$. These contributions are not included since they can be accounted for by redefining the operator coefficients.

The matrix elements of the operator $V^Q + \epsilon\delta V^Q$ between $SU(6)$ symmetric baryon states give the actual values for the baryon charges Q_B including first-order SB. For example, at the physical values $N_c = N_f = 3$, the proton charge reads

$$Q_p + \epsilon\delta Q_p = -\frac{2}{3}a_0 + \frac{4}{3}b_0 + \frac{4}{3\sqrt{3}}\epsilon a + \frac{4}{3\sqrt{3}}\epsilon b - \frac{2}{\sqrt{3}}\epsilon c + \frac{2}{\sqrt{3}}\epsilon g - \frac{7\sqrt{3}}{10}\epsilon h. \tag{20}$$

IOP Conf. Series: Journal of Physics: Conf. Series **912** (2017) 012026 doi:10.1088/1742-6596/912/1/012026

Similar expressions are found for the other members of the baryon octet [2]. When analyzing the expressions for the full baryon charges, it is straightforward to notice that the coefficients of the $(0, \mathbf{27})$ representation are not independent and a new coefficient involving $a_{(2)}^{\mathbf{27}}$ and $a_{(3)}^{\mathbf{27}}$ can be introduced, namely,

$$x_{(2)}^{\mathbf{27}} = a_{(2)}^{\mathbf{27}} + \frac{2}{3}a_{(3)}^{\mathbf{27}}. \tag{21}$$

Since the new coefficient $x_{(2)}^{\mathbf{27}}$ appears in the eight expressions for the baryon charges this reduces the number of unknowns in the linear system by one. Solving the system of linear equations yields

$$c_{(1)}^{\mathbf{8}} = 1, \; c_{(2)}^{\mathbf{8}} = a_{(0)}^{\mathbf{1}} = a_{(1)}^{\mathbf{8}} = a_{(2)}^{\mathbf{8}} = a_{(3)}^{\mathbf{10}+\overline{\mathbf{10}}} = x_{(2)}^{\mathbf{27}} = 0. \tag{22}$$

The above nicely reproduces the Ademollo and Gatto results using the $1/N_c$ expansion formalism. Operator coefficients of (19) are related to the coupling constants introduced in (11) by

$$a_0 = -\frac{1}{2}c_{(1)}^{\mathbf{8}} + \frac{1}{12}c_{(2)}^{\mathbf{8}}, \tag{23a}$$

$$b_0 = \frac{1}{2}c_{(1)}^{\mathbf{8}} + \frac{5}{12}c_{(2)}^{\mathbf{8}}, \tag{23b}$$

$$a = -\frac{1}{4}a_{(1)}^{\mathbf{8}} + \frac{1}{24}a_{(2)}^{\mathbf{8}}, \tag{23c}$$

$$b = \frac{1}{4}a_{(1)}^{\mathbf{8}} + \frac{5}{24}a_{(2)}^{\mathbf{8}}, \tag{23d}$$

$$c = \frac{1}{6}a_{(3)}^{\mathbf{10}+\overline{\mathbf{10}}} \tag{23e}$$

$$g = \frac{3}{2}a_{(0)}^{\mathbf{1}}, \tag{23f}$$

$$h = -\frac{1}{6}x_{(2)}^{\mathbf{27}}. \tag{23g}$$

Notice a matching between the various coefficients and the representations they belong to.

The next step is to incorporate flavor SB at second-order into the $1/N_c$ expansion for the baryon charge operator. The most general expression is [2]

$$\begin{aligned}
\epsilon^2 \delta V^c \; = \; & \epsilon^2 b_{(0)}^{\mathbf{1}} N_c d^{c88} + \epsilon^2 b_{(1)}^{\mathbf{8}} \delta^{c8} T^8 + \epsilon^2 e_{(1)}^{\mathbf{8}} f^{c8e} f^{8eg} T^g + \epsilon^2 g_{(1)}^{\mathbf{8}} d^{c8e} d^{8eg} T^g \\
& + \epsilon^2 h_{(1)}^{\mathbf{8}} \left(i f^{ceg} d^{8e8} T^g - i d^{ce8} f^{8eg} T^g - i f^{c8e} d^{eg8} T^g \right) + \epsilon^2 b_{(2)}^{\mathbf{8}} \frac{1}{N_c} \delta^{c8} \{J^r, G^{r8}\} \\
& + \epsilon^2 e_{(2)}^{\mathbf{8}} f^{c8e} f^{8eg} \frac{1}{N_c} \{J^r, G^{rg}\} + \epsilon^2 g_{(2)}^{\mathbf{8}} d^{c8e} d^{8eg} \frac{1}{N_c} \{J^r, G^{rg}\} \\
& + \epsilon^2 h_{(2)}^{\mathbf{8}} \left(i f^{ceg} d^{8e8} - i d^{ce8} f^{8eg} - i f^{c8e} d^{eg8} \right) \{J^r, G^{rg}\} \\
& + \epsilon^2 b_{(3)}^{\mathbf{10}+\overline{\mathbf{10}}} \frac{1}{N_c^2} d^{c8e} \left(\{T^e, \{J^r, G^{r8}\}\} - \{T^8, \{J^r, G^{re}\}\} \right) \\
& + \epsilon^2 b_{(2)}^{\mathbf{27}} \frac{1}{N_c} \Bigg[d^{c8e} \{T^e, T^8\} - \frac{N_f - 4}{N_f^2 - 4} (N_c + N_f) d^{c8e} d^{8eg} T^g \\
& \qquad - \frac{2N_f}{N_f^2 - 4} d^{c8e} d^{8eg} \{J^r, G^{rg}\} \Bigg] \\
& + \epsilon^2 b_{(3)}^{\mathbf{27}} \frac{1}{N_c^2} \Bigg[d^{c8e} \left(\{T^e, \{J^r, G^{r8}\}\} + \{T^8, \{J^r, G^{re}\}\} \right) \\
& \qquad - \frac{2}{N_f + 2} (N_c + N_f) d^{c8e} d^{8eg} \{J^r, G^{rg}\} - \frac{2}{N_f + 2} d^{c8e} d^{8eg} \{J^2, T^g\} \Bigg] \\
& + \epsilon^2 b_{(3)}^{\mathbf{64}} \frac{1}{N_c^2} \Bigg[\{T^c, \{T^8, T^8\}\} - \frac{N_f - 2}{N_f (N_f^2 - 1)} N_c (N_c + 2N_f) \delta^{88} T^c \\
& \qquad - \frac{1}{2} f^{c8e} f^{8eg} T^g - \frac{N_f - 4}{2(N_f^2 - 4)} (N_c + N_f) d^{c8e} \{T^e, T^8\} \\
& \qquad - \frac{N_f - 4}{2(N_f^2 - 4)} (N_c + N_f) d^{88e} \{T^c, T^e\} - \frac{2}{N_f^2 - 1} \delta^{88} \{J^2, T^c\} \\
& \qquad - \frac{N_f}{N_f^2 - 4} d^{c8e} \{T^8, \{J^r, G^{re}\}\} - \frac{N_f}{N_f^2 - 4} d^{88e} \{T^c, \{J^r, G^{re}\}\} \Bigg]. \qquad (24)
\end{aligned}$$

The matrix elements of $\epsilon^2 \delta V^Q$ between $SU(6)$ symmetric baryon octet states give second-order corrections to the baryon octet charges. For the proton the contribution reads

$$\epsilon^2 \delta Q_p = -\epsilon^2 b_{(0)}^{\mathbf{1}} + \frac{1}{2} \epsilon^2 b_{(1)}^{\mathbf{8}} + \frac{1}{12} \epsilon^2 b_{(2)}^{\mathbf{8}} + \frac{1}{3} \epsilon^2 g_{(1)}^{\mathbf{8}} + \frac{1}{6} \epsilon^2 g_{(2)}^{\mathbf{8}} - \frac{1}{9} \epsilon^2 b_{(3)}^{\mathbf{10}+\overline{\mathbf{10}}} - \frac{1}{15} \epsilon^2 b_{(2)}^{\mathbf{27}} - \frac{2}{45} \epsilon^2 b_{(3)}^{\mathbf{27}} + \frac{13}{120} \epsilon^2 b_{(3)}^{\mathbf{64}}, \qquad (25)$$

Similar expressions for the remaining baryon octet charges are found [2]. Again, the operator coefficients of the $(0, \mathbf{27})$ representation are not independent and a new coefficient that includes $b_{(2)}^{\mathbf{27}}$ and $b_{(3)}^{\mathbf{27}}$ can be defined as

$$y_{(2)}^{\mathbf{27}} \equiv b_{(2)}^{\mathbf{27}} + \frac{2}{3} b_{(3)}^{\mathbf{27}}. \qquad (26)$$

Only eight out of twelve operator coefficients that appear in (24) make up the octet baryon charges. Again, using the argument that the electric charge remains unrenormalized to all orders in perturbation theory, it is possible to solve the resulting linear system for the eight unknowns, which yields

$$b_{(0)}^{\mathbf{1}} = b_{(2)}^{\mathbf{8}} = g_{(2)}^{\mathbf{8}} = y^{\mathbf{27}} = 0, \tag{27a}$$

$$g_{(1)}^{\mathbf{8}} = -\frac{5}{24} b_{(3)}^{\mathbf{10}+\overline{\mathbf{10}}}, \tag{27b}$$

$$b_{(1)}^{\mathbf{8}} = -\frac{13}{72} b_{(3)}^{\mathbf{10}+\overline{\mathbf{10}}}, \tag{27c}$$

$$b_{(3)}^{\mathbf{64}} = \frac{5}{2} b_{(3)}^{\mathbf{10}+\overline{\mathbf{10}}}. \tag{27d}$$

Hence, following the above results, the vector current at second-order in flavor SB can be written in terms of eight operators and five operator coefficients.

The matrix elements of $V^{4\pm i5} + \epsilon^2 \delta V^{4\pm i5}$ between $SU(6)$ baryon states yields the actual expressions for the leading vector form factors at second-order in flavor SB. For the observed $\Delta S = 1$ transitions, one has:

$$\left[\frac{f_1}{f_1^{SU(3)}}\right]_{\Lambda p} = 1 + \frac{3}{4}\tilde{e}_{(1)}^{\mathbf{8}} + \frac{3}{8}\tilde{e}_{(2)}^{\mathbf{8}} + \frac{1}{18} b_{(3)}^{\mathbf{10}+\overline{\mathbf{10}}} - \frac{1}{10} y_{(2)}^{\mathbf{27}}, \tag{28}$$

$$\left[\frac{f_1}{f_1^{SU(3)}}\right]_{\Sigma^- n} = 1 + \frac{3}{4}\tilde{e}_{(1)}^{\mathbf{8}} - \frac{1}{8}\tilde{e}_{(2)}^{\mathbf{8}} - \frac{1}{18} b_{(3)}^{\mathbf{10}+\overline{\mathbf{10}}} - \frac{1}{30} y_{(2)}^{\mathbf{27}} + \frac{1}{30} b_{(3)}^{\mathbf{64}}, \tag{29}$$

$$\left[\frac{f_1}{f_1^{SU(3)}}\right]_{\Xi^- \Lambda} = 1 + \frac{3}{4}\tilde{e}_{(1)}^{\mathbf{8}} + \frac{1}{8}\tilde{e}_{(2)}^{\mathbf{8}} + \frac{1}{18} b_{(3)}^{\mathbf{10}+\overline{\mathbf{10}}} + \frac{1}{10} y_{(2)}^{\mathbf{27}}, \tag{30}$$

$$\left[\frac{f_1}{f_1^{SU(3)}}\right]_{\Xi^- \Sigma^0} = 1 + \frac{3}{4}\tilde{e}_{(1)}^{\mathbf{8}} + \frac{5}{8}\tilde{e}_{(2)}^{\mathbf{8}} - \frac{1}{18} b_{(3)}^{\mathbf{10}+\overline{\mathbf{10}}} + \frac{1}{30} y_{(2)}^{\mathbf{27}} + \frac{1}{30} b_{(3)}^{\mathbf{64}}, \tag{31}$$

where the $\tilde{e}_{(n)}$ coefficients are written in terms of those coefficients that come along with n-body octet operators.

Substituting the obtained expressions for the leading vector form factors (28)-(31) in (9a) and (9b) yields

$$\delta_1^{SB} = \frac{1}{6} y_{(2)}^{\mathbf{27}}, \qquad \delta_2^{SB} = \frac{1}{6} b_{(3)}^{\mathbf{10}+\overline{\mathbf{10}}} + \frac{1}{20} b_{(3)}^{\mathbf{64}}. \tag{32}$$

With the results obtained in (27), sum rules (9a) and (9b) become

$$\frac{1}{4}\left[\frac{f_1}{f_1^{SU(3)}}\right]_{\Xi^- \Sigma^0} + \frac{3}{4}\left[\frac{f_1}{f_1^{SU(3)}}\right]_{\Xi^- \Lambda} - \frac{1}{4}\left[\frac{f_1}{f_1^{SU(3)}}\right]_{\Sigma^- n} - \frac{3}{4}\left[\frac{f_1}{f_1^{SU(3)}}\right]_{\Lambda p} = 0, \tag{33a}$$

and

$$\frac{3}{4}\left[\frac{f_1}{f_1^{SU(3)}}\right]_{\Xi^- \Sigma^0} - \frac{3}{4}\left[\frac{f_1}{f_1^{SU(3)}}\right]_{\Xi^- \Lambda} + \frac{3}{4}\left[\frac{f_1}{f_1^{SU(3)}}\right]_{\Sigma^- n} - \frac{3}{4}\left[\frac{f_1}{f_1^{SU(3)}}\right]_{\Lambda p} = -\frac{1}{24} b_{(3)}^{\mathbf{10}+\overline{\mathbf{10}}}. \tag{33b}$$

There are some interesting results here: Sum rule (33a) vanishes even in the presence of second-order SB, which is equivalent to state that corrections to it arise at third-order SB and should be, in principle, highly suppressed. On the other hand, sum rule (33b) gets corrections mainly from the $\mathbf{10}+\overline{\mathbf{10}}$ flavor representation.

In order to be conclusive about the above findings, sum rules (33) can be tested against theoretical results found in the literature. Several methods have been used to evaluate the effects of SB in the leading vector form factors of HSD. One of the earliest is baryon chiral perturbation theory (BChPT). In this context, Krause [11] included calculations in relativistic BChPT to order $\mathcal{O}(p^2)$ and Anderson and Luty [12] used heavy baryon chiral perturbation theory (HBChPT) up to (partially complete) order $\mathcal{O}(p^3)$ in the chiral expansion. In both works, only spin-1/2 octet baryons were included as baryonic degrees of freedom. In a later work, Villadoro [13] computed (partially) complete corrections up to order $\mathcal{O}(p^3)$ corresponding to subleading in $1/M_B$ terms; octet and decuplet baryon degrees of freedom were included in this work. Also, in the frame of covariant BChPT, Lacour, Kubis, and Meissner [14] used infrared regularization whereas Geng, Camalich, and Vicente-Vacas [15] used the extended on-mass-shell (EOMS) renormalization scheme. Both works included corrections to order $\mathcal{O}(p^3)$; while the former included only spin-1/2 as baryonic degrees of freedom, the latter included both octet and decuplet baryons. Finally, another calculation within large-N_c baryon chiral perturbation theory to order $\mathcal{O}(p^2)$ has been provided in Ref. [16]. In this approach loop graphs with octet and decuplet intermediate states are systematically incorporated into the analysis because both spin-1/2 and spin-3/2 baryons together form an irreducible representation of spin-flavor symmetry.

A comparison with all the different analytical results mentioned above can readily be performed. Thus, sum rule (33a) is by the $f_1/f_1^{SU(3)}$ ratios given in Eqs. (65)–(68) of Ref. [16]. It is interesting to notice that the sum rule is also satisfied when the decuplet fields are not explicitly retained in the effective theory. Therefore, it is fulfilled by all the expressions for the $f_1/f_1^{SU(3)}$ ratios obtained within (heavy) baryon chiral perturbation theory to order $\mathcal{O}(p^2)$ of [11], [13], [14], and [15]. The only exception is found in Ref. [12] where there is a wrong sign in one of the loop diagrams. Terms of subleading order in $1/M_B$ of Refs. [13], [14] and [15] also satisfy sum rule (33a). As for higher-order terms, for now it is not possible to perform a detailed analysis because only partial results are available.

Sum rule (33b), on the other hand, can be checked with the analytical expressions of Ref. [16]. Its right-hand-side yields

$$\frac{3}{2}D^2\left[H\left(m_K,m_\eta\right)-H\left(m_\pi,m_K\right)\right]-\frac{3}{8}\mathcal{C}^2\left[K\left(m_K,m_\eta,\Delta\right)-K\left(m_\pi,m_K,\Delta\right)\right], \tag{34}$$

where D and $\mathcal{C}$ are the conventional $SU(3)$ invariant couplings, the functions $H\left(m_1,m_2\right)$ and $K\left(m_1,m_2,\Delta\right)$ arise from loop integrals and Δ is the decuplet-octet baryon mass difference. The numerical values of the quantities between square parentheses are small [16], of the order of 0.056 and 0.029, which in turn explains the smallness of the right-hand side of sum rule (33b).

Finally, an analysis of the available experimental data [17] can be performed by using the results obtained here. The experimentally measured quantities in HSD are the total decay rate R, angular correlation coefficients $\alpha_{e\nu}$, and angular spin-asymmetry coefficients α_e, α_ν, α_B, A, and B. Often, the data is presented in terms of R and the ratio g_1/f_1 for the decay. The theoretical expressions for the total decay rates and angular coefficients can be found in Ref. [18], where radiative corrections and the four-momentum-transfer contribution to the form factors are also discussed.

For a numerical analysis, it is convenient to rewrite Eqs. (28)–(31) is a more compact way as

$$\left[\frac{f_1}{f_1^{SU(3)}}\right]_{\Lambda p} = 1+\frac{3}{4}v_1+\frac{3}{8}v_2+\frac{11}{18}v_3, \tag{35}$$

$$\left[\frac{f_1}{f_1^{SU(3)}}\right]_{\Sigma^- n} = 1+\frac{3}{4}v_1-\frac{1}{8}v_2+\frac{1}{6}v_3, \tag{36}$$

$$\left[\frac{f_1}{f_1^{SU(3)}}\right]_{\Xi^-\Lambda} = 1+\frac{3}{4}v_1+\frac{1}{8}v_2+\frac{11}{18}v_3, \tag{37}$$

$$\left[\frac{f_1}{f_1^{SU(3)}}\right]_{\Xi^-\Sigma^0} = 1 + \frac{3}{4}v_1 + \frac{5}{8}v_2 + \frac{1}{6}v_3, \tag{38}$$

where $v_1 = \tilde{e}_{(1)}$ and $v_2 = \tilde{e}_{(2)}$ and v_4 comprises the remaining coefficients from the $\mathbf{10} + \overline{\mathbf{10}}$ operator coefficients.

Without further ado, a fit to data using information on R and g_1/f_1 yields

$$v_1 = 0.002 \pm 0.036, \qquad v_2 = -0.095 \pm 0.053, \qquad v_3 = 0.91 \pm 1.01. \tag{39}$$

With these results sum rules (33) are well satisfied. However, future experimental measurements should provide a more stringent test of these sum rules.

6. Closing remarks

The main aspects of the derivation of two novel sum rules involving the ratios of the leading semileptonic vector form factors with SB effects to the ones in the symmetric limit, $f_1/f_1^{SU(3)}$, have been presented. These sum rules, given in (33), reveal some interesting features. The first one, (33a), gets corrections at third-order in SB effects. This result is surprising, in view of the Ademollo-Gatto theorem. The second one, (33b), gets corrections mainly from the flavor $\mathbf{10} + \overline{\mathbf{10}}$ representation. These sum rules can be tested with analytical expressions obtained within baryon chiral perturbation theory, at least at chiral order $\mathcal{O}(p^2)$ and some partial $\mathcal{O}(p^3)$ expressions available. The validity is fulfilled. A complete comparison with the full $\mathcal{O}(p^3)$ expressions is not possible for now. In the near future, lattice QCD results should also be available, and the sum rules could be checked.

Acknowledgments

This work has been partially supported by Consejo Nacional de Ciencia y Tecnología (México) and Fondo de Apoyo a la Investigación (Universidad Autónoma de San Luis Potosí).

References

[1] t Hooft G 1974 *Nucl. Phys.* **B72** 461
[2] Flores-Mendieta R and Padrón-Stevens R 2017 *Phys. Rev.* D **95** 076018
[3] Ademollo M and Gatto R 1964 *Phys. Rev. Lett.* **13** 264
[4] Cabibbo N 1963 *Phys. Rev. Lett.* **10** 531
[5] Pham T N 2013 *Phys. Rev.* D **87** 016002
[6] Leutwyler H and Roos M 1984 *Z. Phys.* **C25** 91
[7] Gervais J L and Sakita B 1984 *Phys. Rev. Lett.* **52** 87; *Phys. Rev. D* **30** 1795
[8] Dashen R F, Jenkins E, and Manohar A V 1995 *Phys. Rev. D* **51** 3697
[9] Jenkins E 1996 *Phys. Rev. D* **53** 2625
[10] Jenkins E and Lebed R F *Phys. Rev. D* **52** 282
[11] Krause A 1990 *Helv. Phys. Acta* **63** 3
[12] Anderson J and Luty M A 1993 *Phys. Rev. D* **47** 4975
[13] Villadoro G 2006 *Phys. Rev. D* **74** 014018
[14] Lacour A, Kubis B, and Meissner U G 2007 *JHEP* 0710 083
[15] Geng L S, Martin-Camalich J, and Vicente-Vacas M J 2009 *Phys. Rev. D* **79** 094022
[16] Flores-Mendieta R and Goity J L 2017 *Phys. Rev. D* **95** 076018
[17] Patrignani C *et al.* (Particle Data Group) 2016 *Chin. Phys. C* **40** 100001
[18] Flores-Mendieta R 2004 *Phys. Rev. D* **70** 114036

IOP Conf. Series: Journal of Physics: Conf. Series **912** (2017) 012027 doi:10.1088/1742-6596/912/1/012027

Contribution of sea quarks to the electromagnetic decay of decuplet baryons

Gustavo Guerrero-Navarro and Roelof Bijker

Instituto de Ciencias Nucleares, Universidad Nacional Autónoma de México, Ciudad de México 04510, México

E-mail: `gustavo.guerrero@correo.nucleares.unam.mx`

Abstract. In this work, we study briefly the electromagnetic dipole transitions from decuplet to octet S-wave baryons, where the current experimental data for the radiative decay width shows that these values are approximately two times larger than the predictions of the constituent quark model. The main aim is to solve this problem with the approach of effective degrees of freedom for quarks in the quark model, considering also the sea quarks, obtaining the valence and the sea quarks contributions. In order to do this, an unquenched quark model has been considered where the effects of the quark-antiquark pairs are incorporated as a perturbation with the quantum numbers of vacuum in the wave function for baryons as baryon-meson intermediate coupled states with the quantum numbers of the initial baryon through the $^3P_0$ model. Also, an extended expression has been obtained to the M1 electromagnetic transition that gives a closer value to the actual experimental data due to the sea quark core contribution.

1. Introduction

In nuclear and particle physics the electromagnetic transitions are a fundamental process for the understanding of the baryon structure. The constituent quark model (CQM) has had a very good achievement by describing the radiative transitions and a good agreement with the spatially excited resonances, the magnetic moments of baryons and charmonium in its non-relativistic approximation [1]-[10], even in the prediction of the decuplet to octet baryon transitions [11]. There have been many attempts in the frame of QCD like effective models of hadrons to predict the experimental data, but there is still a lack of models that can explain the differences between them. The main approach of effective models is taking account degrees of freedom QCD allowed by breaking symmetries. Those can be taken away considering intermediate states of sea quarks that can contribute to the electromagnetic amplitude transitions [12]-[21]. In this context, we consider the S-wave magnetic dipole (M1) transition between decuplet and octet baryons and we investigate the contributions of the sea core to this term.

In the case of the electromagnetic transitions from decuplet to octet, the transition amplitude used, even for the photoproduction or the decay transitions, takes into account the models that describe the wave function of baryons but it can't reproduce the available experimental data. The current values of the electromagnetic decay widths, $\Gamma_{A\to A'\gamma}$, are 703 ± 61 keV, 251 ± 61 keV and 449 ± 78 keV for $\Gamma_{\Delta\to N\gamma}$, $\Gamma_{\Sigma^{*+}\to\Sigma^{+}\gamma}$ and $\Gamma_{\Sigma^{*0}\to\Lambda^{0}\gamma}$ respectively which corresponds to the $(0.60 \pm 0.05)\%$, $(0.70 \pm 0.17)\%$ and $(1.25 \pm 0.13)\%$ of the total Breit-Wigner width [22]. Meanwhile, the result of the CQM is almost one half of the cited experimental data (tables 1

and 2).
On the other hand, we have extensions of the CQM (χQM, MCM, UQM) where the effects of sea quarks are considered. Those are motivated by the Gottfried sum rule violation measurements for the proton that probes the flavor asymmetry on it (a violation of the basic assumption of the CQM). Using the parton distribution functions (PDFs) of quarks and antiquarks, the Gottfried sum rule is given by

$$S_G = \frac{1}{3} + \frac{2}{3}\Delta P \tag{1}$$

where $\Delta P = \left[\int_0^1 dx(\bar{u}_p(x) - \bar{d}_p(x))\right]$ and x the Bjorken variable. This means that if $\bar{d}$ and $\bar{u}$ in the proton are symmetric, the Gottfried sum rule would give 1/3 but the measurements of the NMC experiment at CERN measured this by deep inelastic muon scattering [23] finding a smaller value than 1/3, $S_G = 0.2281 \pm 0.0065$, i.e. $\bar{u} < \bar{d}$ gives the flavor asymmetry. It has also measured (experiment NA51) the x dependence of the radio $(\bar{d}(x)/\bar{u}(x))$ at $0.004 < x < 0.8$ with the Drell-Yan process [24]. Another experiment (E866) at Fermilab measured it at regions of between $0.015 < x < 0.35$ with a final value of $S_G = 0.255 \pm 0.008$ [25] and the Hermes collaboration [26], $S_G = 0.23 \pm 0.02$ at values of $0.020 < x < 0.30$.
By this the understanding of the flavor asymmetry of the antiquarks in the proton can be tested with the EM transitions of baryons by using QCD effective and hadron models such as meson cloud models that contains virtual meson cloud states like $|p\rangle = \alpha|p_0\rangle + \beta|n\pi^+\rangle + \gamma|\Delta^{++}\pi^-\rangle + \ldots$. Those states are described in the Unquenched Quark Model (UQM) [27] through the $^3P_0$ pair quark creation model.

2. The Unquenched Quark Model

The UQM is an extension to the CQM inspired by the $^3P_0$ model originally used to describe the meson strong decays [28] including the degrees of freedom allowed by the QCD. On this formalism, and taking only the effective degrees of freedom from the constituent quarks, the hadron states expands in the different order Fock states as follows

$$|qqq\rangle \rightsquigarrow |qqq\rangle + |qqq\; q\bar{q}\rangle + |qqq\; q\bar{q}q\bar{q}\rangle + \cdots \tag{2}$$

In the UQM the wave states for baryons are obtained from the quark level introducing the $T^\dagger(^3P_0)$ operator [27, 29] to generate the sea pair quarks and coupling this to the zeroth order three quark state to generate the first order Fock states, $|qqq\; q\bar{q}\rangle$. These states are generated as a sum of the quantum number of the possible higher Fock states in the form of the baryon-meson first order state, $|BC\rangle$, coupled with the vacuum quantum numbers from the sea pair and the zeroth order state $|A\rangle$. The UQM state for a baryons reads

$$|\Psi_A\rangle = \mathcal{N}_A\left[|A\rangle + \sum_{BClJ}\int dk d\vec{K} k^2 |BClJ;\vec{K}k\rangle \frac{\langle BClJ;\vec{K}k|T^\dagger|A\rangle}{m_A - E_B - E_C}\right], \tag{3}$$

where the $\mathcal{N}_A$ is the normalization factor, $|A\rangle$ is the valence baryon state which corresponds to the CQM state, $\vec{l}$ is the orbital relative angular momentum of the baryon-meson state, $|BC\rangle$, and $\vec{J} = \vec{J}_B + \vec{J}_C + \vec{l}$ their total angular momentum, and in the same way $\vec{k}$ and $\vec{K}$ are the relative and total momentum of the baryon-meson state, m_A, E_B and E_C are the corresponding energies (A baryon is in the rest frame). The strength factor for the sea quark pair creation within $T^\dagger$ actually can be fitted in the UQM through the flavor content to the proton asymmetry flavor, ΔP (1).

3. Electromagnetic transitions

The decuplet to octet electromagnetic transitions have been widely studied for a variety of quark models [12]-[21]. In the UQM frame to determine the electromagnetic decay width for the $A \to A'\gamma$ process in the form of the matrix element we use the Fermi Golden rule $\Gamma_{i\to f} = 2\pi\phi \left|\langle f|\hat{H}_{em}|i\rangle\right|^2$, where $\hat{H}_{em}$ is the electromagnetic hamiltonian operator and ϕ the invariant phase space factor. A brief calculation give us four possible terms that contains the contributions of the qq quark transition, the sea annihilation $q\bar{q}$, the pair creation $q\bar{q}$ and the sea antiquark $\bar{q}\bar{q}$ transition respectively. An initial calculation taking into account only the first Hamiltonian term involving the only qq transitions results in the non-zero contributions from the valence $\langle A|\hat{H}^{(1)}_{em}|A\rangle$ and the sea coupled $\langle BC|\hat{H}^{(1)}_{em}|BC\rangle$ matrix elements. After a brief reduction to this equation, we have in terms of the matrix elements of transition magnetic moment operator, $\vec{\mu}$, that

$$\Gamma_{A\to A'\gamma} = \frac{\alpha E_{A'} p_\gamma^3}{2m_A m_N^2} \left|\langle \Psi_{A'}|\vec{\mu}|\Psi_A\rangle\right|^2 \tag{4}$$

in unit of the nuclear magneton μ_N, where $\alpha = 1/137$ is the electromagnetic coupling constant, $\vec{p}_\gamma$ is the photon momentum, m_N is the nucleon mass and

$$\vec{\mu} = \vec{\mu}_S + \vec{\mu}_l = \sum_i \mu_i \left(2\vec{s}_i + \vec{l}_i\right), \tag{5}$$

where μ_i, $\vec{s}_i$ and $\vec{l}_i$ are the magnetic moments, the spin operator and the orbital operator of the ith quark respectively, $\vec{\mu}_S$ is the spin part magnetic moment operator and $\vec{\mu}_l$ is the orbital magnetic moment operator where, in the ground state, only contributes the relative orbital momentum between the baryon, B, and the meson, C, taking the expectation value as the matrix element of

$$\vec{\mu}_l = \sum_i \mu_i \vec{l}_i \to \left(\alpha_{BC}\frac{\mu_1+\mu_2+\mu_3}{3} + (1-\alpha_{BC})\frac{\mu_4+\mu_5}{2}\right)\vec{l}, \tag{6}$$

with $\alpha_{BC} = m_C/(m_B+m_C)$ a function of the baryon and meson masses, m_B and m_C, μ_1,μ_2 and μ_3 the magnetic moments of the constituent quarks in the B baryon; in the same way μ_4 and μ_5 corresponds to the quarks in the C meson.

In the UQM the matrix element of the total magnetic moment operator in (4) results as a function of the widths, α_A, α_B, of the baryon and meson state functions as well of the quark-antiquark pair width, α_D, fixed approximately to 0.35 fm. Considering the isospin symmetry for the case of the $\Delta \to N\gamma$ we are able to compute the expectation value for the spin part transition magnetic moment operator, $\vec{\mu}_S$ (5), obtaining an expression in the form

$$\begin{aligned}\langle \Psi_N|\hat{\mu}_S|\Psi_\Delta\rangle &= \mu^{val}_{\Delta N} + \mu^{sea}_{\Delta N\mathrm{spin}} \\ &= \mu_{\Delta N\mathrm{spin}},\end{aligned} \tag{7}$$

where $\mu^{val}_{\Delta N}$ indicates the valence spin contribution that corresponds to the CQM expression including the normalization factor given in (3) and $\mu^{sea}_{\Delta N\mathrm{spin}}$ is the baryon-meson coupled state, $|BC\rangle$, contribution. As mentioned above, another contribution to the transition magnetic moment in the UQM for the ground states of baryons is the orbital relative magnetic momentum from the intermediate states $|BC\rangle$. As always, considering the Jacobi coordinates for the constituent quarks coordinates and the operator $\vec{l}$, the matrix element of (6) results

$$\langle \Psi_N|\hat{\mu}_l|\Psi_\Delta\rangle = \mu^{sea}_{\Delta N\mathrm{orb}}, \tag{8}$$

Table 1. Meson cloud contributions for the decuplet to octet transition magnetic moments (in unit of μ_N). In each group, the first line indicates the valence quark contribution, $\mu^{val}_{BB'}$, and the sum of all meson cloud contributions (boldface), the second line the pion cloud contributions, the third line the kaon cloud, and the fourth the $\eta\eta'$ mixed cloud. The column $\mu_{BB'\text{spin}}$ presents the sum of the valence quark contribution, $\mu^{val}_{BB'}$, and the total spin sea quark core, $\mu^{sea}_{BB'\text{spin}}$. The column $\mu^{sea}_{BB'\text{orb}}$ shows the relative orbital contributions. The column $\mu_{BB'\text{total}}$ shows respectively the total meson cloud contributions and the final results (boldface).

Process	cloud	$\mu^{val}_{BB'}$	$\mu^{sea}_{BB'\text{spin}}$	$\mu_{BB'\text{spin}}$	$\mu^{sea}_{BB'\text{orb}}$	$\mu_{BB'\text{total}}$
$\Delta \to N\gamma$		1.02±0.10	**2.26±0.14**	**3.28±0.04**	**0.33±0.02**	**3.62±0.06**
	π		2.11±0.13		0.33±0.02	2.43±0.15
	K		0.08±0.01		0.01±0.00	0.09±0.01
	$\eta\eta'$		0.08±0.01			0.08±0.01

that contains only the sea contribution given the zero orbital contribution of the ground state for the valence part.

4. Fitting Procedure

In order to calculate the dipole transition magnetic moment and the EM decay width taking into account the experimental data we need to determine the $^3P_0$ sea amplitude coefficients in equation (3) which depends on the experimental masses, the charge radii of proton and pion [22] that depends (through the oscillator harmonic model) of the baryon, meson and sea quark pair widths, $\alpha_B, \alpha_C, \alpha_D$, respectively (into the expression for $\langle BClJ; \vec{K}k|T^\dagger|A\rangle$). The experimental flavor asymmetry in the proton, ΔP [23, 25], fits the strength coupling factor for the operator $T^\dagger$, as well the Color-Spin-Flavor coefficients [27] employing numerical methods. Once the sea coefficients have been fitted to the data, we follow by fitting the magnetic moments of the constituent quarks using the magnetic moments for p, n and Λ^0 octet baryons [22] needed in the equations (7) and (8).

In order to calculate the error propagation from the experimental data to all the parameters into the UQM expressions until the final results, we use a Monte-Carlo simulation [30] using random gaussian distributed values for each experimental data to ensure the correct correlation among parameters into the simulation. The fit was made by computing 10000 random points having enough statistical information finding that, in this particular case, the transition magnetic moment (and the EM decay widths) has also a gaussian frequency distribution. For that reason, we are able to calculate in good approximation the error bars values at the 1σ for this proper gaussian function. In this way, we have been considered all the correlations between the parameters in the UQM.

5. Results

The final fit in the UQM to the data give us the valence and the sea contribution to the transition magnetic moment of the $\Delta \to N\gamma$ process. On the sea quark core contributions, we have the spin and the relative orbital baryon-meson contributions, while the valence contribution is only spin dependent due to the treatment of the ground state of the baryons. In the table 1 we show the numerical values of this giving the sea cloud contribution of each meson, C, corresponding to the intermediate $|BC\rangle$ channel states as showed in (3).

Table 2. Electromagnetic transition of the Δ baryon in the UQM. Error bars are reported at the 1σ and take into account all the correlations among parameters.

$\Delta \to N\gamma$	CQM	UQM *spin*	UQM *total*	Exp [22]
$\mu_{BB'}$ (μ_N)	2.66±0.00	3.28±0.04	3.62±0.06	3.53±0.16
$\Gamma_{B\to B'\gamma}$ (keV)	399±7	608±20	738±29	703±61

In the table 2 we give the corresponding electromagnetic decay width for the transition magnetic moment for the CQM case and the UQM spin contribution following by the spin plus orbital (total) contribution and the experimental comparison. Figures 1 and 2 show the plots where the sea contributions in the UQM increase the values in comparison with CQM, due to the valence contribution is the only one contribution for it.

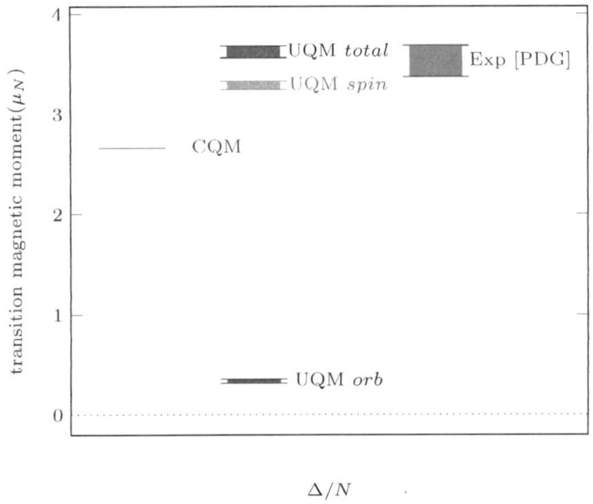

Figure 1. Transition magnetic moment values for the Δ/N case as given in the table 1 for the CQM and the orbital, spin and total contributions in the UQM compared to the experimental data [22].

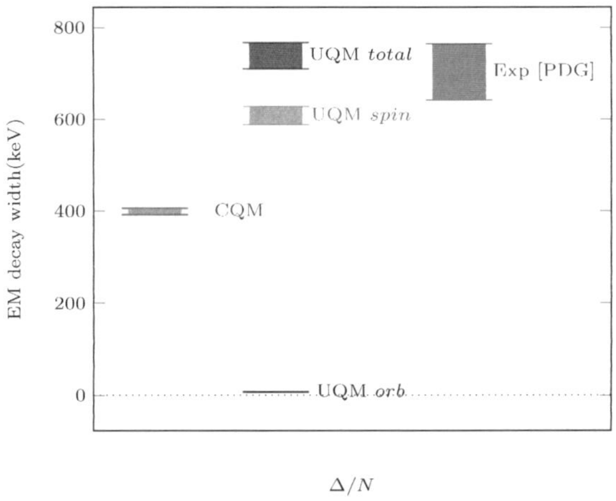

Figure 2. Electromagnetic decay widths values for the $\Delta \to N\gamma$ transition as given in the table 2 for the CQM and the orbital, spin and total contributions in the UQM compared to the experimental data [22].

6. Discussion and conclusions

Our results show a valence and sea core explicit contribution to the transition magnetic moment for Δ/N, as well to the electromagnetic decay width. We can see directly in the figure 3, corresponding to the values in the table 1, that the UQM spin results are closer to the experiment result than the CQM. At the same time in the UQM, the sum of the spin sea cloud contributions become largest than the valence (spin) quark core while the orbital sea contribution is the smaller one. It seems more important the pion cloud contribution to the electromagnetic transition because of the pion smaller mass as we expected. Similarly, the lower sea contribution is given by the Kaon and $\eta\eta'$ cloud with larger mass for these mesons. The orbital sea contribution it's not so important at least it is added to the total spin contribution in the sense that the total, spin plus orbital, contribution reach the experimental data results, as a consequence that the

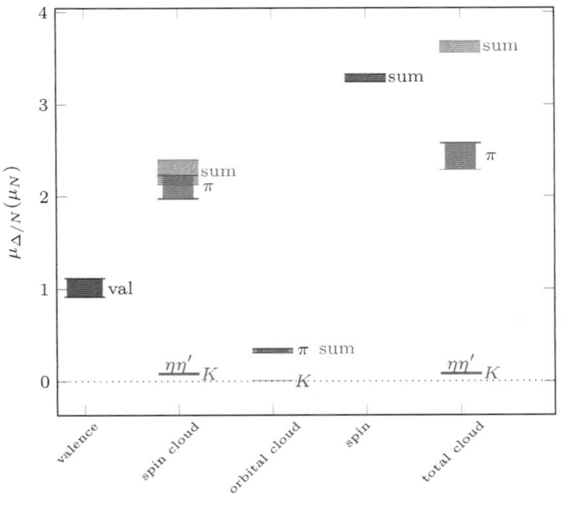

Figure 3. Valence and sea quark contributions for each meson cloud channel as showed in the table 1.

total spin contribution results very close to the experimental value.
It's worth to mention that the valence contribution in the UQM and the CQM value only differs on the normalization factor, although the expressions are in that sense the same. By this, the normalization factor is close to 1/2.
About the total spin contribution, we found that the valence and the sea part are correlated through the $^3P_0$ strength factor giving us a lower error bar for this fit because of the experimental error bar of the proton flavor asymmetry.
Finally, the UQM results given in the table 2 shows the effective contribution of the meson sea cloud by a unquenching in the CQM. Those results are in agreement with the experimental value for the $\Delta \to N\gamma$ and we found that the sea part has an important role as the valence core do it.

Acknowledgments

This work is supported by SNI-CONACYT, Mexico (Exp. Ayte. 13928).

References

[1] Eichten E, Gottfried K, Kinoshita T, B Kogut J, Lane D K and M Yan T 1975 *Phys. Rev. Lett.* **34** 369; Eichten E, Gottfried K, Kinoshita T, D Lane K and M Yan T 1978 *Phys. Rev.* D **17** 3090; 1980 *Phys. Rev.* D **21** 203
[2] Isgur N and Karl G 1978 *Phys. Rev.* D **18** 4187; 1979 *Phys. Rev.* D **19**, 2653; 1979 *Phys. Rev.* D **20** 1191
[3] Godfrey S and Isgur N 1985 *Phys. Rev.* D **32** 189
[4] Capstick S and Isgur N 1986 *Phys. Rev.* D **34** 2809
[5] Ferraris M, Giannini M M, Pizzo M, Santopinto E and Tiator L 1995 *Phys. Lett.* B **364** 231; Santopinto E, Iachello F and Giannini M M 1998 *Eur. Phys.* J. A **1** 307; Santopinto E and Giannini M M 2012 *Phys. Rev.* C **86** 065202; Giannini M M, Santopinto E 2015 *Chin. J. Phys.* **53** 020301; Aiello M *et al* 1996 *Phys.Lett.* B **387**, 215; Aiello *et al* 1988 *J. of Phys.* G **24** 753; Bijker R, Iachello F, Santopinto E 1998 *J. of Phys.* A **31** 9041; M De Sanctis *et al* 2007 *Phys. Rev.* C **6** 062201
[6] Glozman L Y, Riska D O 1996 *Phys. Rept.* **268** 263; 1998 Glozman L Y, Plessas W, Varga K, Wagenbrunn R F, *Phys. Rev.* D **58** 094030
[7] Loring U, Metsch B C and Petry H R 2001 *Eur. Phys.* J. A **10** 395
[8] Santopinto E 2005 *Phys. Rev.* C 72 022201; Ferretti J, Vassallo A and Santopinto E 2011 *Phys. Rev.* C **83**, 065204; De Sanctis M, Ferretti J, Santopinto E and Vassallo A (2014); De Sanctis M *et al* 2011 *Phys. Rev.* C **84** 055201
[9] Galata G and Santopinto E 2012 *Phys. Rev.* C **86** 045202
[10] Bijker R, Iachello F and Leviatan A 1994 *Ann. Phys. (N.Y.)* **236** 69; 1996 *Phys. Rev.* C **54** 1935; 1997 *Phys. Rev.* D **55** 2862; 2000 *Ann. Phys. (N.Y.)* **284** 89
[11] Beg M A B, Lee B W and Pais A 1964 *Phys. Rev. Lett.* **13** 514

[12] Leinweber D B, Draper T and Woloshyn R M 1993 *Phys. Rev.* D **48** 2230
[13] Kim H C, Polyakov M, Praszalowicz M, Yang G S and Goeke K 2005 *Phys. Rev.* D **71** 094023
[14] Ramalho G and Tsushima K 2013 *Phys. Rev.* D **87** 9-093011
[15] Dhir R and Verma R C 2009 *Eur. Phys. J.* A **42** 243
[16] Ramalho G and Tsushima K 2013 *Phys. Rev.* D **88** 053002
[17] Keller D and Hicks K 2013 *Eur. Phys. J.* A **49** 53
[18] Wang L and Lee F X 2009 *AIP: Conf. Proc.* **1182** 532
[19] Hong S T 2007 *Phys. Rev.* D **76** 094029
[20] Jenkins E E 2012 *Phys. Rev.* D **85** 065007
[21] Lebed R F and Martin D R 2004 *Phys. Rev.* D **70** 016008
[22] Patrignani C *et al* 2016 (Particle Data Group) *Chin. Phys.* C **40** 100001
[23] Amaudruz P *et al* 1991 *Phys. Rev. Lett.* **66** 2712; Arneodo M *et al* 1997 *Nucl. Phys.* B **487** 3
[24] Baldit A *et al* 1994 *Phys. Lett.* B **332** 244; Towel R S *et al* 2001 *Phys. Rev.* D **64** 052002
[25] Hawker E A *et al* 1998 *Phys Rev. Lett.* **80** 3715
[26] Ackerstaff K *et al* 1998 *Phys. Rev. Lett.* **81** 5519
[27] Bijker R, Ferretti J, Galata G, Garcia-Tecocoatzi H, Santopinto E 1996 *Phys. Rev.* D **94** 074040
[28] Micu L 1969 *Nucl. Phys.* B **10** 521
[29] Roberts W and Silvestre-Brac B 1992 *Few-Body Syst.* **11**, 171
[30] Fernández-Ramírez C, Danilkin I V, Mathieu V, and Szczepaniak A P (JPAC Collaboration) 2016 *Phys. Rev.* D **93** 074015

IOP Conf. Series: Journal of Physics: Conf. Series **912** (2017) 012028 doi:10.1088/1742-6596/912/1/012028

Use of a non-relativistic basis for describing the low energy meson spectrum

D. A. Amor-Quiroz[1], **T. Yepez-Martinez**[1], **P. O. Hess**[1] **and O. Civitarese**[2]

[1]Instituto de Ciencias Nucleares, Universidad Nacional Autónoma de México Circuito Exterior, C.U., Apartado postal 70-543, 04510 Mexico City, Mexico.
[2]Departamento de Física, Universidad Nacional de La Plata, C.C.67 (1900), La Plata, Argentina.

E-mail: arturo.amor@nucleares.unam.mx, tochtli.yepez@nucleares.unam.mx, hess@nucleares.unam.mx, osvaldo.civitarese@fisica.unlp.edu.ar

Abstract. We justify the use of harmonic oscillator eigenfunctions for expanding the fermionic fields of an effective field theory and evaluate its utility for solving a Hamiltonian inspired by the QCD formalism in the Coulomb Gauge. Since the functions involved in such expansion are non-relativistic, the Talmi-Moshinsky transformations can be used to recover the translational invariance of the center of mass of the mesonic states. Finally, many-body methods and an *a posteriori* flavor mixing procedure are used to compute a preliminary spectrum for the mesons below 1 GeV.

1. Introduction

In the low energy regime (below 1 GeV) the strong interaction coupling constant α_s is of the order of the unit and therefore it is impossible to model the color interaction as a perturbation. Despite the fact that there has been a great effort in describing the non-perturbative regime of QCD (NP-QCD), there is still not a satisfactory solution which fully describes this regime.

There are many phenomenological models that have tried to formulate an effective NP-QCD theory. One of the first to do this was the MIT bag model [1], whose main disadvantage is to be unrealistic in addition to requiring external parameters to the theory.

So far, mainly the Lattice QCD [2] is able to derive non-perturbative results from first principles, for example, it has been able to describe the hadronic states at low energy [3, 4]; as well as to the highly excited mesonic states [5, 6]. However, this approach is intrinsically limited by the rapid rise in computational effort required as one approaches the physically relevant limit of small lattice spacing and large lattice volume. In addition, it has the less important disadvantage that recovering the rotation symmetry is unnecessarily complicated due to the lattice discretization [7].

Another way to approach the NP-QCD calculations is via solving the Dyson-Schwinger equations (DSEs). The main advantage of such formalism is that it represents a continuum method and thus, both the UV and the IR regions can be investigated. In particular, DSEs are also able to describe the dynamical mass generation of the Yang-Mills sector of the theory in a self-consistent way [8, 9, 10]. Nevertheless, the DSEs form an infinite set of coupled integral equations. Therefore the system has to be truncated and cannot be solved completely and there are no clear prescriptions on how to perform such a truncation.

For these reasons, the overall objective is to formulate a model that represents an alternative to perform non-perturbative calculations based on an effective field theory. The present model is meant to offer the following properties:

- That requires less computing power than lattice calculations. This includes the fact that the codes involved are easier to program and can even be run by an standard-performance personal computer.
- That requires the least possible number of free parameters. This is achieved through the use of variational methods to relate the parameters to each other, in such manner that not all of them are free but arise naturally as a function of the theory to be evaluated.
- That offers a more intuitive physical interpretation than the approaches mentioned above. This is expected to be achieved after an *a posteriori* verification of the effectiveness of the assumed ansatz.
- That is simple yet realistic enough, *i.e.*, by simulating the most important features of QCD, it can be expected to represent a good approximation to reality.
- Last but not least, it is expected that the numerical solutions generate a sufficiently rich spectrum to compare with the results of other non-perturbative methods and even with the experimental mesonic spectrum.

2. Characteristics of the model

As mentioned above, one of the goals of the model is to be able to reproduce the important aspects of the theory by combining the following properties:

- One considers an effective QCD Hamiltonian where the quarks interact via the non-relativistic Cornell Potential, which simulates the effect of the gluons [11].
- The main characteristic of the model is that the spatial distribution of the fermionic fields is expanded in the non-relativistic 3D-harmonic oscillator (3D-h.o.) basis in the coordinate space [12, 13], *i.e.*, the angular functions are the spherical harmonics while the radial part is given by

$$\begin{aligned} R_{nl}(r) &= \mathcal{N}_{nl} r^l e^{-\frac{\lambda}{2}r^2} L_n^{l+\frac{1}{2}}(\lambda r^2) \\ \mathcal{N}_{nl} &= \left[\frac{2(n!)}{\Gamma(n+l+\frac{3}{2})}\right]^{\frac{1}{2}} \lambda^{\frac{3}{4}+\frac{l}{2}} \\ n &= \frac{N-l}{2} \ , \end{aligned} \tag{1}$$

  where $L_n^\alpha(x)$ are the associated Laguerre polynomials. This procedure will bring several advantages that are discused in Subsection 2.1.
- Even though this basis is not Lorentz invariant, it is possible to recover the Galilei invariance of the center of mass (CM) with the help of the so called Talmi-Moshinsky (TM) transformations [14, 15].
- Many-body methods such as the Tamm-Dankoff method (TDA) and the Random Phase Approximation (RPA) are used to diagonalize the effective Hamiltonian [16].
- Flavor symmetry breaking is set by introducing an *a posteriori* flavor mixing procedure [17].

IOP Conf. Series: Journal of Physics: Conf. Series **912** (2017) 012028 doi:10.1088/1742-6596/912/1/012028

2.1. The election of a basis

We are not the first group to have implemented the many-body methods for finding the mesonic spectrum related to an effective static potential [18, 19]. Nevertheless, the fundamental difference proposed by the present model is the use of the 3D-h.o. basis for expanding the fermionic fields in the position space.

In principle any complete basis can be used to construct the Fock space, however, the structure of the vacuum will depend on the chosen basis: The better the ansatz is to represent the vacuum of the interacting theory, the more efficiently it will describe the physics involved.

The fields (fermionic or bosonic) are usually expanded in the plane waves basis, which are solutions to the free propagation problem, that is, to the Dirac or Klein-Gordon equation respectively. The Quantum Field Theories have been deeply explored in such formalism that comes very useful when perturbative expansions are made, but little has been done to give formalism to a field theory expanded in 3D-h.o. solutions.

The election of such an initial basis for expanding the fields is due to the following reasons:

- As mentioned in Section 1, lattice QCD calculations are limited by the finite spacing of the grid. Thus, the expansion of the fields in 3D-h.o. eigenfunctions can be interpreted as a discrete mapping of continuous solutions that may avoid the loss of information in the small spacing limit.
- Due to confinement, the fermionic fields are expected to be restricted to a finite volume, forming *localized* hadrons. This is why the oscillator basis is intuitively more adequate than plane waves, since it establishes a characteristic lenght for the fields in the position space: By expanding the fields in such basis, the color charge distribution functions are suppressed over long distances by an exponential $e^{-\frac{\lambda}{2}r^2}$, while the function of the hadron's CM is suppressed by a weight function $e^{-\frac{\lambda}{2}R^2}$.
- Although the 3D-h.o. basis is not Lorentz invariant, it is possible to recover the Galilean invariance of the mesons' CM by using the TM transformations [14]. This means that the individual quarks will not be invariant under spatial translations, however, the important degrees of freedom to be considered in QCD at low energies are light mesons and not their constituent components.
- It is not necessary to introduce any sort of cut-off in the integrals involved. This is because the integrals in the position space are finite for both short and large relative distances. However, a cut-off has to be introduced numerically by truncating the expansion on the functions up to a certain number of oscillator quanta N_{cut}.
- After implementing the transformation to the CM frame, the integrals involved in the matrix elements of the interaction are decoupled, so that they can be expressed in a single one-dimensional integral, which will result to be a *Talmi integral* with analytical results.
- Free-propagation solutions may not be the best approach to a strong interacting theory, while in the CM frame the 3D-h.o. basis implies a significant simplification of the problem as well as fast convergence of the solutions [15].
- This basis seems to be suitable in order to extend the analysis in many directions, like adding gluonic degrees of freedom or evaluating other hadronic states.

3. Construction of the effective Hamiltonian

In the same way as in QED, fixing the Coulomb gauge gives rise to a local and instantaneous potential. Lattice calculations have shown that this potential is approximately of the form $-\frac{\alpha}{r} + \beta r$ [20, 21], where $r = |\boldsymbol{x} - \boldsymbol{y}|$ is the relative distance between two points of the color charge densities. Such interaction is known in the literature as Cornell's potential and its linear component reproduces the phenomenon of confinement, since it makes energetically less favorable to increase the distance between charges. Because of this, the Cornell potential simulates the effect of gluons, which makes possible the treatment of a simplified theory where the only fields involved are fermionic.

This approach to the problem is then equivalent to replacing the Faddeev-Popov operator by the static potential

$$\langle a, \boldsymbol{x}|\frac{1}{\boldsymbol{\nabla}\cdot\boldsymbol{\mathcal{D}}}(-\boldsymbol{\nabla}^2)\frac{1}{\boldsymbol{\nabla}\cdot\boldsymbol{\mathcal{D}}}|a'\boldsymbol{y}\rangle \longrightarrow V(|\boldsymbol{x}-\boldsymbol{y}|) = \frac{-\alpha}{|\boldsymbol{x}-\boldsymbol{y}|} + \beta|\boldsymbol{x}-\boldsymbol{y}| \ . \tag{2}$$

Thus, the effective Hamiltonian to be solved in the present model is

$$\begin{aligned} H &= \boldsymbol{H}_{K_q} + \boldsymbol{H}_{m_q} + \boldsymbol{H}^{q-q}_{Coulomb} \\ \boldsymbol{H}_{K_q} &= \textstyle\int \boldsymbol{\psi}^\dagger(\boldsymbol{x})[-i\boldsymbol{\alpha}\cdot\boldsymbol{\nabla}]\boldsymbol{\psi}(\boldsymbol{x})d\boldsymbol{x} \\ \boldsymbol{H}_{m_q} &= \textstyle\int \boldsymbol{\psi}^\dagger(\boldsymbol{x})[\beta m_0]\boldsymbol{\psi}(\boldsymbol{x})d\boldsymbol{x} \\ \boldsymbol{H}^{q-q}_{Coulomb} &= -\tfrac{1}{2}g^2\delta_{a'a}\textstyle\int \rho^{(q)}_a(\boldsymbol{x})V(|\boldsymbol{x}-\boldsymbol{y}|)\rho^{(q)a}(\boldsymbol{y})d\boldsymbol{x}d\boldsymbol{y} \ , \end{aligned} \tag{3}$$

where $\alpha_i = \gamma^0\gamma^i$ and $\beta = \gamma^0$ are related to the Dirac matrices γ^μ; while m_0 is the bare mass of the quark. On the other hand,

$$\rho^{(q)a} = \sum_{c'c} \boldsymbol{\psi}^\dagger{}_{cf} T^a_{cc'} \boldsymbol{\psi}_{c'f} \ , \tag{4}$$

is the color charge density *restricted* to the quark sector (q) only.

3.1. The free propagation Hamiltonian

The free propagation Hamiltonian is conformed by the kinetic and mass terms restricted to the quark sector only.

$$\boldsymbol{K} \;=\; \boldsymbol{H}_{K_q} + \boldsymbol{H}_{m_q} = \int \boldsymbol{\psi}^\dagger(\boldsymbol{x})[-i\boldsymbol{\alpha}\cdot\boldsymbol{\nabla} + \beta m_0]\boldsymbol{\psi}(\boldsymbol{x})d\boldsymbol{x} \ ,$$

where, in the current status of the model, the bare mass for the u and d quarks is assumed to be degenerated though unequal to the mass of the quark s, i.e.,

$$(m_0)_{YT} = m_{u,d}\delta_{Y,\frac{1}{3}}\delta_{T,\frac{1}{2}} + m_s\delta_{Y,-\frac{2}{3}}\delta_{T,0} \ . \tag{5}$$

The operator $\boldsymbol{K}$ is not diagonal in the 3D-h.o. basis, and thus it is necessary to define a unitary transformation that diagonalizes it.

$$Q^\dagger_{\tau(Nl)jmcf} \;=\; \sum_{\pi k\lambda}\left(\alpha^j_{\tau(Nl)\lambda\pi k}\right)^* q^\dagger_{\lambda\pi kjmcf}\delta_{\pi,(-1)^{\frac{1}{2}-\tau+l}} \ . \tag{6}$$

This will also allow to identify the particle-hole operators $q^\dagger_{\pm\frac{1}{2}}$ with the quark(antiquark) operators $b^\dagger(d)$ via the definitions

$$\begin{aligned} b^\dagger_{\pi kjmcf} &\equiv q^\dagger_{\frac{1}{2}\pi kjmcf} \\ d_{\pi kjmcf} &\equiv q^\dagger_{-\frac{1}{2}\pi kjmcf} \ . \end{aligned} \tag{7}$$

This way, after performing the trasformation of eq. (6), the free propagation Hamiltonian can be written as a one-body operator of the form

$$\boldsymbol{K} \;=\; \sum_{kj}\epsilon_{kjYT}\left(\boldsymbol{b}^\dagger_{kjYT}\cdot\boldsymbol{b}^{kjYT} - \boldsymbol{d}_{kjYT}\cdot\boldsymbol{d}^{\dagger kjYT}\right) \ ,$$

where ϵ_{kjYT} are the eigenenergies obtained by diagonalizating the $\boldsymbol{K}$ in the harmonic oscillator basis.

At this point it is better to make an observation: as the free propagation Hamiltonian is, by construction, diagonal in the basis of the operators b's and d's (denoted as $(\lambda\pi k)$-basis); and knowing that the exact solution for said Hamiltonian are the plane waves, it is to conclude that both basis are closely related.

In fact, due to the completeness of the 3D-h.o. basis, the solutions shall tend to reproduce plane waves in the limit where all the quanta of oscillation are taken into account $N_{cut} \to \infty$. This is the reason why the $(\lambda\pi k)$-basis represents the connection that translates the excitations described by the operators b's and d's with the asymptotic states of a more standard field theory, where said excitations can actually be interpreted as physical quarks and antiquarks respectively.

3.2. The interaction Hamiltonian

Evidently, a one-body operator has a trivial CM, nevertheless, for the interaction Hamiltonian in (3) one obtains a two-body operator whose CM should in principle be removed in order to recover Galilean invariance. For doing so, the so called TM transformations are introduced:

$$[\Psi_{n_a l_a}(\boldsymbol{x}) \otimes \Psi_{n_b l_b}(\boldsymbol{y})]^L_M = \sum_{n_r l_r n_R l_R} \langle n_r l_r, n_R l_R; L | n_a l_a n_b l_b; L\rangle \left[\Psi_{n_r l_r}(\boldsymbol{r}) \otimes \Psi_{n_R l_R}(\boldsymbol{R})\right]^L_M \ , \tag{8}$$

where the Jacobi coordinates $\boldsymbol{R} = \frac{1}{\sqrt{2}}(\boldsymbol{x}+\boldsymbol{y})$ and $\boldsymbol{r} = \frac{1}{\sqrt{2}}(\boldsymbol{x}-\boldsymbol{y})$ refer to the CM and relative motions of two color charge densities distributions originally centered in $\boldsymbol{x}$ and $\boldsymbol{y}$ respectively.

It can be shown [22] that the expansion in the 3D-h.o. basis of an interaction operator for an arbitrary central potential is of the form

$$\boldsymbol{H}^{q-q}_{Coulomb} = -\tfrac{1}{2}\sum_{\{N_i l_i j_i Y_i T_i\} J_0} V^{J_0}_{\{N_i l_i j_i Y_i T_i\}} \times \left[\left[\boldsymbol{b}^\dagger_{(N_1 l_1) j_1 Y_1 T_1} \otimes \boldsymbol{b}_{(N_2 l_2) j_2 \bar{Y}_2 T_2}\right]^{0 J_0 (11) 00} \otimes \left[\boldsymbol{b}^\dagger_{(N_3 l_3) j_3 Y_3 T_3} \otimes \boldsymbol{b}_{(N_4 l_4) j_4 \bar{Y}_4 T_4}\right]^{0 J_0 (11) 00}\right]^{00(00)0}_{0000} \ , \tag{9}$$

where the intermediate coupling refers to pseudospin 0, total angular momentum J_0, color irrep (11) and 00 in flavor isospin and hypercharge. On the other hand, the coefficients $V^{J_0}_{\{N_i l_i j_i Y_i T_i\}}$ are given by

$$\begin{aligned} V^{J_0}_{\{N_i l_i j_i Y_i T_i\}} = & (-1)^{\frac{1}{3}+\frac{Y_1}{2}+T_1}\frac{\sqrt{2T_1+1}}{\sqrt{3}}(-1)^{\frac{1}{3}+\frac{Y_3}{2}+T_3}\frac{\sqrt{2T_3+1}}{\sqrt{3}}\delta_{Y_1Y_2}\delta_{T_1T_2}\delta_{Y_3Y_4}\delta_{T_3T_4} \\ & \hat{j}_1\hat{j}_2\hat{j}_3\hat{j}_4\hat{J}_0 \sum_S \sum_{N_r' N_r l_r N_R l_R J} (-1)^{j_2+j_3+J_0+J}(\hat{l}_r)^2(\hat{J})^2(\hat{S})^2 \\ & \left\{\begin{array}{ccc} j_1 & j_3 & J \\ j_2 & j_4 & J_0 \end{array}\right\}\left\{\begin{array}{ccc} l_1 & \frac{1}{2} & j_1 \\ l_3 & \frac{1}{2} & j_3 \\ l_r & S & J \end{array}\right\}\left\{\begin{array}{ccc} l_2 & \frac{1}{2} & j_2 \\ l_4 & \frac{1}{2} & j_4 \\ l_r & S & J \end{array}\right\} \\ & \langle N_r' l_r, N_R l_R; l_r | N_1 l_1, N_3 l_3; l_r\rangle\langle N_r l_r, N_R l_R; l_r | N_2 l_2, N_4 l_4; l_r\rangle \\ & \int_0^\infty r^2 dr R_{n_r' l_r}(r) V(r) R_{n_r l_r}(r) \ , \end{aligned} \tag{10}$$

where $\hat{j}$ is the shorthand notation for $\hat{j} = \sqrt{2j+1}$. In particular, for the Cornell Potential $V(r) = -\frac{\alpha}{r} + \beta r$, the coefficients $V^{J_0}_{\{N_i l_i j_i Y_i T_i\}}$ involve only integrals of the form .

$$I_{\{n_i, l_i, L\}} = \int R^2 dR \int r^2 dr R_{n_1 l_1}(r) R_{n_2 l_2}(r)\left(\frac{\alpha}{r} + \beta r\right) R_{n_3 l_3}(R) R_{n_4 l_4}(R) \tag{11}$$

which are called *Talmi Integrals* and are analytical and well behaved in both the limits when $r \to 0$ and $r \to \infty$. In addition, the dependence on the CM coordinate R sums out naturally through the orthogonality rules of the TM coefficients, so that the interaction does not generate spurious excitations.

After applying the transformation in (6), each of the intermediate couplings of the equation (9) as can be rewritten as

$$
\begin{aligned}
\left[\boldsymbol{Q}^{\dagger}_{\tau_1(N_1l_1)j_1Y_1T_1} \otimes \boldsymbol{Q}_{\tau_2(N_2l_2)j_2\bar{Y}_2T_2}\right]^{0J_0(11)00} &= \tfrac{1}{\sqrt{2}} \textstyle\sum_{\tau\pi_1k_1\pi_2k_2} \\
\times\Big\{ \left(\alpha^{j_1}_{\tau(N_1l_1)\frac{1}{2}\pi_1k_1}\right)^{*} \left(\alpha^{j_2}_{\tau(N_2l_2)\frac{1}{2}\pi_2k_2}\right) & \left[\boldsymbol{b}^{\dagger}_{\pi_1k_1j_1Y_1T_1} \otimes \boldsymbol{b}_{\pi_2k_2j_2\bar{Y}_2T_2}\right]^{J_0(11)00}_{MC} \\
- \left(\alpha^{j_1}_{\tau(N_1l_1)\frac{1}{2}\pi_1k_1}\right)^{*} \left(\alpha^{j_2}_{\tau(N_2l_2)-\frac{1}{2}\pi_2k_2}\right) & \left[\boldsymbol{b}^{\dagger}_{\pi_1k_1j_1Y_1T_1} \otimes \boldsymbol{d}^{\dagger}_{\pi_2k_2j_2\bar{Y}_2T_2}\right]^{J_0(11)00}_{MC} \\
+ \left(\alpha^{j_1}_{\tau(N_1l_1)-\frac{1}{2}\pi_1k_1}\right)^{*} \left(\alpha^{j_2}_{\tau(N_2l_2)\frac{1}{2}\pi_2k_2}\right) & \left[\boldsymbol{d}_{\pi_1k_1j_1Y_1T_1} \otimes \boldsymbol{b}_{\pi_2k_2j_2\bar{Y}_2T_2}\right]^{J_0(11)00}_{MC} \\
- \left(\alpha^{j_1}_{\tau(N_1l_1)-\frac{1}{2}\pi_1k_1}\right)^{*} \left(\alpha^{j_2}_{\tau(N_2l_2)-\frac{1}{2}\pi_2k_2}\right) & \left[\boldsymbol{d}_{\pi_1k_1j_1Y_1T_1} \otimes \boldsymbol{d}^{\dagger}_{\pi_2k_2j_2\bar{Y}_2T_2}\right]^{J_0(11)00}_{MC} \Big\} \quad , \qquad (12)
\end{aligned}
$$

where the sum over the pseudo-spin λ_1, λ_2 has been written explicitly in order to trace the type of excitation it represents.

To express in a simple manner the result of applying the transformation (6) to the operators involved in the interaction, it turns out convenient to introduce the short notation for operators $\mathcal{A}$ and $\mathcal{B}$, which include only the couplings in the $SU(2)$ subalgebra of flavor.

$$
\begin{aligned}
\mathcal{A}^{\tilde{\Gamma}}_{\pi_1 \mathrm{l}_1 \pi_2 \mathrm{l}_2;\ \tilde{\mu}} &= \frac{1}{\sqrt{2}} \left(\alpha^{j_1}_{\tau(N_1l_1)\frac{1}{2}\pi_1k_1}\right)^{*} \left(\alpha^{j_2}_{\tau(N_2l_2)\frac{1}{2}\pi_2k_2}\right) \left[\boldsymbol{b}^{\dagger}_{\pi_1k_1j_1Y_1T_1} \otimes \boldsymbol{b}_{\pi_2k_2j_2\bar{Y}_2T_2}\right]^{J_0(11)00}_{MC} \\
&- \frac{1}{\sqrt{2}} \left(\alpha^{j_1}_{\tau(N_1l_1)-\frac{1}{2}\pi_1k_1}\right)^{*} \left(\alpha^{j_2}_{\tau(N_2l_2)-\frac{1}{2}\pi_2k_2}\right) \left[\boldsymbol{d}_{\pi_1k_1j_1Y_1T_1} \otimes \boldsymbol{d}^{\dagger}_{\pi_2k_2j_2\bar{Y}_2T_2}\right]^{J_0(11)00}_{MC} \\
\mathcal{B}^{\tilde{\Gamma}}_{\pi_1 \mathrm{l}_1 \pi_2 \mathrm{l}_2;\ \tilde{\mu}} &= \frac{1}{\sqrt{2}} \left(\alpha^{j_1}_{\tau(N_1l_1)-\frac{1}{2}\pi_1k_1}\right)^{*} \left(\alpha^{j_2}_{\tau(N_2l_2)\frac{1}{2}\pi_2k_2}\right) \left[\boldsymbol{d}_{\pi_1k_1j_1Y_1T_1} \otimes \boldsymbol{b}_{\pi_2k_2j_2\bar{Y}_2T_2}\right]^{J_0(11)00}_{MC} \\
&- \frac{1}{\sqrt{2}} \left(\alpha^{j_1}_{\tau(N_1l_1)\frac{1}{2}\pi_1k_1}\right)^{*} \left(\alpha^{j_2}_{\tau(N_2l_2)-\frac{1}{2}\pi_2k_2}\right) \left[\boldsymbol{b}^{\dagger}_{\pi_1k_1j_1Y_1T_1} \otimes \boldsymbol{d}^{\dagger}_{\pi_2k_2j_2\bar{Y}_2T_2}\right]^{J_0(11)00}_{MC} .
\end{aligned}
\qquad (13)
$$

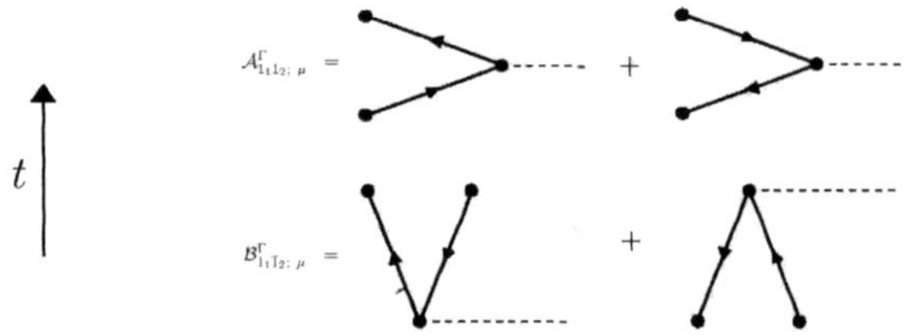

Figure 1. The $\mathcal{A}$ operator involves dispersion diagrams while the $\mathcal{B}$ operator represents the creation or annihilation of a pair.

In this way, the interaction can be expressed in a simple form as

$$
\begin{aligned}
\boldsymbol{H}_{coul} &= -\tfrac{1}{2} \textstyle\sum_{J_0\tau\tau'} \sum_{\{N_il_ik_ij_i\lambda_i\pi_iY_iT_i\}} V'^{J_0}_{\{N_il_ij_iY_iT_i\}} \\
&\times\Big([\mathcal{A}_{12}\mathcal{A}_{34}]^0_0 + [\mathcal{A}_{12}\mathcal{B}_{34}]^0_0 + [\mathcal{B}_{12}\mathcal{A}_{34}]^0_0 + [\mathcal{B}_{12}\mathcal{B}_{34}]^0_0 \Big) . \qquad (14)
\end{aligned}
$$

One can notice that only the combinations $[\mathcal{A}_{12}\mathcal{A}_{34}]$ and $[\mathcal{B}_{12}\mathcal{B}_{34}]$ will have a contribution at both the TDA and RPA level and thus, in this approach one can consider the Hamiltonian of the model to be

$$\begin{aligned}\boldsymbol{H} &= \sum_{kj} \epsilon_{kjYT} \left(\boldsymbol{b}^{\dagger}_{kj} \cdot \boldsymbol{b}^{kj} - \boldsymbol{d}_{kj} \cdot \boldsymbol{d}^{\dagger kj}\right) \\ &-\tfrac{1}{2} \sum_{J_0 TT'} \sum_{\{N_i l_i k_i j_i \lambda_i \pi_i Y_i T_i\}} V^{J_0}_{\{N_i l_i j_i Y_i T_i\}} \Big([\mathcal{A}_{12}\mathcal{A}_{34}]^0_0 + [\mathcal{B}_{12}\mathcal{B}_{34}]^0_0 \Big) \,, \end{aligned} \tag{15}$$

where the coefficients $V^{J_0}_{\{N_i l_i j_i Y_i T_i\}}$ have an analytical expression for the Cornell Potential [22].

4. The diagonalization method

The effective Hamiltonian of the model presented in equation (15) consists of a one-body free propagation term and a two-body interaction mediated by the Cornell static potential. This kind of non-relativistic Hamiltonians of at most two-body operators is a standard problem and there is a lot of literature with several methods of solution. In particular, the Many-body Methods have shown to be a powerful tool in describing quantum systems where interactions between many particles create strong quantum correlations between them.

As mentioned in Section 2, in the present model we have chosen to use two different many-body methods in order to diagonalize our Hamiltonian: the TDA and the RPA methods. The first one does not include ground state correlations, so it would correspond to the vacuum of a free theory and therefore is a naive approximation to a strong interacting theory such as NP-QCD. Pair correlations in the vacuum are introduced by the RPA method and thus only such calculation are reported in Section 5. For further details on the comparison of both methods, see [22].

The main reason for applying the many-body methods relies in the fact that the physical interpretation is direct: On one hand, in the language of nuclear physics any low-energy excitation of the system can be interpreted as the creation of a particle above the Fermi level of the nucleai and a hole below it. On the other hand, in Dirac's scheme, fermionic antimatter can be interpreted as states with negative energy that completely fill the Dirac Sea up to a Fermi energy $\epsilon_F = 0$, so that in a direct analogy to nuclear physics, one can interpret a quark-antiquark excitation as the creation of a pseudo-particle above the Dirac Sea and the creation of its respective hole below it. This allows us to model the light-flavored mesons as pairs that are scalar in color.

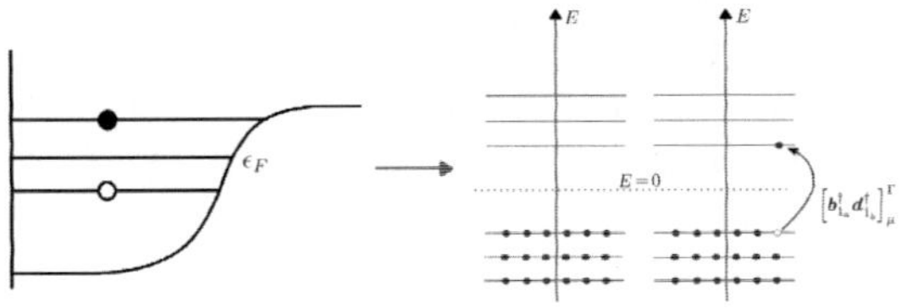

Figure 2. On the left, a particle-hole excitation above the nucleai's Fermi level. On the right, a quark-antiquark excitation above the Dirac sea.

Another reason for using this kind of methods is based in the idea that the appearance of a quark-antiquark vacuum condensate is typically associated with the confining potential that arises when fixing the Coulomb gauge [23, 24]. Thus, it is expected that the pairs involved in both the TDA and RPA methods form a suitable basis for describing the physics involved.

IOP Conf. Series: Journal of Physics: Conf. Series **912** (2017) 012028 doi:10.1088/1742-6596/912/1/012028

4.1. A posteriori flavor mixing

After applying the many-body methods, it is possible to identify η and η'-like mesons, as pure $q\bar{q}$ and $s\bar{s}$ states, which can be mixed by introducing an *a posteriori* interaction via

$$H_{mix} = \begin{pmatrix} M_{q\bar{q}} + 2H_{FM} & \sqrt{2}\,H_{FM} \\ \sqrt{2}\,H_{FM} & M_{s\bar{s}} + H_{FM} \end{pmatrix} . \tag{16}$$

The masses $M_{q\bar{q}}$ and $M_{s\bar{s}}$ of the equation (16) correspond to the lowest eigenvalues at the RPA level for the pure $q\bar{q}$ and $s\bar{s}$ states of the Hamiltonian (3) in the subspace $(J^P)YT = \{0^-, (0,0)\}$.

The interaction in (16) emulates the dynamical mechanism responsible for flavor mixing, which is the annihilation of a quark-antiquark pair into two or three (depending on the angular momentum conservation rules) gluons coupled to color zero that subsequently create a new quark-antiquark colorless pair, but not necessarily with the same flavor [17].

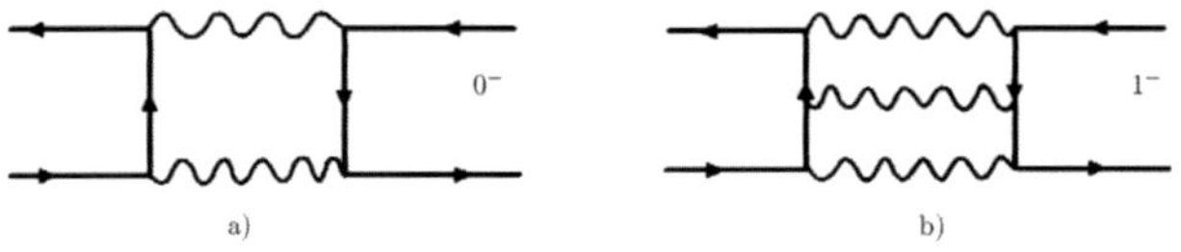

Figure 3. Virtual annihilation of neutral mesons into gluons for a) $J^P = 0^-$ and b) $J^P = 1^-$. Such process is responsible of the dynamical flavor mixing.

The eigenvalues of H_{mix} are given by

$$E_{\pm} = \frac{1}{2}\left(M_{q\bar{q}} + M_{s\bar{s}} + 3H_{FM} \pm \sqrt{9H_{FM}^2 + 2H_{FM}(M_{q\bar{q}} - M_{s\bar{s}}) + (M_{q\bar{q}} - M_{s\bar{s}})^2}\ \right) . \tag{17}$$

Even though H_{FM} should in principle be a function of the energy scale in accordance with asymptotic freedom, in the present case is set to a constant value that could be interpreted as a medium field approximation. This constant is selected in order that the mass of the η' meson is reproduced.

5. Results

Since at the moment no distinction is made between charged and neutral mesons, only a pion-like state is reported, as well as a single kaon-like state.

Here, the parameters for the masses, α and β are adjusted so that the RPA calculation reproduces as well as possible the mass of the π^0 meson ($\approx$ 135 MeV). The parameter H_{FM} is chosen to exactly reproduce the mass of the η', though the election could have been to fit the energy gap between the η and η' mesons.

Without a flavor mixing interaction the pion-like states are degenerated with the η-like state. This would mean that the η state would be too far below its experimental value. With a flavor mixing interaction the η meson is still too low, but in better agreement with the experiments.

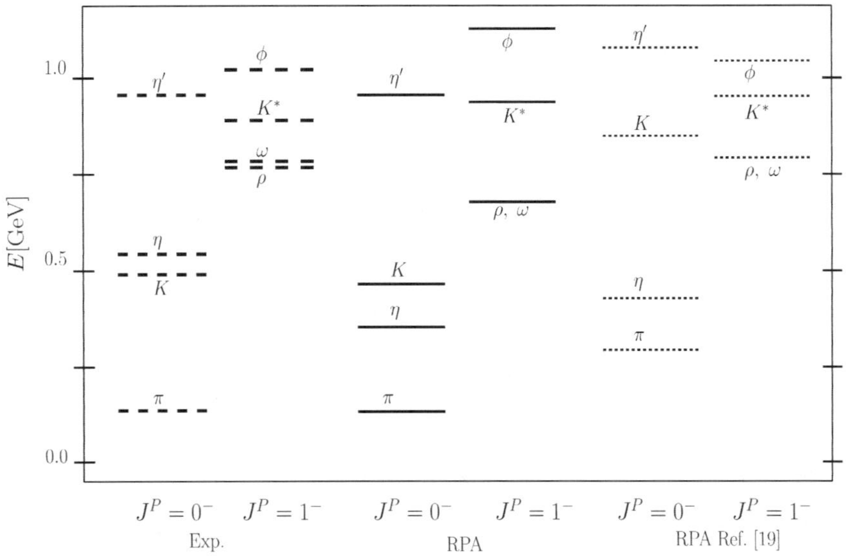

Figure 4. RPA pseudoscalar and vector states after adding the flavor mixing interaction. On the left, the experimental reported values for pseudoscalar and vector mesons up to 1 GeV. In the middle, the results for the present model. To the right, the RPA calculations for a plane wave basis from reference [19].

$m_{u,d}[GeV]$	$m_s[GeV]$	α	$\beta[GeV^2]$	H_{FM}[GeV]
0.05	0.31	0.16	0.40	0.188

Table 1. Parameters used for the RPA calculations shown on Figure 4 after applying the flavor mixing procedure in eq. (16). The optimal oscillator lenght corresponds to 0.75 fm

Last but not least, the spectrum for the pseudoscalar mesons obtained within the present model is in better agreement with the experiment than the results reported in [19], with particularly good results for the K and K^*-like states; while the rest of the spectrum is also relatively well reproduced.

6. Conclusions

The model is not ready yet and a lot of improvements to the spectra shown in Figure 4 can be achieved, for example, by introducing a dynamical flavor mixing.

The recovery of the traslational invariance of the CM with the use of the TM transformations is left for a future work and is expected to greatly improve the current results.

The model's advantages obtained so far are summarized below:

- It is a simple yet semi-realistic model. Even though the spectra can be improved, it's current state is relatively good.
- The 3D-h.o. basis is confining and allows analytic integration, as the integrals involved in the interaction matrix elements are well behaved.
- Rapid convergence of the solutions respect to the number of quanta considered. By introducing a remormalization scheme [22], the solutions barely change after $N_{max} = 8$.

- The calculations involved do not require too much computational power. This is both because of the rapid convergence of the solutions and due to the fact that the integrals are precalculated analically, as mentioned above.
- The coordinate space does not contain any divergences as those found on momentum space for plane wave calculations.
- Unlike lattice QCD, the space-time is intrinsically continuous, introducing no inconvenients due to lattice finite spacing.
- The structure of a correlated vacuum introduced by the RPA method simulates the virtual pairs creation.

Because of the above, it can be concluded that the present model is feasible but still requires further research to be able to compete with the current approaches for NP-QCD calculations.

References

[1] C. E. Carlson, T. H. Hansson y C. Peterson, Phys. Rev. D, 27, 1556-1564, (1983).
[2] T. DeGrand and C. DeTar, *Lattice Methods for Quantum Chromodynamics*, World Scientific (2006).
[3] W. Bietenholz *et al.* QCDSF-UKQCD Collaboration Phys. Rev. D **84**, 054509 (2011).
[4] W. Bietenholz, Int. J. Mod. Phys. E **25**, 1642008 (2016).
[5] J. J. Dudek, Phys. Rev. D**84**, 074023 (2011).
[6] J. J. Dudek *et al*, Phys. Rev. D **83**,111502(R) (2011)
[7] M. Alforda *et al.* Physics Letters B 361 (1995)
[8] C. D. Roberts and A. G. Williams, Prog. Part. NucL Phys., **33**, 477 (1994)
[9] R. Alkofer and L. von Smekal, Phys. Rep. **353**, 281465 (2001).
[10] Adnan Bashir, Chang lei, Ian C. Cloët, et. all., Commun. Theor. Phys. Vol. 58, No. 1 (2012).
[11] Hadron transitions in the Quark Model A. Le Yaouanc, L. Oliver *et al*, Gordon and Breach (1988).
[12] T. Yépez-Martínez, P. O. Hess, Phys. Rev. C 81, 045204 (2010).
[13] T. Yépez-Martínez, A. Amor, *et al*, Int. J. Mod. Phys. E 20, 192 (2011).
[14] Gintautas P. Kamuntavičius, Journal of Mathematical Physics, 55, 042103 (2014).
[15] Algirdas Deveikis, Gintautas P. Kamuntavičius Journal of Modern Physics, 6, 403-413 (2015).
[16] P. Ring, P. Schuck, *The Nuclear Many-Body Problem*, Springer (1980).
[17] A. De Rújula, H. Georgi and S. L. Glashow, Phys. Rev. D **12**, 147 (1975).
[18] F. J. Llanes-Estrada, S. R. Cotanch, Phys. Rev. Lett. 84, 6 (2000).
[19] F. J. Llanes-Estrada and S. R. Cotanch, Nucl. Phys. A **697**, 303 (2002).
[20] K.J. Juge, J. Kuti, and C.J. Morningstar, Nucl. Phys. Proc. Suppl. 63, 326 (1998).
[21] A. P. Szczepaniak, E. S. Swanson, Phys. Rev. D 65, 025012 (2001).
[22] A. Amor-Quiroz, P. O. Hess, *et al*, ArXiv:1704.01947 (to be published).
[23] J. R. Finger and J. E. Mandula, Nucl. Phys. B 199, 168 (1982) ;
[24] S. L. Adler, A. C. Davis, Nucl. Phys. B **244**, 469 (1984)
[25] T. Yépez-Martínez, D. A. Amor-Quiroz, *et al*, 2017 J. Phys.: Conf. Ser. 876 012022

AMDPF2017 IOP Publishing
IOP Conf. Series: Journal of Physics: Conf. Series **912** (2017) 012029 doi:10.1088/1742-6596/912/1/012029

Kähler potential of heterotic orbifolds with multiple Kähler moduli

Yessenia Olguín-Trejo, Saúl Ramos-Sánchez

Instituto de Física, Universidad Nacional Autónoma de México, POB 20-364, Cd.Mx. 01000, México

E-mail: yess.olt@ciencias.unam.mx, ramos@fisica.unam.mx

Abstract.
Aiming at improving our knowledge of the low-energy limit of heterotic orbifold compactifications, we determine at lowest order the Kähler potential of matter fields in the case where more than three bulk Kähler moduli appear. Interestingly, bulk matter fields couple to more than one Kähler modulus, a subtle difference with models with only three Kähler moduli that may provide a tool to address the question of moduli stabilization in these models.

1. Introduction

String theory is possibly the prime candidate for an UV-complete and unified description of particle physics and cosmology. In order to make contact with our observable 4D universe, expecting $\mathcal{N} = 1$ supersymmetry at some scale, the six extra dimensions predicted by the theory must be compactified on Calabi-Yau manifolds or orbifolds. Below the compactification scale, the features of the resulting SUGRA theory are entirely determined by the geometry and topology of the compactification space. Thus, one expects to identify a compactification space that yields the standard models of particle physics and cosmology.

Toroidal orbifold compactifications of the $E_8 \times E_8$ heterotic string[1] have shown to lead to large sets of models that, beyond reproducing properties of the MSSM, provide plausible solutions to problems of particle physics [5, 6, 7, 8, 9, 10]. However, to study the cosmological consequences of this scheme one has still to overcome the hurdles posed by moduli stabilization. Bulk moduli are complex scalar fields parametrizing the size and shape of the compact space; they have a flat potential at tree level in the effective theory of these models, leading thus to undesirable properties, such as unobserved fifth forces. Efforts to solve this issue have been made [11, 12], revealing that it is still unclear whether orbifold compactifications can lead to stable 4D vacua.

Addressing moduli stabilization requires knowing the details of the associated effective field theory. In $\mathcal{N} = 1$ preserving heterotic orbifolds, one then needs to determine the superpotential, the gauge kinetic functions and the Kähler potential. The former two have been computed[2] even considering some non-perturbative contributions [17, 18, 19, 20, 21]. The Kähler potential is less known because, among other things, it is not protected by non-renormalization theorems [22], although it has been explicitly obtained at leading order in some cases [23, 24].

[1] See e.g. [1, 2, 3, 4] for details on heterotic orbifold compactifications in the bosonic formulation.
[2] Yet, despite the great progress on coupling selection rules [13, 14, 15, 16], there remain some details to clarify.

IOP Conf. Series: Journal of Physics: Conf. Series **912** (2017) 012029 doi:10.1088/1742-6596/912/1/012029

In this contribution, after briefly reviewing the origin of bulk moduli in heterotic orbifolds, we reconstruct the Kähler potential for one previously unexplored case: when more than three bulk Kähler moduli arise. To achieve this goal, we present the computation of the relevant scattering amplitudes in the underlying CFT of the heterotic string and then compare them to the SUGRA results, which leads to the Kähler metric and thus the Kähler potential.

2. Untwisted moduli in orbifold compactifications

We focus here on the simplest Abelian toroidal orbifolds, defined by the quotient $\mathbb{T}^6/\mathbb{Z}_n$, with

$$\mathbb{Z}_n = \{\theta = \vartheta^k \ | \ 0 \leq k < n\}, \tag{1}$$

where n is restricted by the crystallography of $\mathbb{T}^6$ to be one of the choices presented in table 1 and the $\mathbb{Z}_n$ generator ϑ acts on the $\mathbb{T}^6$ complex coordinates, z_i, $i = 1, 2, 3$, as

$$\vartheta : \ z_i \ \rightarrow \ z_i e^{2\pi \mathrm{i} v_i}, \quad 0 \leq |v_i| < 1, \ i = 1, 2, 3\,, \tag{2}$$

where the twist vector $\boldsymbol{v} = (v_1, v_2, v_3)$ is subject to the $\mathcal{N} = 1$ condition $\pm v_1 \pm v_2 \pm v_3 = 0$. In table 1 we provide a choice of the twist vectors for all admissible $\mathbb{Z}_n$ orbifolds.

Invariance of the toroidal metric under the orbifold action leads to the *bulk* or *untwisted moduli*. The number of bulk moduli is given by the Hodge numbers $h^{1,1}$ and $h^{1,2}$, counting respectively Kähler moduli T and complex structure moduli U. Table 1 displays the number of modular fields of each kind. Notice particularly that, even though most of these orbifolds have three Kähler moduli T_i, each associated with one of the coordinates z_i, there can appear up to nine of these moduli, labeled as T_{ij}, $i, j = 1, 2, 3$.

Besides bulk moduli, there appear matter states Φ_α of two species: *bulk matter states*, free to move everywhere in 10D, and *twisted* or *localized matter states* that are linked to the singularities of the orbifold, yet free in the uncompactified space. Matter states build the matter content of the emerging field theory and can reproduce the properties of MSSM superfields.

In orbifold compactifications without Wilson lines, the effective field theory is invariant under $SL(2, \mathbb{Z})$ modular transformations of each of the bulk moduli. The resulting target-space modular group $[SL(2, \mathbb{Z})]_T^{h^{1,1}} \times [SL(2, \mathbb{Z})]_U^{h^{1,2}}$ is broken down to some congruent subgroup in semi-realistic models due to the inclusion of Wilson lines [25, 26]. Bulk and twisted matter states Φ_α transform non-trivially under the modular group [27, 28], according to their modular properties encoded in the so-called modular weights n_α and ℓ_α, for T and U symmetries respectively, which are vectors of fractional charges that can be determined from the computation of the Kähler potential contributions involving both bulk moduli and matter fields.

Table 1: Twist vectors and Hodge numbers of $\mathbb{Z}_n$ orbifolds

$\mathbb{Z}_n$	twist vector	$h^{1,1}$	$h^{1,2}$	$\mathbb{Z}_n$	twist vector	$h^{1,1}$	$h^{1,2}$
$\mathbb{Z}_3$	$\frac{1}{3}(1,1,-2)$	9	0	$\mathbb{Z}_8$-I	$\frac{1}{8}(1,2,-3)$	3	0
$\mathbb{Z}_4$	$\frac{1}{4}(1,1,-2)$	5	1	$\mathbb{Z}_8$-II	$\frac{1}{8}(1,3,-4)$	3	1
$\mathbb{Z}_6$-I	$\frac{1}{6}(1,1,-2)$	5	0	$\mathbb{Z}_{12}$-I	$\frac{1}{12}(1,4,-5)$	3	0
$\mathbb{Z}_6$-II	$\frac{1}{6}(1,2,-3)$	3	1	$\mathbb{Z}_{12}$-II	$\frac{1}{12}(1,5,-6)$	3	1
$\mathbb{Z}_7$	$\frac{1}{7}(1,2,-3)$	3	0				

IOP Conf. Series: Journal of Physics: Conf. Series **912** (2017) 012029 doi:10.1088/1742-6596/912/1/012029

3. Kähler potentials for orbifold compactifications

The Kähler potential when *three* bulk Kähler moduli, none or one complex structure modulus (see table 1) and some matter fields are present, is given by [28]

$$K = -\sum_{i=1}^{3} \ln(T_i + \overline{T}_i) - \delta_1^{h^{1,2}} \ln(U + \overline{U}) + \sum_{\alpha} \prod_{i=1}^{3} (U + \overline{U})^{-\ell_\alpha} (T_i + \overline{T}_i)^{-n_\alpha^i} \Phi_\alpha \overline{\Phi}_{\overline{\alpha}} \,. \qquad (3)$$

The modular weights of an untwisted matter field Φ_α are $n_\alpha^i = -q_\alpha^i$ and $\ell_\alpha = -q_\alpha^3$, with q_α denoting the Neveu-Schwarz right-moving momentum of the field (with entries 0 or -1); and for twisted states Φ_α of the θ-sector they are given by

$$\begin{array}{lll} n_\alpha^i = 1 - \eta_\alpha^i - \Delta\tilde{N}_\alpha^i \,, & \ell_\alpha = 1 - \eta_\alpha^3 + \Delta\tilde{N}_\alpha^3 & \text{for } \eta_\alpha^i \neq 0 \,, \\ n_\alpha^i = \eta_\alpha^i \,, & \ell_\alpha = \eta_\alpha^3 & \text{for } \eta_\alpha^i = 0, \quad i = 1,2,3 \,, \end{array} \qquad (4)$$

where $\Delta\tilde{N}_\alpha^i$ denotes the number of holomorphic minus antiholomorphic left-moving oscillator excitations in the z^i complex plane generating the state, and $\eta^i = kv_i \mod 1$, such that $0 \leq \eta^i < 1$, in the $\theta = \vartheta^k$ twisted sector. $\ell_\alpha = 0\ \forall\alpha$ if $h^{1,2} = 0$.

For models with more than three Kähler moduli, we expect K to have a similar form, but the slight differences may be relevant for moduli stabilization. In general, denoting any bulk modulus as M, the Kähler potential of interest can be written as an expansion of the matter fields Φ as

$$K(M, \overline{M}, \Phi, \overline{\Phi}) = \mathcal{K}(M, \overline{M}) + \sum_{\alpha,\beta} K_{,\alpha\bar{\beta}}\, \Phi^\alpha \overline{\Phi}^{\bar{\beta}} + ..., \qquad (5)$$

where $\mathcal{K}$ is a function that has been determined through dimensional reduction of 10D SUGRA theories [29] as it only depends on the effective behavior of the moduli. On the other hand, $K_{,\alpha\bar{\beta}}$ is an unknown function of the bulk moduli that describes the coupling between moduli and matter fields, that carry full information from the original string theory.

In string models, this function can be obtained at first order from four-point scattering amplitudes [23, 24], which are determined by the correlation functions between the asymptotic states $|\Phi\rangle$ and $|\Phi'\rangle$ corresponding to the matter fields Φ and Φ', affected by the interaction with the vertex operators V_M and $V_{M'}$ associated with the moduli M and M'. Schematically, the amplitude to compute reads $A(M, \Phi, (\Phi')^\dagger, (M')^\dagger) := \frac{1}{4\pi} \int dz^2 \langle \Phi' | V_{\overline{M}'} V_M | \Phi \rangle$.

The vertex operator of a Kähler modulus $M_{a\bar{e}}$ with components in the z^a and $\bar{z}^{\bar{e}}$ complex directions, and in the 0-ghost picture, takes the form (we set $\alpha' = 1/2$) [30]

$$V_0(M_{a\bar{e}}, k, z, \bar{z}) = \left(\partial_z X_R^a + \frac{1}{2} \Psi_R^a k \cdot \Psi_R \right) e^{ik\cdot X_R} \partial_{\bar{z}} X_L^{\bar{e}} e^{ik\cdot X_L}, \qquad (6)$$

where X and Ψ are the bosonic and fermionic worldsheet fields and k is the momentum of M, Latin indices are $1, 2, 3$. On the other hand, the Neveu-Schwarz components of the matter states with gauge momentum p_α are given by [31]

$$\begin{array}{llll} |\Phi_\alpha\rangle & = b_{-1/2}^{\alpha_i} |0\rangle_R \otimes |p_\alpha\rangle_L \,, & \alpha_i = 1,2,3 & \text{untwisted sector,} \\ |\Phi_\alpha\rangle & = \Omega_\alpha |0\rangle_R \otimes \widetilde{\Omega}_\alpha |p_\alpha\rangle_L & & \text{twisted sector,} \end{array} \qquad (7)$$

where Ω_α and $\widetilde{\Omega}_\alpha$ are products of right and left-handed oscillators, respectively, in possibly various complex directions. Consequently, in contrast to untwisted states, a twisted state $|\Phi_\alpha\rangle$ can carry more than one index $i = 1, 2, 3$ and many oscillators in various directions.

Using the canonical formalism in CFT, we can now compute the correlation function associated with the scattering of two untwisted matter fields and two Kähler moduli. After a lengthy computation, the result reads

$$A(M_{a\bar{e}}, \Phi_\gamma, (\Phi_\beta)^\dagger, (M_{b\bar{f}})^\dagger) = \frac{1}{32}\delta_{\bar{e}\bar{f}}\left(\delta_{\gamma\beta}\delta_{ab}\frac{us}{t} + s\delta_{\gamma_i a}\delta_{\beta_i b}\delta_{\gamma\beta}\right)\frac{\Gamma(1-u/8)\Gamma(1-s/8)\Gamma(1-t/8)}{\Gamma(1+u/8)\Gamma(2+t/8)\Gamma(1+s/8)}, \quad (8)$$

where s, u, t are the kinematic Mandelstam variables of the process. We are interested in this result at low energies, i.e. when $|s|, |u|, |t| \ll 1$. In this case, (8) takes the form

$$A(M_{a\bar{e}}, \Phi_\gamma, (\Phi_\beta)^\dagger, (M_{b\bar{f}})^\dagger) \approx \frac{1}{32}\delta_{\bar{e}\bar{f}}\left(\delta_{\gamma\beta}\delta_{ab}\frac{us}{t} + s\delta_{\gamma_i a}\delta_{\beta_i b}\delta_{\gamma\beta}\right). \quad (9)$$

Now, in SUGRA one can also determine the analogous amplitude in terms of the fields instead of strings and then express it in the small-moduli limit ($M \ll 1$), by generalizing the methods developed in [32]. A is given thus by

$$A(M_{a\bar{e}}, \Phi_\gamma, (\Phi_\beta)^\dagger, (M_{b\bar{f}})^\dagger) \approx \frac{1}{32}\left(\frac{us}{t}\delta_{ab}\delta_{\bar{e}\bar{f}}\delta_{\gamma\beta} + sG_{\gamma\bar{\beta};a\bar{e},b\bar{f}}(0,0)\right), \quad (10)$$

where $G_{\gamma\bar{\beta};a\bar{e},b\bar{f}}(0,0)$ is the derivative of the Kähler metric with respect to the moduli $M_{a\bar{e}}$ and $(M_{b\bar{f}})^\dagger$ in the small-moduli limit. Contrasting (9) and (10), we can readily conclude that the Kähler metric is

$$G_{\gamma\bar{\beta}} \approx \delta_{\gamma\beta} + \delta_{\gamma\beta}\delta_{\bar{e}\bar{f}}\delta_{\gamma_i a}\delta_{\beta_i b}M_{a\bar{e}}(M_{b\bar{f}})^\dagger, \quad (11)$$

which is, by definition, the function $K_{,\gamma\bar{\beta}}$ in (5).

From (11), we find that, if the in and out states coincide, i.e. $\gamma = \beta$, then $\gamma_i = \beta_i$ and, after the holomorphic field redefinition $M_{i\bar{j}} \to 1 - T_{ij}/1 + T_{ij}$, a Kähler transformation and taking the small-moduli limit, the relevant contribution to the Kähler potential reads

$$K \supset \sum_e \left(1 + M_{\beta_i\bar{e}}(M_{\beta_i\bar{e}})^\dagger\right)|\Phi_\beta|^2 \quad \longrightarrow \quad \prod_e (T_{\beta_i e} + \overline{T}_{\beta_i e})^{-1}|\Phi_\beta|^2. \quad (12)$$

Note that, in contrast to dimensional reductions of 10D SUGRA [29], orbifold compactifications do not yield terms proportional to $\Phi_\beta\overline{\Phi}_{\bar{\gamma}}$, with $\beta \neq \gamma$. The modular weights of Φ_β in the untwisted sector are directly read off (12). They build now the matrix $n_\beta^{ij} = \delta_{\beta_i}^i\ \forall j$, where the values that i, j take depend on the Kähler moduli present in the model. In the $\mathbb{Z}_4$ and $\mathbb{Z}_6$-I orbifolds, only T_{ij} with $i, j = 1, 2$ and T_{33} exist.

The computation for the twisted sectors is analogous. There are several sources of differences though. First, the expressions for X_R^a, Ψ_R^a and $X_L^{\bar{e}}$ in the vertex operator (6) of a modulus $M_{a\bar{e}}$ depend on two components of the twist, η^a and $\eta^{\bar{e}} = \eta^e$. Secondly, the asymptotic states $|\Phi_\alpha\rangle$ carry non-trivial oscillators (with frequencies that depend on the twist) in various directions. Massless states exhibit holomorphic and antiholomorphic oscillators in X_L, whose difference in the compact direction z^i is denoted $\Delta\tilde{N}_\alpha^i$; right-moving bosonic oscillators in Ψ_R of the type b_ν^i, for some frequencies $\nu = \nu(\eta^i)$, affect the correlators; and there are no oscillators from X_R in massless states. These features leave their trace in the expressions of the correlators. Setting first $\Delta\tilde{N}_\alpha^i = 0$, after a lengthy computation, we arrive at

$$A(M_{a\bar{e}}, \Phi_\beta, \Phi_\gamma^\dagger, (M_{b\bar{f}})^\dagger) = \delta_{\gamma\beta}\delta_{ab}\delta_{\bar{e}\bar{f}}\frac{s}{32t}(u + t(1-\eta_\beta^{\bar{e}}))\frac{\Gamma(1-\eta_\beta^{\bar{e}}-u/8)\Gamma(\eta_\beta^a-s/8)\Gamma(1-t/8)}{\Gamma(\eta_\beta^{\bar{e}}+u/8)\Gamma(2+t/8)\Gamma(1+s/8-\eta_\beta^{\bar{e}})}. \quad (13)$$

Including oscillators and taking its low-energy limit (using that $\eta_\beta^a = \eta_\beta^{\bar{e}}$ for non-diagonal moduli) yields

$$A(M_{a\bar{e}}, \Phi_\beta, \Phi_\gamma^\dagger, (M_{b\bar{f}})^\dagger) \approx \delta_{\gamma\beta}\delta_{ab}\delta_{\bar{e}\bar{f}}\frac{s}{32t}\left(u + t(1-\eta_\beta^a + \Delta\tilde{N}_\beta^a)\right). \quad (14)$$

We must now compare this result with the SUGRA analogue, to identify the Kähler metric, which for twisted states turns out to be

$$G_{\gamma\bar{\beta}} \approx \delta_{\gamma\beta} + \delta_{\gamma\beta}\delta_{ab}\delta_{\bar{e}\bar{f}}(1 - \eta^a_\beta + \Delta\tilde{N}^a_\beta) M_{a\bar{e}} (M_{b\bar{f}})^\dagger \,. \tag{15}$$

As in the untwisted case, by performing the proper transformations and considering the small-moduli limit, we finally obtain

$$K \supset \sum_{a,e} \left(1 + n^{ae}_\beta M_{a\bar{e}}(M_{a\bar{e}})^\dagger\right) |\Phi_\beta|^2 \quad \longrightarrow \quad \prod_{a,e}(T_{ae} + \overline{T}_{ae})^{-n^{ae}_\beta} |\Phi_\beta|^2 \,, \tag{16}$$

where we have identified the twisted modular weights as $n^{ae}_\beta = 1 - \eta^a_\beta + \Delta\tilde{N}^a_\beta \;\forall e$. We realize that (12) and (16) are very similar, but they have two differences. The first one is the expression for the modular weights, given under these equations. The second one is that the Kähler contribution of the twisted fields prescribes a kinetic interaction between these fields and all Kähler moduli, whereas the untwisted fields interact only with a subset of the moduli. The latter is also an interesting difference with respect to compactifications with only three bulk moduli: each bulk matter field interacts only with one Kähler modulus in those models.

From our discussion, we find that the Kähler potential at leading order in heterotic orbifold compactifications endowed with more than the three diagonal Kähler moduli takes the form

$$K = -\ln\det(T_{ij} + \overline{T}_{ij}) - \delta^{h^{1,2}}_1 \ln(U + \overline{U}) + \sum_\beta \prod_{i,j} (U + \overline{U})^{-\ell_\beta} (T_{ij} + \overline{T}_{ij})^{-n^{ij}_\beta} \Phi_\beta \overline{\Phi}_{\bar{\beta}} \,, \tag{17}$$

where the Kähler modular weights are defined below (12) and (16), and the complex structure modular weights are given in (4). The first contribution in K is a result of dimensional reduction in SUGRA [29], which applies here because bulk moduli of heterotic orbifolds behave just as in a SUGRA compactification.

4. Discussion and forthcoming research

This proceedings contribution has been devoted to ameliorating our understanding of the SUGRA limit of supersymmetric heterotic orbifold compactifications. We have reconstructed the Kähler potential at lowest order in the case where bulk and twisted matter fields interact with more than the three usual bulk Kähler moduli. To do so, we computed the associated four-point scattering amplitudes within the formalism of the underlying CFT of the heterotic string and compared them with the known analogous result in SUGRA.

Our findings are summarized in eq. (17). In the presence of multiple bulk Kähler moduli in orbifolds, the structure of the couplings between matter fields and bulk matter fields becomes richer than in models with only diagonal Kähler moduli. Especially, each bulk matter field is coupled to more than one Kähler modulus. This feature may become instrumental to address the longstanding issue of moduli stabilization, which might only find a solution in a model-dependent basis, in promising $\mathbb{Z}_4$, $\mathbb{Z}_6$-I heterotic orbifolds [10] with more than three moduli. From there, one can aim at gaining some insight into the string cosmology of these constructions.

Our result also reveals further details about the modular properties of the effective fields in heterotic orbifolds. Provided that the modular weights can be regarded as charges of the target space modular symmetries, on may ask oneself if they can play a role in some 4D particle phenomenology, as it has been sometimes suggested (see e.g. [33]).

Our results are only the starting point of a long quest, whose progress shall be reported elsewhere. We should be able to compute further corrections of the Kähler potential and thereby try to identify some symmetries governing its structure. The final goal would be to configure the final shape of the effective SUGRA theory in these simple compactifications.

Acknowledgments
We thank P.K.S Vaudrevange and M. Ratz for interesting discussions on this subject. This work was partly supported by DGAPA-PAPIIT grant IN100217 and CONACyT grant F-252167. S. R-S. would like to thank the ICTP for the kind hospitality and support received through its Junior Associateship Scheme during the realization of this work.

References

[1] Dixon L J, Harvey J A, Vafa C and Witten E 1985 *Nucl. Phys.* **B261** 678–686
[2] Bailin D and Love A 1999 *Phys. Rept.* **315** 285–408
[3] Ramos-Sánchez S 2009 *Fortsch. Phys.* **10** 907–1036 (*Preprint* `0812.3560`)
[4] Vaudrevange P K S 2008 Ph.D. thesis Bonn U. (*Preprint* `0812.3503`)
[5] Kobayashi T, Raby S and Zhang R J 2005 *Nucl. Phys.* **B704** 3–55 (*Preprint* `hep-ph/0409098`)
[6] Buchmüller W, Hamaguchi K, Lebedev O and Ratz M 2006 *Phys. Rev. Lett.* **96** 121602 (*Preprint* `hep-ph/0511035`)
[7] Kim J E, Kim J H and Kyae B 2007 *JHEP* **06** 034 (*Preprint* `hep-ph/0702278`)
[8] Nilles H P, Ramos-Sánchez S, Ratz M and Vaudrevange P K S 2009 *Eur. Phys. J.* **C59** 249–267 (*Preprint* `0806.3905`)
[9] Lebedev O, Nilles H P, Ramos-Sánchez S, Ratz M and Vaudrevange P K S 2008 *Phys. Lett.* **B668** 331–335 (*Preprint* `0807.4384`)
[10] Nilles H P and Vaudrevange P K S 2015 *Mod. Phys. Lett.* **A30** 1530008 (*Preprint* `1403.1597`)
[11] Parameswaran S L, Ramos-Sánchez S and Zavala I 2011 *JHEP* **01** 071 (*Preprint* `1009.3931`)
[12] Dundee B, Raby S and Westphal A 2010 *Phys. Rev.* **D82** 126002 (*Preprint* `1002.1081`)
[13] Kobayashi T, Parameswaran S L, Ramos-Sánchez S and Zavala I 2012 *JHEP* **05** 008 [Erratum: JHEP12,049(2012)] (*Preprint* `1107.2137`)
[14] Cabo Bizet N G, Kobayashi T, Mayorga Peña D K, Parameswaran S L, Schmitz M and Zavala I 2013 *JHEP* **05** 076 (*Preprint* `1301.2322`)
[15] Nilles H P, Ramos-Sánchez S, Ratz M and Vaudrevange P K S 2013 *Phys. Lett.* **B726** 876–881 (*Preprint* `1308.3435`)
[16] Cabo Bizet N G, Kobayashi T, Mayorga Peña D K, Parameswaran S L, Schmitz M and Zavala I 2014 *JHEP* **02** 098 (*Preprint* `1308.5669`)
[17] Burwick T T, Kaiser R K and Müller H F 1991 *Nucl. Phys.* **B355** 689–711
[18] Erler J, Jungnickel D, Spalinski M and Stieberger S 1993 *Nucl. Phys.* **B397** 379–416 (*Preprint* `hep-th/9207049`)
[19] Choi K S and Kobayashi T 2008 *Nucl. Phys.* **B797** 295–321 (*Preprint* `0711.4894`)
[20] Dixon L J, Kaplunovsky V and Louis J 1991 *Nucl. Phys.* **B355** 649–688
[21] Mayr P and Stieberger S 1993 *Nucl. Phys.* **B407** 725–748 (*Preprint* `hep-th/9303017`)
[22] Dine M and Seiberg N 1986 *Phys. Rev. Lett.* **57** 2625
[23] Bailin D, Gandhi S K and Love A 1992 *Phys. Lett.* **B275** 55–62
[24] Kaplunovsky V and Louis J 1995 *Nucl. Phys.* **B444** 191–244 (*Preprint* `hep-th/9502077`)
[25] Bailin D, Love A, Sabra W A and Thomas S 1994 *Mod. Phys. Lett.* **A9** 1229–1238 (*Preprint* `hep-th/9312122`)
[26] Love A and Todd S 1996 *Nucl. Phys.* **B481** 253–288 (*Preprint* `hep-th/9606161`)
[27] Ferrara S, Lüst D and Theisen S 1989 *Phys. Lett.* **B233** 147–152
[28] Ibáñez L E and Lüst D 1992 *Nucl. Phys.* **B382** 305–361 (*Preprint* `hep-th/9202046`)
[29] Ferrara S, Kounnas C and Porrati M 1986 *Phys. Lett.* **B181** 263
[30] Hamidi S and Vafa C 1987 *Nucl. Phys.* **B279** 465–513
[31] Bailin D and Love A 1992 *Phys. Lett.* **B288** 263–268
[32] Dixon L J, Kaplunovsky V and Louis J 1990 *Nucl. Phys.* **B329** 27–82
[33] Feruglio F 2017 (*Preprint* `1706.08749`)

Searching for heavy Majorana neutrinos in $\tau^- \to \pi^+\mu^-\mu^-\nu_\tau$ at Belle II

D. Rodriguez Perez, P.L.M.Podesta Lerma, Isabel Dominguez Jimenez

Universidad Autónoma de Sinaloa

E-mail: est.david.rope@uas.edu.mx, podesta.pedro@uas.edu.mx and isadoji@uas.edu.mx

Abstract. A feasibility study is done concerning the lepton number violation decay channel $\tau^- \to \pi^+\mu^-\mu^-\nu_\tau$. Monte Carlo events are produced using phase space and the response of the detector is simulated. The observation will be a proof of the Majorana nature of the neutrinos.

1. Introduction

The SM is the current theory of fundamental particles and their interactions, which was created based on observations and theoretical predictions through several years. In this theory all the elementary matter particles are fermions, interacting among themselves by the exchange of gauge bosons and the mass is generated by the Higgs mechanism.

But, despite its tremendous success in describing the basic forces of nature many fundamental questions remain unanswered within the SM. Thus, considering the open questions that in the SM remain unanswered, it is fair to conclude that the present theory is an extremely successful but phenomenological description of subatomic processes at the energy scales up to O(1 TeV).

The future experiments in high energy physics are designed to address the above questions about New Physics using complementary approaches. One approach is at the energy frontier with the main representatives being the ATLAS and the CMS experiments at the Large Hadron Collider (LHC) at CERN [1]. The second is the rate/precision frontier, exemplified by the LHCb experiment at the LHC and the Belle II experiment at SuperKEKB [2].

At the energy frontier, the LHC experiments will be able to discover new particles produced in proton-proton collisions at a center-of-mass energy of up to 14 TeV. Since the constituent gluons or quarks interact in these collisions, only a fraction of the center-of-mass energy is available to produce such new particles, and the mass reach is limited to O(1 TeV/c^2). Sensitivity to the direct production of a specific new particle depends on the cross section and on the size of the data sample.

At the rare/precision frontier, the observable signatures of new particles or processes can be obtained through measurements of flavor physics reactions at lower energies and evidence of a deviation from the SM prediction. An observed discrepancy can be interpreted in terms of NP models. This is the approach of Belle II.

2. Lepton number violation

Experiments with neutrinos [3][4] have questioned the SM since it considers that neutrinos are massless. Solar neutrino experiments [5] show that they oscillate from a definite flavor to another through long distances, thus neutrinos should have mass. However, these neutrino oscillations don't give information about absolute neutrino masses. Different mechanisms exist to obtain neutrino masses. The requirement of lepton number violation is necessary to generate neutrino masses in the case of Majorana nature.

Neutrinoless double beta decay [6] is the most studied process that violates the lepton number by two units. Until now, it has not been observed. Besides the neutrinoless double beta decay, other processes have been searched in LHCb at CERN [1], BaBar at SLAC [7] and Belle at KEKB [8]. They have imposed limits in the branching fraction to different decays, Figure 1.

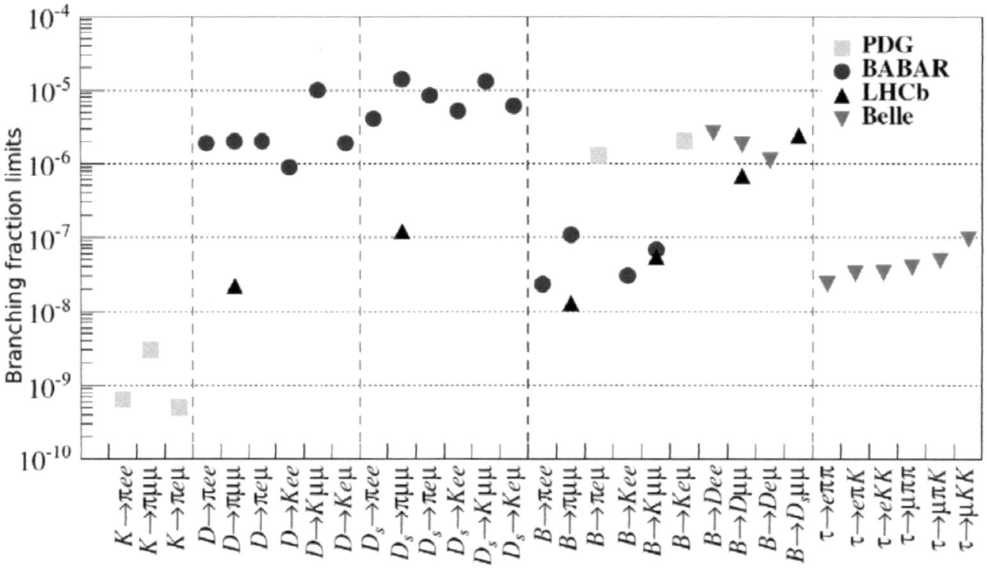

Figure 1. Branching fraction limits in channels where $\Delta L = 2$ [9].

However, the Majorana neutrino mass (m_N) is not determined and we can consider it in the range of ∼(MeV to few GeV). Then, the branching fractions can be enhanced for decays to four bodies (for more details see [9]). We focus in semi-leptonic decays, specifically on $\tau^- \to \pi^+\mu^-\mu^-\nu_\tau$ channel. We will study the possibility to search this channel in the Belle II detector.

Decay processes that violate the total lepton number $\Delta L = 2$ can be induced by the exchange of Majorana neutrinos. We consider a scenario where these decays are dominated by the exchange of only one heavy neutrino, which produces an enhancement of the decay amplitude via the resonant mechanism and the high luminosity in the Belle II experiment allows to study the $\tau^- \to \pi^+\mu^-\mu^-\nu_\tau$ channel. The resonant contribution of a heavy Majorana neutrino to $\Delta L = 2$ four-body decays of τ lepton provide additional information and complementary constrains on the parameter space of Majorana neutrino.

3. Belle II

The B factory experiment Belle II at the SuperKEKB [2] accelerator in Tsukuba, Japan, has confirmed the Kobayashi-Maskawa mechanism of CP violation in the SM. Furthermore, it has a rich physics program of τ leptons. The aim of e^+e^- colliders is to accumulate 50 ab^{-1},

IOP Conf. Series: Journal of Physics: Conf. Series **912** (2017) 012030 doi:10.1088/1742-6596/912/1/012030

corresponding to about 47 billion $\tau^+\tau^-$ events, by the year 2022. This correspond to 50 times the luminosity of its predecessor KEKB.

3.1. Accelerator

SuperKEKB uses the same two rings that were used by KEKB: a low-energy ring (LER) for positrons and a high-energy ring (HER) for electrons. These are located side by side 11 meters below the ground level in the TRISTAN tunnel, which has a circumference of around 3 km. However, the beam energies have been changed from the present values of 3.5 and 8.0 GeV to 4.0 and 7.0 GeV. In SuperKEKB the Nano-Beam scheme is used [2], in which the emittance growth due to intra-beam scattering and the short beam lifetime due to the Touschek effect and both are serious problems, particularly in the LER. The increase in the beam energy of the LER from 3.5 to 4.0 GeV helps mitigate these problems. The decrease in the beam energy of the HER from 8.0 to 7.0 GeV allows a lower emittance.

In the interaction region, the 7 GeV electrons in the HER and the 4 GeV positrons in the LER collide at one interaction point (IP) with a non-zero crossing angle of 83 mrad.

4. Luminosity

The center-of-mass (CM) energy corresponds to the mass of the $\Upsilon(4S)$ resonance,

$$E_{CM} = 2\sqrt{E_{HER}E_{LER}} = 10.58GeV \sim M_{\Upsilon(4S)} \quad (1)$$

the production cross section for $\Upsilon(4S)$ (bound $b\bar{b}$ state) at the CM is 1.1 nb, 1.3 nb for $c\bar{c}$ pair production, and 2.1 nb for $q\bar{q}$ pair production. Where q is a u, d, or s quark. Cross sections of various processes are summarized in Table 1. The rate of interaction in e^+e^- collisions is proportional to the interaction cross section, $\sigma_{interaction}$:

$$\frac{dN_{\text{interaction}}}{dt} = \mathcal{L}_{e^+e^-}\sigma_{\text{interaction}} \quad (2)$$

where the coefficient $\mathcal{L}_{e^+e^-}$ is the luminosity of the accelerator and is a measure of the number of particles per unit area per unit time at the IP,

$$\mathcal{L}_{e^+e^-} = \nu n \frac{N_{e^+}N_{e^-}}{A} \quad (3)$$

here, n is the number of bunches in the storage ring composed of N_{e^+} positrons and N_{e^-} electrons that collide ν times per second and A is the overlapping area of the two colliding beams. The luminosity goal of SuperKEKB is $8 \times 10^{35}\text{cm}^{-2}\text{s}^{-1}$, which is around 40 times as large as the peak luminosity achieved by the KEKB collider.

5. Analysis

We produced the signal using the KKMC generator [10], $e^+e^- \to \tau^+\tau^-(\gamma)$, with τ decaying to the signal mode $\tau^- \to \pi^+\mu^-\mu^-\nu_\tau$, and the other decaying to all experimentally measured SM decay modes of the τ, approximately 42 decays. A total of 50,000 events were generated for this decay.

Samples for a range of background events were produced by the Belle II collaboration; the generators used for each background process are given in Table 1. Initial state radiation (ISR) and final state radiation (FSR) of multiple photons is generated by KKMC and BABAYAGA@NLO[11] generators. Beam backgrounds are also taken into account and include radiative Bhabha, Touschek scattering and beam-gas interactions.

Table 1. Samples for a range of background events were produced by the Belle II collaboration; the generators used for each background process

Background process	Cross-section [nb]	Generator
$e^+e^- \to B^+B^-$	0.525	EvtGen, PYTHIA
$e^+e^- \to B^0\bar{B}^0$	0.525	EvtGen, PYTHIA
$e^+e^- \to u\bar{u}$	1.61	KKMC
$e^+e^- \to d\bar{d}$	0.40	KKMC
$e^+e^- \to s\bar{s}$	0.38	KKMC
$e^+e^- \to c\bar{c}$	1.30	KKMC
$e^+e^- \to \tau\bar{\tau}$	1.30	KKMC

5.1. Reconstruction

Reconstruction of events can be separated into three sections: tracking, calorimeter reconstruction and particle identification. Tracking of charged particle is achieved using the innermost detectors, the VXD and the CDC. Pattern recognition algorithms are to collect all detector hits belonging to a single track, then a track candidate is created. Impact parameters can be determined through the VXD, and hence vertex positions of decaying particles can be calculated. A helical fit is applied to the track candidate, using track impact parameters combined with highly accurate measurements of charged track momenta from the CDC. The track is extrapolated from the CDC to the point-of-closest approach [2].

5.2. Event selection

Some selection criteria were applied to the reconstructed events removing the more obvious background components.

- $0.5 < M_{\pi^+\mu^-\mu^-} < 2.0$ GeV
- For particle identification,
  - $\mathcal{L}(\pi) > 0.9$
  - $\mathcal{L}(\mu) > 0.9$
- Tag side: One charged track.
- Signal side: Three charged tracks.

Taus are unstable particles, and will decay before reaching any of the Belle II detectors. Therefore, they must be reconstructed from the detection of their decay products. However, the neutrino can not be detected and it will be missing mass. The hemispheres are defined in the CM by the plane perpendicular to the thrust axis $\hat{n}$, defined as the unit vector in the direction of the thrust $T = \max\left[\sum_i |\hat{n} \cdot \vec{p_i}| / \sum_i |\vec{p_i}|\right]$, where $\vec{p_i}$ is the momentum of the i-th particle.

These selection criteria will allow to have the control of background sample. For this purpose, we observe the two dimensional distribution of the ΔE and the missing mass of the τ candidate, where ΔE is the difference between the energy of τ candidate and the half energy of the system in the center-of-mass frame, as shown in Fig. 2.

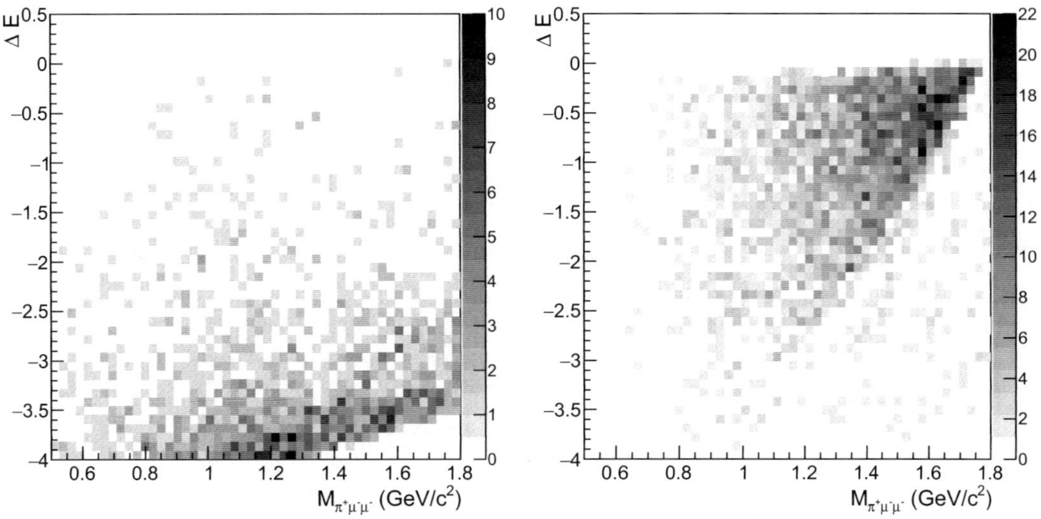

Figure 2. (left) All background. (right) Signal

6. Conclusions

We simulated the channel

$$\tau^- \to \pi^+ \mu^- \mu^- \nu_\tau$$

which is a decay to four bodies and its branching fraction is enhanced with respect to decays to three bodies. This mode has not been measured before, we also simulated the possible source of background and from this study, we have found that is reasonable to look for this decay mode.

Forthcoming Research

Selection criteria could be optimised to increase signal efficiency through use of multivariate analysis techniques. Additional selection variables could be investigated and signal region analysis can be extended to utilise Mass and ΔE resolutions and provide a statistics-motivated signal region.

References

[1] Brüning O S, Collier P, Lebrun P, Myers S, Ostojic R, Poole J and Proudlock P 2004 *LHC Design Report* (Geneva: CERN)
[2] Abe T *et al.* (Belle-II) 2010 (*Preprint* `1011.0352`)
[3] Mann A K and Primakoff H 1977 *Phys. Rev. D* **15**(3) 655–665
[4] Frati W, Gaisser T K, Mann A K and Stanev T 1993 *Phys. Rev. D* **48**(3) 1140–1149
[5] Kuo T K and Pantaleone J 1990 *Phys. Rev. D* **41**(12) 3842–3845
[6] Bilenky S M 690–715 ISSN 1063-7796 (*Preprint* `1001.1946`)
[7] Aubert B *et al.* (BaBar) 2002 *Nucl. Instrum. Meth.* **A479** 1–116 (*Preprint* `hep-ex/0105044`)
[8] 1996 *Il Nuovo Cimento A* **109** ISSN 0369-3546
[9] Quintero N 2014 *PhD thesis, Departamento de Física, Cinvestav*
[10] Jadach S, Ward B F L and Was Z 2000 *Comput. Phys. Commun.* **130** 260–325 (*Preprint* `hep-ph/9912214`)
[11] Balossini G, Carloni Calame C M, Montagna G, Nicrosini O and Piccinini F 2006 *Nucl. Phys.* **B758** 227–253 (*Preprint* `hep-ph/0607181`)

IOP Conf. Series: Journal of Physics: Conf. Series **912** (2017) 012031 doi:10.1088/1742-6596/912/1/012031

Potential of CCDs for the study of sterile neutrino oscillations via Coherent Neutrino-Nucleus Elastic Scattering

Marisol Chávez-Estrada and Alexis A. Aguilar-Arevalo

Instituto de Ciencias Nucleares. Universidad Nacional Autónoma de México. Circuito Exterior, Ciudad Universitaria, Apartado postal 70-543, 04510 Ciudad de México, México

E-mail: `fismarisol@gmail.com`, `alexis@nucleares.unam.mx`

Abstract. We study the potential of a detector based on CCD sensors (CONNIE experiment) to study neutrino oscillations to sterile states using reactor neutrinos. We calculate the number of events expected in a 1 kg detector and determine the sensitivity to oscillations $\nu_e \to \nu_s$ in the Δm^2_{41} vs. $\sin^2\theta_{es}$ parameter space for various exposures. The sensitivity is compared with the regions excluded by the Daya Bay experiment under the assumption $\theta_{24} = \theta_{34} = 0$. This work was carried out independently of the CONNIE Collaboration using published information, and its results are not official.

1. Introduction

While the majority of neutrino oscillation measurements fit well in the 3 massive neutrino model, a handful of anomalous results [1] suggest the existence of at least one *sterile* neutrino with a mass splitting $\Delta m^2 \sim 1\text{eV}^2$. Models considering sterile neutrinos (3+N) have been considered but different observations impose severe constraints. The 3+1 scenario can be probed with an experiment sensitive to all 3 active flavors $(\nu_e, \nu_\mu, \nu_\tau)$ via a neutral current interaction.
In the **C**oherent **E**lastic **N**eutrino-**N**ucleus **S**cattering (**CENNS**) [2], a neutrino, or antineutrino, scatters off a nucleus as a whole. When the 4-momentum transfer is small compared to the reciprocal of the nucleus size ($Q^2 < 1/R^2$), the interaction with the constituent nucleons coherently enhances the cross section. The differential cross section is:

$$\frac{d\sigma}{dE_{rec}} = \frac{G_F^2}{8\pi}[Z(4\sin^2\theta_W - 1) + N]^2 M \left(2 - \frac{E_{rec}M}{E_{\bar{\nu}_e}^2}\right)|f(q)|^2 \quad (1)$$

where E_{rec} is the nuclear recoil energy, G_F is the Fermi constant, M is the mass of the nucleus, Z, and N are the number of protons and neutrons in the nucleus respectively, θ_W is the weak mixing angle, and $f(q)$ is the nuclear form factor, which is a function only of the 4-momentum transfer q. Since $\sin^2\theta_W \sim 0.231$, the term involving Z nearly cancels numerically and the differential cross section is approximately proportional only to N^2.
CENNS is a Standard Model (SM) process that has never been observed experimentally (see **Note** in [3]) primarily because the nuclear recoil energies E_{rec} are very small ($\lesssim$ 15 keV), and escape the capabilities of most detectors.

Charge Coupled Devices (CCD) with very low thresholds (~10 eV) and relatively large sensitive mass (~5 g per CCD), have been successfully used in the DAMIC experiment [4] at SNOLAB to perform direct DM searches in the low mass region < 10 GeV/c^2, demonstrating the unique capabilities of CCDs for the detection of low-energy recoils.
This channel could be used as a probe to study oscillations between active and sterile states [5].

2. The CONNIE experiment

The **Co**herent **N**eutrino **N**ucleus **I**nteraction **E**xperiment (**CONNIE**) aims to detect for the first time the CENNS using an array of CCDs installed at 30 m from the Angra-II reactor at the Angra dos Reis nuclear power plant in Brazil. A prototype [6] was installed in 2014. The detector was upgraded to 80 g of mass (14 CCDs) in 2016.

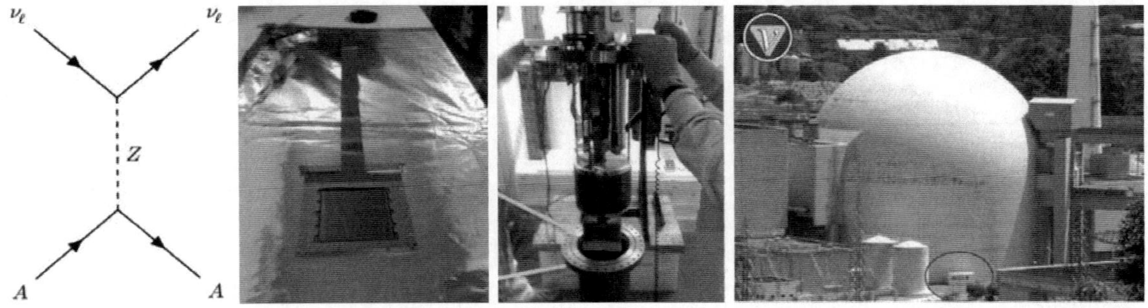

Figure 1: (a) Feynmann diagram for CENNS. (b) 4kx4k pixels CCD on Si substrate and copper frame. Wire-bonded to flex circuit. (c) The CCDs are stacked inside a vacuum vessel (~ 10^{-7} torr). (d) Localization of the CONNIE detector in Angra II.

The Angra-II reactor is a 3.95 GW$_{\rm th}$ PWR reactor that generates ~8.7×10^{20} $\bar{\nu}_e$/s and a flux at the detector location of ~$7.7\times10^{12}\bar{\nu}_e$ cm^{-2} s^{-1}.

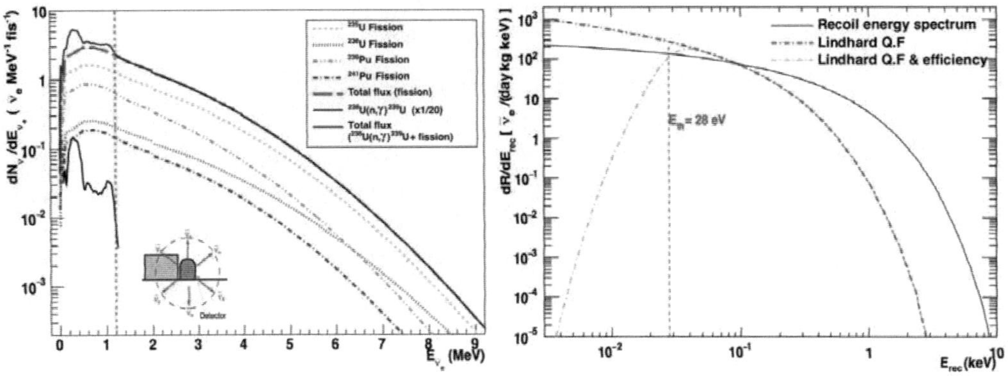

Figure 2: Left: Reactor neutrino flux (normalized for Angra-II). The vertical dotted line at ~1.2 MeV indicates the value for the maximum neutrino energy that contributes to the neutron capture process. Right: Expected CENNS energy spectrum in a Si detector.

3. Expected event rates

Assuming a 1 kg detector, an energy threshold of $\sim$ 28 eV, and a Lindhard quenching factor gives an expected rate of $\sim$16.1 events kg^{-1} day^{-1} [7].

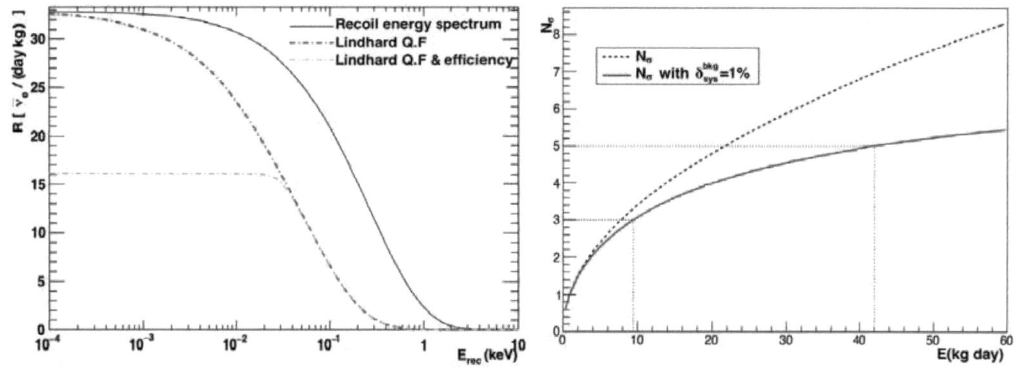

Figure 3: Left: Total CENNS event rate in a 1 kg Si detector as a function of the threshold. The effect of the quenching factor and efficiency are shown. Right: Significance of a CENNS signal as a function of the exposure E.

It is possible to achieve a 5σ significance (figure 3, right) with an exposure of 42 kg day: a 50 g detector operating for 840 days or a 100 g detector operating for 420 days or a 1 kg detector for 42 days.

4. Sterile neutrino oscillations. 3+1 Model

With a fourth light sterile neutrino (3+1 model), the mix between flavor eigenstates ν_α ($\alpha = e, \mu, \tau, s$) and mass eigenstates ν_i ($i = 1, 2, 3, 4$) is $\nu_\alpha = \sum_i U_{\alpha i}\nu_i$, where the mixing matrix is now 4×4, including new mixings with the sterile state $U_{\alpha i}$.
Under the aproximation $1\text{eV}^2 \sim |\Delta m^2_{41}| >> |\Delta m^2_{31}| >> |\Delta m^2_{21}|$, one can choose L/E such that $\Delta m^2_{31}/2E$, and $\Delta m^2_{21}/2E$ are $<< 1$. The survival probability of electron neutrinos in this experiment is:

$$P(\nu_e \to \nu_{\text{active}}) = 1 - P(\nu_e \to \nu_{\text{s}}) \tag{2}$$

$$= 1 - \sin^2 2\theta_{es} \sin\left(\frac{1.27\Delta m^2 L}{E}\right) \tag{3}$$

where $\sin^2 2\theta_{es} = 4|U_{e4}|^2|U_{s4}|^2$. In the case $\theta_{24} = \theta_{34} = 0$, $\sin^2 2\theta_{es} = \sin^2 2\theta_{14}$. This is the relevant probability since the detector is sensitive to all three active neutrino flavors through the CENNS.

5. Analysis

We divided the energy spectrum (figure 5, left) in 20 logarithmical uniform bins from 0.01 keV to 0.5 keV and defined the statistic:

$$\chi^2(\sin^2 2\theta, \Delta m^2) = \sum_i^{nb} = \frac{(D_i - P_i(\sin^2 2\theta, \Delta m^2))^2}{(\sigma_i^{\text{stat}})^2 + (\sigma_i^{\text{sys}})^2} \tag{4}$$

where D_i are the observed events and P_i are the predicted events. The observed and predicted events have a signal and background component:

$$D_i = S_i^D + B_i^D, \quad P_i(\sin^2 2\theta, \Delta m^2) = S_i^P(\sin^2 2\theta, \Delta m^2) + B_i^P \tag{5}$$

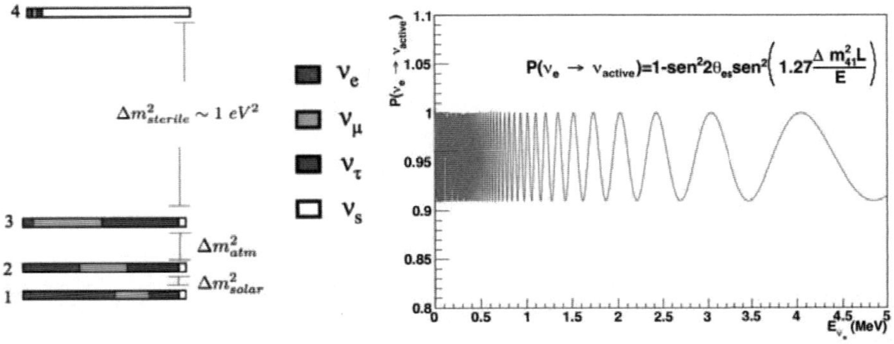

Figure 4: Left: 3+1 model direct mass ordering scheme. Right Survival Probability $\nu_e \rightarrow \nu_{\text{active}}$

The background was assumed to be uniform in energy at ~600 events kg^{-1} day^{-1}, demonstrated for CCDs at sea level. We consider two sources of systematic error σ_i^{sys}: an uncertainty on the background level B_i^P, δ_B^{sys}, and a contribution from the flux prediction δ_{flux}^{sys} into the signal S_i^P. We assumed a δ_B^{sys} around ~0.5%, and a δ_{flux}^{sys} around ~5%. (fig. 5 right).

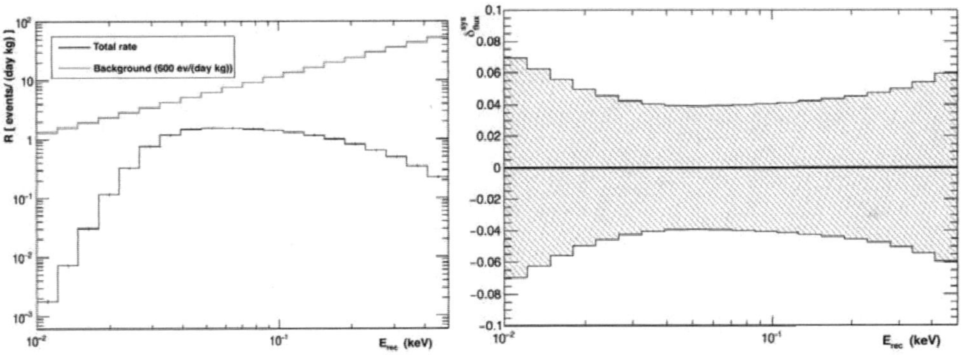

Figure 5: Left: Signal and background energy spectra used in the analysis. Right: Effect of the systematic error δ_{flux}^{sys} on the signal spectrum.

We constructed a reticule of 50×50 points in the parameter space. To calculate the sensitivity of a 1kg CONNIE detector we made the assumption that:

$$S_i^D = S_i^P(0,0), \quad B_i^D = B_i^P$$

which makes $\chi^2_{min} = 0$. The 90% C.L region was extracted from the excursion $\chi^2 = 4.6$.

6. Results

The figure 6 shows the expected sensitivity for different exposures. In the case $\theta_{24} = \theta_{34} = 0$ this can be compared with the Daya Bay and Bugey limits [8].

7. Conclusions

The sensititivity of a CCD experiment like CONNIE to sterile neutrino oscillations in a 3+1 model was calculated. A 10 kg year exposure is competitive in the region $\Delta m^2 > 1\text{eV}$ with the

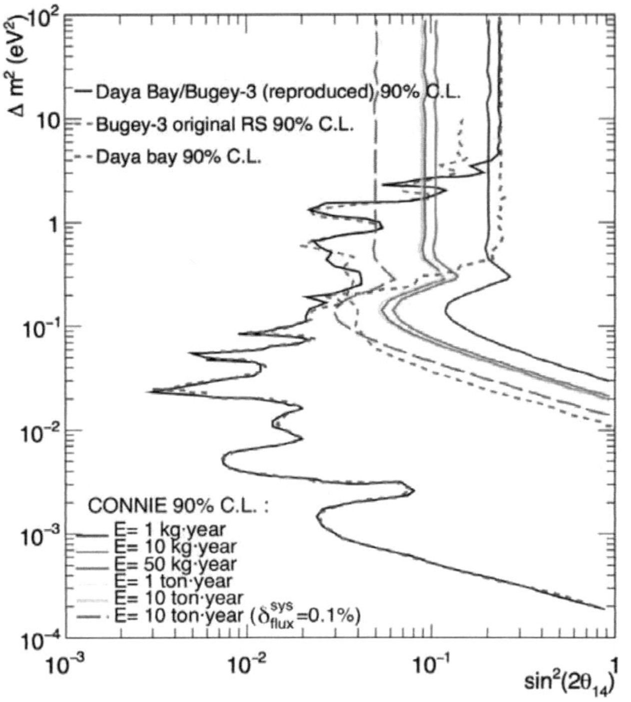

Figure 6: 90% C.L. sensitivity for a CONNIE experiment with exposure E of 1, 10, 50 kg·year and 10 ton·year assuming $\theta_{24} = \theta_{34} = 0$. The Daya Bay and Bugey limits are also shown.

recent Daya Bay limits assuming $\theta_{24} = \theta_{34} = 0$. This result can be combined with ν_μ and $\bar{\nu}_\mu$ like MINOS to constraint $\nu_\mu \to \nu_e$ and $\bar{\nu}_\mu \to \bar{\nu}_e$ transitions. This will be subject of a future work.

Acknowledgments

We acknowledge support from CONACYT through project No. 240666 and grant 398495 (scholarship holder No: 570620), and from DGAPA-UNAM, PAPIIT grant No. IN108917.

References

[1] K.N. Abazajian *et al.*, 2012. Light Sterile Neutrinos: A White Paper , FERMILAB-PUB-12-881-PPD (2012)
[2] D. Z. Freedman, 1974. Coherent neutrino nucleus scattering as a probe of the weak neutral current, *Phys. Rev.* D9, 1389-1392.
[3] **Note:** Since the time of this meeting the COHERENT Collaboration has reported the first observation of this process in D. Akimov *et al.*, 2017. Observation of coherent elastic neutrino-nucleus scattering, Science, 10.1126/science.aao0990
[4] A. Aguilar-Arevalo et al. [DAMIC Collaboration], 2016. Search for low-mass WIMPs in a 0.6 kg day exposure of the DAMIC experiment at SNOLAB.*Phys. Rev.* D **94** 082006
[5] M. Chávez-Estrada 2017 *Potencial de los dispositivos CCD para el estudio de oscilaciones a neutrinos estériles a través de la dispersión elástica coherente neutrino-núcleo.* Tesis de Maestría. Posgrado en Ciencias Físicas. (México: Universidad Nacional Autónoma de México.)
[6] A. Aguilar-Arevalo et al. [CONNIE Collaboration], 2016. Results of the engineering run of the Coherent Neutrino Nucleus Interaction Experiment (CONNIE). JINST 11 (2016) no.07, P07024
[7] G. Fernandez Moroni et al. 2015. Charge Coupled Devices for detection of coherent neutrino-nucleus scattering. *Phys.Rev.* D **91** 7, 072001
[8] P. Adamson et al. [Daya Bay, MINOS Collaboration], 2016. Limits on Active to Sterile Neutrino Oscillations from Disappearance Searches in the MINOS, Daya Bay, and Bugey-3 Experiments. *Phys. Rev. Lett.* **117**, 151801

Higgs decay mediated by top-quark with flavor-changing neutral scalar interactions

J. A. Orduz-Ducuara

División de Matemáticas e Ingeniería,
FES-Acatlán, UNAM
C.P. 53150,
Estado de México, México.

E-mail: `jaorduz@ciencias.unam.mx`

Abstract. We explore the flavor-changing parameters mediated by a Higgs boson within the THDM-III context. In particular, the $h \to t^*c$ processes, and check the high suppression for the FC in the THDM-III context for the low t_β parameters. Our exploration in the $\chi^u_{ij} - \chi^d_{ij}-$parameter space shows the allowed regions for different t_β values. We explored different modes for Higgs decays, considered the experimental constraints to get scattering plots for the FC parameters and some relevant decay modes. We expect future results to figure out the FC and its implications in the scalar sector.

Higgs, Flavor-Changing Neutral Scalar Interactions, Multi-Higgs

1. Introduction

Different experiments have examined the Standard Model observables which have good agreement with the theoretical results. Currently LHC explores the nature at energy scale of order TeV's and the SM works very well to analyze the structure of the matter. However we have some questions to solve: the matter-antimatter asymmetry, the CP violation and flavor-changing neutral currents (FCNC) mediated by gauge and scalar bosons. In this document, we shall discuss the last one where the neutral scalar boson can change the fermion flavor.

We shall discuss the Flavor-Changing Neutral Scalar Interactions (FCNS) in the Two-Higgs Doublet Model type III (THDM-III) context considering different scenarios for the parameters of the model. Previous studies have considered different versions for this THDM-III [1]. In fact, the total Higgs decay is given by [2]:

$$\Gamma_{total} \approx \Gamma(h \to b\bar{b}) + \Gamma(h \to c\bar{c}) + \Gamma(h \to WW^*) + \Gamma(h \to ZZ^*) + \Gamma(h \to c\bar{b}W^-); \qquad (1)$$

however the purpose of this paper is focused on flavor-changing mediated by Higgs bosons, in particular, we explore the $hct-$vertex, which is interesting for the last term in eq. (1).

Several reports have shown the following theoretical and experimental limit values for the branching ratios (Br) [3]: $\text{Br}\,(t \to uh) = 5.5 \times 10^{-6}$, $\text{Br}\,(t \to ch) = 1.5 \times 10^{-3}$, $\text{Br}\,(t \to u\gamma) \sim 10^{-6}$, $\text{Br}\,(t \to c\gamma) \sim 10^{-6}$, $\text{Br}\,(t \to uZ) \sim 10^{-7}$, $\text{Br}\,(t \to cZ) \sim 10^{-7}$.

Our calculation introduces the Passarino-Veltman functions implemented on Looptools, FeynCalc, FeynArts and SARAH [4–6].

IOP Conf. Series: Journal of Physics: Conf. Series **912** (2017) 012032 doi:10.1088/1742-6596/912/1/012032

This letter is organized as follows: section 2 shows the models and methods in the THDM-III context, section 3 we expose a results for the $h \to t^*c$ process and section 4 contains the conclusions.

2. Models and Methods

We shall consider the THDM-III where the most general potential is [7]

$$\begin{aligned} V(\Phi_1\Phi_2) &= m_{11}^2\Phi_1^\dagger\Phi_1 + m_{22}^2\Phi_2^\dagger\Phi_2 - [m_{12}^2\Phi_1^\dagger\Phi_2 + \text{h.c.}] + \tfrac{1}{2}\lambda_1(\Phi_1^\dagger\Phi_1)^2 + \tfrac{1}{2}\lambda_2(\Phi_2^\dagger\Phi_2)^2 \\ &+\lambda_3(\Phi_1^\dagger\Phi_1)(\Phi_2^\dagger\Phi_2) + \lambda_4(\Phi_1^\dagger\Phi_2)(\Phi_2^\dagger\Phi_1) + \Big\{\tfrac{1}{2}\lambda_5(\Phi_1^\dagger\Phi_2)^2 + \big[\lambda_6(\Phi_1^\dagger\Phi_1) \\ &+\lambda_7(\Phi_2^\dagger\Phi_2)\big]\Phi_1^\dagger\Phi_2 + \text{h.c.}\Big\}. \end{aligned}$$

In a general way, the Yukawa sector for the THDM-III is given by [8]

$$\begin{aligned} \mathcal{L}_n^{THDM-III} &= \frac{g}{2}\left(\frac{m_i}{m_W}\right)\bar{d}_i\left[\frac{\cos\alpha}{\cos\beta}\delta_{ij} + \frac{\sqrt{2}\sin(\alpha-\beta)}{g\cos\beta}\left(\frac{m_W}{m_i}\right)\left(\tilde{Y}_2^d\right)_{ij}\right]d_jH^0 \\ &+ \frac{g}{2}\left(\frac{m_j}{m_W}\right)\bar{d}_i\left[-\frac{\sin\alpha}{\cos\beta}\delta_{ij} + \frac{\sqrt{2}\cos(\alpha-\beta)}{g\cos\beta}\left(\frac{m_W}{m_i}\right)\left(\tilde{Y}_2^d\right)_{ij}\right]d_jh^0 \\ &+ \frac{ig}{2}\left(\frac{m_i}{m_W}\right)\bar{d}_i\left[-\tan\beta\delta_{ij} + \frac{\sqrt{2}}{g\cos\beta}\left(\frac{m_W}{m_i}\right)\left(\tilde{Y}_2^d\right)_{ij}\right]\gamma^5 d_jA^0 \\ &+ \frac{g}{2}\left(\frac{m_i}{m_W}\right)\bar{u}_i\left[\frac{\sin\alpha}{\sin\beta}\delta_{ij} + \frac{\sqrt{2}\sin(\alpha-\beta)}{g\sin\beta}\left(\frac{m_W}{m_i}\right)\left(\tilde{Y}_2^u\right)_{ij}\right]u_jH^0 \\ &+ \frac{g}{2}\left(\frac{m_u}{m_W}\right)\bar{u}_i\left[-\frac{\cos\alpha}{\sin\beta}\delta_{ij} + \frac{\sqrt{2}\cos(\alpha-\beta)}{g\sin\beta}\left(\frac{m_W}{m_i}\right)\left(\tilde{Y}_2^u\right)_{ij}\right]u_jh^0 \\ &+ \frac{ig}{2}\left(\frac{m_u}{m_W}\right)\bar{u}_i\left[-\cot\beta\delta_{ij} + \frac{\sqrt{2}}{g\sin\beta}\left(\frac{m_W}{m_i}\right)\left(\tilde{Y}_2^u\right)_{ij}\right]\gamma^5 u_jA^0. \end{aligned} \tag{2}$$

Where the (2) is the Lagrangian density for the fermion-fermion-ϕ, and $\phi - A^0, H^0, h^0$ which are the pseudoscalar, heavy Higgs and standard model Higgs, and the superscript (zero) labels the mass eigenstates.

One can rewrite the Yukawa couplings as $\widetilde{Y}_1^f = \sqrt{2}\frac{M_f}{v\cos\beta} - \widetilde{Y}_2^f\ t_\beta$; because we explore a simple model where the flavor-changing is due to the χ_f^i-parameters, and it is simple if we reduce the number of those parameters, which are associated to the Yukawa couplings; namely $Y_{ij}^f = \sqrt{2}\frac{\sqrt{m_im_j}}{v}\chi_{ij}^f$, where the m_i and m_j are the fermion masses and χ_{ij}^f are the dimensionless parameters that will probe the flavor-changing mediated by scalar bosons.

We explore the THDM-III because it is possible to have mixing between the fermion flavors at tree level. For the purpose of this paper, we introduce $B \to X_s\gamma$ processes considering $\Gamma(B \to X_s\gamma) \simeq \Gamma(b \to s\gamma)$, since the non-perturbative effects are small [9]. We consider the constraints coming from $t \to cV, b \to s\gamma, h \to l_il_j, h \to \gamma Z$, to show the correlation between branching ratios $b \to s\gamma$ and $h \to \gamma Z$, and found excluded regions (e. g. figs. 3-4).

3. Results

Figs. 3 and 4 show a correlation between parameters of the model, considering different constraints coming from t and b-quark and h decays [10].

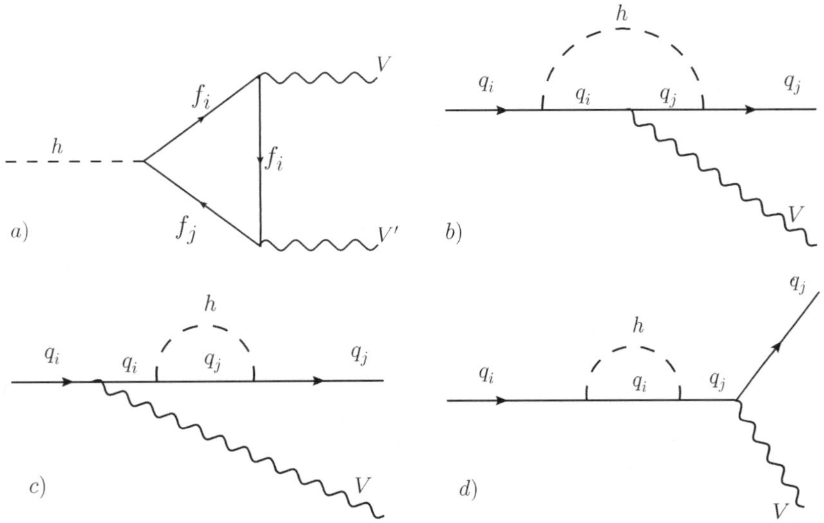

Figure 1. The fig. a) shows the Feynman diagrams for $h \to VV'$ considered to constrain our parameter analysis. The figs. c)-d) are the Feynman diagram for the $q_i q_j V$ process at one-loop level with a scalar as a flavor-changing mediator.

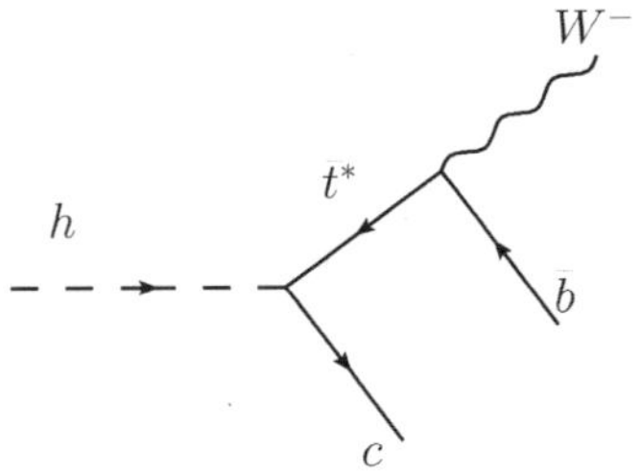

Figure 2. The Feynman diagram for the $h \to t^* c$.

The process that is represented in fig. 2 was proposed by ref. [2] to explore the flavor-changing modes.

Our analysis shows that it could be possible to have FC if we consider the quark type separately, since the parameter space is wider for the u-quark type. We find that if $t_\beta \lesssim 10^{-3}$ then $-200 \lesssim \chi_{ij}^u \lesssim 200$, while $t_\beta \lesssim 10^{-4}$ then $-200 \lesssim \chi_{ij}^d \lesssim 200$. Those figs. (3 and 4) show that $\chi_{ij}^d - t_\beta$ is more suppressed than the $\chi_{ij}^u - t_\beta$ for the lowest t_β values.

In figs 5-6, the dark regions represent the highly allowed region. We generated randomly the parameter set, as follow, $-200 \leq \chi_{ij}^{u,d} \leq 200, 0 \leq t_\beta \leq 100, 350 \text{ GeV} \leq m_{H,H^\pm} \leq 1000 \text{ GeV}$, as well considering the experimental bounds for the t, b and h branching ratios at tree and one-loop level.

If we consider the W^- decays to $\nu_l l$ then we obtain $\text{Br}(h \to llq\bar{q})$ of order 10^{-4}, therefore our results are interesting if we compare to the experimental results $\left(\text{Br}(h \to llq\bar{q}) \sim 10^{-2} - 10^{-3}\right)$ as is shown in ref. [11] for $m_h = 125$ GeV. We found $\text{Br}(h \to t^* c) \sim 10^{-3}$ for the $1 \lesssim t_\beta \lesssim 20$ (fig. 7), which is a very interesting channel to explore in the LHC or the next generation of colliders.

Fig. 7 shows an interesting channel ($\text{Br}(h \to t^* c)$) to probe new physics and we expect the

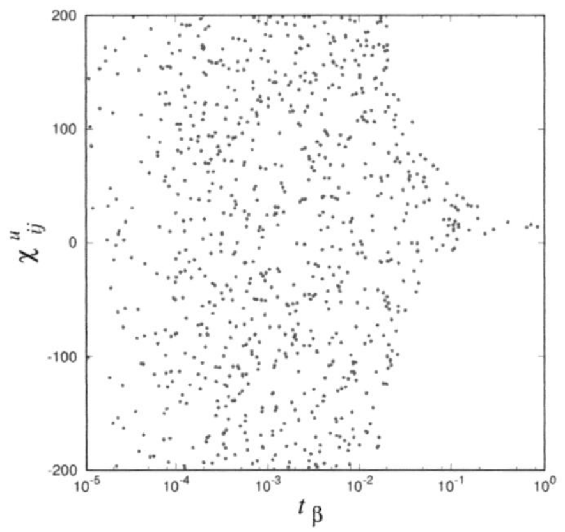

Figure 3. Scattering plot for the flavor parameters χ^u_{ij} versus t_β.

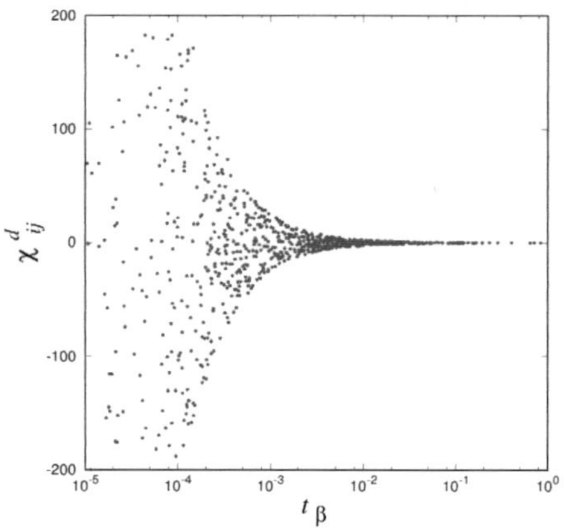

Figure 4. Scattering plot for the flavor parameters χ^d_{ij} versus t_β.

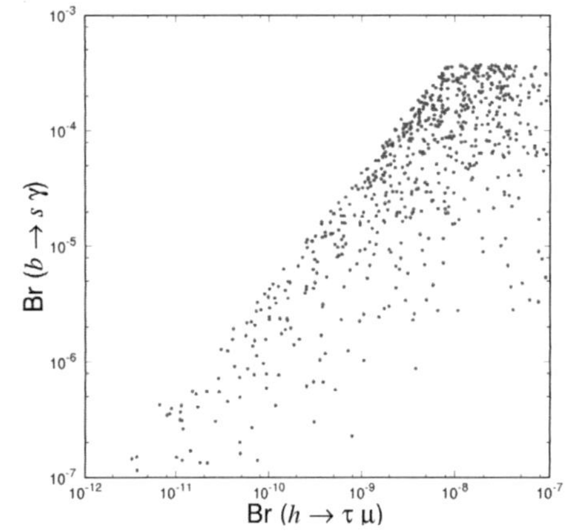

Figure 5. Scattering plot for the branching ratios of the processes $b \to s\gamma$ versus $h \to \tau\mu$.

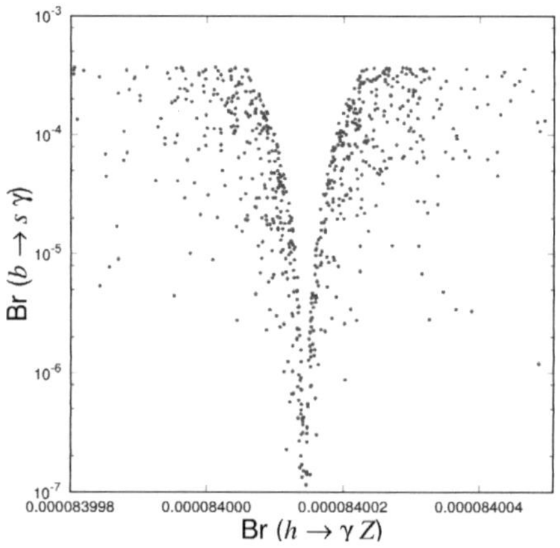

Figure 6. Scattering plot for the branching ratios of the processes $b \to s\gamma$ versus $h \to \gamma Z$.

next experimental results to test the THDM-III as the simplest SM extension.

4. Conclusions

We have explored the $h \to t^*c$ processes, and checked the high suppression considering FC mediated by Higgs boson.

Our exploration in the $\chi^u_{ij} - \chi^d_{ij}$−parameter space showed the allowed regions for different t_β values (see figs. 3-4). Besides we explore the different modes for Higgs decays to (see figs. 5-6). Our results showed $\mathrm{Br}(h \to \mu\tau) \lesssim 10^{-5}$ and $\mathrm{Br}(h \to \gamma Z) \sim 10^{-6}$ as long as $\mathrm{Br}(b \to s\gamma) \lesssim 10^{-4}$ as is shown in figs. 5-6, those engaging values for exploring in LHC. Besides we predicted $\mathrm{Br}(h \to t^*c) \sim 10^{-3}$ for $1 \lesssim t_\beta \lesssim 20$, this a feasible channel to explore our model in the LHC.

Acknowledgments

I acknowledge support from CONACYT-SNI (Mexico).

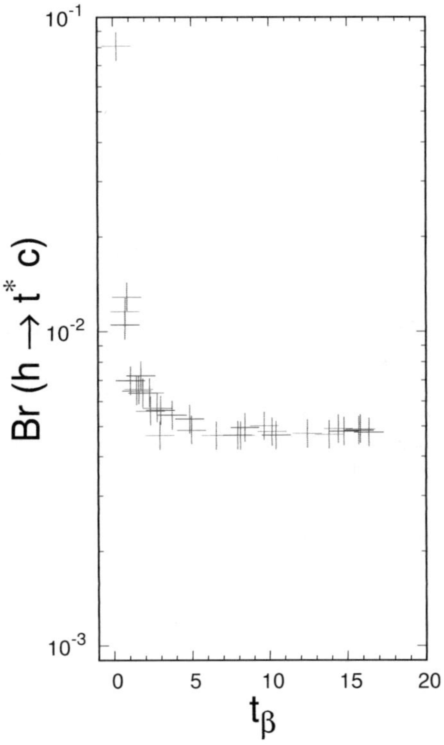

Figure 7. The Br for the process $h \to t^{*}c$ that we have predicted.

References

[1] Arroyo M, Diaz-Cruz J L, Diaz E and Orduz-Ducuara J A 2016 *Chin. Phys.* **C40** 123103 (*Preprint* 1306.2343)
[2] Diaz-Cruz L, Diaz-Furlong A, Gaitán-Lozano R and Montes de Oca J 2012 (*Preprint* 1203.6893)
[3] collaboration T A (ATLAS) 2013 ATLAS-CONF-2013-081
[4] Hahn T and Perez-Victoria M 1999 *Comput. Phys. Commun.* **118** 153–165 (*Preprint* hep-ph/9807565)
[5] Hahn T 2001 *Comput. Phys. Commun.* **140** 418–431 (*Preprint* hep-ph/0012260)
[6] Staub F 2015 *Adv. High Energy Phys.* **2015** 840780 (*Preprint* 1503.04200)
[7] Gunion J F and Haber H E 2003 *Phys. Rev.* **D67** 075019 (*Preprint* hep-ph/0207010)
[8] Diaz-Cruz J L, Noriega-Papaqui R and Rosado A 2005 *Phys. Rev.* **D71** 015014 (*Preprint* hep-ph/0410391)
[9] El-Khadra A X and Luke M 2002 *Ann. Rev. Nucl. Part. Sci.* **52** 201–251 (*Preprint* hep-ph/0208114)
[10] Gaitán R, Montes de Oca J and Orduz-Ducuara J 2017 *Progress of Theoretical and Experimental Physics* **2017** 073B02
[11] Dittmaier S *et al.* (LHC Higgs Cross Section Working Group) 2011 (*Preprint* 1101.0593)

Minimal supersymmetric B – L extension of the standard model, heavy *H* and light *h* Higgs boson production and decay at future e^+e^- linear colliders

F Ramírez-Sánchez[1], A Gutierrez-Rodríguez[2] and M A Hernández-Ruiz[3]

[1] Facultad de Física, Universidad Autónoma de Zacatecas, Apartado Postal F-630, 99000 Fresnillo, Zac. México.
[2] Facultad de Física, Universidad Autónoma de Zacatecas, Apartado Postal C-580, 98060 Zacatecas, Zac. México.
[3] Facultad de Ciencias Químicas, Universidad Autónoma de Zacatecas, Apartado Postal C-580, 98060 Zacatecas, Zac. México.

E-mail: [1]frank.ramirez@fisica.uaz.edu.mx, [2]alexgu@fisica.uaz.edu.mx and [3]mahernan@uaz.edu.mx

Abstract. We study the phenomenology of the light h and heavy H Higgs boson production and decay in the context of a $U(1)_{B-L}$ extension of the standard model with an additional Z' boson at future e^+e^- linear colliders with center-of-mass energies of $\sqrt{s} = 500 - 3000$ GeV and integrated luminosities of $\mathcal{L} = 500 - 2000$ fb^{-1}. The study includes the processes $e^+e^- \rightarrow (Z, Z') \rightarrow Zh$ and $e^+e^- \rightarrow (Z, Z') \rightarrow ZH$, considering both the resonant and non-resonant effects. We find that the total number of expected Zh and ZH events can reach 10^6 and 10^5, respectively, which is a very optimistic scenario allowing us to perform precision measurements for both Higgs bosons h and H, as well as for the Z' boson in future high-energy and high-luminosity e^+e^- colliders.

1. Introduction

The $U(1)_{B-L}$ model [1–5] is one of the simplest extensions of the Standard Model (SM) with an extra $U(1)$ local gauge symmetry [6], where B–L represents baryon minus lepton number. This symmetry plays an important role in various physics scenarios beyond the SM: a) The gauge $U(1)_{B-L}$ symmetry group is contained in the Grand Unification Theory (GUT) described by a $SO(10)$ group [1]. b) The scale of the B–L symmetry breaking is related to the mass scale of the heavy right-handed Majorana neutrino mass terms and provides the well-known see-saw mechanism [7–11] to explain light left-handed neutrino mass. (c) The B–L symmetry and the scale of its breaking are tightly connected to the baryogenesis mechanism through leptogenesis [12]. The model also contains an extra gauge boson Z' corresponding to B–L gauge symmetry and an extra SM singlet scalar (heavy Higgs boson H). This may change the SM phenomenology. Therefore, another Higgs factory besides the LHC, such as the ILC (International Linear Collider) and CLIC (Compact Linear Collider), which can precisely determine the properties of the Higgs bosons h and H, can be an important future step in high-energy and high-luminosity physics exploration.

The Higgs-Strahlung process $e^+e^- \to Zh$ [13-17] is one of the main production mechanisms of the Higgs boson in e^+e^- linear colliders. After the discovery of the Higgs boson, detailed experimental and theoretical studies are necessary for checking its properties [18-21]. It is possible to search for the Higgs boson in the framework of the B–L model; however, the existence of a new gauge boson could also provide new Higgs particle production mechanisms that could prove its non-standard origin.

We examine a variety of h, H decay channels while we concentrate on the $e^+e^- \to Zh$ and ZH production modes, including the possibility of Z' mediation, which could be resonant, as we allow for Z/Z' mixing. The B–L model [22-23] is attractive due to its relatively simple theoretical structure. The crucial test of the model is the detection of the new heavy neutral Z' gauge boson and the new Higgs boson H.

2. Brief review of the $U(1)_{B-L}$ theoretical model

The solid evidence for the non-vanishing neutrino masses has been confirmed by various neutrino oscillation phenomena [32-33] and indicates the evidence of new physics beyond the SM. The most attractive idea to naturally explain the tiny neutrino masses is the see-saw mechanism [8-10], in which the right-handed (RH) neutrinos singlet under the SM gauge group is introduced. The gauged $U(1)_{B-L}$ model based on the gauge group $SU(3)_C \times SU(2)_L \times U(1)_Y \times U(1)_{B-L}$ [7] is an elegant and simple extension of the SM in which the RH heavy neutrinos arise, associated with the $U(1)_{B-L}$ gauge symmetry breaking.

We consider a $SU(3)_C \times SU(2)_L \times U(1)_Y \times U(1)_{B-L}$ model, which is one of the simplest extensions of the SM [4, 5, 7, 22], where $U(1)_{B-L}$, represents the additional gauge symmetry. The gauge invariant Lagrangian of this model is given by:

$$\mathcal{L} = \mathcal{L}_s + \mathcal{L}_{YM} + \mathcal{L}_f + \mathcal{L}_Y \tag{1}$$

where $\mathcal{L}_s$, $\mathcal{L}_{YM}$, $\mathcal{L}_f$, $\mathcal{L}_Y$ are the scalar, Yang-Mills, fermion and Yukawa sector, respectively.

The model consists of one doublet Φ and one singlet χ and we briefly describe the lagrangian including the scalar, fermion and gauge sector. The Lagrangian for the gauge and scalar sector is given by [4, 24]:

$$\mathcal{L}_g = -\frac{1}{4}W^a_{\mu\nu}W^{a\mu\nu} - \frac{1}{4}B_{\mu\nu}B^{\mu\nu} - \frac{1}{4}B'_{\mu\nu}B'^{\mu\nu} \tag{2}$$

where we see the corresponding field strength tensors for $SU(2)_L \times U(1)_Y \times U(1)_{B-L}$ respectively.

The Lagrangian for the scalar sector of the model is:

$$\mathcal{L}_s = (D^\mu\Phi)^\dagger(D_\mu\Phi) + (D^\mu\chi)^\dagger(D_\mu\chi) - V(\Phi,\chi) \tag{3}$$

here we can clearly see the added χ field of our model, and where our new potential term is [25]:

$$V(\Phi,\chi) = m^2(\Phi^\dagger\Phi) + \mu^2|\chi|^2 + \lambda_1(\Phi^\dagger\Phi)^2 + \lambda_2|\chi|^4 + \lambda_3(\Phi^\dagger\Phi)|\chi|^2 \tag{4}$$

with Φ and χ as the complex scalar Higgs doublet and singlet fields, respectively. The covariant derivative is given by [25-27]:

$$D^\mu = \partial_\mu + ig_s\, t^a G^a_\mu + i\left[gT_a \boldsymbol{W}^a_\mu + g_1 Y B_\mu + (\tilde{g}Y + g'_1 Y_{B-L})B'_\mu\right] \tag{5}$$

where g_s, g, g_1 and g'_1 are the $SU(3)_C \times SU(2)_L \times U(1)_Y \times U(1)_{B-L}$ couplings with t^a, T_a, Y and Y_{B-L} being their respective group generators. The electromagnetic charges of the fields are the same as those of the SM and the new "hypercharges" Y_{B-L} are 1/3 for quarks, -1 for leptons, 0 and 2 for the Φ and χ fields respectively [4-5, 25-27], to preserve the gauge invariance of the model. The doublet Φ and singlet χ scalar fields before and after spontaneous symmetry breaking are given by:

$$\Phi = \begin{pmatrix} G^\pm \\ \frac{v+\phi^0+iG_Z}{\sqrt{2}} \end{pmatrix} \to \begin{pmatrix} 0 \\ \frac{v+\phi^0}{\sqrt{2}} \end{pmatrix} \qquad \chi = \left(\frac{v'+\phi'^0+iz'}{\sqrt{2}}\right) \to \left(\frac{v'+\phi'^0}{\sqrt{2}}\right) \tag{6}$$

where v' is the B–L symmetry breaking scale constrained by the electroweak precision measurements data whose value is assumed to be at least of the order of TeV.

After minimizing the potential around these values we see that the mass eigenstates are linear combinations of Φ^0 and Φ'^0 and written:

$$\begin{pmatrix} h \\ H \end{pmatrix} = \begin{pmatrix} cos\alpha & -sin\alpha \\ sin\alpha & cos\alpha \end{pmatrix} \begin{pmatrix} \phi^0 \\ \phi'^0 \end{pmatrix} \tag{7}$$

where h is the SM-like Higgs boson, H is an extra Higgs boson and the scalar mixing angle α ($-\pi/2 \leq \alpha \leq \pi/2$).

The extension we are studying is in the Abelian sector of the SM gauge group, so that the charged gauge bosons $W^{\pm}$ will have masses given by their SM expressions related to the $SU(2)_L$ factor only, the other gauge boson masses are not so simple to identify because of mixing. In fact, analogously to the SM, the fields of definite mass are linear combinations of W_μ^3, B^μ, and B'^μ, the relation between the neutral gauge bosons and the corresponding mass eigenstates is given by [22, 23, 26, 27];

$$\begin{pmatrix} B^\mu \\ W^{3\mu} \\ B'^\mu \end{pmatrix} = \begin{pmatrix} cos\theta_w & -sin\theta_w cos\theta_{B-L} & sin\theta_w sin\theta_{B-L} \\ sin\theta_w & cos\theta_w cos\theta_{B-L} & -cos\theta_w sin\theta_{B-L} \\ 0 & sin\theta_{B-L} & cos\theta_{B-L} \end{pmatrix} \begin{pmatrix} A^\mu \\ Z^\mu \\ Z'^\mu \end{pmatrix} \tag{8}$$

with θ_{B-L} ($-\pi/4 \leq \theta_{B-L} \leq \pi/4$).

3 The Higgs strahlung process $e^+e^- \to Z\,h$ and $e^+e^- \to Z\,H$ in the B - L model

We calculate the Higgs production cross section σ via the Higgs-strahlung process in the context of the B − L model considering both, a Z and a Z' mediator.

The Feynman diagrams contributing to the processes $e^+e^- \to (Z, Z') \to Zh$ and $e^+e^- \to (Z, Z') \to ZH$ are given in figure 1.

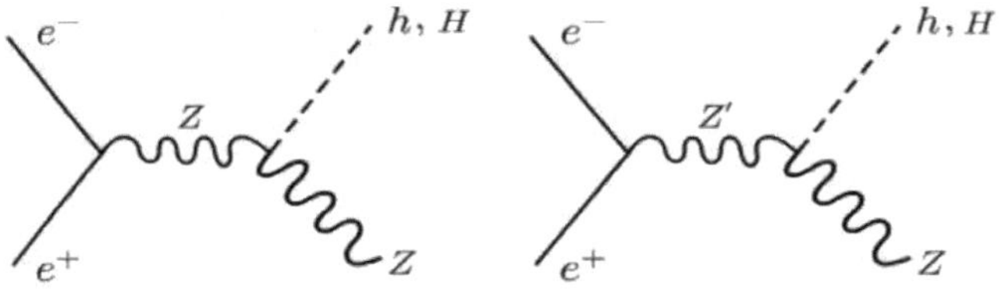

Figure 1
Feynman diagrams for the processes $e^+e^- \to (Z, Z') \to Zh$ and $e^+e^- \to (Z, Z') \to ZH$

Here we present the cross sections σ for the standard model Higgs boson h;

$$\sigma_Z(e^+e^- \to Zh) = \frac{G_F^2 M_Z^4 cos^2\alpha (g_v^{e2} + g_a^{e2}) s\sqrt{\lambda}}{\left(24\pi (s - M_Z^2)^2 + M_Z^2 \Gamma_Z^2\right)} \left(\lambda + \frac{12 M_Z^2}{s}\right) \tag{9}$$

$$\sigma_{(Z')}(e^+e^- \to Zh) = \frac{G_F^2 M_Z^6}{24\pi}((g_v'^e)^2 + (g_a'^e)^2) \frac{s\sqrt{\lambda}[\lambda + 12 M_{Z'}^2/s]}{M_{Z'}^2\left[\left[s - M_{Z'}^2\right]^2 + M_{Z'}^2 \Gamma_{Z'}^2\right]} \; [f(\theta')cos\alpha + g(\theta')sen\alpha]^2 \tag{10}$$

$$\sigma_{(z,z')} = \frac{G_F^2 M_Z^6 cos\alpha}{6\pi}(g_v'^e g_v^e + g_a'^e g_a^e) s\sqrt{\lambda}\left[\frac{1}{M_Z^2}\left(\lambda + 12\frac{M_Z^2}{s}\right) + \frac{1}{M_{Z'}^2}(\lambda + 6 + \frac{s\lambda}{8 M_Z^2 M_{Z'}^2}\left(\lambda - 12\frac{M_Z^2}{s}\right)\right] \frac{\left[(s - M_Z^2)(s - M_{Z'}^2) + M_Z M_{Z'} \Gamma_Z \Gamma_{Z'}\right]}{\left[(s - M_Z^2)^2 + M_Z^2 \Gamma_Z^2\right]\left[\left(s - M_{Z'}^2\right)^2 + M_{Z'}^2 \Gamma_{Z'}^2\right]} [f(\theta')cos\alpha + g(\theta')sen\alpha] \tag{11}$$

With

$$\lambda(M_Z^2, s, M_h^2) = \left(1 - \frac{M_Z^2}{s} - \frac{M_h^2}{s}\right)^2 - 4\frac{M_Z^2 M_h^2}{s^2} \tag{12}$$

the usual two-particle phase space function.

The expression given in equation (9) corresponds to the cross section with the exchange of the Z boson, while equations (10) and (11) come from the contributions of the B–L model and of the interference, respectively.

4 Results and conclusions

We evaluate the total cross section σ of the Higgs-strahlung process $e^+e^- \to (Z, Z') \to Zh$ in the B-L model using these values for our computation [31] ; $\sin^2\theta_W = 0.23126 \pm 0.00022$, $m_\tau = 1776.82 \pm 0.16$ MeV, $m_b = 4.18 \pm 0.06$ GeV, $m_t = 172.44 \pm 0.13$ GeV, $M_W = 80.389 \pm 0.023$ GeV, $M_Z = 91.1876 \pm 0.0021$ GeV, $\Gamma_Z = 2.4952 \pm 0.0023$ GeV, $M_h = 125.09 \pm 0.4$ GeV. If we consider the most recent limit from $\frac{M_{Z'}}{g'_1} \geq 6.9$ TeV [28-30], it is possible to obtain a direct bound on the B–L breaking scale v', and take $v' = 3.45$ TeV and $\alpha = \pi/9$ in our numerical analysis. We will assume $\sqrt{s}$, $M_{Z'}$, g'_1, $\theta_{B\text{-}L}$ as free parameters.

In figure 2, we show the cross section σ $(e^+e^- \to Zh)$, for different contributions as a function of the center-of-mass energy $\sqrt{s}$ for $\theta_{B-L} = 10^{-3}$ and $g'_1 = 0.290$: the solid line corresponds to the SM and the dashed line corresponds to σ_Z where the $U(1)_{B-L}$ model contributes to the couplings g^f_V and g^f_A of the SM gauge boson Z to electrons. The dot-dashed line corresponds to $\sigma_{z'}$, which is only the B-L contribution, the dot dot-dashed line corresponds to the interference $\sigma_{z,z'}$. Finally, the dotted line corresponds to the total cross section, $\sigma_{Tot.}$. We can see that the cross section corresponding to $\sigma_Z(e^+e^- \to Zh)$ decreases for large $\sqrt{s}$, whereas the cross section of the B-L model and the total cross section, there is an increase for large values of the center-of-mass energy, reaching its maximum value at the resonance of Z', $\sqrt{s} = 2000$ GeV.

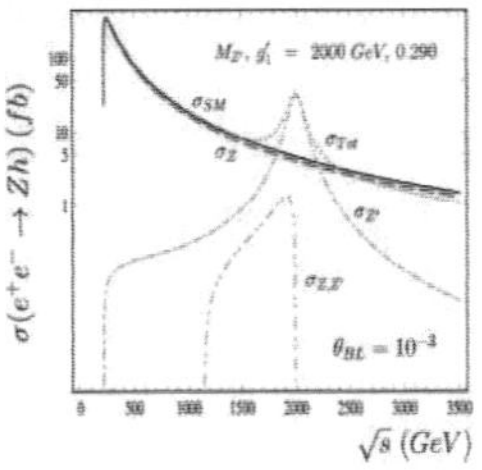

Figure 2

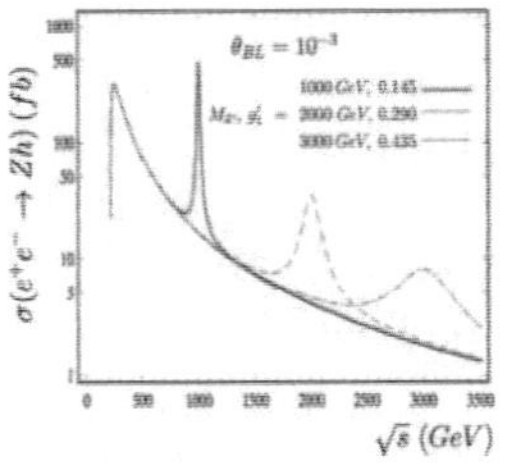

Figure 3

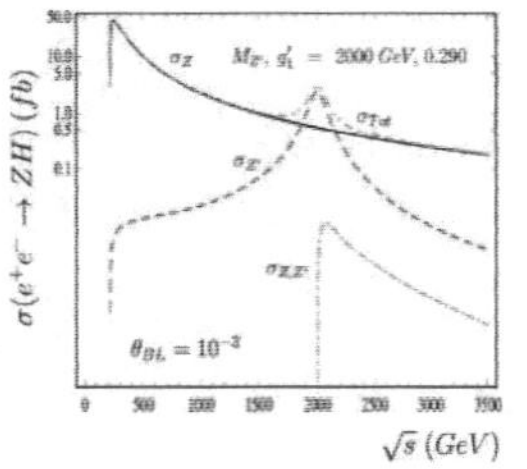

Figure 4

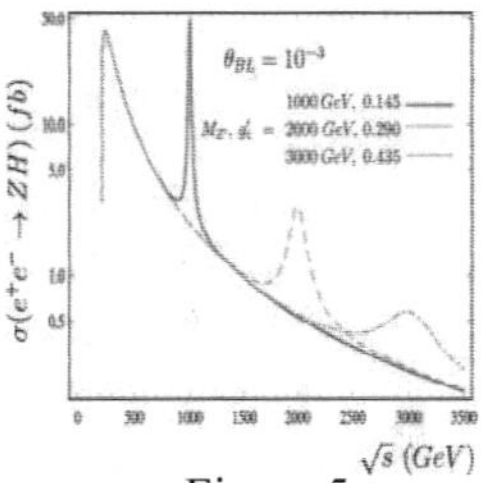

Figure 5

Total cross sections of the processes, $e^+e^- \to Zh$ (2, 3), and $e^+e^- \to ZH$ (4, 5) as a function of the center of mass energies $\sqrt{s}$, with $\theta_{B-L} = 10^{-3}$ for different values of $M_{Z'}$ and g'_1, including both the resonant and non-resonant effects

We plot the total cross section of the reaction $e^+e^- \to Zh$ in figure 3 as a function of the center-of-mass energy $\sqrt{s}$, for values of the heavy gauge boson mass $M_{Z'} = 1000, 2000, 3000$ GeV and $g'_1 = 0.145, 0.290, 0.435$, respectively. $M_{Z'}$ and g'_1 maintain the previous relationship. Here we see that the cross section is sensitive to the free parameters and also that the height of the resonance peaks for the boson Z' changes depending on the value of $\sqrt{s} = M_Z'^2$. In addition, the resonances are broader for larger g'_1 values, as the total width of the Z' boson increases with g'_1.

The total cross section for the production processes $e^+e^- \to ZH$ as a function of the collision energy for $M_h = 125$ GeV, $M_H = 800$ GeV, $M_{\nu R} = 300$ GeV, $M_Z' = 2000$ GeV and $g'_1 = 0.290$ is shown in figure 4. Here the curves are for $\sigma_Z(e^+e^- \to ZH)$ (solid line), $\sigma_{Z'}(e^+e^- \to ZH)$ (dashed line), $\sigma_{Z,Z'}(e^+e^- \to ZH)$ (dot-dashed line), and the dot dot-dashed line corresponds to the total cross section of the process $\sigma_{\text{Tot}}(e^+e^- \to ZH)$.

To see the effects of $\theta_{B\text{-}L}$, g'_1, $M_{Z'}$, the free parameters of the B-L model, we plot the total cross section of the process $e^+e^- \to ZH$ in figure 5 as a function of the center-of mass energy $\sqrt{s}$ for the values of the heavy gauge boson mass of $M_Z' = 1000$ GeV with $g'_1 = 0.145$, $M_Z' = 2000$ GeV with $g'_1 = 0.290$ and $M_{Z'} = 3000$ GeV with $g'_1 = 0.435$, preserving the relationship between $M_{Z'}$ and g'_1. In this figure we observed that for $\sqrt{s} = M_Z'$ the resonant effect dominates, the cross section is sensitive to the free parameters. We also observe that the height of the resonance peaks for the Z' boson change depending on the value of $\sqrt{s} = M^2_{Z'}$, and in addition, we see that the resonances are broader for larger g'_1 values, as the total width of the Z' boson increases with g'_1.

We see that the expected Z_h and Z_H events can reach a number of $\mathcal{O}$ (10^6 and 10^5), respectively, which is a very optimistic scenario and it would be possible to perform precision measurements for both Higgs bosons h and H, for the Z' heavy gauge boson, as well as for the parameters of the model θ_{B-L}, g'_1 and α in future high-energy and high-luminosity e^+e^- colliders experiments. In addition, the SM expression for the cross section of the reaction $e^+e^- \to Zh$ can be obtained in the decoupling limit, when θ_{B-L}, $g'_1 = 0$ and $\alpha = 0$. Our study complements other studies on the B−L model and on the Higgs-strahlung processes $e^+e^- \to (Z, Z') \to Zh$ and $e^+e^- \to (Z, Z') \to ZH$.

References

[1] Buchmuller W, Greub C and Minkowski P 1991 *Phys. Lett.* B **267** 395
[2] Marshak R and Mohapatra R N 1980 *Phys. Lett.* B **91** 222
[3] Mohapatra R N and Marshak R 1980 *Phys. Rev. Lett.* **44** 1316
[4] Khalil S 2008 *J. Phys. G: Nucl. Part. Phys.* G **35** 055001
[5] Eman W and Khalil 2007 S *Eur. Phys. J. C* **52** 625
[6] Carlson E D 1987 *Nucl. Phys* B **286** 378
[7] Mohapatra Rabindra N and Senjanovic G 1980 *Phys. Rev. Lett.* **44** 912
[8] Minkowski P 1977 *Phys. Lett.* B **67** 421
[9] Van Nieuwenhuizen P and Freedman D Z 1979 *Proc. Supergravity Conf.* (Stony Brook University) ed P Van Nieuwenhuizen and D Z Freedman (Amsterdam, Netherlands: North Holland Publishing Co.) p 341
[10] Yanagida T 1979 *Proc. Workshop on the Baryon Number of the Universe and Unified Theories* (Tsukuba, Japan, 13–14 February) p 95
[11] Gell-Mann M, Ramond P and Slansky R 1979 *Proc. Supergravity Conf.* (Stony Brook University, New York) ed P Van Nieuwenhuizen and D Z Freedman (Amsterdam, Netherlands: North Holland Publishing Co.) pp 315–21
[12] Fukugita M and Yanagida T 2003 *Physics of Neutrinos and Applications to Astrophysics* (Berlin: Springer)
[13] Ellis J, Gaillard M K and Nanopoulos D V 1976 *Nucl. Phys.* B **106** 292
[14] Ioffe B L and Khoze V A 1978 Sov. *J. Part. Nucl.* **9** 50
[15] Lee B W, Quigg C and Thacker H B 1977 *Phys. Rev.* D **16** 1519
[16] Bjorken J D 1976 *Proc. Summer Institute on Particle Physics* SLAC Report 198
[17] Barger V D *et al* 1994 *Phys. Rev.* D **49** 79
[18] Ellis J 2013 arXiv:1312.5672
[19] Dawson S *et al* 2013 arXiv:1310.8361
[20] Klute M *et al* 2013 *Europhys. Lett.* **101** 51001
[21] Behnke T *et al* 2013 arXiv:1306.6327 [physics.acc-ph]
[22] Basso L *et al* 2009 *Phys. Rev.* D **80** 055030
[23] Basso L *et al* 2009 *J. High Energy Phys.* JHEP10(2009)006
[24] Basso L 2011 arXiv:1106.4462 [hep-hp]
[25] Basso L, Moretti S and Pruna G M 2010 *Phys. Rev.* D **82** 055018
[26] Basso L *et al* 2011 *Eur. Phys. J.* C **71** 1613
[27] Basso L, Moretti S and Pruna G M 2012 *J. Phys. G: Nucl. Part. Phys.* G **39** 025004
[28] Carena M, Daleo A, Dobrescu B A and Tait T M 2004 *Phys. Rev.* D **70** 093009
[29] Heeck J 2014 *Phys. Lett.* B **739** 256
[30] Cacciapaglia G, Csaki C, Marandella G and Strumia A 2006 *Phys. Rev.* D **74** 033011
[31] Olive K A *et al* (Particle Data Group) 2014 *Chin. Phys.* C **38** 090001
[32] Fogli G L, Lisi E, Marrone A, and Palazzo A 2006 *Prog. Part. Nucl. Phys.* **57** 742
[33] Fogli G L *et al* 2007 *Phys. Rev.* D **75** 053001

IOP Conf. Series: Journal of Physics: Conf. Series **912** (2017) 012034 doi:10.1088/1742-6596/912/1/012034

Invariant mass reconstruction of the η meson in the decay channel $\pi^+\pi^-\pi^0$ with the REDTOP experiment

S. D. Escobar Martínez[1,2], B. Fabela Enríquez[2,3], M. I. Pedraza Morales[2] on behalf of the REDTOP collaboration

[1]Universidad Nacional Autónoma de Honduras, Honduras, [2]Benemérita Universidad Autónoma de Puebla, Mexico, [3] Universidad Autónoma de Zacatecas, Mexico.

E-mail: `sdem2393@gmail.com`

Abstract. REDTOP is a novel experiment proposed at the Delivery Ring of Fermilab with the intent of producing more than 10^{13} η mesons per year to detect possible rare η decays which can be a clear evidence of the existence of Physics Beyond the Standard Model. Such statistics are sufficient for investigating several discrete symmetry violations, searching for new particles and interactions and to perform precision studies. One of the golden processes to study is the $\eta \to \pi^+\pi^-\pi^0$ decay [7], where π^0 decays promptly into two photons. In the context of the Standard Model, the dynamics of the charged pions is symmetric in this process. Thus, any mirror asymmetry in the Dalitz plot would be a direct indication of C and CP violation. We present a study on the performance of the REDTOP experiment detector by reconstructing the invariant mass of the final state $\pi^+\pi^-\gamma\gamma$ using Monte Carlo samples.

1. Introduction

The *Rare Eta Decays with a Time Projection Chamber (TPC) for Optical Photons* (REDTOP) experiment is a project in its proposal stage. It intends to produce about 10^{13} η mesons in one year of running. The $\eta \to \pi^+\pi^-\pi^0$ decay channel is used to study the performance of the detector based on the smearing of the energy response of the calorimeter.

2. $\eta \to \pi^+\pi^-\pi^0$ process

The η meson is a pseudo-Goldstone boson symmetrically constrained by its QCD dynamics; it is an eigenstate of the C, P, CP and G operators: $I^G J^{PC} = 0^+0^{+-}$. Thus, the η meson can be used to test C and CP invariance and all of its possible main decays are forbidden in lowest order by P and CP invariance, G-parity conservation as well as by isospin and charge symmetry invariance.

The η decays are mostly constituted by three-body decays; the eta decay to three pions ($\eta \to \pi\pi\pi$) does not conserve isospin and the partial widths are similar to the second order electromagnetic processes. However, the contribution of these processes are suppressed and the decays are driven by an isospin violation in the QCD lagrangian proportional to $m_d - m_u$. Thus, it is ideal for testing Chiral Perturbation Theory (ChPT), CP and C invariance.

3. The REDTOP experiment

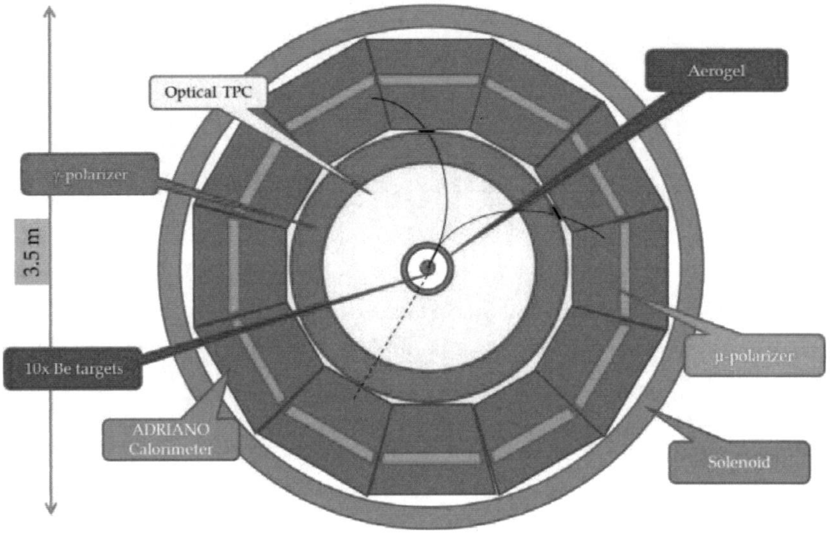

Figure 1. Schematics of the REDTOP detector.

REDTOP is a high intensity fixed target experiment, which is planned to use a continuous proton beam of 1.8 GeV to produce η mesons via the hadronic production from inelastic scattering of protons in beryllium (Be) targets. The target system consists of ten round foils of beryllium; each foil will be of $\frac{1}{3}$ mm thick. The target system will be held inside the beam pipe made of carbon-fiber or beryllium. The expected production of η mesons in one year of running is about 2×10^{13} η, considering an average intensity of the beam of 10^{11} protons/s.

The *Optical Time Projection Chamber* (OTPC) is designed to measure the momentum and position in space of a charged track by using a deflecting solenoidal field of 0.6 T. The OTPC is based on the Cerenkov effect instead of the conventional gas ionization; the pattern of Cerenkov light emitted by the particle will be detected by an array of photo-sensors surrounding the radiator (approximately 500,000 SiPM).

The two Cerenkov radiators present in the OTPC are:

- A double aerogel cylinder of about 3 cm of thickness at the inner wall supported by the beam pipe. The innermost aerogel has a refractive index of 1.2 and the outermost has a refractive index of 1.3.
- A low-pressure methane gas filling the rest of the volume of the OTPC. The pressure is adjusted in order for its refractive index to be very close to 1.0.

The OTPC will be only sensitive to muons, electrons/positrons and fast pions, that are generated in the decays of the η meson. Hadrons and slower pions will be under the Cerenkov threshold and will not contribute to the recorded hits.

The *ADRIANO calorimeter* (A Dual-Readout Integrally Active Non-segmented Option) is a dual-readout calorimeter based on the two simultaneous measurements of the energy deposited by a hadronic or electromagnetic showers into two media with different properties. Typically, the first medium is a plastic scintillator; any charged particle will produce a scintillation signal

IOP Conf. Series: Journal of Physics: Conf. Series **912** (2017) 012034 doi:10.1088/1742-6596/912/1/012034

there. The second medium is a heavy glass with high refractive index and high density where only the charged electromagnetic component of the shower will emit light via the Cerenkov effect; the high density of the medium will be ideal to absorb all the particles impinging the detector.

The *Muon and Photon Polarimeters* are made from an array of plastic scintillators that will count the number of electron and positrons emitted when a muon is stopped inside the calorimeter or when a photon is converted into a e^+e^- pair. The difference in the counting performed inside the solenoidal magnetic field will provide an estimate of the polarization of the particle.

4. Invariant mass reconstruction of the η meson

The Monte Carlo samples with the process $\eta \to \pi^+\pi^-\pi^0 \to \pi^+\pi^-\gamma\gamma$ are generated using GenieHad and the decay is processed in the REDTOP experimental configuration using Geant4. The corresponding process is:

$$p + Be \to \eta X \quad \text{where} \quad \eta \to \pi^+\pi^-\pi^0 \tag{1}$$

The proton beam energy is of 1.8 GeV and the Ultra Relativistic Quantum Molecular Dynamics (UrQMD) scattering model is used. In this study 4390 events were analyzed. Using the Monte Carlo information of the particles passed through the detector, the π^0 is reconstructed by summing up the Lorentz vectors of pairs of photons in each event; likewise the η is reconstructed adding the 4-momentum vectors of the π^+, π^- and π^0 in the event. It is important to mention that in a particular event, the presence of π^+, π^- and π^0 which are not produced by the decay of an η meson is possible, and for that reason, we can have more than one combination which does not correspond to a true η.

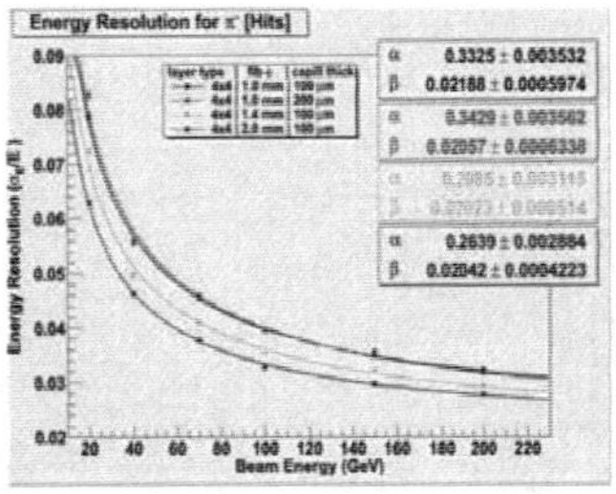

Figure 2. Hadronic resolution of the ADRIANO calorimeter in dual readout mode.

To test the performance of the REDTOP detector in this task, a smearing in the energy response of the ADRIANO calorimeter is done, using values based on simulations for the hadronic resolution, that is given by:

$$\frac{\sigma_E}{E} = \frac{\alpha}{\sqrt{E}} \oplus \beta. \tag{2}$$

Here, the values of the α and β parameters are varied and the goal is to show the differences on the invariant mass distributions. After obtaining the invariant mass distribution for the η meson, a Gaussian fit is performed and the values of the η mass and the resolution given by the σ of the distribution are calculated using the Java Analysis Studio 3 (JAS3) framework.

5. Results

- Values used for the α and β parameters and the obtained η invariant mass (and corresponding sigmas) and obtained distributions.

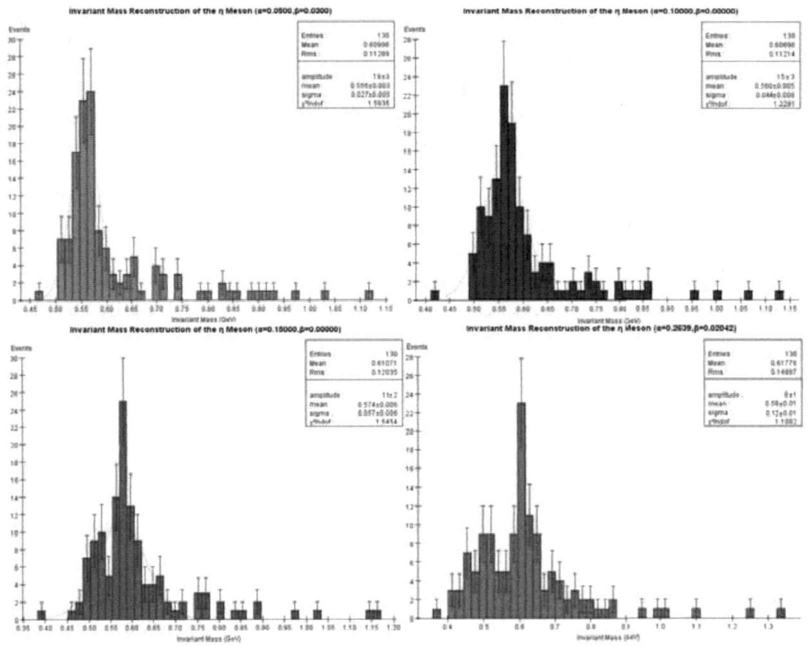

Figure 3. Reconstructed invariant mass distributions for the η meson.

α(GeV $^{1/2}$)	β	m(η) (GeV)	σm(η)(GeV)	σm(η)/m(η) (%)
0.050	0.030	0.556 $\pm$ 0.003	0.028 $\pm$ 0.005	5.0 $\pm$ 0.9
0.100	0.000	0.560 $\pm$ 0.005	0.044 $\pm$ 0.008	7.8 $\pm$ 1.4
0.150	0.000	0.574 $\pm$ 0.006	0.057 $\pm$ 0.006	9.9 $\pm$ 1.0
0.191	0.040	0.579 $\pm$ 0.009	0.081 $\pm$ 0.012	14.0 $\pm$ 1.9
0.264	0.020	0.580 $\pm$ 0.013	0.117 $\pm$ 0.014	20.2 $\pm$ 2.0
0.298	0.020	0.590 $\pm$ 0.014	0.126 $\pm$ 0.014	21.4 $\pm$ 1.8
0.333	0.022	0.572 $\pm$ 0.028	0.179 $\pm$ 0.031	31.3 $\pm$ 3.9
0.343	0.021	0.564 $\pm$ 0.033	0.194 $\pm$ 0.038	34.4 $\pm$ 4.7

Table 1. Values of the invariant mass of the meson and its resolution for each value of the energy resolution for the calorimeter considered.

6. Conclusions and future work

The invariant mass of the meson was reconstructed, smearing on the energy resolution of the ADRIANO calorimeter, the obtained value $m(\eta) = 547.86 \pm 0.02$ MeV agrees with the previous reports[6]. The best energy resolution is achieved when $\alpha < 0.15$. The plan is to increase the statistics, and smear the size of the calorimeter cells and the parameters of the particle trajectories in the OTPC to refine and confirm the results presented here.

IOP Conf. Series: Journal of Physics: Conf. Series **912** (2017) 012034 doi:10.1088/1742-6596/912/1/012034

Acknowledgments
We thank to the REDTOP collaboration and specially to Joseph Comfort at Arizona State University and Corrado Gatto at INFN sezione di Napoli and Northern Illinois University for their advice and support in the simulation work as well as in the preparation of this poster; to the local organizing committee of the XXXI Annual Meeting of the Division of Particles and Fields of the Mexican Physical Society for providing us the full scholarship to attend the event and present this work.

References
[1] Kupsc, A. (2007, November). What is interesting in η and η' Meson Decays?. In A. Kupsc, P. Moskal and M. Wolke (Eds.), *AIP Conference Proceedings*, Vol. 950, No. 1, pp. 165-179. AIP.
[2] Ambrosino, F., Antonelli, A., Antonelli, M., Archilli, F., Bacci, C., Beltrame, P., and Boccheta, S. (2008). Determination of $\eta \to \pi^+\pi^-\pi^0$ Dalitz plot slopes and asymmetries with the KLOE detector. *Journal of High Energy Physics*, 2008 (05), 0006.
[3] The physics. The REDTOP experiment, `https://redtop.fnal.gov/the-physics/`. Date of consultation: May 21th, 2017.
[4] Gatto, C., Fabela, B. and Pedraza, M. I. (2016). The REDTOP project: Rare Eta Decays with a TPC for Optical Photons.
[5] Gatto, C. "ADRIANO: A Dual Readout Integrally Active Non-segmented Option for Future Colliders", `https://indico.cern.ch/event/105142/contributions/26738/attachments/16788/24318/Adriano-LHeC20101112.ppt`. Date of consultation: May 21th, 2017.
[6] K. A. Olive, *et al.* (Particle Data Group), *Chin. Phys. C*, 38, 090001 (2014) and 2015 update.
[7] Colangelo, G., Lanz, S., Leutwyler H. (2017, January) The decay $\eta \to 3\pi$: study of the Dalitz plot and extraction of the quark mass ratio Q. *Phys. Rev. Lett.* 118, 022001 (2017). `arXiv:1610.03494v2 [hep-ph]`.

Design, construction and characterization of a three channels detector of cosmic rays

Julian Félix

Department of Physics, Division of Science and Engineering Campus Leon, Universidad de Guanajuato

E-mail: felix@fisica.ugto.mx

Oscar Moreno

Department of Physics, Guanajuato University, León, UT 37150

E-mail: morenopo2012@licifug.ugto.mx

Everardo Granados

Department of Physics, Guanajuato University, León, UT 37150

E-mail: granadosve2012@licifug.ugto.mx

Luis Arceo

Department of Physics, Guanajuato University, León, UT 37150

E-mail: miquel@fisica.ugto.mx

Abstract. Cosmic rays are particles produced by Astrophysical sources, currently they are being studied to obtain information about the sources and their properties. Cosmic rays research can be used to improve new technology in science, for instance, spectroscopy for material distinction. A three-channel detector was built using three photo-multipliers, two scintillation plastic and common materials water, air, oil, aluminum, and others-. They were allocated in a vertical position, where materials channel is between scintillation plastics channels to validate the signal captured and study the interaction of cosmic rays with these materials. The characterization process is reported achieving positive results in the materials distinction using cosmic rays.

1. Introduction

Since Victor Hess discovered cosmic rays around 1912, measuring the dependence of air ionization with the height [1], its properties have been studied to obtain information about them, solving puzzles and therefore improving our universe understanding. In addition, new technology could be developed due to the continuous natural radiation on the Earth. The design, construction and characterization of a cosmic rays detector are presented. Complementary studies could be developed like new materials for paticle detection.

IOP Conf. Series: Journal of Physics: Conf. Series **912** (2017) 012035 doi:10.1088/1742-6596/912/1/012035

2. Desing and construction

The prototype was made in the Laboratory of Elemental Particles of the Universidad de Guanajuato. It was built with Aluminum due to his light spectrum reflectance property[2]. The photomultipliers were located inside, attached to plastic scintillators and different materials, in the figure 1 the detector structure and components are shown. The plastic scintillators were covered by mirrors and the material channel cavity was polished to mirror level.

Figure 1. Left to right: Three channel detector internal structure, aluminum tube where the photomultiplier is introduced, plastic Scintillator Saint-Gobain BC-408 [3] and Sens Tech photomultiplier P30CW5[4].

Each photomultiplier operates with two voltages, the ranges are 5 V to 8 V and 0.3 V to 1.8 V, to avoid use three dual-voltage sources, a six channels electronic plate was built, therefore only one voltage source is needed for the proper operation. Every photomultiplier output was connected to a discriminator plate to convert the analog signal into a digital signal, this is connected to a data acquisition system (Compact RIO 9025) to collect all the digital signals and to save the information in a data file through a program developed in LabView. In figures 2 and 3, these elements are shown.

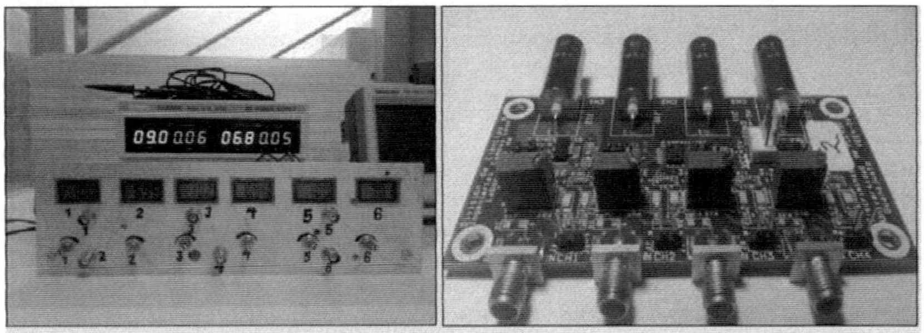

Figure 2. Left to right: Six-channel electronic plate test, four-channels discriminator plate used.

The assembled detector is shown the figure 4.

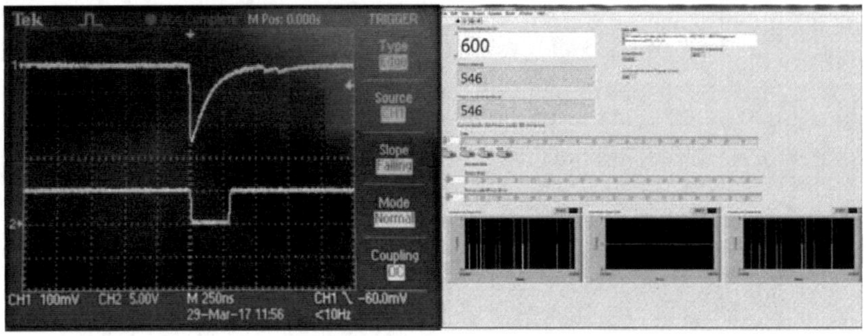

Figure 3. Left to right: analogic and digital signals observed by an oscilloscope and graphical interface of Data collected in LabView.

Figure 4. Detector connected to the six channel plate, discriminator plate and system data acquisition.

3. Characterization

3.1. Voltage operation

To find the plateau region [3] of each photomultiplier, the range voltage of operation was varied in a combination of (5-8) V and (0.5-1.5) V. Each sampling took ten minutes. In figure 5 the numbers of events versus voltage are shown for the three channels.

After graphs analysis the selected operation voltages are showed in Table 1.

Table 1. Voltages selected by the characterization.

Channel	V1 (V)	V2 (V)
1	6.00	0.60
2	7.00	1.05
3	6.00	0.70

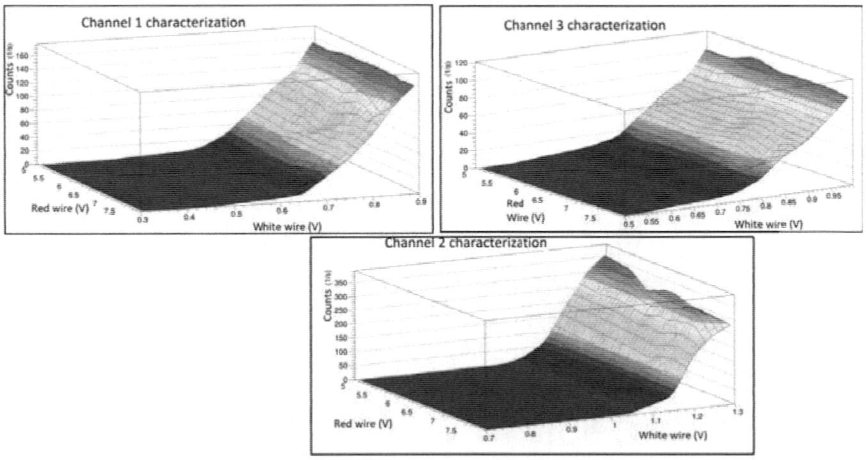

Figure 5. Characterization graphs for each photomultiplier.

3.2. Dark counts
The optical part and the tube behind were isolated from light with sticky aluminum as is showed in figure 6.

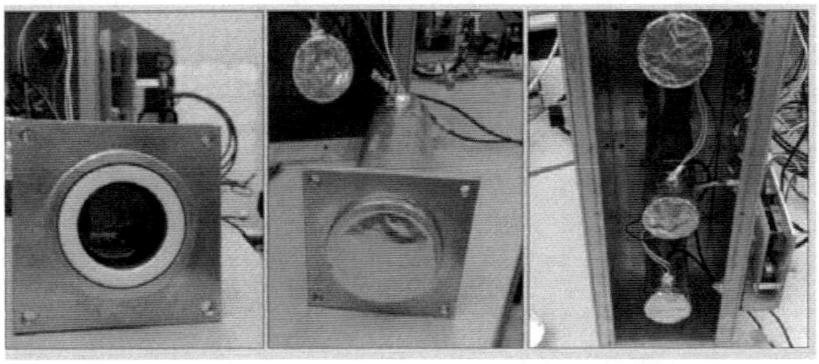

Figure 6. Left to right: Optical part of the photomultiplier, optical part isolated, tube behind isolated from light.

Data were taken during five hours to see the dark counts rate. The distribution frequencies for each photomultiplier are shown in figures 7. the dark counts rate for the channel two is greater due to the voltage selected.

4. Results and conclusions
The frequencies distribution was plotted with the dark counts rested. To compare each material, the distributions were normalized to one, in the fig. 8 the histograms are shown for every channel. Channel 1 corresponds to the lower one. The material tested was mineral oil, adhesive glue, tridistilled water, air, aluminum, asphalt, liquid detergent, quarry and marble.

The channel one and three correspond to plastic scintillators when given material was tested, for that reason it is hope see similar distributions between channel one and three.

It is possible to recognize, identify, different materials using cosmic rays. Tests should continue to collect more data, improve the statistics and observe interactions in all year seasons. Other parameters should be compared like the whether, the position on the Earth, altitude, and others.

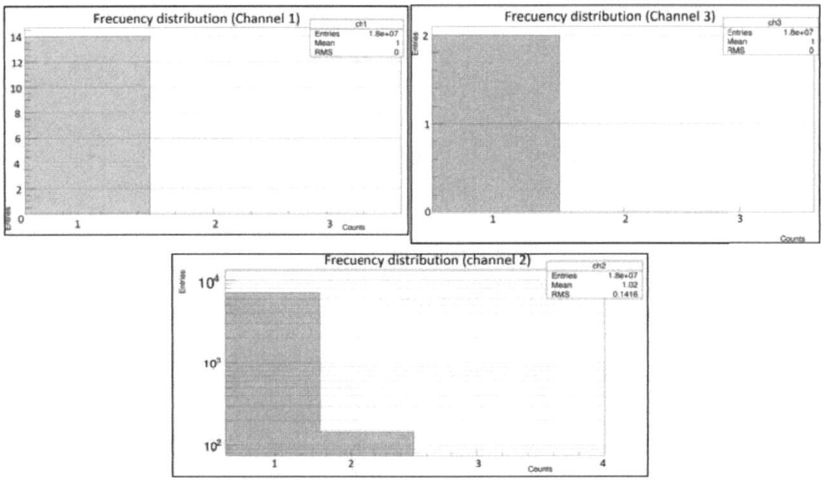

Figure 7. Frequency distribution (Dark counts).

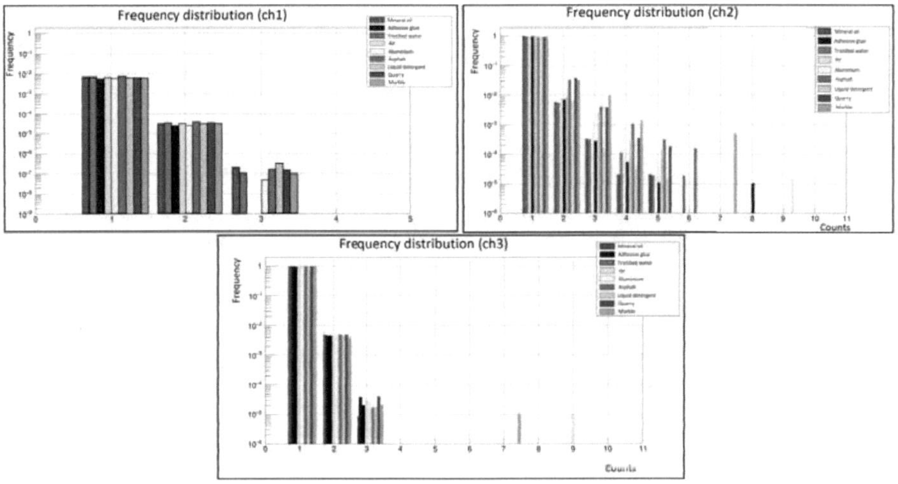

Figure 8. Frequencies distributions.

5. Acknowledgments

The group members of the Laboratory of Elementary Particles. CONACYT grant Fondo I0017 Fondo SEP CONACYT 223179.

6. References

[1] J.J. Beatty, J. Matthews, S.P. Wakely (2016), *Cosmic Rays*, from Data Particle Group.
[2] M. Bass, E.W. Van Stryland, D.R. Williams, and W.L. Wolfe (2001), *Handbook of optics*, vol. **2**, Mc Graw-Hill New York, U.S.
[3] Crystals Saint-gobain, Premium Plastic Scintillators BC-400, BC-404,BC-408,BC-412,BC-416. *http : //www.crystals.saint − gobain.com/sites/imdf.crystals.com/files/documents/sgc − bc400 − 404 − 408 − 412 − 416 − data − sheet.pdf* (accessed March, 2017)
[4] Sens-Tech sensors electronics, photodetector module P30CW5 series. *http : //cosmic.lbl.gov/documentation/P30CW5$_i$ss01.pdf* (accessed March,2017).

Cu Hybrid 4 Channel Cosmic Ray Detector

F. J. Rosas Torres[1], M. A. Hernández Morquecho[2], L. Arceo[3], J. Félix[4]

Laboratorio de Partículas Elementales, Departamento de Física, Campus León, Universidad de Guanajuato, Loma del Bosque 103, León 37150, Guanajuato, México

rosastf2012@licifug.ugto.mx[1] , hernandezmm2012@licifug.ugto.mx[2], miquel@física.ugto.mx[3], felix@física.ugto.mx[4]

Abstract. There are, in the universe, several sources that produce very energetic cosmic rays that interact with the Earth´s atmosphere and create new low energy particles. To detect them there are different methods, according to the interaction with a medium such as the ionization of a material and Cerenkov radiation, among others. In this work a hybrid cosmic ray detector of 4 channels was designed, built and tested at the Laboratorio de Partículas Elementales of the Universidad de Guanajuato. A Copper bar was used as detection material, both smaller area faces have an ionization and a Cerenkov radiation detection channel. To detect the Cerenkov radiation, Hamamatsu silicon photodiodes were used, and for the ionization channels an RC circuit was developed to measure the signal. The ionization channels were tested simultaneously, observing the analogic signal on an oscilloscope. The RC circuit and discriminator were designed to be on the same board; with the discriminator we can digitize the analogic signal. Details of the design, construction and testing of the ionization channel are presented.

1. Introduction

Cosmic rays consist of particles that come from the outer space; large number of them interacts with the atmosphere of the Earth. These particles are classified into primary and secondary, the primary particles are those created in an astrophysical source. The secondary are those that are generated by the interaction between primary cosmic rays and the atmosphere of the Earth [1].

To detect them, they have to interact with a medium, such is the case in the detection method of ionization of a material and Cerenkov radiation. Ionization takes place when a particle has enough energy to ionize an atom of a material, which results in an electron and an ion generating a flow of current. This flow of current is amplified if it is immersed in an electric field.

For the detection of Cerenkov radiation a photodiode [2] is used. A photodiode is a device that transforms light into voltage or current. For the detection by ionization we use an RC circuit that operates when a pulse is produced by a particle interacting with the material immersed in an electric field.

2. Design

The complete design consists of one copper bar and two plastics scintillator, all with a length of 1 meter. The copper bar is among the plastics scintillator, they are isolated with an aluminum cube, and the copper bar with electrical tape. Using this design the signal of the copper bar can be validated (see Figure 1).

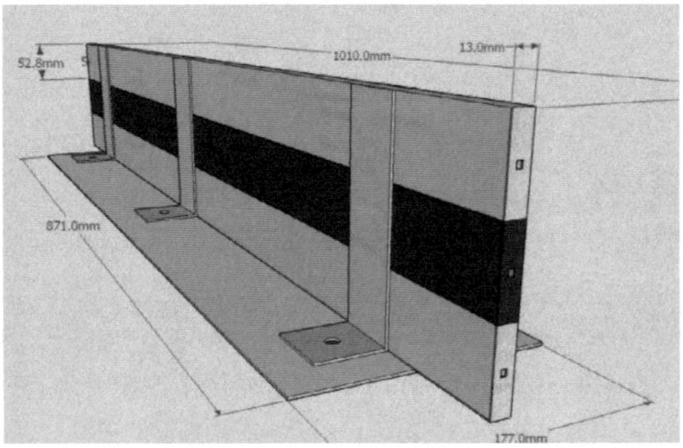

Figure 1. Design of the three bars in SketchUp software.

On the smaller faces of the copper bar, a board is placed with the soldered photodiode in the center and the connections for the electric field, and the collectors of signals of the ionization channel in the bottom and top of the board (see Figure 2).

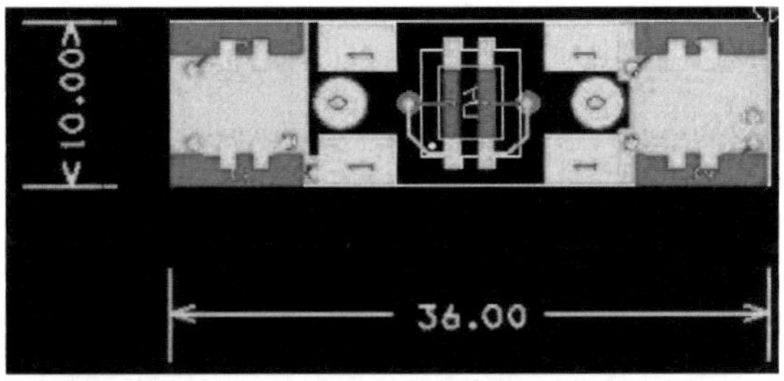

Figure 2. Design of the photodiode board

The board is connected to a second board that has the RC circuit, the discriminator and the power supply connections of the system (see Figure 3).

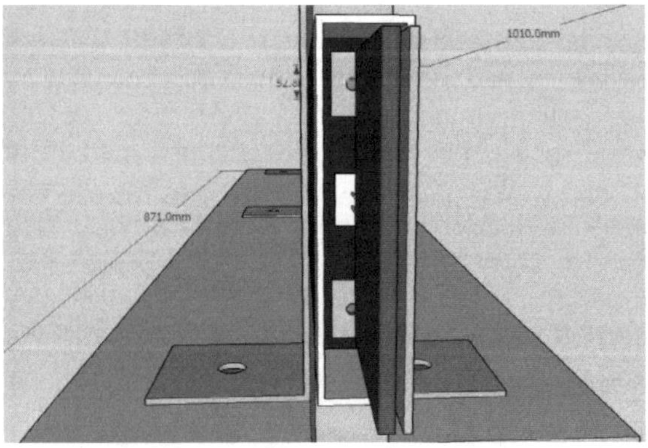

Figure 3. Design of the three bars in SketchUp software.

The board with the RC circuit is connected forming a 90 degrees angle with respect to the photodiode board. These boards were designed using OrCAD software [3]. On the left side, the connections with the photodiode board are shown. The analog-digital comparator is located in the center. The power supply and signal output connections to the data acquisition system are shown on the right. (see Figure 4).

IOP Conf. Series: Journal of Physics: Conf. Series **912** (2017) 012036 doi:10.1088/1742-6596/912/1/012036

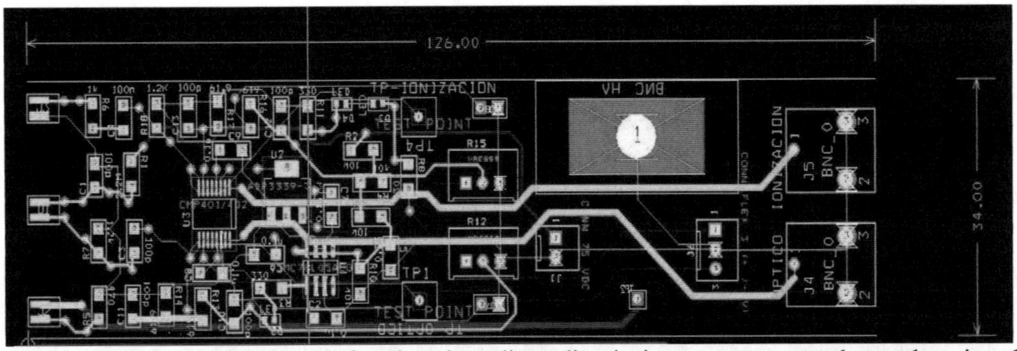

Figure 4. Board with RC circuit for signal reading, discriminator to convert the analog signal to digital pulses, BNC connectors for high voltage of the power supply and regular BNC for signals.

3. Construction

The copper bar was cut and cleaned of impurities to have the following dimensions: 101 cm X 3.8 cm X 1.2 cm. Holes were drilled in the smaller faces of the bar to put two screws that will hold the photodiode board in place. A small subsidence was made in the center of the face to place the photodiode (see Figure 5).

Figure 5. Drillings and subsidence in the bar. Clean copper bar. Screws set in the bar.

The bar was then cleaned again to put electrical insulating tape. To make the arrangement for the electric field, copper tape was used, and to have the connections to obtain the signal in the ionization channels (see Figure 6).

Figure 6. Copper bar with final configuration.

IOP Conf. Series: Journal of Physics: Conf. Series **912** (2017) 012036 doi:10.1088/1742-6596/912/1/012036

A test board was built to activate the ionization channels. It has 4 pins to make the connections, two for the signal and two for the high voltage to feed the system. Cables are connected to the pins that go to the high voltage source and to the RC circuit (see Figure 7).

Figure 7. Test board to set on copper bar.

The detector was built in the Laboratorio de Partículas Elementales of the Visión de Ciencias e Ingenierías, Campus León of the Universidad de Guanajuato.

4. Testing

For the first tests, a high voltage source, Ultravolt 3M24-P1 M SERIES [4], is connected to the test board to activate the electric field. The voltage used was 1500V. The RC circuit was connected in a breadboard to change the values of the resistors and capacitors until finding the best values. The output signal was observed on an oscilloscope (seeFigure 8).

Figure 8. Copper bar connected to RC circuit and power supply.

The RC circuit is made up of two resistors and a capacitor. A manifold of the copper bar is connected to a capacitor, followed by a resistor. In this route the oscilloscope tip is connected. The other collector connects a resistor which is then grounded and at that point the oscilloscope reference is placed (see Figure 9).

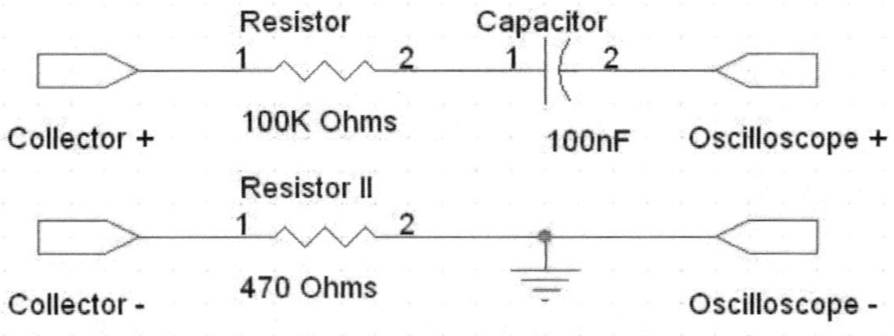

Figure 9. Schematic design of the RC circuit.

With this RC circuit and with a fixed input voltage of 2000 V to generate the electric field, signals with decaying exponential form were obtained. The pulse amplitude is 300 mV and with a discharge time of 1 μs. This was the best configuration to reduce the pulse discharge time without totally deforming the signal (see Figure 10).

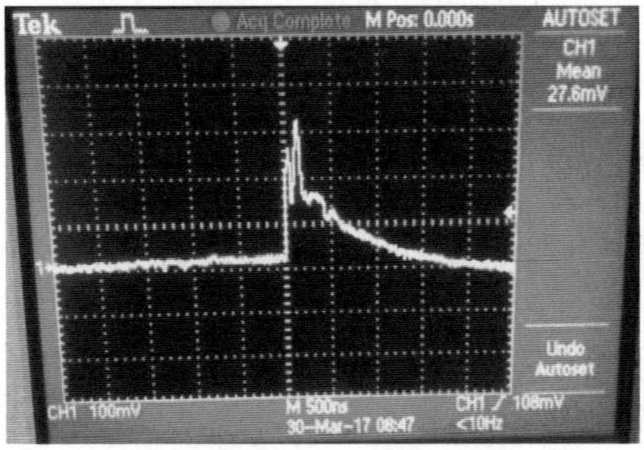

Figure 10. Ionization channel signal.

5. Conclusions

The copper bar was set to activate the ionization channels. The capacitance of the copper bar was measured, which is 3 nF. It was possible to operate the ionization channels and to observe the signal on an oscilloscope. The values of the resistors and capacitors in the RC circuit were varied until the best performance of the ionization channels was achieved.

Acknowledgments

To Jazmín Sánchez Sánchez for previous contributions, and CONACYT grant Fondo I0017 Fondo SEP – CONACYT 223179.

References

[1] Beatt J J, Mat thews J and Wakely S P. *Cosmic Ray* [Internet]. http://pdg.lbl.gov/2015/reviews/rpp2015-rev-cosmic-rays.pdf (accessed 21 June 2016).

[2] Hamamatsu phototonics K.K. and "Opto-semiconductor Handbook" Editorial Committee. *Opto-semiconductor Handbook*. : Hamamatsu phototonics K.K. and Solid State Divition; 2014. https://www.hamamatsu-news.de/hamamatsu_optosemiconductor_handbook/#362/z (accessed 21 June 2016).

[3] OrCAD Cadence PCB Solutions. *OrCAD PCB Designer* [Internet]. http://www.orcad.com/products/orcad-pcb-designer/overview (accessed 16 July 2016).

[4] Advanced Energy Industries, Inc. Ultravolt M Series Minature, *Micro-Sized High Voltage Biasing Supplies* [Internet]. http://www.ultravolt.com/uv_docs/MSeries.pdf (accessed 21 June 2016).

Design and construction of a uniform magnetic field generator for a 32 channel cosmic ray detector.

K N Herrera-Guzman, R A Gutierrez-Sanchez, J Felix, L J Arceo, C Araujo

Department of Physics Engineering, University of Guanajuato, Leon, GTO 37150
felix@fisica.ugto.mx

Abstract. The trajectory of a particle can be measured if some points of its track are known. This is applied to any kind of particle, including cosmic rays. We have designed and built a device for this purpose. We present the design, construction and characterization of a uniform magnetic field generator system in a finite volume. An array of Cerenkov detectors will be placed inside of it for determining the cosmic rays charge and to reconstruct their trajectories.

1. Introduction

Charged cosmic rays at sea level are mostly muons [1]. The main goal of this work is to build a base to place a particle detector inside the magnetic field. It is very important that both materials, of the detector and base, do not distort the magnetic field. We built a system, with the requirements described above, for an arrangement of detectors of 8 in X 8 in X 8 in.

2. Analytic description

Cosmic Rays are very energetic, therefore it is necessary to consider a relativistic its deflection due to the presence of a magnetic field. We start with the Lorentz's principle [2] and the second Newton's principle with the relativistic correction,

$$\frac{d\vec{p}}{dt} = q\vec{v}x\vec{B}, \tag{1}$$

where $p = \gamma m v$ is the relativistic momentum, γ is the Lorentz factor, $\vec{v}$ and m are the velocity and the mass of the particle respectively.

In this case, $\vec{B} = (0, B, 0)$ and $\vec{v} = (v_x, 0, v_z)$ where $\alpha = \frac{\gamma m}{qB}$ and solving (1):

$$x = \alpha\left[Asen\left(\frac{t}{\alpha}\right) - Bcos\left(\frac{t}{\alpha}\right)\right] + K \tag{2}$$
$$z = -\alpha\left[Acos\left(\frac{t}{\alpha}\right) + Bsen\left(\frac{t}{\alpha}\right)\right] + C$$

and

$$v_x = Acos\left(\frac{t}{\alpha}\right) + Bsen\left(\frac{t}{\alpha}\right) \tag{3}$$
$$v_z = Asen\left(\frac{t}{\alpha}\right) - Bcos\left(\frac{t}{\alpha}\right)$$

Applying the initial conditions $x = 0$, $z = z_0$ and $v_x = 0$, $v_z = v_{z0}$ in $t = 0$ to (2) and (3) the solution is

$$x = \alpha\, v_{z0} cos\left(\frac{t}{\alpha}\right) - \alpha\, v_{z0} \tag{5}$$

and

$$z = \alpha\, v_{z0} sen\left(\frac{t}{\alpha}\right) + z_0. \tag{6}$$

3. Design

The prototype design consists of Helmholtz coils with 210 turns and a ferric nucleus. Both coils are joined by ferric bars to close the magnetic field. An aluminum base, which will be used to place detectors inside the magnetic field, was designed too. Figure 1 shows a drawing of the coils and the aluminum base.

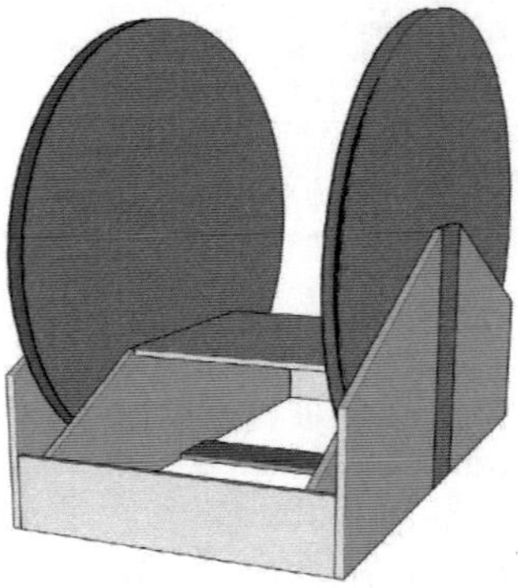

Figure 1. System perspective view.

Because this prototype is supposed to hold the electronic cards, few aluminum plates must be added. Those extra aluminum plates, and their location inside the coils, are shown in Figure 2.

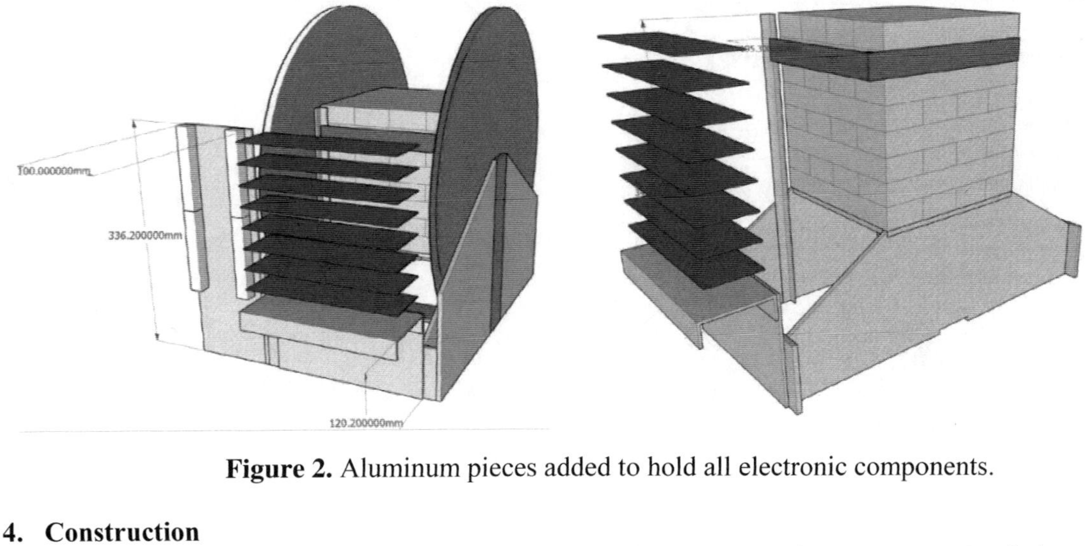

Figure 2. Aluminum pieces added to hold all electronic components.

4. Construction

All pieces were joined together with nuts, bolts and aluminum angles when necessary. Steel plates were glued together to form bars, resulting two bars of 27.5 cm X 5.0 cm X 0.6 cm and one of 23.4 cm X 5.0 cm X 0.6 cm (for future references this pieces are called L and K pieces respectively). Drilled discs form

2 reels, each one containing two 23.4 cm radio discs and eleven 21.4 cm radio discs. The discs were glued together by the perimeter.
The steel reels were winded with 210 turns each with 19 AWG wire. Then, the reels were painted red color and the L and K pieces matte matt black color. Figure 3 shows the assembled system with and without electric connection.

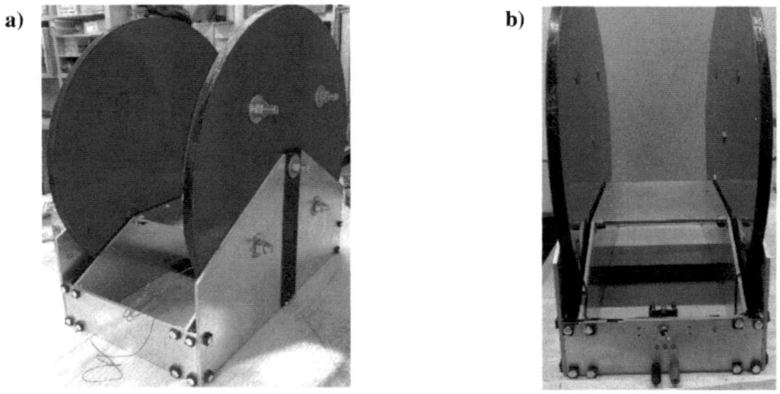

Figure 3. Assembled system a) without and b) with electric connection.

In order to map the magnetic field, a MakeBlock XY Plotter was adapted [3]. A graduated acrylic tube was used instead of a pencil. With this, it was possible to map the magnetic field in a volume of 20.00±0.05 X 19.00±0.05 X 20.0±0.1 cm in *x*, *y*, *z* directions respectively. Inside the acrylic tube, a Vernier VGM was placed. Figure 4 shows the measurement system on the prototype mentioned above. Each surface is mapped in 20 lines from left to right; each line is made of points 20. The volume mapped is made of 20 flat surfaces 1 cm apart.

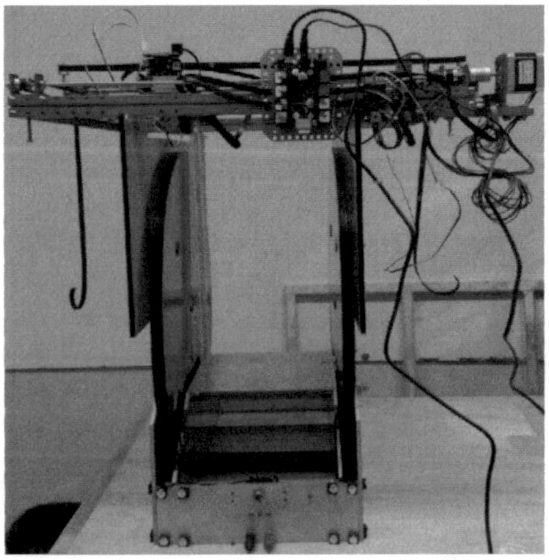

Figure 4. Magnetic field measurement system on the prototype.

5. Results

A model of the coils was simulated in Poisson Superfish [4]. Figure 5 shows the generated graph by Poisson Superfish.

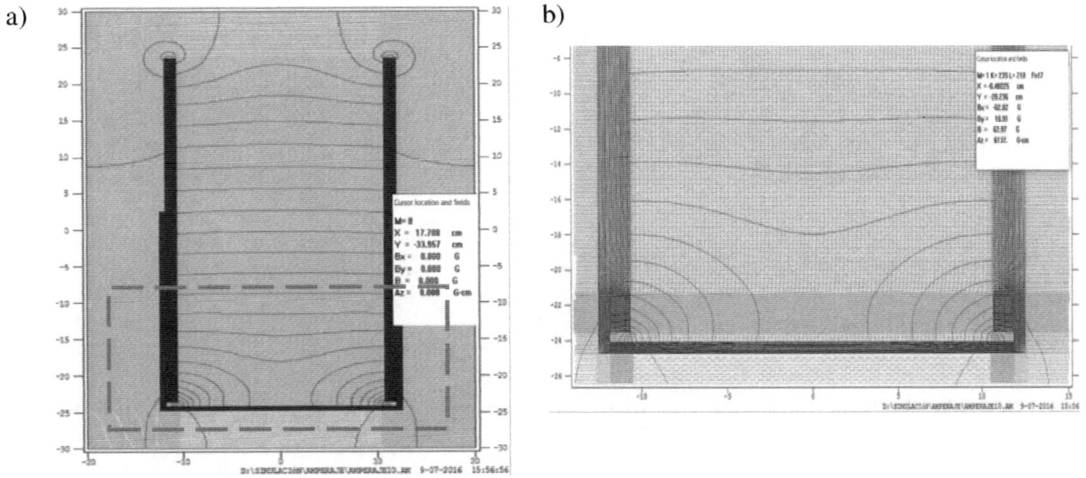

Figure 5. a) Magnetic field lines simulated and b) zoom of the simulated lines inside the red rectangle.

Equations (5) and (6) have to be solved to find the maximum and minimum particle energy in which the incident particle will be deflected in the desired way.
To solve this equations it is necessary to give B, m, q y z_0 values, those are the magnetic field generated by the coils, the muon mass, the muon charge and the point in which the particle enters in the magnetic field. Those values are: $B = 29.44$ mT (Poisson Superfish for a 20 Amperes), $m = 1.08838x10^{-28}$ kg, $z_0 = 23.52$ cm and $q = \pm 1.0602x10^{-19}$ C. Defining $\beta = \frac{m}{qB}$ we have:

$$x = \frac{\pm \beta xc}{\sqrt{c^2 - v_{z0}{}^2}} \cdot v_{z0} \left[cos\left(\frac{\sqrt{c^2 - v_{z0}{}^2}}{\beta xc} \cdot t \right) \mp 1 \right], \tag{7}$$

$$z = \frac{\pm \beta xc}{\sqrt{c^2 - v_{z0}{}^2}} \cdot v_{z0} \left[sen\left(\frac{\sqrt{c^2 - v_{z0}{}^2}}{\beta xc} \cdot t \right) \mp 0.2352 \right].$$

Lower bound; case in which the particle leaves the array of detectors before crossing its half:

$$x = \pm 0.1016 \text{ m}, \quad z = 0.0254 \text{ m}.$$

Upper bound; case in which the particle leaves the array of detectors without a detectable deflection (the width of one detector is 5.08 cm):

$$x = \pm 0.0508 \text{ m}, \quad z = -0.1016 \text{ m}.$$

The energy bounds found are:

$$E_{max} = 106.31 \text{ MeV}, \quad E_{min} = 105.85 \text{ MeV},$$

which were calculated with $E^2 = \gamma^2 m^2 c^2 + m^2 c^4$.
Table 1 shows the initial and final current and temperatures in which the measurements took place for some planes.

Table 1. Values of current and temperature between each plane was mapped.

Plane number	I_i (A)	I_f (A)	T_i $(°C)$	T_f $(°C)$
0	4.77	4.69	72.5	77.0
10	4.66	4.66	80.3	80.2
20	4.70	4.69	77.0	78.5

Figure 6 shows the graphs of some planes of the magnetic field generated by the coils. Figure 6.4 shows the maximum and minimum value obtained for each plane.

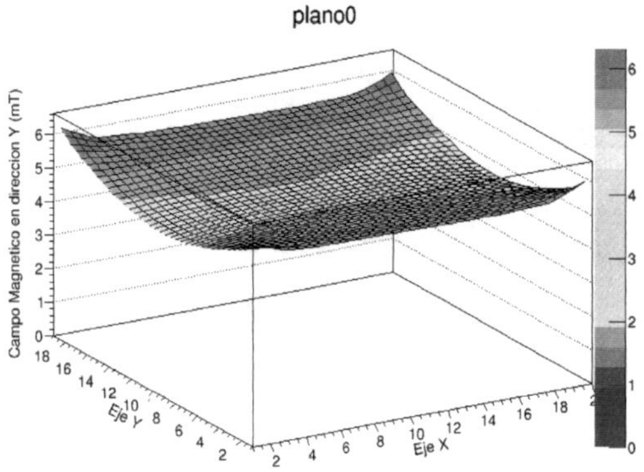

Figure 6.1. Magnetic field map in plane 0.

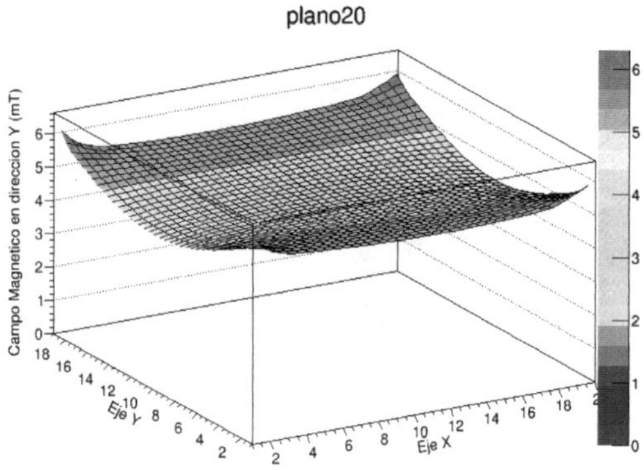

Figure 6.2. Magnetic field map in plane 20.

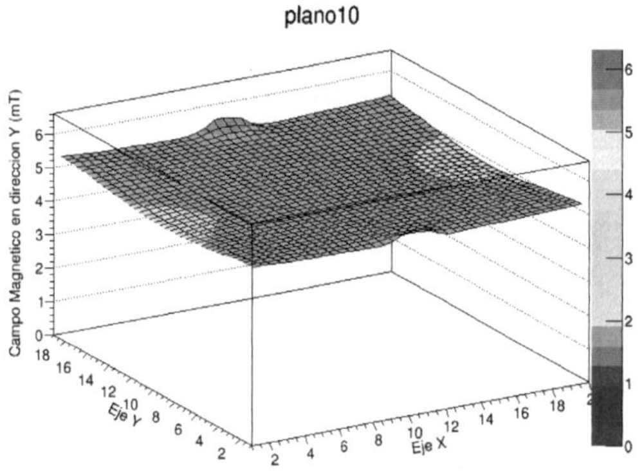

Figure 6.3. Magnetic field map in plane 10.

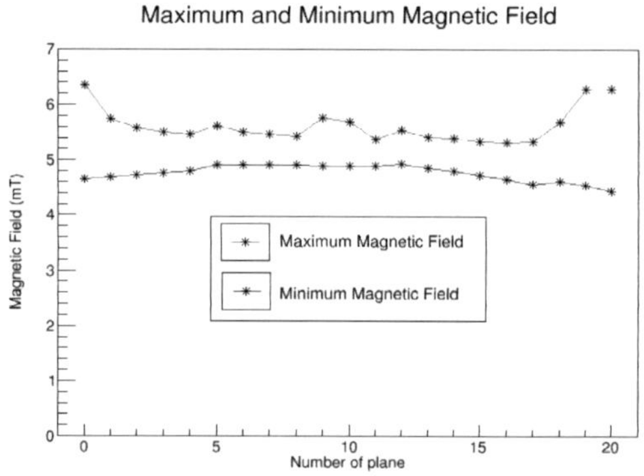

Figure 6.4. Maximum and minimum values of the magnetic field.

6. Conclusions

We simulated the magnetic field as a function of electric current; we have measured the magnetic field as a function of the applied electric current. Incident cosmic particles with energy between E_{max}=106.31 MeV and E_{min}=105.86 MeV must be detected by this spectrometer. Loss of energy from particles passing through the detectors is not considered. Magnetic field percent variation decreases as the sensor approaches to the axis of the coils and increases as they move away from it. Also it is observed that this variation increases faster in the upper planes than in the lower planes.

From figure 6.4 can be seen that the magnetic field uniformity is higher at the center planes and that the maximum difference is rapidly decreasing at the top and bottom planes.

Aknowledgments

CONACYT Project 223179, CB-2013-01 and UGTO Project.

References

[1] Beatt J J, Mat thews J and Wakely S P. Cosmic Ray [Internet]. http://pdg.lbl.gov/2015/reviews/rpp2015-rev-cosmic-rays.pdf
[2] R. A. Serway y R J Beichner, *2001 Fisica para Ciencias e Ingenierías. Tomo II* , Mexico/McGraw-Hill, Chapter 29, pag 922.
[3] Makeblock website. http://learn.makeblock.com/xy-plotter-robot-kit/
[4] Poisson Superfish Website. http://laacg.lanl.gov/laacg/services/download_sf.phtml

IOP Conf. Series: Journal of Physics: Conf. Series **912** (2017) 012038 doi:10.1088/1742-6596/912/1/012038

Status Report on the R&D of a 5 T/m Normal Conducting Quadrupole Magnet for the 10-MeV Beam Line of the Electron Linac of the Mexican Particle Accelerator Community

D Chavez [1], G H I Maury[1], M Napsuciale[1], C B Ortiz[2], J Peña[1], C Galvan-Duarte[3], C A Valerio[3] and A I Sattarov[4].

[1]Universidad de Guanajuato, Loma del Bosque 103, Lomas del Campestre CP. 37150 León, Gto., México.
[2]DNyN CINVESTAV- IPN, Av. IPN 2508, Col. San Pedro Zacatenco, Delegación Gustavo A. Madero, CP. 07360, México D.F. México.
[3]Universidad Autónoma de Sinaloa, Calle Gral. Ángel Flores Pte. S/N Colonia Centro, CP. 80000, Culiacán, Sinaloa, México.
[4]Texas A&M University, 3380 University Dr. East/ Magnet Lab. CP. 77845, College Station, Texas, U.S.A.
.

Email: dchavez@fisica.ugto.mx

Abstract. The Mexican Particle Accelerator Community (CMAP) has endeavoured the development of a 100 MeV electron Linac (eLINAC) that will have three beam lines at 10, 60 and 100 MeV. The Linac, at present under design, will make use of four normal conducting quadrupole magnets for the 10-MeV beam line. The quadrupoles follow a standard type geometry and they will operate using Cu conductors at room temperature. This work reports the progress on the overall quadrupole design, analytic, magnetic, thermal and mechanical studies. In addition, the selection of varied materials for the Yoke and how affects the multipole content and gradient uniformity is explored. A winding strategy is discussed as well as the CMAP plan for the development of the first prototype.

1. Introduction

A preliminary design on the first Mexican electron Linac (eLINAC) is being developed by the Mexican Particle Accelerator Community (CMAP). The eLINAC consists in a sequence of three 2-m in length focusing defocusing cells (FODO) composed of four quadrupoles. Three radiofrequency cavities (RF) cavities are interspersed among the three cells. To minimize the beam size along the line, a 60° phase advance per FODO cell is selected, thereby minimizing the maximum value of the beta functions; the emittances are assumed to be constants.

The 10-MeV beam line of the Mexican eLINAC will be the first step of the accelerator complex of the 100-MeV eLINAC. Magnetic gradients for the three pairs of quadrupoles for the 10, 60 and 100 MeV lines are computed based on the required magnetic flux density at the tip of the poles, and the aperture radius. The 10-MeV line requires a magnetic flux density of 0.09 T at the tip of the pole, assuming a 0.02 m bore radius, the desired gradient is 4.5 T/m.

The quadrupoles were designed bearing in mind infrastructure, materials, cost-effective development techniques (air cooled design) and equipment available at the University of Guanajuato. The current design makes use of regular copper wire (AWG 12) for the windings and allows future energy (current) upgrades, ranging from 4 T/m up to 19 T/m while preserving field quality and thermal stability. In the following sections, we are going to describe the quadrupole design for the 10-MeV beam line. However, thanks to the optimized geometry and the parametrized design, small modifications will be needed for the quadrupoles to be employed in the 60-MeV and 100-MeV beam lines. A thermal model driven by Joule-Lenz law and heat conduction in solids was modeled using the commercial software COMSOL Multiphysics ® [1] for finite element analysis. A mechanical model of the Lorentz force distribution was studied as well as the fringe field effects at the end of the quadrupole.

2. Quadrupole design

The 10-MeV beam line has an aperture radius of 0.02 m and requires a flux density of 0.09 T at the tip of the pole, i.e. a 4.5 T/m gradient. The use of standard copper wire AWG 12 for winding suggested the use of a "standard-type geometry" for the quadrupole design. 126 turns per pole were used and a 6.6 A current was applied. The pole shape was modeled following an optimization method described in [2, 3]. In this method, the pole geometry follows a truncated hyperbole and field quality is related to pole thickness. Parallel simulations using COMSOL Multiphysics® were done to determine the optimum value for pole thickness vs. field quality (see Figure 1). According to beam dynamics requirements, the 1% uniformity in field quality is achieved when pole thickness, which can be expressed in terms of the distance from the pole to the axis *"A"*, is A=0.0061 m. Analogously, a shimming strategy was added to the pole tip geometry to boost field uniformity [3]. Main quadrupole parameters are listed in Table 1.

Table1. List of main design Parameters

Parameter	Value	Units
Wire OD [m]	0.002	m
Iop [A]	6.60	A
Turns per pole	126	
Bore Radius [m]	0.02	M
Bnorm at Pole Tip	0.09	T
Gradient	4.71	T/m
Power Dissipated	31.8	W
Gradient upgrade suggested limit	15.0	T/m
Current upgrade recommended limit	19.00	A
Quadrupole length	0.10	m
Quadrupole mass	20.85	m

Magnetic relaxation calculations where done using Finite Element Analysis software COMSOL Multiphysics®. Figure 1 shows the optimized geometry. For assembly purposes, the poles will be wind separately and installed at the iron yoke. This process allows us to use the available inter-pole space.

2.1. Thermal model

Normal-conducting magnets use non-superconductor materials for windings, because of their characteristic non-vanishing resistivity ($\rho=RA/L$, ρ resistivity [Ωm], R electrical resistance [Ω], A cross-section are of conductor [m^2], and L the conductor length [m]) the Joule-Lenz Law takes place, increasing the temperature of the system proportionally to the square of the current passing through the conductor.

Considering standard copper wire AWG 12, the number of turns per pole, the length of the quadrupole and the resistivity of pure copper (1.68×10^{-8} Ωm [4]), we estimated 31.8 W of dissipated power for a 4 T/m gradient. Bearing in mind thermal properties of pure iron as the "iron yoke" material, and using heat equation, it is possible to estimate the increment in temperature produced by that amount of heat. The finite element analysis simulation estimates a temperature rise of 2 °C, assuming the quadrupole is in a 20 °C room. The gradient upgrade case was studied, the amount of current needed to produce a 19.27 T/m gradient is 19 A per wire, which produces a 263 W power. For that case, the temperature increments 3.6 °C, which does not affect the actual air cooling method.

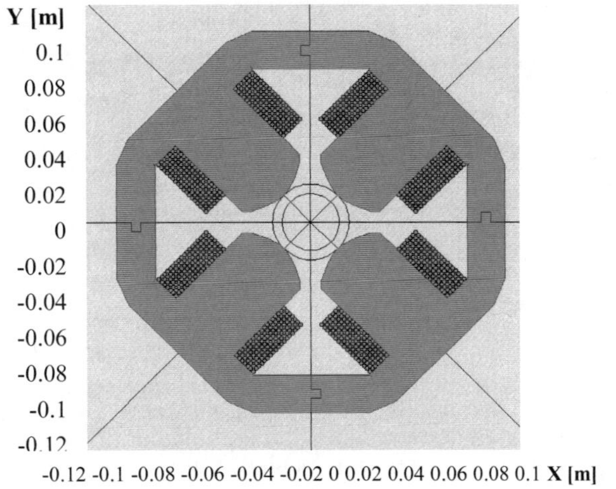

Figure 1. Optimized quadrupole design. Four hyperbolic truncated poles house 126 turns per pole of standard copper wire AWG 12. 18 layers, 7 wires each are wound around each pole. The pole tip was optmized varying pole thicknes as a function of multipole content. A shimming strategy allow to boost the gradient uniformity at the fard end of the aperture.

2.2. *Mechanical model*

If Lorentz forces are large, they could cause permanent damage or even affect magnet performance. The present magnetic requirements are small to produce large Lorentz forces. Although this statement is true, it must be validated. A finite element analysis study was performed to estimate the magnetic forced produced while energizing the quadrupole at full field.

The model considers contributions from each turn at the pole, assuming no-ends effects, which in a first approximation is valid. In the 2D model, the wires are aligned parallel to beam direction. The estimated overall contribution along the total length of the quadrupole (0.1m) is 1.2 N toward the base of the pole. This amount of force gets redistributed along the iron yoke and does not represent a damage to the wires nor the mechanical structure of the quadrupole.

2.3. *Materials for Iron Yoke and winding strategy*

Original field relaxation model assumed LHC type steel [5], which has an extended performance over the 2 T saturation limit. This is related to the carbon content in the alloy. For low field purposes, as in this case (0.09 T at pole tip), an evaluation of the multipole content must be study for candidate materials to determine how smaller magnetic permeabilities affect quadrupoles overall performance.

A field relaxation study will be done varying different materials, i.e. changing the B-H curves at room temperature, see Figure 2. The study of the materials will consider the most common steel alloys that can be readily obtained in Mexico [6, 7]. The materials used were: Pure Iron, Pure Iron Annealed, Steel 1018, Steel 1020, Cast Iron, Cold drawn Low Carbon Steel. Hot Rolled Low Carbon Steel, Mu-Metal, Magnet Steel and LHC Steel. The goal of the study will be to preserve the field quality requirements (less than 1 unit for the allowed multipoles) at full field in both cases, the desired operation gradient and for future upgrades, in which the maximum recommended gradient will be 19 T/m.

Analytic, thermal, mechanical and material studies confirm that, for the present field requirements, the current design satisfies the expectations. The next step is to develop a winding strategy for a model quadrupole. A barrel-winding procedure [8] will be used to wind the 18 layers (7 wires each) on each of the poles, starting inside-out of the base of the pole. Winding tables and special tooling is being develop for the first prototype.

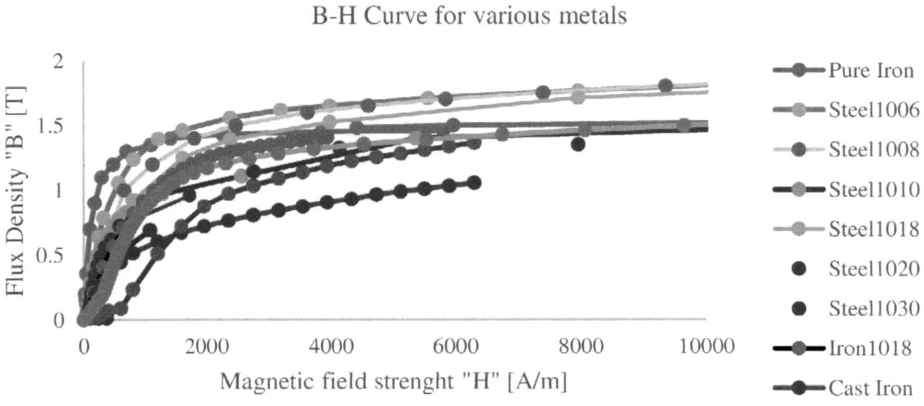

Figure 2. Magnetic permeabilities for different materials. Notice the saturation point for most materials falls behind the 2 T limit. The desired effect is to reach saturation at higher fields. For our purposes, the field at the tip of the pole is 0.09 T, which implies that almost every material could be use. Taking into account future energy upgrades, it is recommendable to select a cost-effective material with high saturation field. Pure Iron, or Steel 1020 are good candidates.

2.4. *Fringe field for sensitive electronics*

Fringe field lines were computed on a 3D model at several magnetic flux density values, starting at 5 G (five times earth's magnetic flux density field). Due to small field values and quadrupole geometry, the field at the body of the magnet is completely enclosed by the iron yoke. Naturally, the remaining component is at the ends of the magnet, in which there is no iron yoke. According to the International commission on non-ionizing radiation protection (ICNIRP, 2009) [9], prudent practice requires posting warnings at the 5 G line, and limit access to areas with more than 10 to 20 G to knowledgeable staff. The 5, 20, 50 and 100 G lines were computed, (see Figure 3). The estimated 5 G line is located at 0.14 m from the centre of the quadrupole while the 20 and 50 G lines are at 0.11 and 0.09 m respectively.

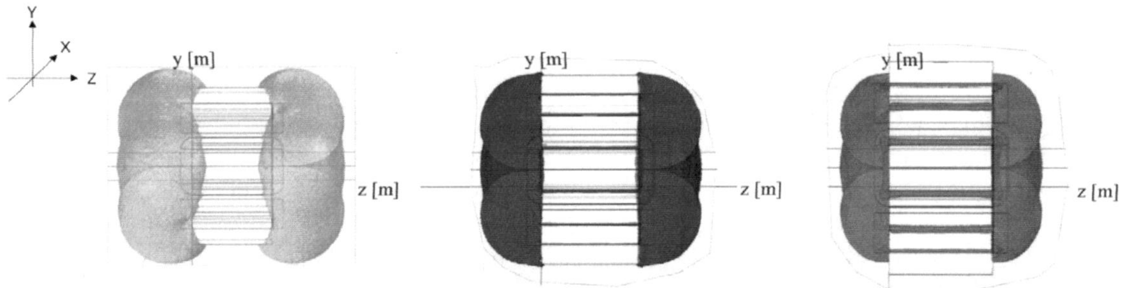

Figure 3. Fringe field distribution along the axis of the quadrupoles. The quadrupoles axis is oriented along the Z axis. In A, B and C), the gauss lines equipotential surfaces were computed for the 5, 20 and 100 G. The 5-G line is located at 0.14 m from the center of the quadrupole, while the 20 and 100-G are at 0.11 and 0.09 m respectively.

IOP Conf. Series: Journal of Physics: Conf. Series **912** (2017) 012038 doi:10.1088/1742-6596/912/1/012038

3. Conclusions

An optimized geometry and design for the quadrupoles of the 10-MeV beam line of the CMAP eLINAC has been presented. A basic analytic model, based on general requirements and limitations, showed that a 637 A current must be applied to each pole to produce a ~4.5 T/m gradient in a 0.02 m bore radius. The quadrupole is designed in such a way that allows future energy updates, ranging from 4 up to 19 T/m while preserving gradient uniformity. To improve field quality and keep gradient uniformity below 1%, the end of the hyperbolic poles was strategically shimmed. The mechanical model showed that a net force of 1.2 N is directed towards the base of each pole. The basic thermal model revealed that when a current at 6.6 A per wire is applied, the power dissipated barely increases the copper temperature by about a 2°C. The 19 A scenario was modelled, producing a temperature increment of 3.6 °C over the 6.6 A model. In both cases, the quadrupoles can be cooled using natural air convection.

The iron yoke fully encloses the magnetic field at the body of the quadrupole, adding no contribution to the fringe field. At the end of the quadrupole, there is a small region in which the winding around the pole is not covered by the iron yoke. This region contributes to the fringe field that extends out of the quadrupole. According to the International Commission on Non-Ionizing Radiation (ICNIRP, 2009), prudent practice requires posting warnings at the 5 G line. The magnetic model computes the 5, 10 and 50 G lines, being the first at 0.14 m from the centre of the quadrupole. Special tooling is being designed and it will be developed shortly

Thermal, mechanical and electromagnetic studies have shown that the quadrupole fulfil the requirements for the quadrupoles of eLINAC, low-cost production and air-cooled. In addition, thanks to the parametrized design only small modifications will be needed for the quadrupoles to be employed in the 60-MeV and 100-MeV beam lines.

References

[1] COMSOL Multiphysics® Software link: https://www.comsol.com
[2] H. Wiedemann. "Particle Accelerators for physics" Third Edition. Springer Ed. 1993.
[3] Jack T. Tanabe. "Iron Dominated Electromagnets: Design, Fabrication, assembly and Measurements". World Scientific Publishing Company, Incorporated. 2005.
[4] W.J.de Haas. .J.de Boer. and.G.J.van dën Berg."The electrical resistance of gold, copper, and lead at low temperatures".Physica. Volume 1, Issues 7-12, May 1934. Pages 1115-1124.
[5] F. Bertinelli ; S. Comel ; P. Harlet ; G. Peiro ; A. Russo ; A. Taquet "Production of low carbon magentic steel for the LHC superconducting dipole and quadrupole magnets". IEEE TRANSACTIONS ON APPLIED SUPERCONDUCTIVITY, VOL. 16, NO. 2, JUNE 2006
[6] Aceros Mexico. http://acerosmexico.com.mx/la-empresa/
[7] Aceros del Toro Mexico. http://acerosmexico.com.mx/la-empresa/
[8] Martin N. Wilson. "Superconducting Magnets". Oxford Science Publications. NY.1983.
[9] ICNIRP Guidelines on limits of exposure to static magnetic fields. HEALTH PHYSICS 96(4):504-514; 2009.

IOP Conf. Series: Journal of Physics: Conf. Series **912** (2017) 012039 doi:10.1088/1742-6596/912/1/012039

ELECTROMAGNETIC DESIGN OF A RADIOFREQUENCY CAVITY

G.R Montoya Soto, Carlos Duarte Galvan[1], Ildefonso Leon Monzon[1], Pedro Luis manuel Podesta Lerma[1], C.A Valerio-Lizarraga[1].

[1]Facultad de Ciencias Físico-Matemáticas, Universidad Autónoma de Sinaloa, Av. De las Américas y Blvd. Universitario, Cd. Universitaria, 80000 Culiacán, Sinaloa, México.
ricams.fcfm@uas.edu.mx
cvalerio@uas.edu.mx

Abstract. Electromagnetic and mechanical studies have been performed with the aim of build a RF cavity in the S-Band (2998 MHz), the design takes into consideration the relativistic change in the electron velocity through the acceleration cavity. Four cavity cases were considered at different input energies, 50 KeV, 100 KeV, 150 KeV and 200 KeV, with output energies from 350 KeV to 5 MeV, the designs show good acceleration efficiency and beam coherence comparable to the one created in the cathode.

1. Introduction

In recent years, Mexico has promoted the development of particle accelerators and radiation detectors. The lack of accelerator facilities in the country forced the researchers to perform most of the testing outside the country, one possible solution is to promote the design and construction os accelerators within the country aiming also for several industrial applications. A normal RF Linac (RFL) is under desig and considered for accelerating an electron beam at 2998 MHz using the $\pi/2$ TM010 mode to accelerate in a stable condition, this design takes in to account the particle velocity within the system to set the cavity length to improve the acceleration efficiency.

The behavior of electromagnetic waves and how they resonates inside conducting walls depends strongly of the geometry itself, we designed a RF cavity geometry that enhance the particle acceleration using resonant electric fields, the code used to find the appropriated electric field to accelerate a single electron was POISSON SUPERFISH [1], with the obtained field we performed multiparticle simulations to quantify the electric field effect in the beam particle distribution including the energy gain. The goal was to find the field that preserve the beam emittance when it passes through the cavity. Therefore, an appropriated brightness is achieved when the beam is used for research activities like spectrometry of like secondary emission microscopy.

2. Cavity Dimensions

3. The electromagnetic resonance frequency f in the cylindrical cavity can be obtained from the angular frequency ω:

IOP Conf. Series: Journal of Physics: Conf. Series **912** (2017) 012039 doi:10.1088/1742-6596/912/1/012039

$$\omega = k_r c = \frac{2.405c}{R} \quad (1)$$

Where R is the radius of the cavity, and $\omega = 2\pi f$, and c the speef of light by setting the frequency f as fix we can calculate R [2].

The number of resonant modes will depend of the number of coupled cavities; the coupling is done ; the coupling is made by a small hole between them called the iris with a radius r_i. The iris radius needs to be smaller than the cavity radius in order to induce the different resonants in the cavity.

It is intended to acquire an RF generator in the S-band, 2998 MHz, with a 2 MW output power, setting this frequency as input, we can calculate the necessary radius for the cavity using the equation (1), then

$$R = \frac{2.405c}{2\pi f} \quad (2)$$

It is important that have the same input frequency as the resonant frequency because is the one that maximize the power transmission to the cavity. There are another aspect that can impact in the generated electric field in the cavity as length and iris aperture between the multiple cavities.

In order to set an appropriated cavity length to accelerate electrons, it is necessary to take into account the relativistic speed of the electrons at the input of each cavity and take in to account the time dependence of the generated electric field within cavity to find the maximum electron acceleration.

For example, by considering two parallel infinite planes separated by a distance L with an applied sinusoidal voltage, taking z=0 as the cavity start, also assuming the electric field E in the planes to be uniform (we ignore the iris holes effect), then the field is directed in the z direction and is given by:

$$E_z(t) = E_0 \cos(\omega t + \varphi) \quad (3)$$

Where φ is the temporal phase at which the electron enters to the acceleration cavity[2],

Then the gained energy in the acceleration gap for a charge particle q that crosses it is given by the equation

$$\Delta E = qE_0 \int_0^L \cos\,[\omega t(z) + \varphi]\; dz \quad (4)$$

Assuming that the energy gain is small enough that there is and negligible change in the electron speed through the cavity, then the total traveling time of the particle through the cavity is given by $t(z) = \frac{z}{v}$, and then $\omega t(z) = \omega \frac{z}{v} = 2\pi f \frac{z}{\beta c}$, now there is a direct connection between the time and the cavity frequency phase $\omega t(z)$ of the wave that subsists inside the cavity.

In the case of three cavities coupled together there will be three resonance frequencies the 0, $\frac{\pi}{2}$, and $\frac{\pi}{1}$ modes. The selected resonant acceleration mode is the $\frac{\pi}{2}$mode, then synchrony of the particles with the RF will be $\omega t(z) = \frac{\pi}{2}$. So, if we choose the length of the cavity in the direction of z, $z = L$, we can calculate the width of our cavity, if we know the energy of the electrons coming to the cavity

$$z = L = \frac{\beta c}{4f} \tag{5}$$

Where β is the relativistic factor v/c, once the basic cavity length is set the next steps is to make the electromagnetic calculations using simulations codes [3].

4. Electromagnetic Design

A 3-cell RF cavity has been designed to accelerate in the $\frac{\pi}{2}$ resonant mode because it is the one that presents the higher gradient accelerations. The electric field map density simulations results are shown in the Figure 2 where the iris connection between the cavities is set to 10 mm.

Figure 1. Electric Field effect in the RF Cavity for an input energy of 100 KeV, it can be observed that there is a large concentration of field in the lower corners of the first and third cavity.

Once we found the correct input phase φ that maximize the time dependent component of the RF electromagnetic field that will interact with an electron traveling in the middle of the cavity(r=0) we can set the maximum energy gain for that cavity design, Figure 3 shows the electric field experimented by the electron traveling in the cavity, as we can see the electric field is always positive leading to an optimum acceleration or energy gain.

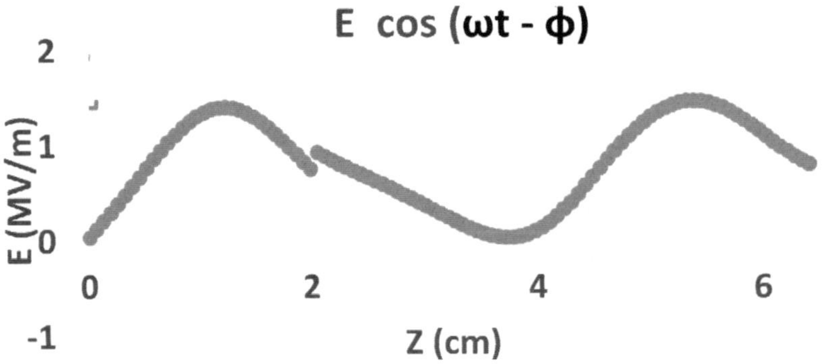

Figure 2. Electric Field Distribution in the cavity center (r=0). The phase φ is chosen in such way that the particles are accelerated by the greater part of the RF, if φ is not chosen correctly the particles can be accelerated towards the source.

5. Beam Dynamics

Once the correct set of acceleration parameters are set for a single electron traveling inside the cavity the next step consist in improve the design to accelerate particle distributions, another important feature of an RF cavity is to minimize the growth of the beam longitudinal and transversal phase space area(figure 5,6), if the phase space area increase beam losses and the minimum beam size that can be focused increase, we perform a campaign of studies to improve the design and accelerate particle distributions instead.
The input beam for simulations is a 100 keV Gaussian profile (Figure 4) of 3 mm 1 σ r.m.s allowing to have particles close to the iris edges inside the cavity. We use the code Travel [4] to perform all the beam dynamics simulations.

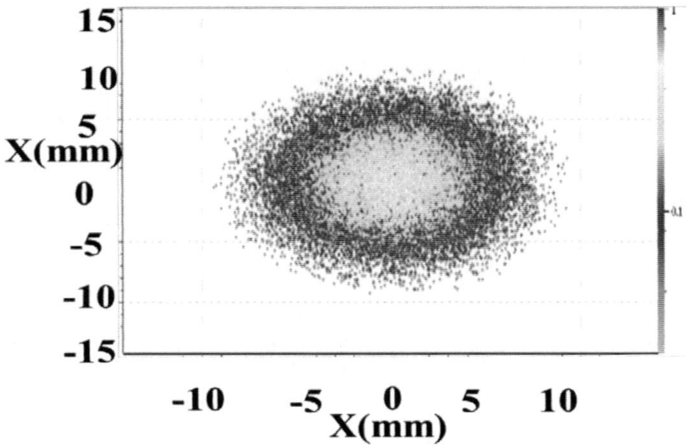

Figure 3. Cross section of the Beam at the end of the cavity. The hole in our cavity is approximately 1 cm in radius, which means that the particles do not collide with cavity walls.

The final transverse phase space shown in figure 5 has an emittance of 2.7796 1 σ r.m.s Pi.mm.mrad with some particles showing divergence bigger than 50 mrad, this effect is due the iris edge effect than can be seen in (Figure 2), then by limit our input beam to 2 mm 1 σ r.m.s we can keep the particles divergences within our needs to have a homogeneous beam at the end of the acceleration system

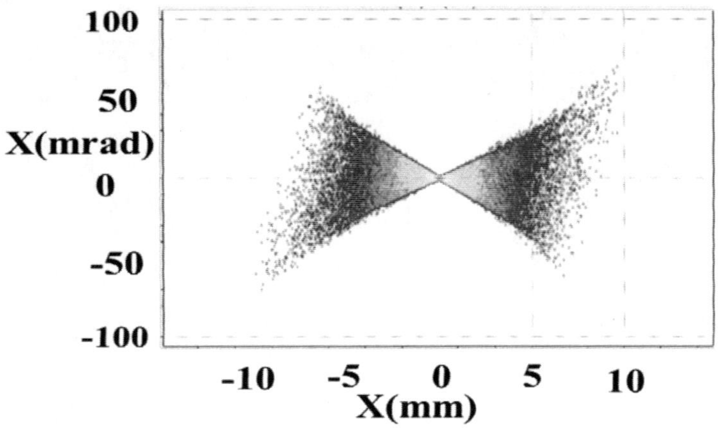

Figure 4. Phase space result of the simulation. The beam divergence within the 5 mm radius shows a divergence below the 50 mrad indicating that the beam is suitable to our desired aplications[5].

In the longitudinal dynamics the cavity a can accelerate the particles properly the particles ±10 degrees within the energy spread of 2% that is the energy spread at the moment on introduce the particles to the acceleration cavity.

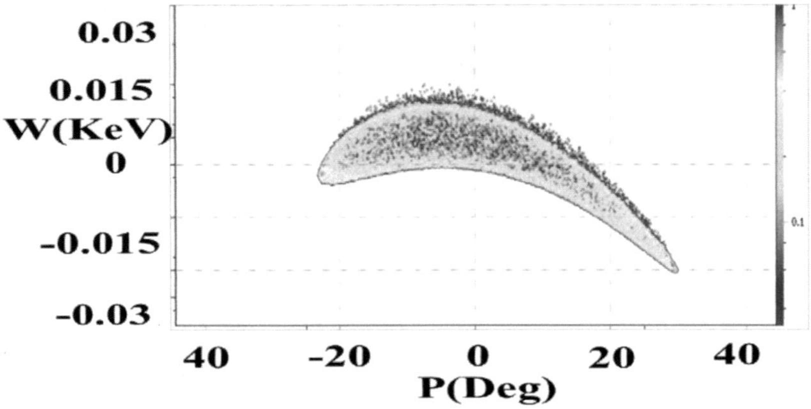

Figure 5. Phase space energy distribution in the longitudinal plane (vertical axis centered 500 KeV). The energy spread is 2% in the center of the acceleration bucket similar to the input beam energy spread.

With the obtained results from simulations we can set the transversal and longitudinal limits to for our input beam and also it works as a feedback loop to reshape the cavity in several iterations. Now the design need to be adapted to the machining tools that are available in the university workshop.

6. Mechanical Design

The cavity mechanical design was performed in the code Inventor AutoCAD[6] which allows us to study the mechanical strength and thermal loads of the cavity once the final design was set see Figure 6, the code also can generate G-code for the pieces to do the machining in a CNC 3-Axis Vertical Milling.

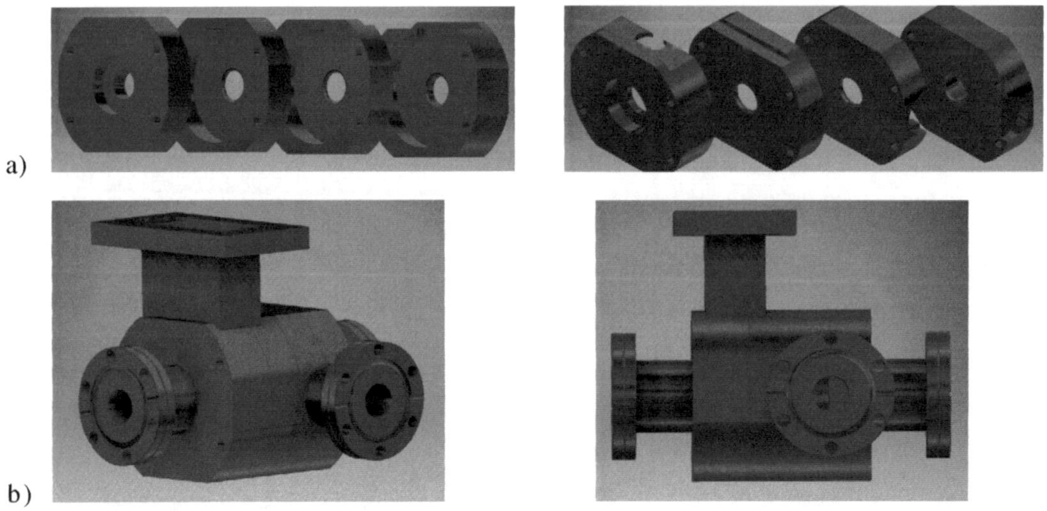

Figure 6. a) Design of individual pieces and b) Assembly of parts. The pieces will be machined in copper, it is thought to make the assembly by welding.

IOP Conf. Series: Journal of Physics: Conf. Series **912** (2017) 012039 doi:10.1088/1742-6596/912/1/012039

At the moment of this workshow, we are fininshing the machining of the first RF cavity and and is foreseeing to finish all the cavity machining during this year.(Figure 7).

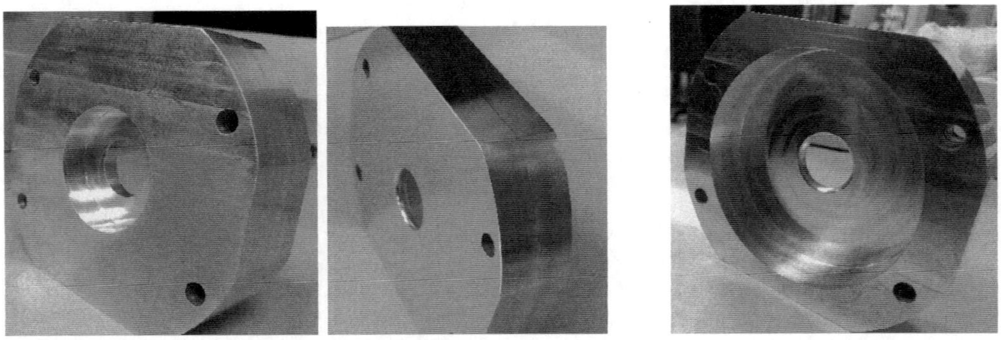

Figure 7. Progress in the RF cavity machining, the pieces were created in a CNC 3-Axis vertical milling center, the polish procedure still missing.

7. Final Remarks

By taking into consideration the relativistic change in the electron speed ty through the acceleration cavity we can optimize the acceleration efficiency of the single electron beam an latter improve the design to enhance the beam emittance in the cavity by reshaping the cavity taking in to account the effect of the cavity in the beam distribution.

We are close to build the first RF cavity in Mexico, first step in the construction of the first High Energy RF accelerator (from 300KeV up to 5 MeV), with a high-brillance beam.

References

[1] Los Alamos National Laboratory, Poisson SuperFish Program
[2] Thomas P. Wangler, 2008, RF Linear Accelerators, Editorial Wiley-VCH.
[3] Virgilio Acosta, Clyde Cowan, Billie J. Graham, 1999, Oxford University Press
[4] Travel v4.07, User Manual, April 2007, CERN
[5] Wiedemann, 2007, Particle Accelerator Physics, Editorial Springer
[6] Autodesk Inventor, https://www.autodesk.mx/products/inventor/overview

IOP Conf. Series: Journal of Physics: Conf. Series **912** (2017) 012040 doi:10.1088/1742-6596/912/1/012040

Decay of a pseudoscalar into two photons

J. I. Aranda[1], E. Cruz-Albaro[1], D. Espinosa-Gómez[1], J. Montaño[1,2], F. Ramírez-Zavaleta[1], E. S. Tututi[1]

[1] Facultad de Ciencias Físico Matemáticas, Universidad Michoacana de San Nicolás de Hidalgo, Avenida Francisco J. Mújica S/N, 58060, Morelia, Michoacán, México.
[2] CONACYT.

E-mail: feramirez@umich.mx

Abstract. The existence of new particles whose masses could be found in the region of TeVs are predicted by extensions of the Standard Model, one of these is the Littlest Higgs model, where it is proposed the existence of a new neutral massive pseudoscalar particle. In this context we analyze the $\Phi^P \to \gamma\gamma$ decay, where γ symbolizes the photon and Φ^P is the pseudoscalar particle predicted by this model.

1. Introduction

Currently, the ATLAS [1] and CMS [2] Collaborations continue with the search for new particles, such as the Randall-Sundrum spin-2 boson or new heavy scalar particles [3, 4]. In this sense, there are several models that predict the existence of more scalar particles, such as the two-Higgs doublet models (2HDMs) [5, 6], three-Higgs doublet model (3HDM) [7], Higgs-singlet extension model [8], little Higgs models (LHMs) [9], etc. Among the large variety of LHMs, we will focus on the simplest model, which is the Littlest Higgs model (LTHM) because it does not contain new degrees of freedom below the TeVs scale. Particularly, the LTHM has a reduced particle spectrum and does not consider additional discrete symmetries. Within this model, it is proposed to study the $\Phi^P \to \gamma\gamma$ decay at one loop level, since the coupling $\Phi^P\gamma\gamma$ is not present at the tree level. In particular, we are interested in estimating the branching ratio of this process as a function of the parameter f. In the same context, a rough analysis for the production cross section of the Φ^P boson via gluon fusion at LHC is presented.

2. The LTHM model

The LTHM model is constituted by a nonlinear sigma model with $SU(5)$ global symmetry together with the gauge group $[SU(2)_1 \times U(1)_1] \times [SU(2)_2 \times U(1)_2]$ [10, 11], where the $SU(5)$ group is spontaneously broken to SO(5) at the energy scale f, which is of the order of TeV. At the same time, the $[SU(2)_1 \times U(1)_1] \times [SU(2)_2 \times U(1)_2]$ group is also broken to its subgroup $SU(2)_L \times U(1)_Y$, which is identified as the electroweak gauge group of the Standard Model (SM) [11, 12]. Due to the global symmetry breaking 14 Goldstone bosons are generated, which are transformed under the $SU_L(2) \times U_Y(1)$ group as a real singlet $\mathbf{1}_0$, a real triplet $\mathbf{3}_0$, a complex doublet $\mathbf{2}_{\pm 1/2}$ and a complex triplet $\mathbf{3}_{\pm 1}$ [11]. The real singlet and the real triplet are absorbed by longitudinal components of the gauge bosons at the energy scale f, while the complex doublet

and the complex triplet remain massless [11].
On the other hand, the mass of the new pseudoscalar particle Φ^P is given by [11]

$$m_\Phi = \frac{\sqrt{2} m_H}{\sqrt{1-y_v^2}} \frac{f}{v}, \tag{1}$$

where $y_v = 4v'f/v^2$, being v and v' the vacuum expectation values of the doublet and triplet, respectively. This mass expression is definite positive if the following relation is satisfied

$$\frac{v'^2}{v^2} < \frac{v^2}{16 f^2}\,. \tag{2}$$

3. The $\Phi^P \to \gamma\gamma$ process

The Feynman rules involved in the $\Phi^P \to \gamma\gamma$ decay are summarized in Table 1 [11].

$\Phi^P \overline{u} u$	$-\frac{m_u}{\sqrt{2} v}(\frac{v}{f} - \sqrt{2} s_p)\gamma^5$
$\overline{u} u A_\mu$	$-ieQ_u \gamma^\mu$

Table 1. LTHM and SM couplings.

The Feynman diagrams corresponding to the $\Phi^P \to \gamma\gamma$ decay are shown in Fig. 1, where inside the loops are circulating quarks of the SM.

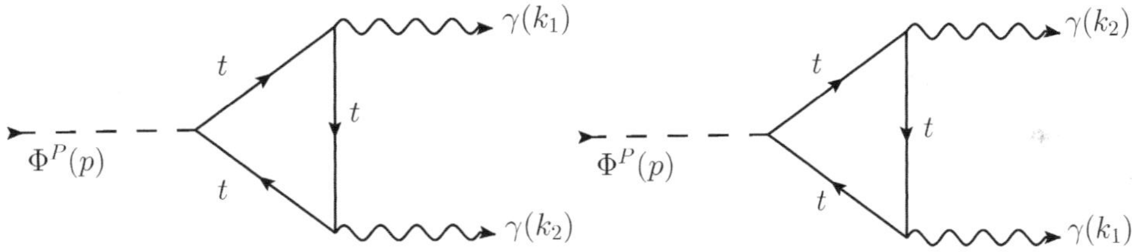

Figure 1. Decay $\Phi^P \to \gamma\gamma$.

The kinematic conditions of this decay are: $k_1^2 = k_2^2 = 0$, $k_1 \cdot k_2 = m_{\Phi^P}^2/2$ and $p^2 = m_{\Phi^P}^2$, where p is the four-moment of Φ^P, k_1 and k_2 are the four-moments of the photons. Considering the above and after the application of the Passarino-Veltaman reduction scheme, the amplitude of the decay takes the form

$$\mathcal{M}(\Phi^P \to \gamma\gamma) = A^{\gamma\gamma} \epsilon^{\mu\nu\alpha\beta} k_{1\alpha} k_{2\beta} \epsilon_\mu^*(k_1) \epsilon_\nu^*(k_2), \tag{3}$$

where the Levi-Civita tensor is a distinctive feature of the pseudoscalar nature of this process [14]. The form factor is

$$A^{\gamma\gamma} = \frac{-g^2\, N_C\, s_W^2}{9\sqrt{2}\, f\, \pi^2} m_t^2 C_0(0,0,m_{\Phi^P}^2, m_t^2, m_t^2, m_t^2), \tag{4}$$

where $C_0(0,0,m_{\Phi^P}^2, m_t^2, m_t^2, m_t^2)$ is the three-point Passarino-Veltman scalar function, m_t is the top quark mass, $N_C = 3$ is the color factor due to quarks circulating in the loop, and $s_W \equiv \sin\theta_W$ being θ_W the weak mixing angle. The form factor is free of ultraviolet divergences and its corresponding Lorentz structure satisfies gauge invariance.

4. Analysis of the results

After using the decay width formula [13], and after performing some algebraic operations we find that the decay width for our decay is

$$\Gamma(\Phi^P \to \gamma\gamma) = \frac{1}{64\pi} |A^{\gamma\gamma}|^2 m_{\Phi^P}. \tag{5}$$

In this work we use the LoopTools [15] software for the numerical evaluation of the decay width $\Gamma(\Phi^P \to \gamma\gamma)$. Before continuing the analysis of results, we recall that m_{Φ^P} is a function of the energy scale f in which the global symmetry is broken. Note that this energy scale is the only free parameter for testing. This symmetry breaking scale is constrained by the experimental data to $2 \leq f \leq 4$ TeV [16]. Thus, the $m_{\Phi^P}(f)$ quantity is a monotonous increasing function, being restricted to $1.66 \leq m_{\Phi^P}(f) \leq 3.32$ TeV. In Fig. 2(a) is shown the behavior of $\Gamma(\Phi^P \to \gamma\gamma)$ as a function of the energy scale f, we can clearly see that the decay width is entirely of the order of 10^{-7} in the interval $2 \leq f \leq 4$ TeV. In Fig. 2(b) is presented the corresponding branching ratio as a function of the energy scale f, it is observed that the branching ratio is of the order of 10^{-7}, for interval 2 TeV $\leq f \leq 2.7$ TeV. It should be noted that the total decay width Γ_{Φ^P} is conformed by the channels ZH, WWZ, WWH, tt.

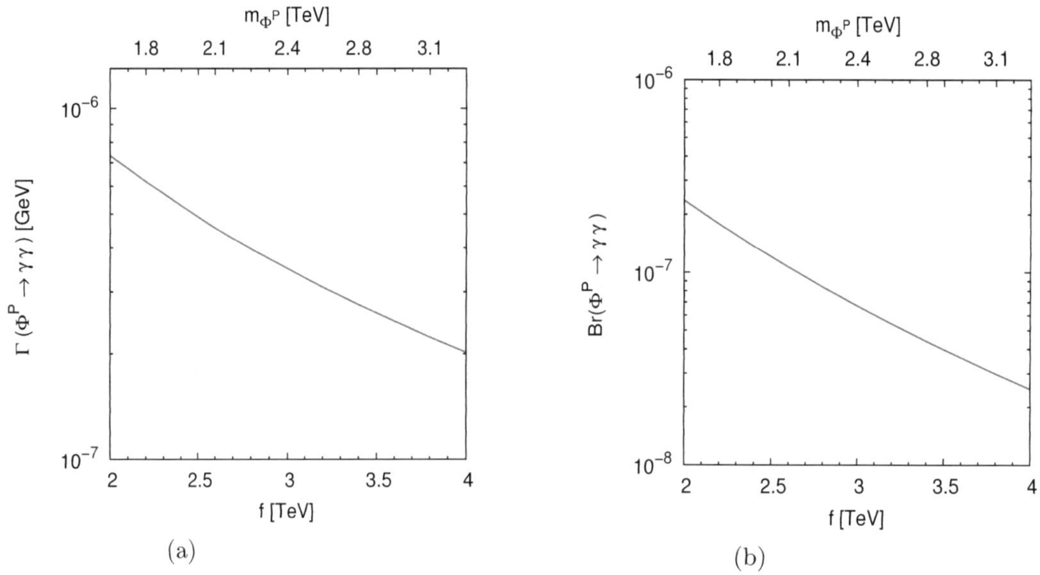

Figure 2. (a) $\Gamma(\Phi^P \to \gamma\gamma)$ and (b) $\mathrm{Br}(\Phi^P \to \gamma\gamma)$ as a function of the f parameter.

4.1. Production of the heavy neutral pseudoscalar boson

Here we develop a rough study for the production cross section of the scalar Φ^P at LHC. To carry out this, we use the Breit-Wigner resonant cross section [17]. In this way, the production cross section via gluon fusion can be computed through the branching ratios $\mathrm{Br}(\Phi^P \to gg)$ and $\mathrm{Br}(\Phi^P \to \gamma\gamma)$. Therefore, our Breit-Wigner cross section is written as

$$\sigma(gg \to \Phi^P \to \gamma\gamma) = \frac{\pi}{12}\frac{\mathrm{Br}(\Phi^P \to gg)\ \mathrm{Br}(\Phi^P \to \gamma\gamma)}{m^2_{\Phi^P}}, \tag{6}$$

where $\sigma(gg \to \Phi^P \to \gamma\gamma)$ is estimated at the resonance of the Φ^P boson [17]. The branching ratio of the $\Phi^P \to gg$ process is

$$\mathrm{Br}(\Phi^P \to gg) = \frac{1}{8\pi}\frac{|A^{gg}|^2 m_{\Phi^P}^3}{\Gamma_{\Phi^P}}, \tag{7}$$

with $A^{gg} = \frac{g_s^2}{8\sqrt{2}\pi^2 f} m_t^2 C_0(m_{\Phi^P}^2, 0, 0, m_t^2, m_t^2, m_t^2)$, being Γ_{Φ^P} the total decay width of Φ^P.

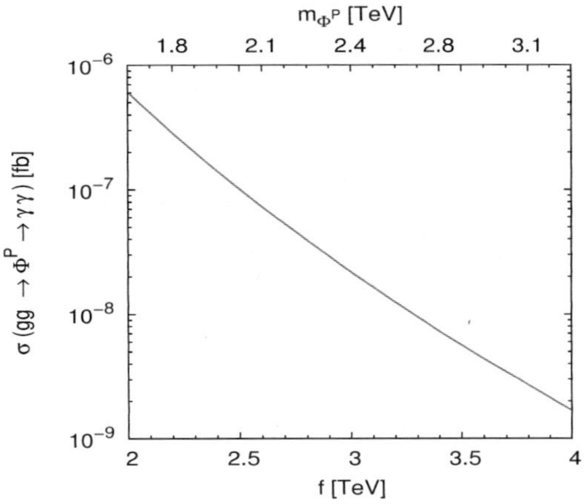

Figure 3. Cross section of $gg \to \Phi^P \to \gamma\gamma$ as a function of f.

In Fig. 3 we show the Breit-Wigner cross section for the Φ^P production as a function of the f parameter from 2 TeV to 4 TeV. The expected integrated luminosity of the LHC at the last stage of operation is planned to be around 3000 fb^{-1} [18]. In this context, from our results shown in Fig. 3, the corresponding $\sigma(gg \to \Phi^P \to \gamma\gamma)$ is 6.077×10^{-7} fb for $f = 2$ TeV, which implies that this observable related with the $\Phi^P \to \gamma\gamma$ is very suppressed and it would be difficult to observe at the LHC.

5. Conclusions

The LTHM is constituted by a nonlinear sigma model together with a $SU(5)$ global symmetry and a subgroup of gauge $[SU(2)_1 \times U(1)_1] \times [SU(2)_2 \times U(1)_2]$, where the presence of new particles with masses of the order of several TeVs is predicted. From all new particles, a new neutral pseudoscalar particle, Φ^P, is the object of analysis in this work. We have established a study region from 2 TeV to 4 TeV for the energy scale f, which represents a mass range for m_{Φ^P} between 1.66 TeV and 3.32 TeV, respectively. In this energy range, the width decay found for the $\Phi^P \to \gamma\gamma$ process is of the order of 10^{-7} GeV, while the branching ratio computed is of the order of 10^{-7} on the range 2 TeV$< f <$ 2.7 TeV. Finally, an estimate for the production cross section of the Φ^P boson via gluon fusion was implemented. For $f = 2$ TeV, the $\sigma(gg \to \Phi^P \to \gamma\gamma) \sim 10^{-7}$ fb, which implies that this observable is very suppressed and it would be difficult to observe at the LHC.

Acknowledgments
This work has been partially supported by SNI-CONACYT and CIC-UMSNH (México).

IOP Conf. Series: Journal of Physics: Conf. Series **912** (2017) 012040 doi:10.1088/1742-6596/912/1/012040

References

[1] G. Aad *et al.* [ATLAS Collaboration], Phys. Lett. B716, 1 (2012).
[2] S. Chatrchyan *et al.* [CMS Collaboration], Phys. Lett. B716, 30 (2012).
[3] CMS Collaboration, CMS-PAS-EXO-15-004; The CMS Collaboration, Phys. Rev. Lett. 117, 051802 (2016).
[4] The ATLAS collaboration, ATLAS-CONF-2015-081; The ATLAS collaboration, JHEP 1609, 001 (2016).
[5] T. D. Lee, Phys. Rev. D8, 1226 (1973).
[6] G. C. Branco, P. M. Ferreira, L. Lavoura, M. N. Rebelo, M. Sher, and J. P. Silva, Phys. Rept. 516, 1 (2012).
[7] I. P. Ivanov, V. Keus, and E. Vdovin, J. Phys. A: Math. Theor. 45, 215201 (2012); I. P. Ivanov and E. Vdovin, Eur. Phys. J. C 73, 2309 (2013); V. Keus, S. F. King, and S. Moretti, JHEP 01, 052 (2014).
[8] G. M. Pruna and T. Robens, Phys. Rev. D88, 115012 (2013).
[9] N. Arkani-Hamed, A. G. Cohen, and H. Georgi, Phys. Lett. B513, 232 (2001); N. Arkani-Hamed, A. G. Cohen, T. Gregoire, and J. G. Wacker, JHEP 08, 020 (2002); N. Arkani-Hamed, A. G. Cohen, E. Katz, A. E. Nelson, T. Gregoire, and J. G. Wacker, JHEP 08, 021 (2002); I. Low, W. Skiba, and D. Smith, Phys. Rev. D66, 072001 (2002).
[10] N. Arkani-Hamed, A. G. Cohen, E. Katz, and A. E. Nelson, JHEP 07, 034 (2002).
[11] T. Han, H. E. Logan, B. McElrath, and L. -T. Wang, Phys. Rev. D67, 095004 (2003).
[12] J. I. Aranda, I. Corts-Maldonado, F. Ramírez-Zavaleta, and E. S. Tututi, J. Phys. G 43, 025002 (2016).
[13] K. A. Olive *et al.*, Chin. Phys. C38, 090001 (2014).
[14] B. A. Kniehl, Nucl. Phys. B 357, 439 (1991).
[15] T. Hahn and M. Perez-Victoria, Comput. Phys. Commun. 118, 153 (1999).
[16] J. Reuter and M. Tonini, JHEP 1302, 077 (2013).
[17] C. Patrignani *et al.*, Chin. Phys. C 40, 100001 (2016).
[18] ATLAS Collaboration, arXiv:1307.7292.

IOP Conf. Series: Journal of Physics: Conf. Series **912** (2017) 012041 doi:10.1088/1742-6596/912/1/012041

Decays $\Phi^P \to \gamma Z, ZZ$ in the context of the Littlest Higgs Model

J. I. Aranda[1], **E. Cruz-Albaro**[1], **F. Ramírez-Zavaleta**[1], **J. Montaño**[1,2], **E. S. Tututi**[1]

[1] Facultad de Ciencias Físico Matemáticas, Universidad Michoacana de San Nicolás de Hidalgo, Avenida Francisco J. Mújica S/N, 58060, Morelia, Michoacán, México
[2] CONACYT

E-mail: `feryuphy@yahoo.com.mx`

Abstract. In this study we analyze the decay of the neutral pseudoscalar Φ^P predicted by the Littlest Higgs Model (LTHM), an extension of the Standard Model (SM) that seeks to solve the hierarchy problem. We focus our research in the decays $\Phi^P \to \gamma Z, ZZ$ at the one-loop level. Here, we analyze the branching ratios of these decays and the production cross section via gluon fusion as function of the new scale of energy f, which characterizes the scale of energy of the breaking of the global symmetry and whose values are restricted to the range of 2 to 4 TeV.

1. Introduction

The experimental collaborations in the LHC [1, 2] are seeking for a wide variety of exotic particles, in particular, new heavy scalar bosons. The LTHM is one of the most simple extensions of the Little Higgs models, since it only presents new degrees of freedom [3, 4, 5] at the scale of TeVs. In addition, it predicts a reduced spectrum of new scalar particles [6], which is not the case in other versions of these type of models. The Little Higgs models are interesting because they offer a possible explanation to the hierarchy problem of the SM. The LTHM represents another approach to the electroweak symmetry breaking pattern based on the dimensional deconstruction [6].

2. The Littlest Higgs Model

The LTHM is built using a nonlinear sigma model with $SU(5)$ global symmetry together with a gauged group $[SU(2)_1 \otimes U(1)_1] \otimes [SU(2)_2 \otimes U(1)_2]$ [6, 7]. Firstly, the $SU(5)$ group is spontaneously broken to the $SO(5)$ group at the energy scale f. Simultaneously, the $[SU(2)_1 \otimes U(1)_1] \otimes [SU(2)_2 \otimes U(1)_2]$ group is broken to its subgroup $SU_L(2) \otimes U_Y(1)$, being the latter the electroweak gauge group of the SM. At the scale f, the spontaneous global symmetry breaking is generated by the vacuum expectation value (VEV) of the Σ field, identified as Σ_0. The structure of the Σ field is given as follows

$$\Sigma = e^{i\Pi/f}\Sigma_0 e^{i\Pi^T/f}, \tag{1}$$

with

$$\Sigma_0 = \begin{pmatrix} \mathbf{0}_{2\times2} & \mathbf{0}_{2\times1} & \mathbf{1}_{2\times2} \\ \mathbf{0}_{1\times2} & 1 & \mathbf{0}_{1\times2} \\ \mathbf{1}_{2\times2} & \mathbf{0}_{2\times1} & \mathbf{0}_{2\times2} \end{pmatrix}, \tag{2}$$

and Π being the Goldstone boson matrix [7]. On the other hand, the effective Lagrangian invariant under the $[SU(2)_1 \otimes U(1)_1] \otimes [SU(2)_2 \otimes U(1)_2]$ group is composed of [7]

$$\mathcal{L}_{LTHM} = \mathcal{L}_G + \mathcal{L}_F + \mathcal{L}_\Sigma + \mathcal{L}_Y - V_{CW}, \tag{3}$$

where $\mathcal{L}_G$ represents the gauge bosons kinetic contributions, $\mathcal{L}_F$ contains the fermion kinetic contributions, $\mathcal{L}_\Sigma$ includes the nonlinear sigma model contributions of the LTHM and $\mathcal{L}_Y$ encompasses the Yukawa couplings of fermions and pseudo-Goldstone bosons. The V_{CW} term is the Coleman-Weinberg potential [8]. For the nonlinear sigma sector, we have the Lagrangian density [7]

$$\mathcal{L}_\Sigma = \frac{f^2}{8}\mathrm{tr} \mid \mathcal{D}_\mu \Sigma \mid^2, \tag{4}$$

where the covariant derivative is

$$\mathcal{D}_\mu \Sigma = \partial_\mu \Sigma - i \sum_{j=1}^{2} \left[g_j \sum_{a=1}^{3} W_{\mu j}^a (Q_j^a \Sigma + \Sigma Q_j^{aT}) + g_j^{'} B_{\mu j} (Y_j \Sigma + \Sigma Y_j^T) \right], \tag{5}$$

being $W_{\mu j}^a$ the gauge fields of the $SU(2)$ group, the $B_{\mu j}$ are the $U(1)$ gauge fields, Q_j^a and Y_j are the generators of $SU(2)$ and $U(1)$ groups, respectively. The quantities g_j and $g_j^{'}$ are the coupling constants of the $SU(2)$ and $U(1)$ groups. After the spontaneous symmetry breaking around Σ_0, the mass eigenstates of the order of the energy scale f are generated. All the details of the LTHM and the corresponding Feynman rules can be found in Ref. [7].

3. Decays $\Phi^P \to \gamma Z, ZZ$ at one-loop level

Because the main subject of this work is to analyze the one-loop level decays $\Phi^P \to \gamma Z, ZZ$, we present some samples of Feynman rules involved in the calculation of the amplitudes [7]:

$$\begin{aligned} \Phi^P \bar{u} u &\to -\frac{m_u}{\sqrt{2} v}(\frac{v}{f} - \sqrt{2} s_P)\gamma^5, \\ Z \bar{u} u &\to \frac{i g \gamma_\mu}{2 c_W}(g_V^f - g_A^f \gamma^5), \\ A \bar{u} u &\to i e Q_f \gamma_\mu, \end{aligned} \tag{6}$$

where u denotes the up quark, g_V^f and g_A^f are the weak charges given by

$$\begin{aligned} g_V^f &= T_3^f - 2 Q_f s_W^2, \\ g_A^f &= T_3^f, \end{aligned} \tag{7}$$

where $T_3^f = -\frac{1}{2}$ for the charged leptons and down type quarks, and $T_3^f = \frac{1}{2}$ for the up type quarks. Here $s_W \equiv \sin\theta_W$ and $c_W \equiv \cos\theta_W$, being θ_W the weak angle. In addition, Q_f is the charge of the fermion in units of e.

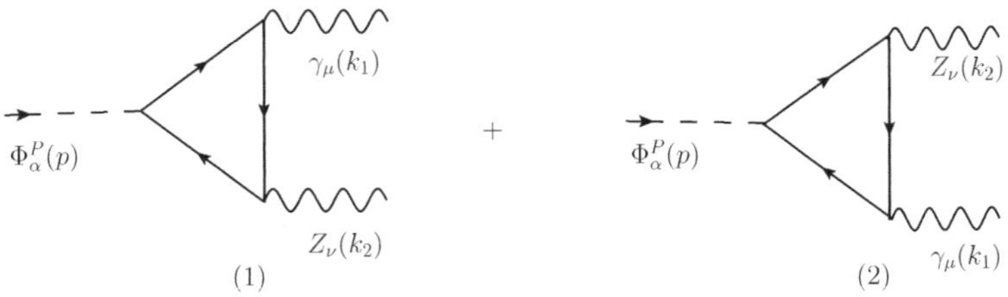

Figure 1: Feynman diagrams contributing to the $\Phi^P \to \gamma Z$ process at the one-loop level.

For the $\Phi^P \to \gamma Z$ process that involves fermions, this decay occurs through two triangle diagrams, as shown in Fig. 1. Thus, the amplitude is obtained as a result of the sum of the two contributions. Here the top quark provides the dominant contribution to the quantum fluctuation. So, for the $\Phi^P \to \gamma Z$ decay, the amplitude is

$$\mathcal{M}(\Phi^P \to \gamma Z) = A^{\gamma Z} \epsilon^{\mu\nu\alpha\beta} k_{1\alpha} k_{2\beta}\ \epsilon^*_\mu(k_1)\epsilon^*_\nu(k_2), \tag{8}$$

where the form factor $A^{\gamma Z}$ is finite,

$$A^{\gamma Z} = \frac{g^2 N_C\ s_W(3 - 8\ s_W^2)}{72\sqrt{2}\ \pi^2 c_W f} m_t^2 C_0(1), \tag{9}$$

with $C_0(1) = C_0(m_Z^2, m_{\Phi^P}^2, 0, m_t^2, m_t^2, m_t^2)$ the Passarino-Veltman scalar function, N_C is the color factor, 3 for quarks and 1 for leptons; f is the scale of the global symmetry breaking, m_t is the mass of the top quark. Thus, the decay width is

$$\Gamma(\Phi^P \to \gamma Z) = \frac{1}{32\ \pi\ m_{\Phi^P}^3}(m_{\Phi^P}^2 - m_Z^2)^3 \mid A^{\gamma Z} \mid^2 . \tag{10}$$

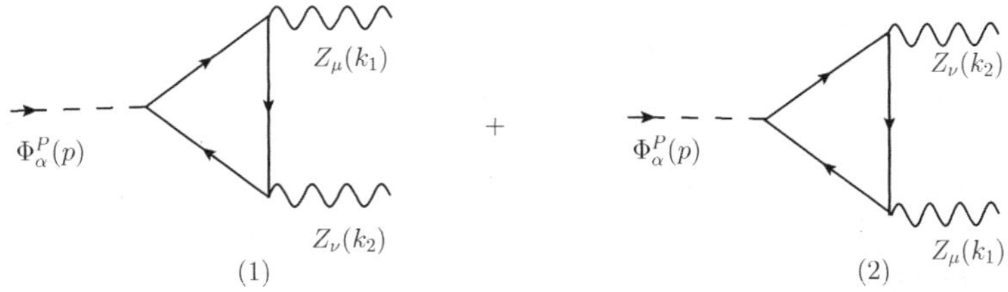

Figure 2: Feynman diagrams that contribute to the $\Phi^P \to ZZ$ decay at one-loop level.

Concerning to the $\Phi^P \to ZZ$ (see Fig. 2) decay, this also has two contributions, the amplitude is

$$\mathcal{M}(\Phi^P \to ZZ) = A^{ZZ} \epsilon^{\mu\nu\alpha\beta} k_{1\alpha} k_{2\beta}\ \epsilon^*_\mu(k_1)\epsilon^*_\nu(k_2), \tag{11}$$

with the form factor A^{ZZ} is

$$A^{ZZ} = \frac{g^2 N_C m_t^2}{144\sqrt{2}\ \pi^2\ c_W^2(4m_Z^2 - m_{\Phi^P}^2) f}\Bigg[\Big(4(3 - 4s_W^2)s_W^2 m_{\Phi^P}^2 + (3 - 8s_W^2)^2\ m_Z^2\Big)C_0(2) + 9\big(B_0(1) - B_0(2)\big)\Bigg], \tag{12}$$

being $C_0(2) = C_0(m_Z^2, m_Z^2, m_{\Phi^P}^2, m_t^2, m_t^2, m_t^2)$, $B_0(2) = B_0(m_Z^2, m_t^2, m_t^2)$ and $B_0(1) = B_0(m_{\Phi^P}^2, m_t^2, m_t^2)$. This amplitude is free of ultraviolet divergences, because the B_0 functions appear as a difference from each other in the equation (12), while C_0 is finite. The gauge structure for this process also satisfies electromagnetic gauge invariance. The decay width is

$$\Gamma(\Phi^P \to ZZ) = \frac{1}{64\,\pi}(m_{\Phi^P}^2 - 4m_Z^2)^{3/2} \mid A^{ZZ} \mid^2 . \tag{13}$$

In this work we also present a study for the production cross section for the pseudoscalar Φ^P in the context of the LTHM at LHC, decaying into final states, γZ and ZZ. To carry out this task, we employ the Breit-Wigner resonant cross section [9]. The production cross section via gluon fusion can be computed by means of the branching ratios $\mathrm{Br}(\Phi^P \to gg)$ and $\mathrm{Br}(\Phi^P \to Y)$, where $Y = \gamma Z, ZZ$. Thus, our Breit-Wigner cross section is

$$\sigma(gg \to \Phi^P \to Y) = \frac{\pi}{12}\frac{\mathrm{Br}(\Phi^P \to gg)\mathrm{Br}(\Phi^P \to Y)}{m_{\Phi^P}^2}, \tag{14}$$

where $\sigma(gg \to \Phi^P \to Y)$ is estimated at the resonance of the pseudoscalar Φ^P.

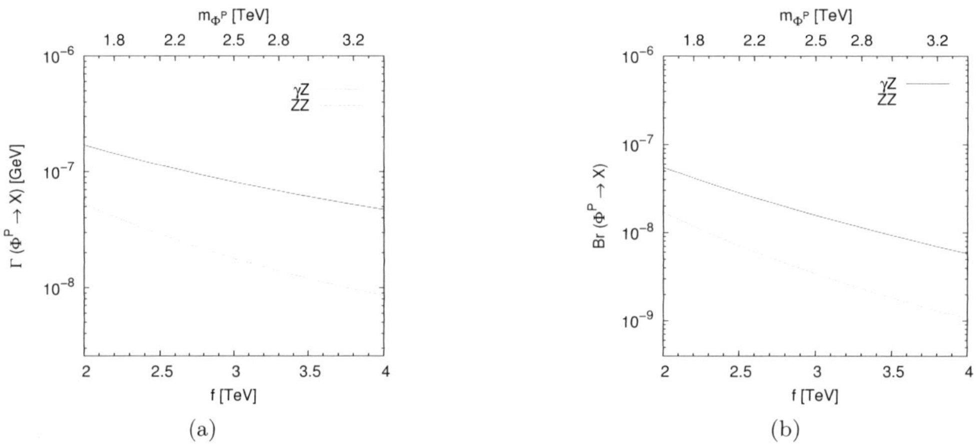

Figure 3: (a) Decay widths for the $\Phi^P \to \gamma Z, ZZ$ processes as a function of the f energy scale. (b) Branching ratios for the same decays on the f energy scale.

4. Results and conclusions

The numerical evaluation of the $\Phi^P \to \gamma Z, ZZ$ processes was carried out using the programs FeynCalc [10] and LoopTools [11]. We stress that the energy scale f is the only free parameter with which we can play. Thus, in strict agreement with the electroweak precision observables of the SM, the energy scale f is found around $3 - 4$ TeV at 99% C.L. [12]. However, we present our numerical analysis from $f = 3$ TeV to $f = 4$ TeV. For the decay widths of each process we have found: $\Gamma(\Phi^P \to \gamma Z) \sim 10^{-7}$ GeV and $\Gamma(\Phi^P \to ZZ) \sim 10^{-8}$ GeV for $f = 2$ TeV, as shown in Fig. 3. In addition, we have calculated its corresponding branching ratios that turned out to be $\mathrm{Br}(\Phi^P \to \gamma Z) \sim 10^{-7}$ and $\mathrm{Br}(\Phi^P \to ZZ) \sim 10^{-8}$ for f around 2 TeV (see Fig. 3). For the cross section, its behavior is depicted in the Fig. 4. In specific, $\sigma(gg \to \Phi^P \to \gamma Z) \sim 10^{-7}$ fb and $\sigma(gg \to \Phi^P \to ZZ) \sim 10^{-8}$ fb for $f = 2$ TeV.

It should be noted that our work presents preliminary results which in the future will serve us as a basis for a more complete study of the various modes of decay of Φ^P in the context of the LTHM.

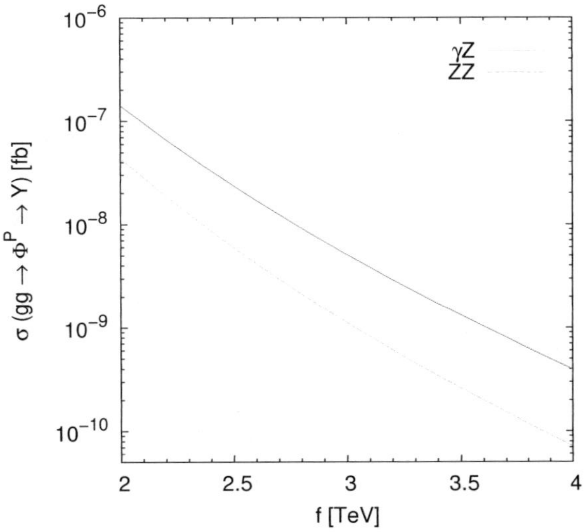

Figure 4: Cross section $\sigma(gg \to \Phi^P \to \gamma\gamma)$ as a function of the f energy scale.

Acknowledgments
This work has been partially supported by CONACYT.

References

[1] G. Aad et al. [ATLAS Collaboration], Phys. Lett. B **716**, 1 (2012).
[2] S. Chatrchyan et al. [CMS Collaboration], Phys. Lett. B **716**, 30 (2012).
[3] T. D. Lee, Phys. Rev., D**8**, 1226 (1973).
[4] G. C. Branco, P.M Ferreira, L. Lavoura, M.N. Rebelo, M. Sher, and J.P. Silva, Phys. Rept. **516**, 1 (2012).
[5] I. P. Ivanov, V. Keus, and E. Vdovin, J. Phys. A: Math. Theor. **45**, 215201 (2012); I. P. Ivanov and E. Vdovin, Eur. Phys. J. C **73**, 2309 (2013); V. Keus, S. F. King, and S. Moretti, JHEP **01**, 052 (2014).
[6] N. Arkani-Hamed, A. G. Cohen, E. Katz, and A. E. Nelson, JHEP **07**, 034 (2002).
[7] T. Han, H.E. Logan, B. McElrath, and L. -T. Wang, Phys. Rev. D**67**, 095004 (2003).
[8] S. R. Coleman and E. J. Weinberg, Phys. Rev. D**7**, 1888 (1973).
[9] C. Patrignani, et al., Chin. Phys. C **40**, 100001 (2016).
[10] R. Mertig, M. Bohm, and A. Denner, Comput. Phys. Commun. **64**, 345 (1991); V. Shtabovenko, R. Mertig and F. Orellana, arXiv:1601.01167.
[11] T. Hahn and M. Perez-Victoria, Comput. Phys. Commun. **118**, 153 (1999).
[12] J. Reuter and M. Tonini, JHEP **02**, 077 (2013).

Detecting physics beyond the Standard Model with the REDTOP experiment

D González[1], D León[1], B Fabela[1] and M I Pedraza[1]
[1]Benemérita Universidad Autónoma de Puebla, Puebla, México
E-mail: dario.gonzalez@alumno.buap.mx

Abstract. REDTOP is an experiment at its proposal stage. It belongs to the High Intensity class of experiments. REDTOP will use a 1.8 GeV continuous proton beam impinging on a fixed target. It is expected to produce about 10^{13} η mesons per year. The main goal of REDTOP is to look for physics beyond the Standard Model by detecting rare η decays. The detector is designed with innovative technologies based on the detection of prompt Cherenkov light, such that interesting events can be observed and the background events are efficiently rejected. The experimental design, the physics program and the running plan of the experiment is presented.

1. Introduction

Rare Eta Decays with a TPC for Optical Photons (REDTOP) is a new proposed experiment to search for physics beyond the Standard Model [1].
The experiment will have a branching ratio sensitivity level better than 10^{-10}, an estimated production of 2×10^{13} η mesons per year using a continuous proton beam with an average intensity of about 10^{11} protons per second impinging on fixed beryllium targets. These are the necessary requirements to produce enough statistical information about η decays. This experiment is expected to study C and CP violation as well as C, T, CP and CPT violations, lepton flavor violation; in addition, it will search for new scalar particles, new baryonic forces and new gauge bosons.

2. Why η meson?

The η meson is a pseudo Goldstone boson and an eigenstate of the C, P, CP and G operators, making it a very special particle and a good option to test C and CP invariance of the strong and electromagnetic interactions. Most of the symmetry violations in η and η' decays can be detected at a level of 10^{-5} or higher; In consequence, REDTOP will increase the η flux to reach the essential levels needed to detect interesting events.

3. Experimental apparatus

The detector

The REDTOP detector consists of the following components: the Optical Time Projection Chamber (OTPC), the ADRIANO Dual-Readout Calorimeter, the Muon Polarimeter and Photon Polarimeter. It will be sensitive to electrons, positrons, muons, photons and charged pions, all of them above the Cherenkov thresholds.

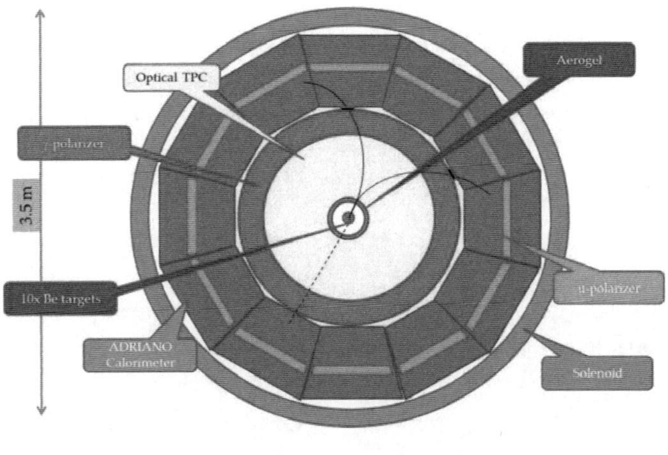

Figure 1. Transverse view of the detector.

The beam and target system
The target system is composed of ten beryllium foils with a thickness of 1/3 mm and 1 cm of diameter, it is located at the center of the beam pipe which is composed of carbon fiber or beryllium. The probability of a proton with an energy of 1.8 GeV interacting by inelastic scattering in any foil of beryllium is 0.5% and the probability to produce an η meson is of approximately 0.4%. Monte Carlo (MC) simulations show that the η decays are not affected by the beryllium foils. The experiment will use a beam with an energy of 1.8 GeV, provided by a cavity added to the Delivery Ring (DR) which will decelerate the beam coming from the Booster synchrotron.

The Optical Time Projection Chamber
The Optical Time Projection Chamber will measure the momentum and position of a charged track using the Cherenkov effect instead of the gas ionization. The momentum is reconstructed from the detection of the photons radiated and the photo sensors will detect them. The OTPC will be very sensitive to leptons and fast pions while hadrons and slower pions will not be recorded. Two Cherenkov radiators are present: a double aerogel cylinder of about 3 cm thick at the inner wall formed by an innermost and outermost aerogel with $n_D = 1.22$ and $n_D = 1.35$, respectively and a low-pressure nitrogen gas will fill the rest of the volume; in consecuence, pions and muons will be discriminated by dual refractive system while electrons and positrons with large momentum will radiate in the aerogel as well as in the gas.

The ADRIANO Dual-Readout Calorimeter
The ADRIANO (A Dual-Readout Integrally Active Non-segmented Option) calorimeter is based on two simultaneous measurements of the energy deposited by an hadronic or electromagnetic shower into two media with different properties. The first is a plastic scintillation, in which any charged particle going through it will generate scintillation. The second medium is a heavy glass with a refactive index of $n_D \geq 1.8$ and a density of $\rho > 5.5\ g/cm^3$, it will record signals of charged particles and absorb all the particles impinging on the detector. The measurements of scintillation and Cherenkov signals will provide the total energy and by comparing them, identify the initial particle. The main advantage of using a dual readout calorimeter to detect the photons from the η decays is that a large amount of background can be easily rejected.

IOP Conf. Series: Journal of Physics: Conf. Series **912** (2017) 012042 doi:10.1088/1742-6596/912/1/012042

The Muon and Photon Polarimeters
The polarimeters are composed of plastic scintillators inserted in specific areas of the detector. The muon polarimeter is inserted between the inner and the outer shells of ADRIANO calorimeter: it will count the e^+ and e^- emitted when a muon is stopped inside the polarimeter while the photon polarimeter is inserted inside the volume of the OTPC: it will count the number of photons decaying to e^+ e^- pairs. The asymmetry of the counting will provide an estimate of the polarization of muons and photons.

4. Physics program

The REDTOP experiment will focus on many areas of research, including: symmetry and conservation laws violations, searches for new particles and forces and searches for new baryonic forces.

Symmetry and conservation violations

- *CP Violation (Type I: P and T odd, C even):* This type of CP violation considers a parity of the final state is reversed while charge conjugation is unchanged. This happens when the **P**=-1 and the η meson decays into an even number of pions.
- *CP Violation (Type II: C and T odd, P even):* The charge of the final state is reserved and parity is not. This will just be possible in certain classes of decays:

$$\eta \to \pi^0 l^+ l^- \qquad \text{and} \qquad \eta \to 3\gamma$$

- *CP Violation via the polarization of the muons:* In the $\eta \to \mu^+\mu^-$ decay, the CP invariance does not allow the muons to be longitudinally polarized. For observing non-zero longitudinal polarization of the muons, an extra Higgs boson must be considered. The muon polarization would be as high as 10^{-2}.
- *CP and C Violation with Dalitz plot studies:* The CP invariance has to be present in the $\eta \to \pi^+\pi^-\pi^0$ decay considering the dynamics of charged pions are symmetric. Essentially, any mirror-asymmetry in the Dalitz plot would be an indication of CP and C violation.
- *T violation via the polarization of the muons:* Detecting T invariance implies that tranverse polarization of the muons be null in the following decays:

$$\eta \to \pi^0\mu^+\mu^- \quad \text{and} \qquad \eta \to \gamma\mu^+\mu^-$$

  Observing any polarization is a direct evidence of violation of T invariance.
- *CPT Violation:* CPT invariance can be test analyzing two processes:

$$\eta \to \pi^+\mu^-\nu^- \quad \text{vs} \qquad \eta \to \pi^-\mu^+\nu$$

  In case of CPT violation, the transverse polarization of the μ^- and μ^+ is reversed. The $\eta \to \pi\mu\nu$ is not detected yet.
- *CPT Violation with non-hermitian lagrangians:* In the following processes,

$$\eta \to \gamma\gamma \quad \text{and} \qquad \pi^0 \to \gamma\gamma$$

  the CPT invariance of the hermitian lagrangians implies that the photons have null circular polarization. The main reason to verify this case is that if the non-null circular polarization of the photons exist [4], Redtop will observe it.

IOP Conf. Series: Journal of Physics: Conf. Series **912** (2017) 012042 doi:10.1088/1742-6596/912/1/012042

- *Lepton flavor violation:* Violation of the leptonic number will be detected in the process:

$$\eta \to \mu^+ e^- \gamma \quad \text{and} \qquad \eta \to \pi^0 \mu^+ e^- + c.c.$$

where current PDG limits are lower than 5.7×10^{13} and CL=90%.

- *Double lepton flavor violation:* Double violation of the leptonic number could be observed through the following process:

$$\eta \to \mu^+ \mu^+ e^- e^- + c.c.$$

this event is considered a rare process but it occurs within the sensitivity of the experiment. Its unique signature make it easy to detect.

- *CP Violation with gamma polarization measurements or in $\eta \to \pi^+\pi^- e^+ e^-$ decays:* For investigating the CP violation we will take a look to the photon in the decay $\eta \to \pi^+\pi^-\gamma$, this needs the photon polarization which will be measured with a gamma polarimeter. Furthermore, It is possible study CP violation considering a virtual photon decaying into a e^- e^+ pair.

$$\eta \to \pi^+\pi^-\gamma^* \qquad \text{with} \qquad \gamma^* \to e^+ + e^-$$

The CP invariance vanish the next asymmetry:

$$A_\Phi = \frac{N(\sin[\Phi]\cos[\Phi] > 0) - N(\sin[\Phi]\cos[\Phi] < 0)}{N(\sin[\Phi]\cos[\Phi] > 0) + N(\sin[\Phi]\cos[\Phi] < 0)}$$

in which Φ is the angle between the decay planes of the electron-positron pair and the two charged pions. The REDTOP experiment will improve on the systematic error by almost two orders of magnitude compared with the best present measurements of such asymmetry performed by WASA collaboration [2].

Searches for new particles and forces

- *The proton radius anomaly:* The processes implicated are

$$\eta \to \gamma e^+ e^- \quad \text{and} \qquad \eta \to \gamma \mu^+ \mu^-$$

This anomaly is a discrepancy of several sigma between the electron and the muon probes, such processes occur mainly through the exchange of one virtual photon. In the case of $\eta \to \gamma l^+ l^-$ can occur either via one photon or two photons. The existence of a light scalar particle S coupling differently to electrons and muons could produce this type of η decays and a precise measurement of the branching ratios of the latter might explain the proton radius anomaly.

- *Search for true muonium:* The proccess implicated is:

$$\eta \to \gamma(\mu^+\mu^-)|_{2M_\mu} \to \gamma e^+ e^-$$

The production of true muonium is associated with a μ^+ μ^- pair that determines a mass of about $2M_\mu$ [3] and it is hoped to decay into a $e+$ e_- pair with a large branching ratio.

- *Indirect searches for dark photons:* The mass of the dark photon particle is expected to be in GeV scale, considering its mass is below the mass of the η meson it can be observed by:

$$\eta \to \gamma A' \to \gamma + l^+ l^-$$

where A' correspond to dark photon is a U(1) vector boson. For two η and A' mass restrictions will help to reject the background.

IOP Conf. Series: Journal of Physics: Conf. Series **912** (2017) 012042 doi:10.1088/1742-6596/912/1/012042

- *Searches for new baryonic forces* Following the same sample search for dark photon case, but without the mass constraint on the resonance, shall be deemed the vector boson B (coupling to baryonic matter) mediating that force is below mass of the η it can be measured via:

  $$\eta \to \gamma B \quad \text{with} \quad B \to \gamma\pi^0 \qquad \text{and} \qquad \eta \to \gamma B \quad \text{with} \quad B \to \gamma + l^+l^-$$

  The expected new forces in the Mev-GeV range are associated with dark matter in some models, among them [5], [6], [7] y [8].

Low and intermediate energy studies
The studies of energy physics using η mesons could lead the next topics:

- Test of nuclear models
- Non-perturbative QCD
- Chiral perturbation theory
- $\pi\pi$ interactions
- Electromagnetic transition form-factors
- Octet-singlet mixing angle
- Isospin breaking due to the u-d quark mass difference

5. Running Plan
The experiment REDTOP is expected to do four runs, one of them is still considered as optional.

Run I - η factory
The η meson factory will be the most area focused on REDTOP and will yield η mesons by using a proton beam with an energy of 1.8 GeV. The aim is to produce 10^{13} η mesons per year.

Run II - η' factory
The REDTOP η' factory will use a proton beam with an energy betwen 3.5 to 4.9 GeV, aiming to collect 10^{11} η' mesons.

Run III - μ scattering experiment (optional)
The beam required is composed of muons with an energy between 0.2 and 0.8 GeV, aiming to collect 10^{12} μ scattering events. In this stage the proton radius anomaly would be explorered but the detector operation is still under investigation keeping Run III as optional. However, the REDTOP detector will need extra requirements, one of them changing the position of the graphite target.

Run IV - Rare kaon decay experiment
The main physics topic will be the study of the CP violating process $k^+ \to \pi^+\nu\nu$. This run will use a kaon beam provided as a secondary from the primary 8 GeV proton beam with expected π/k production radio of 13. Nevertheless, the target has to be replaced with an active plastic target while the rest of the detector will stand unchanged.

6. Conclusions
Studying the decays of mesons η/η' can lead to the discovery of new physics. Using new and innovative technologies the REDTOP experiment aims to reach new levels of intensity and sensitivity for this type of decays obtaining measurements with low background comming from the Standard Model Events.

References

[1] Fabela B, Pedraza M I , Gatto C 2016. The REDTOP project: Rare Eta Decays with a TPC for Optical Photons.
[2] Adlarson P, Augustyniak W, Bardan W, Bashkanov M, Bergmann F S, Berowski M, ... and Ciepa, I. (2016). Measurements of branching ratios for decays into charged particles. Physical Review C, 94(6), 065206.
[3] Brodsky S J and Lebed R F 2009. Production of the smallest QED atom: true muonium ($\mu^+ \mu^-$). Physical review letters, 102(21), 213401.
[4] Okun L B 2002. C, P, T are Broken. Why Not CPT?. arXiv preprint hep-ph/0210052.
[5] Arkani-Hamed N, Finkbeiner D P, Slatyer T R and Weiner, N. (2009). A theory of dark matter. Physical Review D, 79(1), 015014.
[6] Tulin S, Yu H B and Zurek K M (2013). Beyond collisionless dark matter: particle physics dynamics for dark matter halo structure. Physical Review D, 87(11), 115007.
[7] Savage C, Gelmini G, Gondolo P and Freese K (2009). Compatibility of DAMA/LIBRA dark matter detection with other searches. Journal of Cosmology and Astroparticle Physics, 2009(04), 010.
[8] Feng J L (2010). Dark matter candidates from particle physics and methods of detection. Annual Review of Astronomy and Astrophysics, 48, 495-545.

Global extraction of the parton-to-kaon fragmentation functions at NLO in QCD

R. J. Hernández-Pinto[1]*, M. Epele[2], D. de Florian[3], R. Sassot[4] and M. Stratmann[5]

[1]Facultad de Ciencias Físico-Matemáticas, Universidad Autónoma de Sinaloa, Ciudad Universitaria, CP 80000, Culiacán, Sinaloa, México,
[2] Instituto de Física La Plata, UNLP, CONICET Departamento de Física, Facultad de Ciencias Exactas, Universidad de La Plata, C.C. 69, La Plata, Argentina,
[3] International Center for Advanced Studies (ICAS), UNSAM, Campus Miguelete, 25 de Mayo y Francia, (1650) Buenos Aires, Argentina,
[4] Departamento de Física and IFIBA, Facultad de Ciencias Exactas y Naturales, Universidad de Buenos Aires, Ciudad Universitaria, Pabellón 1 (1428) Buenos Aires, Argentina,
[5] Institute for Theoretical Physics, University of Tübingen, Auf der Morgenstelle 14, 72076 Tübingen, Germany

E-mail: roger@uas.edu.mx

Abstract. In this document, we present the global QCD analysis of parton-to-kaon fragmentation functions at next-to-leading order accuracy using the latest experimental information on single-inclusive kaon production in electron-positron annihilation, lepton-nucleon deep-inelastic scattering, and proton-proton collisions. An extended analysis of this work can be found in Ref. [1].

1. Introduction
Parton-to-hadron fragmentation functions (FFs) parametrize how quarks and gluons that are produced in hard interactions at high energies confine themselves into hadrons measured and identified in experiment. This information is beyond the reach of perturbative Quantum Chromodynamics (pQCD) and must therefore be inferred from the wealth of data on identified hadron production under the theoretical assumption that the relevant non-perturbative dynamics of FFs factorizes in a universal way from the calculable hard partonic cross sections up to small corrections which can be usually neglected. A precise knowledge on FFs is vital for the quantitative description of a wide variety of hard scattering processes designed to probe the spin and flavor structure of nucleons and nuclear matter and their interpretation at the most elementary and fundamental level. In this review, we present the basic features of the updated parton-to-kaon FFs.

2. Functional form and fit parameters
Similar to the updated pion FFs [2], we found that the functional form adopted in Ref. [3] is still flexible enough to accommodate also the wealth of new experimental information included

* Speaker

in the present fit. Specifically, we parametrize the hadronization of a parton of flavor i into a positively charged kaon K^+ at an initial scale of $Q_0 = 1$ GeV as

$$D_i^{K^+}(z, Q_0) = \frac{N_i z^{\alpha_i}(1-z)^{\beta_i}[1+\gamma_i(1-z)^{\delta_i}]}{B[2+\alpha_i, \beta_i+1] + \gamma_i B[2+\alpha_i, \beta_i+\delta_i+1]}. \quad (1)$$

Here, $B[a, b]$ denotes the Euler Beta-function, and the N_i are chosen in such a way that they represent the contribution of $zD_i^{K^+}$ to the momentum sum rule. The corresponding FFs for negatively charged kaons are obtained by charge conjugation symmetry. In order to reduce the number of parameters in the fit we impose, without changing the total χ^2, the following constraints: $\beta_g = \beta_{\bar{u}}$, $1\gamma_{s+\bar{s}} \simeq \gamma_{\bar{u}}$, $\delta_{s+\bar{s}} \simeq \delta_{\bar{u}}$, $\gamma_{c+\bar{c}} = \gamma_{b+\bar{b}} = 0$, $\alpha_{c+\bar{c}} = \alpha_{b+\bar{b}} = 0$. In total we now have 20 free fit parameters describing our updated FFs for quarks, antiquarks, and gluons into a positively charged kaon. These parameters are then determined from data by a standard χ^2-minimization procedure that includes a χ^2-penalty from computing the optimum relative normalization of each experimental set of data analytically as was outlined in the DSS 14 [2].

3. Data selection

In addition to the data sets already used in Ref. [3] for SIA, we now introduce the new results from BaBar [4] and Belle [5] in SIA at a c.m.s. energy of $\sqrt{s} = 10.5$ GeV. Both sets are very precise and reach all the way up to kaon momentum fractions z close to one. We analyze both sets with $n_f = 4$ active, massless flavors using the standard expression for the SIA cross section at NLO accuracy. As is customary, we limit ourselves to data with $z \geq 0.1$ to avoid any potential impact from kinematical regions where hadron mass corrections, proportional to $M_K/(s\,z^2)$, might become of any importance. For all previous SIA data, taken at higher $\sqrt{s}$, we use $n_f = 5$ and also $z \geq 0.1$.

In case of SIDIS, we replace the preliminary multiplicity data from Hermes by their final results [6]. More specifically, we use the data for charged kaon multiplicities in four bins of z as a function of both momentum transfer Q^2 and the target nucleon's (proton or deuteron) momentum fraction x. The kinematical ranges of average values of Q^2 and x covered by these data are from about 1.1 GeV^2 to 7.4 GeV^2 and 0.064 to 0.277, respectively, for the $z - Q^2$ projections and from about 1.19 GeV^2 to 10.24 GeV^2 and 0.034 to 0.45, respectively, for the $z - x$ projections, with $0.2 \leq z \leq 0.8$. In addition, we include for the first time multiplicity data for $K^\pm$ production from the Compass Collaboration [7], which are given as a function of z in bins of inelasticity y and the initial-state momentum fraction x. Experimental information is available for $0.004 \leq x \leq 0.7$ and $1.2 \leq Q^2 \leq 60$ GeV^2.

Finally, we update and add new sets of data for inclusive high-p_T kaon production in *pp* collisions with respect to those included in the DSS 07 analysis. Most noteworthy are the first results for the kaon-to-pion ratio from the ALICE Collaboration [8] at CERN-LHC, covering unprecedented c.m.s. energies of up to 2.76 TeV. In addition, we include Star data taken at $\sqrt{s} = 200$ GeV for charged kaon production and for the K^-/K^+ ratio from the Star Collaboration [9]. As was discussed in detail in Ref. [2] in the context of pion FFs, it turns out that a good global fit of RHIC and LHC *pp* data, along with all the other world data, can only be achieved if one imposes a cut on the minimum transverse momentum of the produced hadron of about 5 GeV. We maintain this cut also for the present global analysis. Such a p_T−cut eliminates some old *pp* data sets included in the previous DSS 07 analysis from the fit.

4. Results

4.1. Parton-to-kaon fragmentation functions

The new set of parton-to-kaon FFs at NLO evolved at $Q^2 = 10$ GeV^2 is shown as a funcion of z in Figure 1. The differences with respect to the DSS 07 results are mainly driven by the

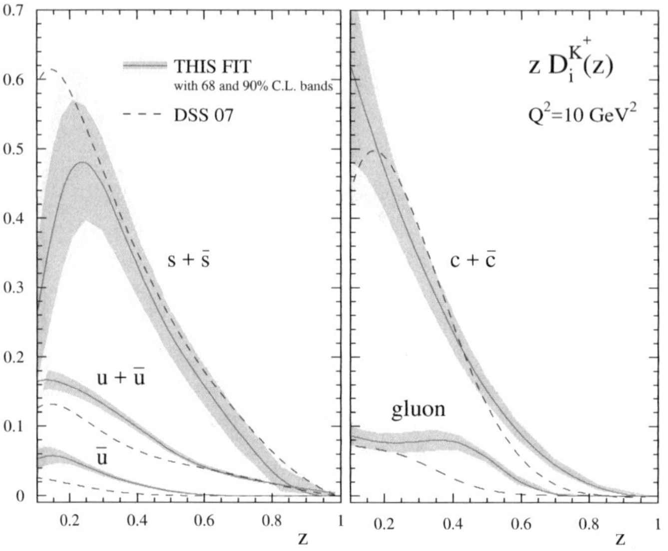

Figure 1. The individual FFs for positively charged kaons $zD_i^{K^+}(z, Q^2)$ at $Q^2 = 10$ GeV^2.

newly added Belle and Babar data at high z, by the $z - x$ projections of the multiplicities both from Hermes and Compass, and by the K^-/K^+ ratios measured in pp collisions by Star and the recent multiplicities released by the ALICE Collaboration. A single parameterization for $D_{\bar{u}}^{K^+} = D_d^{K^+} = D_{\bar{d}}^{K^+} = D_s^{K^+}$ is still the most economical choice to reproduce the data, as was the case in the original DSS 07 analysis. The overall quality of the fit is summarised in Table 1. It is also worth mentioning that there is a more than five-fold increase in the number of available

experiment	# data in fit	χ^2
TPC [10]	12	13.4
SLD [11]	48	81.9
ALEPH [12]	13	29.7
DELPHI [13]	36	31.0
OPAL [14]	25	100.1
BABAR [4]	45	30.6
BELLE [5]	78	15.6
HERMES [6]	288	389.3
COMPASS [7]	618	550.9
STAR [9]	16	7.6
ALICE [8]	15	21.6
TOTAL:	1194	1271.7

Table 1. Data sets used in our NLO global analysis.

data points as compared to the original DSS 07 analysis. Secondly, the overall quality of the global fit has improved dramatically from $\chi^2/d.o.f. = 1.83$ for DSS 07 to $\chi^2/d.o.f. = 1.08$ for the current fit. The biggest improvement concerns the SIDIS multiplicities from Hermes which are described rather well by the updated fit. Also, the charged kaon multiplicities from Compass and the new SIA data from BaBar and Belle integrate very nicely into the global QCD analysis. In the present fit, the set of PDFs has have been upgraded to the recent MMHT 2014 analysis, that gives a much more accurate description of sea-quark parton densities on which the analysis of SIDIS multiplicities depends strongly. It is worth mentioning that a better χ^2 have found recently when MMHT 14 is replaced by NNPDF3.0 [15].

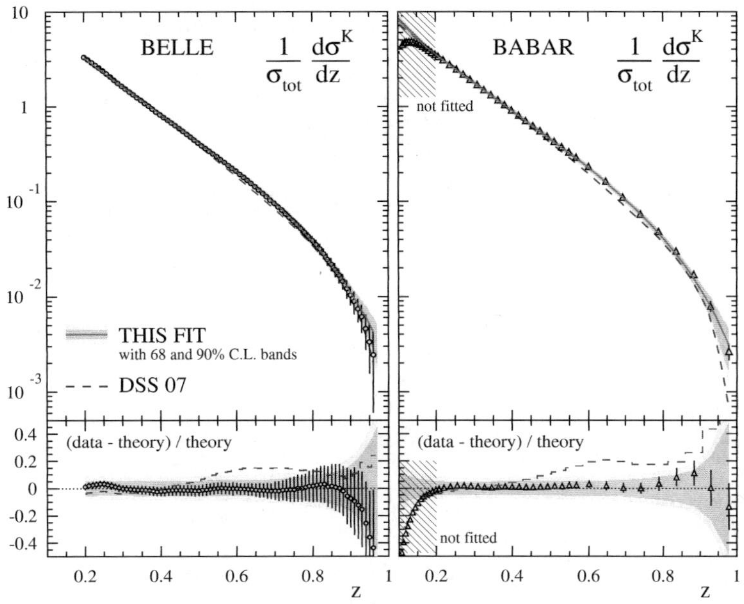

Figure 2. Results of the global QCD fit at NLO for SIA data points. On the left panel the results for Belle while in the right panel the results for BaBar.

4.2. Experimental data and the global QCD fit

In this short review, we only comment on the data provided by the Belle and BaBar experiments, the multiplicities from Hermes and the measurements from the ALICE experiment. However, the reader is encouraged to read the complete analysis in Ref. [1].

In Figure 2, 3 and 4, we present a comparison of the result of the fit at 68% and 90% C.L. with the SIA, SIDIS and proton-proton collisions. Regarding to Figure 2, the agreement of the fit is excellent in the entire energy and z-range covered by Belle and BaBar experiments. The new fit improves very significantly the description of the newly added Belle and BaBar data as can be best seen from the "(data-theory)/theory". The estimated uncertainty bands at 68% and 90% C.L. reflect the accuracy and kinematical coverage of the fitted and data sets and, hence, increase towards both small and large values of z.

Turning into the SIDIS data, in Figure 3 we plotted the multiplicities $M^{K^\pm}_{\ell,p(d)}$ which are defined as the ratio of the inclusive kaon yield and the total DIS cross section in the same x and Q^2 bin in the lepton-deuteron scattering:

$$M^{K^\pm}_{\ell,p(d)} = \frac{d\sigma^{K^\pm}_{\ell,p(d)}/dx\,dQ^2\,dz}{d\sigma_{\ell,p(d)}/dx\,dQ^2}. \qquad (2)$$

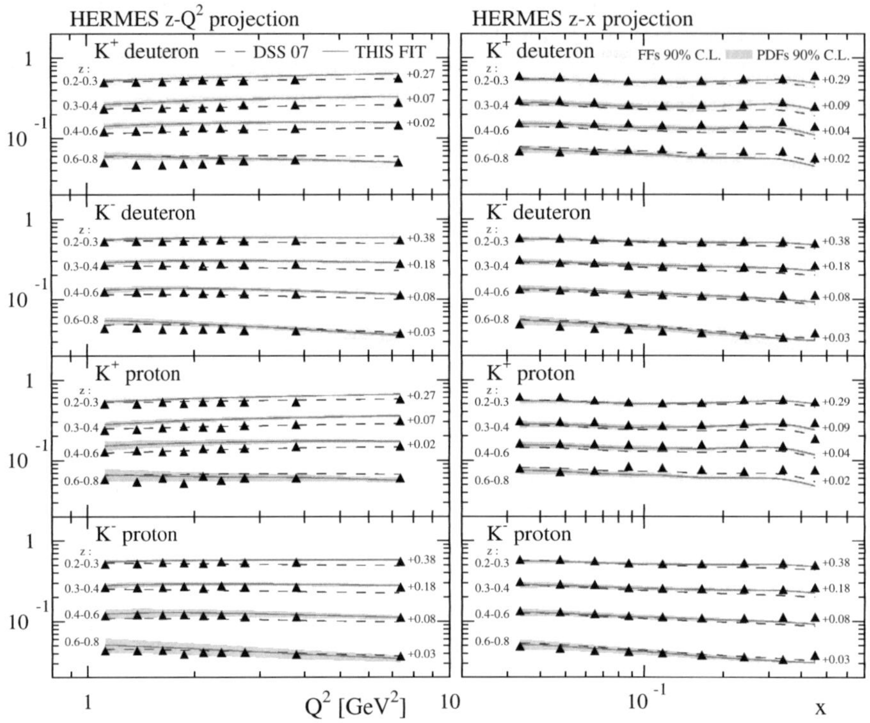

Figure 3. Results of the global QCD fit at NLO for SIDIS data points. On the left panel the results for the projection $z - Q^2$ and the right panel presents the projection $z - x$ both for the Hermes data.

In the global fit we consider the two-dimensional projections of the three-dimensional multiplicity data ontho the $z - Q^2$ dependence and also the $z - x$ dependence. A good agreement among all data points, demonstrate that the low-energy Hermes charged kaon multiplicity data can be described correctly and, equally important, without spoiling the agreement with SIA results.

The third and final main ingredient in our global analysis is the experimental information coming from hadron-hadron collisions. In Figure 4, we show the charged kaon to charged pion cross section ratio as a function of the transverse momentum p_T as measured by the ALICE collaboration in pp collisions at midrapidity at a c.m.s. energy of 2.76 TeV. As can be seen, the current description of the data is much better than the one achieved bt the previous DSS 07 sets of pion and kaon FFs which turns out to be way too small in the entire range of p_T. One reason is the much reduced gluon-to-pion FF in the DSS 14 set as compared to DSS 07, which pushes the kaon-to-pion ratio up.

5. Conclusions

We have presented a new, comprehensive global QCD analysis of parton-to-kaon fragmentation functions at NLO accuracy including the latest experimental information. The very satisfactory and simultaneous description of all data sets within the estimated uncertainties strongly supports the validity of the underlying theoretical framework based on pQCD and, in particular, the notion of factorization and universality for parton-to-kaon fragmentation functions.

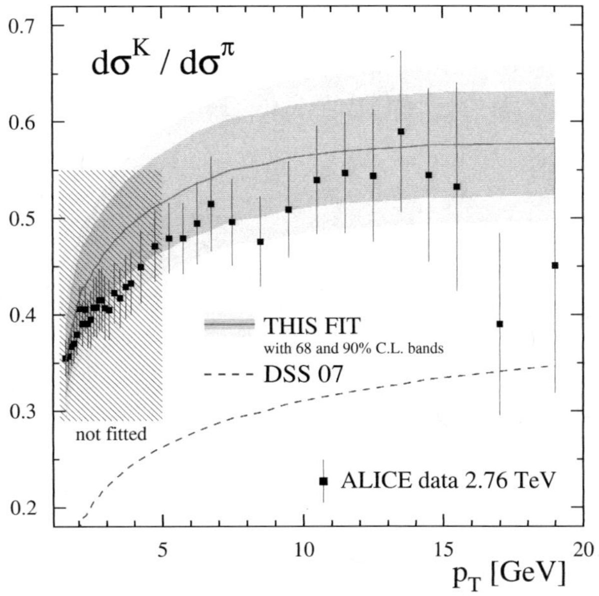

Figure 4. Results of the global QCD fit at NLO for ALICE data points.

6. Acknowledgments

The work of R. J. H. P. is supported by CONACyT and PROFAPI 2015 Grant No. 121. This work is also supported in part by CONICET, ANPCyT, and the Institutional Strategy of the University of Tübingen (DFG, ZUK 63).

References

[1] D. de Florian, R. Sassot, M. Epele, R. J. Hernandez-Pinto, and M. Stratmann, Phys. Rev. D **95**, 094019 (2017).
[2] D. de Florian, R. Sassot, M. Epele, R. J. Hernandez-Pinto, and M. Stratmann, Phys. Rev. D **91**, 014035 (2015).
[3] D. de Florian, R. Sassot, and M. Stratmann, Phys. Rev. D **75**, 114010 (2007).
[4] J. P. Lee *et al.*, (BaBar Collaboration), Phys. Rev. D **88**, 032011 (2013).
[5] M. Leitgab *et al.*, (Belle Collaboration), Phys. Rev. Lett **111**, 062002 (2013).
[6] A. Airapetian *et al.*, (Hermes Collaboration), Phys. Rev. D **87**, 074029 (2013).
[7] C. Adolph *et al.*, (Compass Collaboration), Phys. Lett. B **767**, 133 (2017).
[8] B. Abelev *et al.*, (ALICE Collaboration), Phys. Lett. B **736**, 196 (2014).
[9] G. Agakishiev *et al.*, (Star Collaboration), Phys. Rev. Lett. **108**, 072302 (2012).
[10] H. Aihara *et al.*, (TPC/Two Gamma Collaboration), Phys. Lett. B **184**, 299 (1987); Phys. Rev. Lett. **61**, 1263 (1988); X.-Q. Lu, Ph.D. thesis, Johns Hopkins University, UMI-87-07273, 1987.
[11] K. Abe *et al.*, (SLD Collaboration), Phys. Rev. D **59**, 052001 (1999).
[12] D. Buskulic *et al.*, (ALEPH Collaboration), Z. Phys. C **66**, 355 (1995).
[13] P. Abreu *et al.*, (DELPHI Collaboration), Eur. Phys. J. C **5**, 585 (1998).
[14] G. Abbiendi *et al.*, (OPAL Collaboration), Eur. Phys. J. C **16**, 407 (2000).
[15] I. Borsa, R. Sassot, M. Stratmann, `arXiv:1708.01630 [hep-ph]`.

Chromomagnetic and chromoelectric dipole moments in the 4GTHDM

A. Hernández-Juárez and G. Tavares

Facultad de Ciencias Físico Matemáticas, Benemérita Universidad Autónoma de Puebla, Apartado Postal 1152, Puebla, Pue., México.

A. Moyotl

Departamento de Física, CINVESTAV, Apdo. Postal 14-740, 07000 México, D. F., México.

Abstract. The contributions to the chromomagnetic (CMDM) and chromoelectric (CEDM) dipole moments of the top quark are calculated in framework of the two-Higgs doublet model with a sequential fourth family. We only consider the contributions arising from the neutral scalar bosons H^0 and A^0, whereas the contribution of the charged scalar will be analyzed elsewhere. For the scalar boson masses we consider the degenerate case, with 200 GeV $< m_\phi <$ 500 GeV, while for the mass of the fourth family up quark we consider the interval 400 GeV $< m_{t_4} <$ 600 GeV. It is found that the CMDM and CEDM are specially sensitive when m_ϕ is about 600 GeV and can reach the sizes of $|a_t| \sim 10^{-5}$ and $|d_t| \sim 10^{-24} e$ cm, respectively.

1. Introduction

Since its discovery, the top quark has played a special role in the standard model (SM) phenomenology due to its large mass, which is of the same order as the electroweak symmetry breaking scale. Furthermore, it is the only quark that decays semi-weakly, and its Yukawa coupling is almost of the order of the unity. In addition, it has a very short life time, therefore it does not hadronize. At the CERN Large Hadron Collider (LHC), the top quark is produced mainly in pairs through the processes $q\bar{q} \to t\bar{t}$ and $gg \to t\bar{t}$, where at $\sqrt{s} = 14$ TeV about 90% of the top quark production arises from gluon-gluon fusion and the remainder from $q\bar{q}$ annihilation.

In the context of the SM there are many unsolved problems, one of the most interesting is the baryon asymmetry of the universe. According to Sakharov's criteria, CP violation is a necessary requirement for this phenomena. CP violation is included in the SM via the complex phase of the CKM matrix, which is not enough to explain the baryon asymmetry and so it is worth studying new sources of CP violation arising beyond the SM. A chromoelectric dipole moment (CEDM) violates time-reversal and parity invariance, which due to the CPT theorem is equivalent to CP violation. So a non-zero CEDM would represent a signal of new sources of CP violation. The top quark chromomagnetic dipole moment (CMDM) and CEDM have been studied in several models such as the SM [1], the two-Higgs doublet model (THDM) [2], the minimal supersymmetric standard model (MSSM) [3, 4], technicolor models, 331 models [5] and models with vectorlike multiplets [6].

The top quark anomalous couplings to a gluon pair can be defined via the following

Lagrangian:

$$\mathscr{L} = g_s \bar{t} T^a \left(\gamma_\mu + \frac{i}{2m_t} \sigma_{\mu\nu} \left(a_t + i d_t \gamma^5 \right) q^\nu \right) t G^\mu_a \tag{1}$$

where a_t and d_t are the CMDM and CEDM respectively, while G^μ_a is the gluon field, T^a are the color generators and q is the outgoing gluon four-momentum. In the SM the top quark CMDM is induced at one loop level [5], whereas the CEDM appears only at the three-loop level [7] and thus the SM prediction is negligibly small. The most recent bounds to CMDM and CEDM are $-0.016 < a_t < 0.008$ and $|d_t| < 4.8 \times 10^{-16} e \cdot cm$ respectively [8].

In this work we study the CMDM and CEDM of the top quark in the THDM with a sequential fourth family (4GTHDM), which was proposed by Bar Shalom *et al.* in 2011 [9]. In 2012 with the discovery of a Higgs boson consistent with the SM, these authors revisited their model [10, 11] and constrained the fourth family mass and some parameters of the model. A fourth family was introduced in the past in the SM4 [12]. Unfortunately, this model is not consistent with the Higgs boson production in the LHC [13, 14]. In the 4GTHDM the fourth family interacts only with the heavy scalar bosons, and thus the theoretical prediction for Higgs boson production remains unchanged. The contributions to the magnetic dipole moment have been calculated in [15] and were reviewed in [10], also different decay modes have been studied in [15]. In this work we only consider the neutral scalar boson contributions to the CMDM and CEDM of the top quark, whereas a discussion on the contribution of the charged scalar boson mass will be presented elsewhere.

2. THDM with a sequential fourth generation

There are many motivations for the study of THDMs. In particular the MSSM scalar sector is composed by two Higgs doublets and axions models also require two Higgs doublets. THDMs have been introduced in order to explain the large top-quark mass through the inequality between the two vacuum expectation values (VEVs) $v_1 \gg v_2$. The Higgs sector of these models is composed of two doublet scalar fields Φ_1 and Φ_2 with VEVs v_1 and v_2 respectively. We use the usual definitions $v \equiv \sqrt{v_1^2 + v_2^2}$ and $\tan\beta \equiv v_2/v_1$. The fourth family can be introduced in three different scenarios [9]: (i) Φ_1 gives masses only to the fermions of the fourth family and Φ_2 generates masses for all other fermions, (ii) Φ_1 generates the masses of both the third and fourth families, while Φ_2 generates masses for all the other families, (iii) Φ_1 is only coupled to fermions with masses at the EW scale. In this work we only consider the case (i), which is compatible with the 125 GeV Higgs [10, 11].

In the 4GTHDM flavor changing neutral currents arise at the tree level and are associated with the heavy scalar bosons. The Yukawa interactions of the neutral scalar bosons are given by:

$$\mathscr{L}_{\rm FV} = \frac{g}{2m_W} f^\phi_\beta \bar{\ell}_i \left[g^\phi_s + g^\phi_p \gamma_5 \right] \ell_j \phi, \tag{2}$$

where $\phi = H^0$, A^0, whereas the coupling constants f^ϕ_β, g^ϕ_s and g^ϕ_p acquire values according to Table 1. The subscripts i, j refers to the flavor of the quarks, but we are interested in non-diagonal couplings ($i \neq j$). On the other hand, since the SM is reproduced in the limit $\sin(\beta - \alpha) = 1$, it is convenient to introduce a parameter χ defined as $\chi \equiv \frac{\pi}{2} - (\beta - \alpha)$, where $\chi = 0$ corresponds to the SM limit. Then, we expect tiny deviations from the SM, i.e. we take $\chi \simeq 0$.

Table 1. Coupling constant f^ϕ_β, g^ϕ_s and g^ϕ_p for the neutral scalar bosons H^0 and A^0. I_q is the weak isospin,whereas α and β are mixing angles. We use the short-hand notation $s_\alpha \equiv \sin\alpha$, $c_\alpha \equiv \cos\alpha$ and $t_\beta \equiv \tan\beta$. Additionally, $\Sigma^{u,d}_{ij}$ is a complex element of the new mixing 4×4 matrix in the 4G2HDM.

ϕ	f^ϕ_β	g^ϕ_s	g^ϕ_p
H^0	$c_\alpha/c_\beta - s_\alpha/s_\beta$	$m_{\ell_i}\Sigma^q_{ij} + m_{q_j}\Sigma^{q*}_{ji}$	$m_{\ell_i}\Sigma^q_{ij} - m_{\ell_j}\Sigma^{q*}_{ji}$
A^0	$2iI_q(t_\beta + 1/t_\beta)$	$m_{\ell_i}\Sigma^q_{ij} - m_{\ell_j}\Sigma^{q*}_{ji}$	$m_{\ell_i}\Sigma^q_{ij} + m_{\ell_j}\Sigma^{q*}_{ji}$

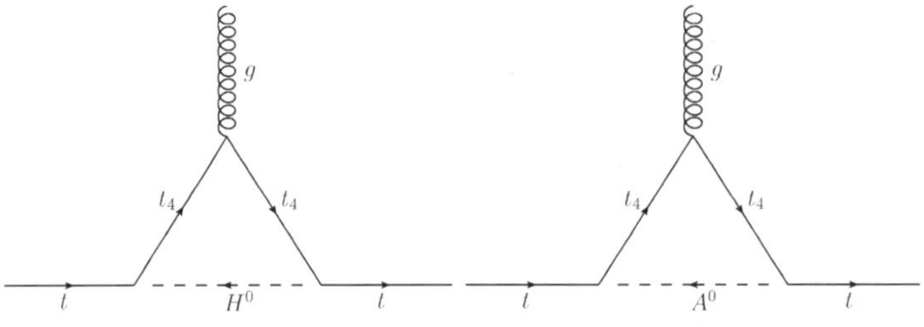

Figure 1. One loop contribution to the CMDM and CEDM of the top quark in the 4GTHDM. In the loop circulate the fourth generation up quark t_4 and the heavy scalar bosons $\phi = H^0$, A^0.

3. Top quark CMDM and CEDM

The neutral scalar bosons contributions to the CMDM and CEDM arise through the Feynman diagrams of Figure 1, where we have a heavy quark t_4 into the loop. We have used Feynman parametrization to solve the loop integrals. The resulting expressions can be written as follows

$$a^\phi_t = \left(\frac{g}{2m_W}\right)^2 \frac{f^2_\phi}{(4\pi)^2}\int_0^1 dx \frac{x^2\left[|g^\phi_p|^2(x-1-r_{t_4}) + |g^\phi_s|^2(x-1+r_{t_4})\right]}{xr^2_{t_4} - (x-1)(x-r^2_\phi)} \tag{3}$$

$$d^\phi_t = -\text{Im}\left(G^\phi_s G^{\phi*}_p\right)\left(\frac{g}{2m_W}\right)^2 \frac{m^2_t f^2_\phi r_{t_4}}{(4\pi)^2}\int_0^1 \frac{x^2 dx}{xr^2_{t_4} - (x-1)(x-r^2_\phi)} \tag{4}$$

where $r_k = m_k/m_t$ and $G^\phi_{s,p} = g^\phi_{s,p}/m_t$. Furthermore, we note that d^ϕ_t is proportional to $\text{Im}\left(G^\phi_s G^{\phi*}_p\right)$, so a complex phase is necessary to obtain a non-zero result.

4. Analysis

To simplify the analysis we will assume degenerate masses for the neutral scalar bosons, and we will consider the interval 200 GeV $< m_\phi <$ 500 GeV. Furthermore, from the analysis of the $\tau \to \mu\gamma$ decay [15], assuming that it is also valid for quarks, we have taken $|\Sigma_{34}| = |\Sigma_{43}| \sim 10^{-3/2}$. We will also take $\tan(\beta) \sim \mathcal{O}(1)$, which is in accordance with the most recent bounds from experimental data. Finally, in Ref. [11], the mass of the fourth family up quark has been constrained to 400 GeV$< m_{t_4} <$ 600 GeV. In this scenario, the lightest scalar boson h is consistent with the 125 GeV Higgs boson signals measured at the LHC. Thus, in the following analysis we will use this interval. In Figure 1 we show the behavior of the CMDM (left plot) and the CEDM (right plot) as functions of the scalar boson mass m_ϕ for $m_{t_4} = 400$ GeV. Additionally we have used $\tan\beta = 5$ and $\chi = 0.1$. We show the partial contributions of the H^0 (dotted line) and A^0 (dashed line) scalar bosons as well as the total contribution (solid line). As for the CEDM, the contributions of both the H^0 and A^0 scalar bosons have the same order of magnitude but opposite signs, therefore the sum of both contributions is of the order $|d_t| \sim 10^{-24} e$ cm. As far as the CMDM is concerned, the total contribution is $|a_t| \simeq 10^{-5}$ for $m_\phi = 200$ GeV, and $|a_t| \simeq 10^{-6}$ for $m_\phi = 500$ GeV.

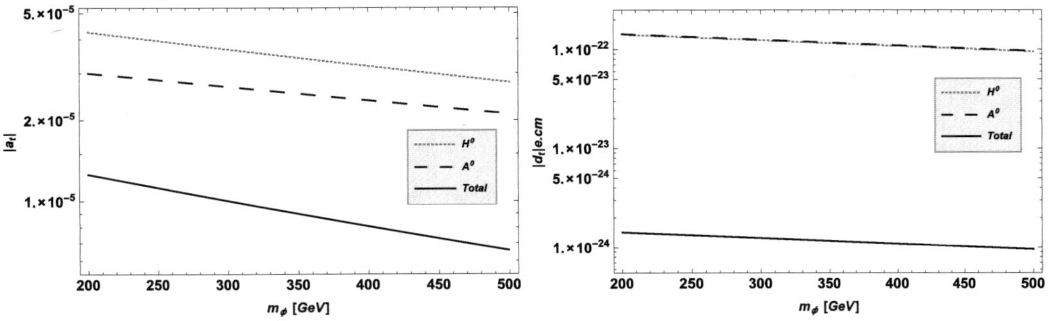

Figure 2. The CMDM (left plot) and CEDM (right plot) as functions of scalar mass m_ϕ. We have used $m_{t_4} = 400$ GeV, $\tan\beta = 5$ and $\chi = 0.1$. We show the partial contributions of H^0 (dotted line) and A^0 (dashed line) as well as the total contribution (solid line).

In Figure 3 we show the behavior of the total contribution to the top quark CDMD and CEDM for $m_{t_4} = 400$ GeV (dashed line), 500 GeV (solid line) and 600 GeV (dotted line). In general we have maximal contributions for a heavier t_4 quark, however, for a relative light scalar boson mass, the difference between the distinct contributions is negligible. For the CMDM we have $|a_t| \simeq 1.25 \times 10^{-5}$ for $m_\phi = 200$ GeV and $|a_t| \simeq 8.2 \times 10^{-6}$ for $m_\phi = 600$ GeV when $m_{t_4} = 600$ GeV, whereas $|a_t| \simeq 6.4 \times 10^{-6}$ for $m_{t_4} = 400$ GeV. On the other hand, for the CEDM we have $|d_t| \simeq (1.4 - 1.5) \times 10^{-24} e$ cm for $m_\phi = 200$ GeV and $|d_t| \simeq 1.18 \times 10^{-24} e$ cm for $m_\phi = 600$ GeV when $m_{t_4} = 600$ GeV, whereas $|d_t| \simeq 9.2 \times 10^{-25} e$ cm for $m_{t_4} = 400$ GeV.

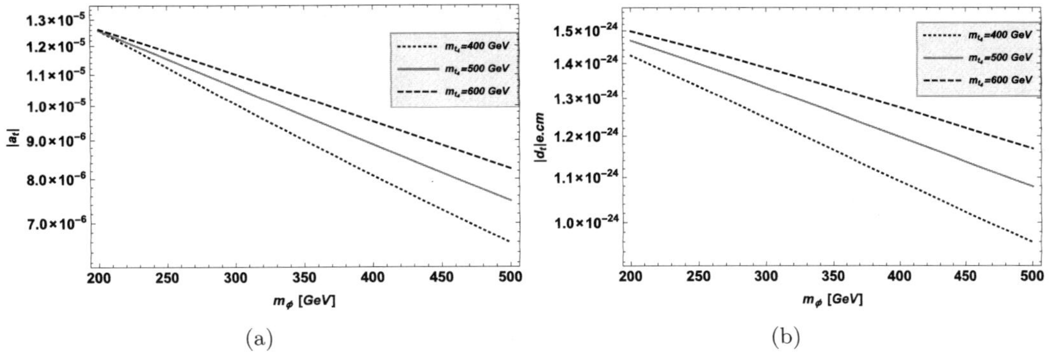

Figure 3. The same as in Figure 2, but for the total contribution to the top quark CMDM and CEDM for tree distinct values of m_{t_4}.

5. Conclusions

We have presented the contributions to both the chromomagnetic and chromoelectric dipole moments of the top quark in the framework of the two-Higgs doublet model with a sequential fourth family. We only discussed the contributions of the neutral scalar bosons. A more comprehensive analysis, including the contribution of the charged scalar boson, will be presented elsewhere. We have found that in general we have maximal contributions for a heavier up quark of the fourth family. We also found that the CMDM and CEDM are sensitive to heavy scalar bosons. In particular, if $m_{t_4} = 600$ GeV $|a_t| \simeq 1.25 \times 10^{-5}$ and $|d_t| \simeq (1.4 - 1.5) \times 10^{-24} e$ cm for $m_\phi = 200$ GeV.

Acknowledgement

We are grateful to the conference organizers for their kind invitation to present our research work and their kind hospitality.

References

[1] Atwood D, Kagan A and Rizzo T G 1995 *Phys. Rev. D* **52**(11) 6264–6270 URL https://link.aps.org/doi/10.1103/PhysRevD.52.6264
[2] Gaitán R, Garcés E A, de Oca J H M and Martinez R 2015 *Phys. Rev. D* **92**(9) 094025 URL https://link.aps.org/doi/10.1103/PhysRevD.92.094025
[3] Martinez R and Rodriguez J A 2002 *Phys. Rev. D* **65**(5) 057301 URL https://link.aps.org/doi/10.1103/PhysRevD.65.057301
[4] Aboubrahim A, Ibrahim T, Nath P and Zorik A 2015 *Physical Review D* **92** 035013– URL https://link.aps.org/doi/10.1103/PhysRevD.92.035013
[5] Martinez R, Perez M A and Poveda N 2008 *Eur. Phys. J.* **C53** 221–230 (*Preprint* hep-ph/0701098)
[6] Ibrahim T and Nath P 2011 *Physical Review D* **84** 015003– URL https://link.aps.org/doi/10.1103/PhysRevD.84.015003
[7] Czarnecki A and Krause B 1997 *Physical Review Letters* **78** 4339–4342 URL https://link.aps.org/doi/10.1103/PhysRevLett.78.4339
[8] Hayreter A and Valencia G 2013 *Physical Review D* **88** 034033– URL https://link.aps.org/doi/10.1103/PhysRevD.88.034033
[9] Bar-Shalom S, Nandi S and Soni A 2011 *Phys. Rev.* **D84** 053009 (*Preprint* 1105.6095)
[10] Bar-Shalom S, Geller M, Nandi S and Soni A 2013 *Adv. High Energy Phys.* **2013** 672972 (*Preprint* 1208.3195)
[11] Geller M, Bar-Shalom S, Eilam G and Soni A 2012 *Phys. Rev.* **D86** 115008 (*Preprint* 1209.4081)
[12] Holdom B, Hou W, Hurth T, Mangano M L, Sultansoy S and Ünel G 2009 *PMC Physics A* **3** 4 URL http://dx.doi.org/10.1186/1754-0410-3-4
[13] Eberhardt O, Herbert G, Lacker H, Lenz A, Menzel A, Nierste U and Wiebusch M 2012 *Physical Review D* **86** 013011– URL https://link.aps.org/doi/10.1103/PhysRevD.86.013011
[14] Eberhardt O, Lenz A, Menzel A, Nierste U and Wiebusch M 2012 *Physical Review D* **86** 074014– URL https://link.aps.org/doi/10.1103/PhysRevD.86.074014
[15] Bar-Shalom S, Nandi S and Soni A 2012 *Phys. Lett.* **B709** 207–217 (*Preprint* 1112.3661)

On a Three-Channel Cosmic Ray Detector based on Aluminum Blocks

L Arceo[1], **J Félix**[1]

[1]Laboratorio de Partículas Elementales - División de Ciencias e Ingenierías campus León - Universidad de Guanajuato, Loma del Bosque,103, Lomas del Campestre, 37150, Leon, Guanajuato, MX

E-mail: `miquel@fisica.ugto.mx`, `felix@fisica.ugto.mx`

Abstract. There are many general purpose cosmic ray detectors based on plastic scintillators and electronic boards from the market. This is a new cosmic ray detector designed on three 2.54 cm X 5.08 cm X 20.32 cm Aluminum blocks in stack arrangement, and three Hamamatsu S12572-100P photodiodes. The photodiode board, the passive electronic board, and the discriminator board are own designed. The electronic signals are stored with a CompactRIO -cRIO- by National Instruments. It is presented the design, the construction, the data acquisition system algorithm, and the preliminary physical results.

1. Introduction

Currently there are many reports around the word on cosmic rays detectors [1]. The electronics has a key role in cosmic ray detection. In Mexico, particularly, the students and professors from Laboratory for elementary particles (laboratorio de partículas elementales) design, construct, and test their own cosmic ray detectors [2]. This report, focuses on the design, construction and test of electronic board to setup the avalanche Hamamatsu photodiode model S12572-100P [3] (the figure 1 display photodiode) for a cosmic ray detector based on Aluminum blocks.

2. Design and construction

Figure 1. S12572-100P photodiode.

Figure 2. Photodiode board top layer.

2.1. Photodiode board

Its function is to support mechanically the electrical connection (anode and cathode) of the photodiode to attach it to the Aluminum block on one of its polished 2.54 cm X 5.08 cm ends. The figure 2 displays the top layer. The figure 3 displays one front end 2.54 cm X 5.08 cm Aluminum block polish and finally the figure 4 displays the Aluminum block attached with a photodiode board and isolated optically with four layers of Aluminum tape 3311.

Figure 3. Aluminum block polished.

Figure 4. Aluminum block and photodiode optically isolated.

2.2. Passive electronic board

Provides high voltage to the photodiode enabling it; for the readout of the signal was implemented a RC circuit. The photodiode requires reverse polarity for working. The figure 5 displays passive electronic board diagram schematic and figure 6 displays the passive electronic board.

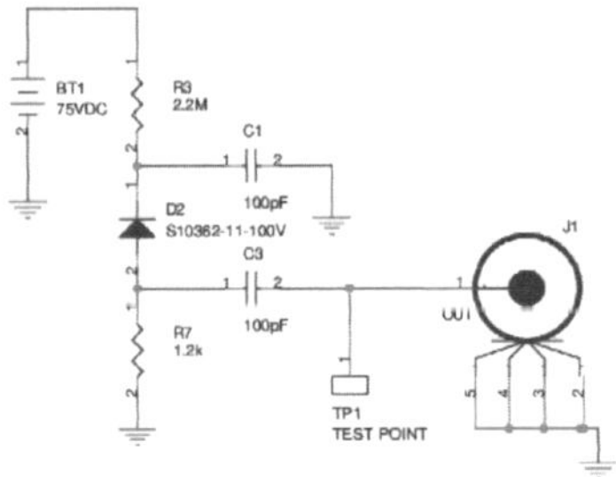

Figure 5. Passive electronic board diagram.

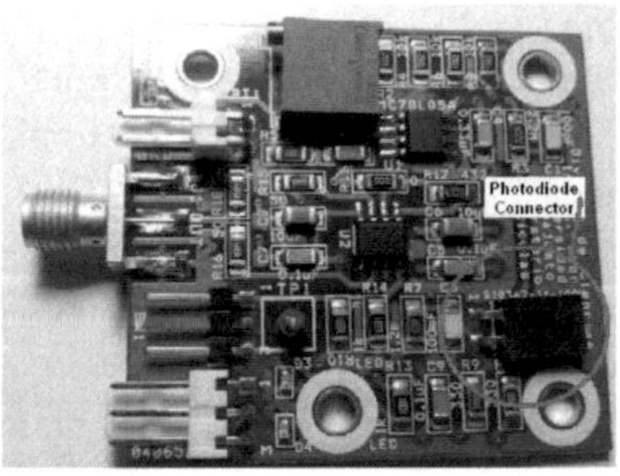

Figure 6. Passive electronic board with photodiode connector.

Figure 7. Four channel discriminator board.

2.3. Discriminator board

Contains a single integrated circuit CMP401 of Analog Device [4], with 23 ns propagation delay, quad comparators and compatible with 5 volts logic. Its function is to compare analogue signal coming from the passive electronic board, with a fixed trigger voltage defined by the final user and give out the digital signal. If the input signal is higher in amplitude than the trigger, the discriminator turned on -one logical state-, otherwise, turns off -zero logical state. The figure 7 displays the four channel discriminator board.

Each channel cosmic ray detector requires one Aluminum block with photodiode board attach and one passive electronic board, the three channels from passive electronic board are connected to the discriminator board, finally, all channels are connected to the data acquisition system, the figure 8, displays detector final assembly.

Figure 8. Three channel detector final assembly.

2.4. Data Acquisition System

The cRIO is assembled with 9025 embedded controller [5], eight NI-9402 [6] interchangeable C modules of four channels of digital input or output and one rack. The cRIO is connected to the PC HOST via Ethernet cable with one number port. The cRIO and PC-HOST required LabVIEW-FPGA and LabVIEW program, respectively.

3. Characterization

The output counts as function of applied high voltage for each channel are displayed in figures 9, 10 and 11. The recording interval began at 60 volts and finished at 100 volts in steps of 5 volts each. Nine ten minute text files were generated. In the three channels for voltages between 60 and 70 volts the result is zero counts. For 75 volts and higher the number the counts increases lineally with applied voltage.

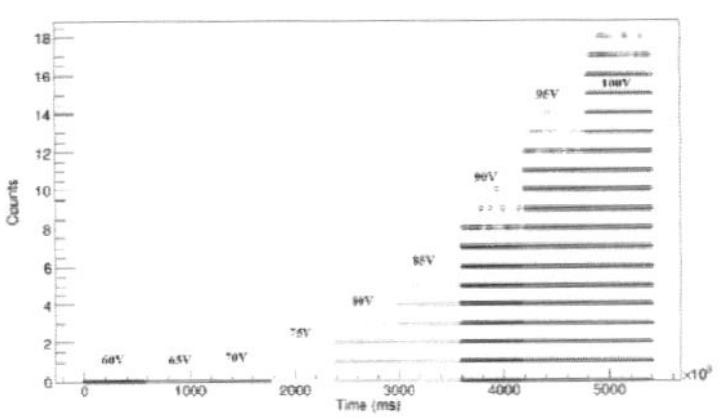

Figure 9. Top Aluminum block.

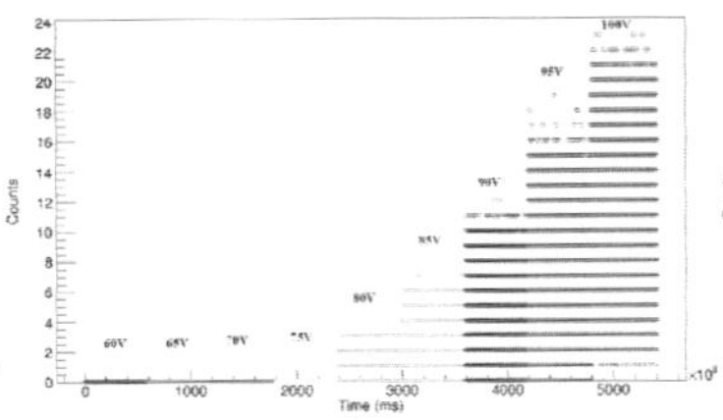

Figure 10. Middle Aluminum block.

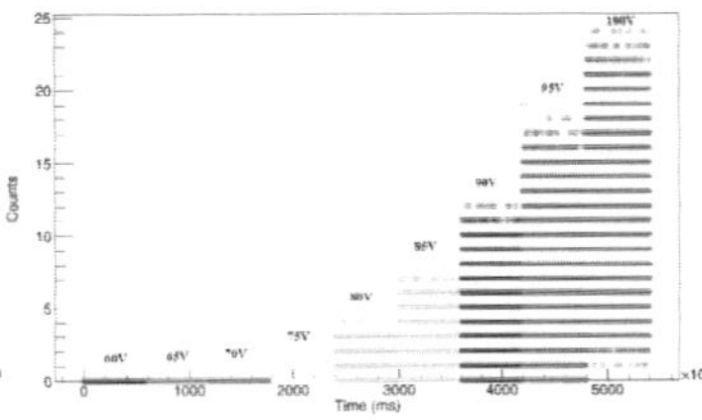

Figure 11. Bottom Aluminum block.

4. Results

The analogue signal noise was 50 mVpp and the configuration parameters were 75 Volts for operation voltage and 100 mV of threshold for the discriminator board.

The figure 12, 14 and 16 display the number of counts vs time for top, middle and bottom channels, respectively. The figure 13, 15 and 17 display frequency vs counts histograms for top, middle and bottom channels, respectively. The recording time length was 30 minutes for each file.

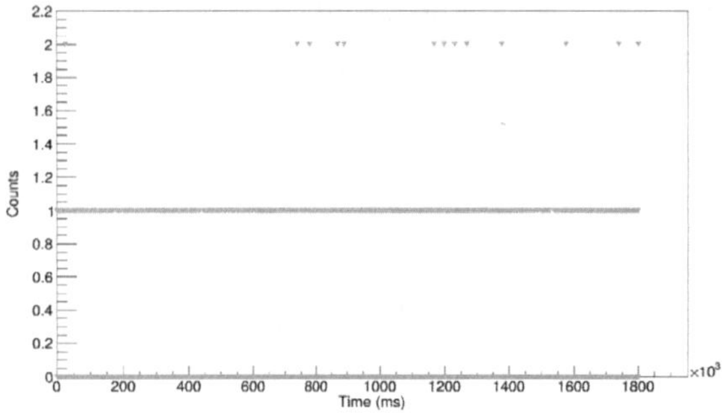

Figure 12. Counts vs time of top Aluminum block.

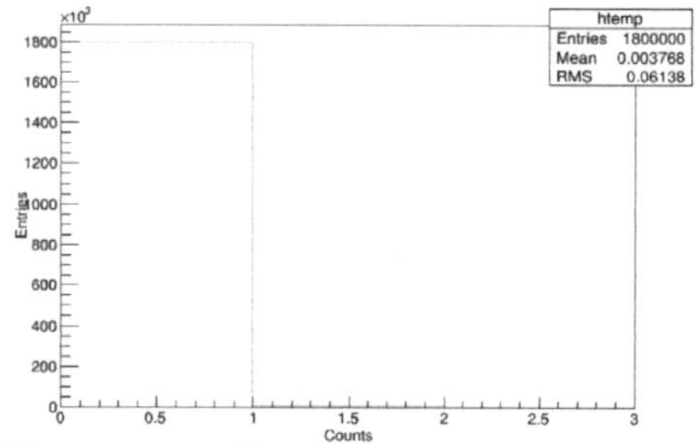

Figure 13. Frequency vs counts of top Aluminum block.

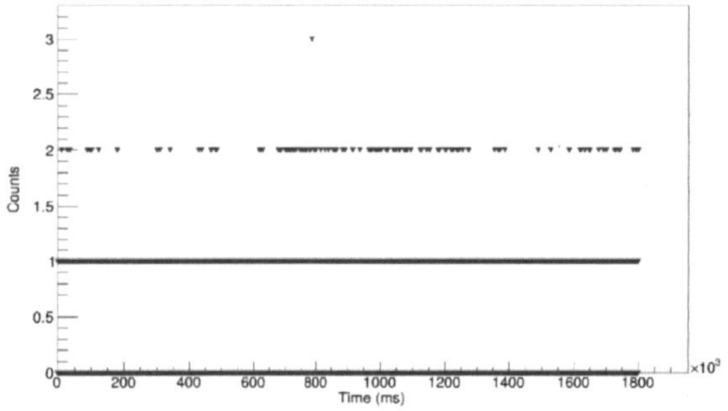

Figure 14. Counts vs time of middle Aluminum block.

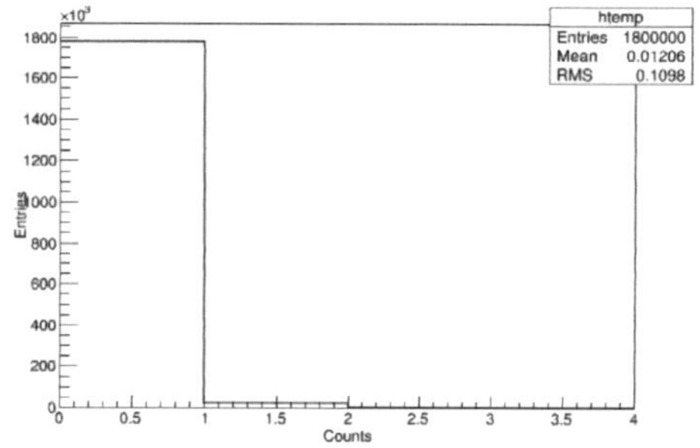

Figure 15. Frequency vs counts of middle Aluminum block.

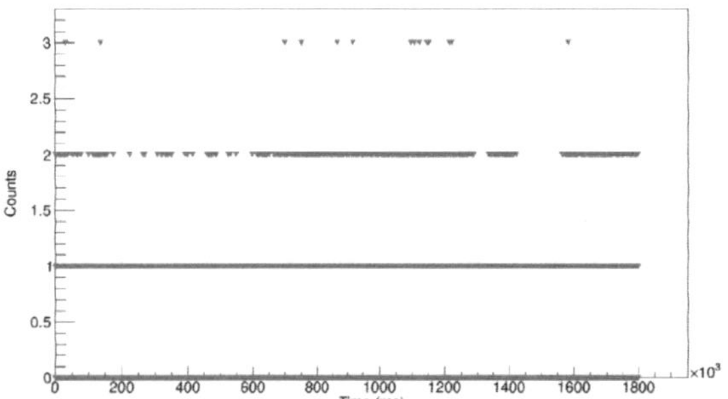

Figure 16. Counts vs time of bottom Aluminum block.

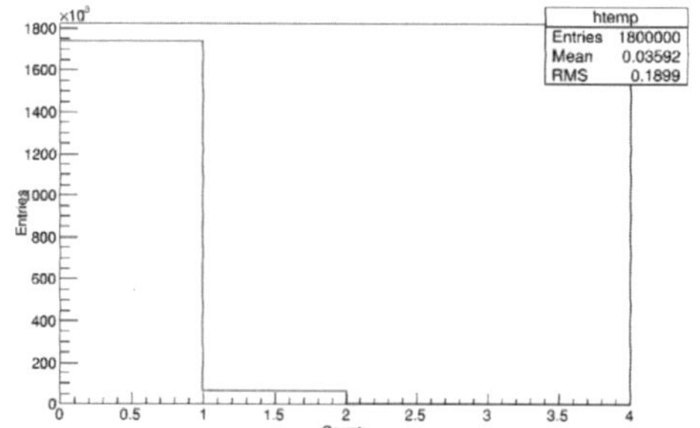

Figure 17. Frequency vs counts of bottom Aluminum block.

5. Conclusions

The electronic board, the data acquisition system and photodiode with Aluminum blocks work properly. It was characterized the cosmic ray detector to obtain a linear function of the counts vs applied high voltage. The distributions of counts vs time are almost flat, for the three channels of the cosmic ray detector.

6. Acknowledgments

CONACYT for the support, project 223179, 10017, CB-2013-01 "Fondo Sectorial de Investigación para educación SEP- CONACYT".

References

[1] Morello C 2010 Cosmic Ray Detectors *PoS* PoS(CRA School)027 can be found on `http://inspirehep.net/record/912798/files/CRA%20School_027.pdf`.
[2] Felix J 2016 Design construction, and operation of small cosmic rays detectors at Universidad de Guanajuato Mexico *ICHEP2016* PoS(ICHEP2016)726 can be found on `https://indico.cern.ch/event/432527/contributions/1072033/attachments/1321179/1981335/posterICHEP2016JFelix.pdf`.
[3] Hamamatsu datasheet December 2015 can be found on `http://www.hamamatsu.com/jp/en/S12572-100P.html`.
[4] Analog Devices datasheet Rev A 2002 can be found on `http://www.analog.com/media/en/technical-documentation/data-sheets/CMP401_402.pdf`.
[5] National Instruments October 2015 user manual and specifications NI cRIO-9025. can be found on `http://www.ni.com/pdf/manuals/375490d.pdf`.
[6] National Instruments January 2016 datasheet, NI-9402. can be found on `http://www.ni.com/pdf/manuals/374614a_02.pdf`.

Simulation of a stripline beam position monitor of a single electrode for educational purposes.

F.Pérez-Suárez and H. Maury-Cuna

División de Ciencias e Ingenierías, Universidad de Guanajuato Loma del Bosque 103, Col. Lomas del Campestre, León, Guanajuato, México, C.P. 37150

E-mail: `f.perezsuarez@ugto.mx` / `fermin.ugfimee@gmail.com`

Abstract. A beam position monitor is a device used in particle accelerators which determine the position of the beam. In this work, the results obtained by computational simulation and the characterization of a stripline beam position monitor using a single electrode under the action of an electric potential are shown. The electric potential is generated along a copper cable, concentrically located on the monitor, to emulate the particle beam. The simulations were made in COMSOL®, and the response of the electric potential, on the electrode, is shown when varying the position of the copper cable on the vertical axis of the monitor. In addition, the teaching value of the device is discussed.

Keywords: Accelerator, particles, simulation, COMSOL®.

1. Introduction

Particle accelerators are instruments that use electromagnetic fields to increase the energy of the charged particles for a BPM of one pair of electrodes as shown in 1 or even a BPM of two pairs of electrodes. These machines produce beams of particles with specific energies and intensities that can be used in many areas, ranging from medical, industrial and scientific applications.

In the 1960s, the first particle beam position monitors were built. The importance of having this information is relevant since the path of a beam along the accelerator was unknown. The most common way to monitor the position of the charged particle beam is by coupling the BPM to the electromagnetic field. A typical arrangement for position monitors consists of a pair of electrodes or two pairs and it is common that the BPMs placed along the tube are located before and after the electromagnets (dipoles, quadrupole, etc.). Subsequently, this signal is analyzed and translated in terms of position. [1]. Beam position measurements provide an enormous amount of diagnostic information. The access to this information can vary from periodic revisions by the operators to a data acquisition in a high bandwidth [2].

Simulation of a stripline beam position monitor.

The main goal of this work is to simulate a single-electrode beam position monitor (BPM). To understand the concept of such a device, we can cite Peter Forck: "A beam position monitor is a non-invasive device that has a low cut-of frequency, i.e., the behavior of a direct current beam cannot be monitored" [3]. A beam position monitor will then be a device able to monitor alternate signal frequencies. The important features of a BPM are position redundancy, and temporal resolution, for example for the BPM system at the Shanghai Synchrotron Radiation Facility (SSRF) the orbit feedback system requires an information of the closed orbit of high resolution in the range of 1 μm at a rate of approximately 1 ms [6]. There are different types of BPMs. The kind of monitor to be chosen depends heavily on the task. The most used are the stripline and radiofrequency (RF) cavity BPMs.

For this project, we focus on a stripline type beam position monitor. A BPM typically has a pair of copper electrodes or even 2 pairs of them, these pick up a voltage signal, depending on the location of the particle a current is induced at each of these electrodes. This phenomenon is known as image currents, when the particles pass through the vacuum chamber, an electromagnetic field is generated that induces a signal on the electrode in the form of a current. Therefore, a pulse extends from end to end at the electrode of length L [1].

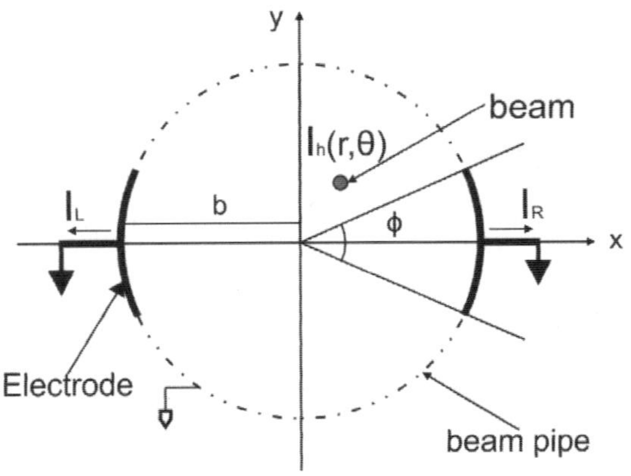

Figure 1: Stripline Beam Position Monitor.

Figure 1 shows a cross-section of a beam position monitor, the outer circular section is known as vacuum chamber, a pair of electrodes in the middle left and right is shown, usually made of copper. It also has ground terminals, where commonly at one end of the electrodes there is a load or impedance which allows us to obtain a signal reflected from it. A particle or bundle of particles following a path outside the intersection of the X and Y axis depending on the radial and angular position (r, θ) is appreciated in Figure 1. Two electrodes (L and R for left and right) of angular width ϕ are placed at 0 and 180 as shown in Figure 1.

Simulation of a stripline beam position monitor.

2. Methodology

A single-electrode stripline BPM (see Figure 2) have been built at the Science and Engineering Division of the University of Guanajuato. Simulations of this BPM geometry were performed using the software COMSOL Multiphysics®.

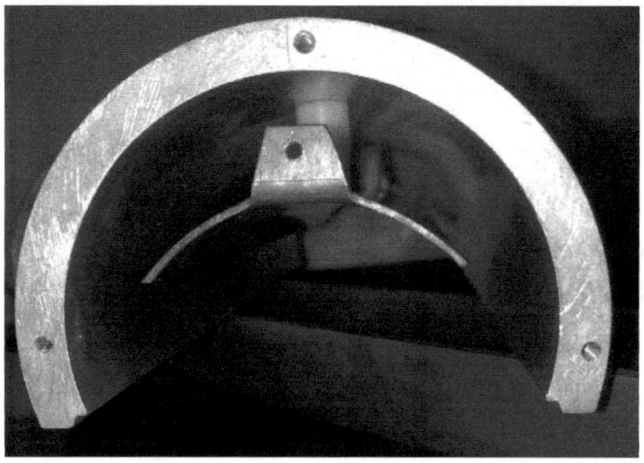

Figure 2: Section of a stripline beam position monitor of a single electrode.

We draw the stripline single-electrode BPM geometry using the CAD environment of COMSOL Multiphysics® as shown in Figure 3.

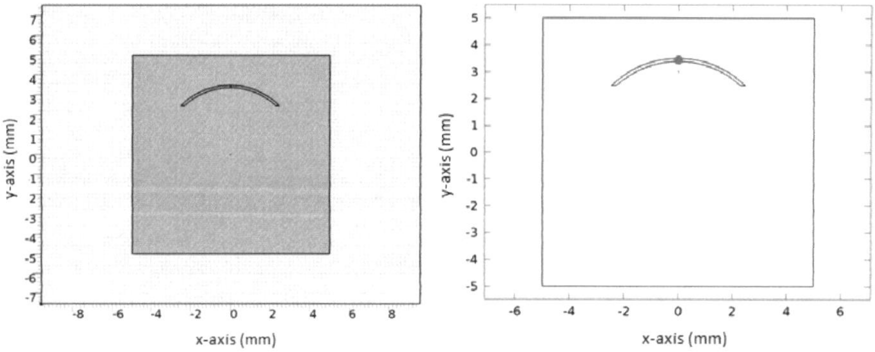

Figure 3: Left: Section of a stripline beam position monitor of an electrode in CAD design. Right: Section with a point probe to measure the electric potential at this point.

We performed 2D electrostatic simulations. We emulated the particle beam using a small charge sphere (representing the beam charge). The analyzed signal was the electrostatic potential (V) induced by the charged sphere on the electrode a variation of the vertical distance between the sphere and the electrode top edge from 0 mm to 3 mm, in intervals of 0.5 mm was made.

If we use a cartesian coordinate system to represent the two-dimensional distribution, we will obtain a set of points known with the scatter diagram, whose analysis allows to

Simulation of a stripline beam position monitor.

study qualitatively the relationship between both variables as seen in Figure 4 (Right). The determination of the functional dependence between the two variables V (voltage) and y(displacement) that best fits the two-dimensional distribution is a linear regression of the form $Y{=}aX{+}b$. Voltage depends on the value of displacement over y axis.

Once the results of the measurements in the simulation were obtained, a equation is determined by the regression as shown in figure 4 (Left). This equation brought us information about the resulting electric potential meanwhile the position is changing.

3. Results

Table 1 shows the results of the electrical potential obtained by simulating a beam of particles moving from the origin (0,0) towards the copper electrode in the vertical direction.

Table 1: Data of Distance and Electric Potential.

Distance(mm)	Electric Potential(V)
0	3.60682x10^6
0.5	3.97706x10^6
1	4.3472x10^6
1.5	4.71374x10^6
2	5.06928x10^6
2.5	5.40403x10^6
3	5.70847x10^6

These results were plotted to observe the behavior of the electric potential as a function of distance. In Figure 4, the values from Table 1 are plotted.

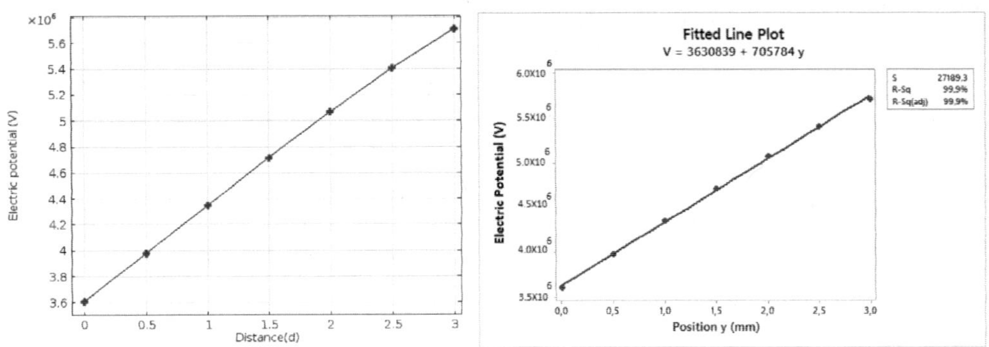

Figure 4: Rigth: COMSOL data plotted from simulation. Left: Linear regression.

Figure 5 shows the voltage influence of the simulated particle beam on the copper electrode.

Simulation of a stripline beam position monitor.

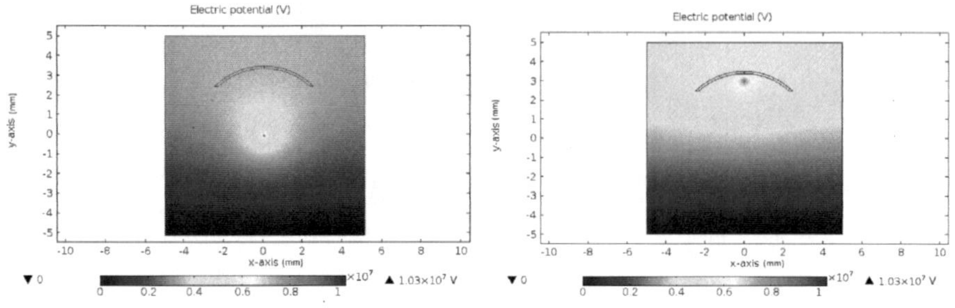

Figure 5: Image of a COMSOL simulation of the particle beam and the copper electrode [Left: At origin, Right: At (0,3) point].

4. Conclusions

In this work we shown that the COMSOL 2D electrostatic simulations yield good results, corroborated by the statistical results. Moreover, the Stripline BPM being built for educational purposes can be constructed from material that is easily obtained and relatively easy manufactured at university machine-tool workshops. When the device is finished, the idea will be to compare the simulation with the measurements obtained by the device and this could be used as a training tool for students and technicians.

5. Future Work:

There is intent to carry out simulations that contain more specific information, as well as to implement a system that simulates a signal that can be monitored by the electrode. In the incoming months, when the prototype and 2D electrostatic simulations are ready, we are going to simulate the 3D system with a time-varying signal instead of an electrostatic punctual charge.

References

[1] Robert E. Shafer, Beam Position Monitoring, *Los Alamos National Laboratory, 1989.*
[2] T.J. Shea, Interfacing to accelerator instrumentation, *Brookhaven National Laboratory.*
[3] L.A. Law and L.Holcom, The Cambridge electron accelerator beam position monitors, *Cambridge Electron Accelerator, Harvard University, 1965.*
[4] V. Sargsyan, Comparison of Stripline and Cavity Beam Position Monitors, *2004.*
[5] A.J. Claus, R.B. Blackman, E.G. Halline and W.C. Ridgway, Orbit determination and prediction, and computer programs, *1963.*
[6] Zhichu Chen,et.al., Beam position monitor design for a third generation light source *2014.*
[7] COMSOL Multiphysics®

IOP Conf. Series: Journal of Physics: Conf. Series **912** (2017) 012047 doi:10.1088/1742-6596/912/1/012047

Scalar dark matter in inert doublet model with scalar singlet

L. G. Cabral-Rosetti

Centro Interdisiplinario de Investigación y Docencia en Educacin Técnica (CIIDET), Unidad de Ciencias Básicas, Av. Universidad No. 282 Pte., Col. Centro, C.P. 76000, Querétaro, Qro., México

E-mail: cabralrosetti@gmail.com

R. Gaitán[1], J.H. Montes de Oca[2], R. Osorio Galicia

Departamento de Física, FES-Cuautitlán, UNAM, C.P. 54770, Estado de México, México

E-mail: [1]rgaitan@unam.mx, [2]josehalim@gmail.com

E. A. Garcés

Departamento de física, CINVESTAV, C.P. 07360, Ciudad de México, México.

E-mail: egarces@fis.cinvestav.mx

Abstract. The simplest scalar representations under $SU(2)_L$ symmetry are considered to extend the scalar sector of the Standard Model. The main motivation is including a dark matter candidate, which could arise from the additional scalar fields. We also include an additional U(1) symmetry in order to control the stability of the dark matter candidate.

1. Introduction

The most convincing evidence for Dark Matter (DM) came from the observation that luminous objects such as stars, gas clouds, globular clusters, or entire galaxies move faster than one would expect if they only felt the gravitational attraction of other visible objects [1]. DM is estimated to constitute about 23 % of the total matter in the universe. However, the origin of DM still remains a mystery. The existence of Dark matter (DM) is now essentially established [2]. The Standard Model (SM) of particle physics successfully explains experimental results observed in colliders, but none of the SM particles can be a good candidate for the dark matter. One of the main reasons is the lack of particles with scarce or null interactions with other SM particles. We assume that dark matter candidates are weakly interacting massive particles (WIMP). The DM relic density is around the observed value (Planck Collaboration) [3, 4]:

$$\Omega h^2_{CDM} = 0.1199 \pm 0.0027. \tag{1}$$

In the case of scalars in different representations under $SU(2)_L$ as DM candidates in models beyond the SM the simplest version includes an additional singlet or doublet [5, 6, 7, 8, 9, 10, 11].

2. The model

In this section the field content and gauge bosons interactions are introduced. We will follow closely the notation introduced by Lee and Sher [12, 13, 14]. We consider the extension to SM with gauge group defined as $G_{IDMU_X} = G_{SM} \otimes U(1)_X$ where G_{SM} is the gauge group symmetry of the SM. Basically, this model is a Inert Doublet Model with $U(1)_X$. The coupling constant associated to $U(1)_X$ is denoted as g_X.

A mixing among the kinetic energy terms for the gauge fields, $U(1)_Y$ and $U(1)_X$, is allowed given the gauge invariance in G_{IDMU_X}. This mixing term is parametrized by the weak mixing angle θ_W and a new parameter denoted by ε [12, 13, 14]. The kinetic terms are written explicitly as

$$\mathcal{L}_{Kin} = -\frac{1}{4}\hat{B}_{\mu\nu}\hat{B}^{\mu\nu} + \frac{1}{2}\frac{\varepsilon}{\cos\theta_W}\hat{B}^{\mu\nu}\hat{Z'}_{0\mu\nu} - \frac{1}{4}\hat{Z'}_{0\mu\nu}\hat{Z'}^{\mu\nu}_{0}, \tag{2}$$

where, $\hat{B}^{\mu\nu}$ and $\hat{Z'}^{\mu\nu}_{0}$ are the field strength tensors of the $U(1)_Y$ and $U(1)_X$ gauge bosons, respectively. The gauge field of Z' is redefined by the rotation

$$\begin{pmatrix} Z'_{0\mu} \\ B_\mu \end{pmatrix} = \begin{pmatrix} \sqrt{1-\varepsilon^2/\cos^2\theta_W} & 0 \\ -\varepsilon/\cos^2\theta_W & 1 \end{pmatrix} \begin{pmatrix} \hat{Z'}_{0\mu} \\ \hat{B}_\mu \end{pmatrix}, \tag{3}$$

and therefore the mixing term in Equation (2) is cancelled. The mixing parameter ε will appear in all terms where Z' is redefined. The magnitude of ε has been constrained to $\varepsilon \leq 10^{-3}$ [13, 15, 16].

In the IDM with U_X the scalar field content is given in the two doublets and one singlet defined as follows

$$\begin{aligned} \Phi_1 &= \begin{pmatrix} \phi_1^+ \\ \frac{1}{\sqrt{2}}(v + \phi_1 + i\eta_1) \end{pmatrix}, \\ \Phi_2 &= \begin{pmatrix} \phi_2^+ \\ \frac{1}{\sqrt{2}}(\phi_2 + i\eta_2) \end{pmatrix}, \\ \Phi_x &= \frac{1}{\sqrt{2}}(v_x + \phi_x + i\eta_x), \end{aligned} \tag{4}$$

where v and v_x are the vacuum expectation values (VEV), with $v = 246$ GeV. The SSB is achieved as $G_{IDMU_X} \xrightarrow{\langle\Phi_x\rangle} G_{SM} \xrightarrow{\langle\Phi_1\rangle} SU(3)_C \otimes U(1)_{EM}$, where $\langle\Phi_1\rangle^T = (0, v_1/\sqrt{2})$ and $\langle\Phi_X\rangle = v_x/\sqrt{2}$. The charges and representations for the scalar fields under the symmetry group G_{IDMU_X} can be written as follows

$$\begin{aligned} \Phi_1 &\sim (1, 2, 1/2, x_1), \\ \Phi_2 &\sim (1, 2, 1/2, x_2), \\ \Phi_x &\sim (1, 1, 0, x), \end{aligned} \tag{5}$$

where two first entries denote the representation under $SU(3)_C$ and $SU(2)_L$ meanwhile the hypercharge and charge $U(1)_X$ are written in the last two entries.

The interactions between the scalar and gauge bosons are given by

$$\mathcal{L}_{\text{scalar}} = |D_\mu\Phi_1|^2 + |D_\mu\Phi_2|^2 + |D_\mu\Phi_x|^2, \tag{6}$$

where the covariant derivative D_μ is defined as

$$D_\mu = \left(\partial_\mu + ig'Y\hat{B}_\mu + igT_3\hat{W}_{3\mu} + ig_xQ'\hat{Z}'_{0\mu}\right), \tag{7}$$

here Q' and g_x denote the $U(1)_X$ charges for fermions and the gauge coupling constant, respectively. After the right spontaneous symmetry breaking SSB, additionally to the mass terms, the mixing term between $Z_{0\mu}$ and $Z_0'^{\mu}$ arises. This new mixing term is proportional to $\Delta^2 = \frac{1}{2}g_{Z'}g_Z v^2 \cos^2\beta + \frac{1}{4}\frac{\varepsilon}{\cos\theta_W} g_Z g' v^2$. In order to cancel the mixing term the following rotation is required

$$\begin{pmatrix} Z \\ Z' \end{pmatrix} = \begin{pmatrix} \cos\xi & -\sin\xi \\ \sin\xi & \cos\xi \end{pmatrix} \begin{pmatrix} Z^0 \\ Z'^0 \end{pmatrix}, \tag{8}$$

where the mixing angle ξ satisfy the expression $\tan 2\xi = \frac{2\Delta^2}{m^2_{Z^0} - m^2_{Z'^0}}$, and has been constrained to $|\xi| < 10^{-3}$ [17]. Therefore the Z' mass is

$$m^2_{Z'} = g^2_{Z'}(v^2\cos^2\beta + v_S^2) + \frac{\varepsilon}{\cos\theta_W} g_{Z'}g'v^2\cos^2\beta + \frac{1}{4}\left(\frac{\varepsilon}{\cos\theta_W}\right)^2 g'^2 v^2, \tag{9}$$

meanwhile the Z mass retains the same value set by the SM,

$$m_Z^2 = \frac{1}{2}gv^2. \tag{10}$$

3. Scalar Sector

The most general, renormalizable and gauge invariant potential is

$$\begin{aligned} V &= \mu_1^2\Phi_1^\dagger\Phi_1 + \mu_2^2\Phi_2^\dagger\Phi_2 + \mu_x^2\Phi_x^*\Phi_x + \left[\mu_{12}^2\Phi_1^\dagger\Phi_2 + h.c.\right] + \lambda_x\left(\Phi_x^*\Phi_x\right)^2 \\ &+ \lambda_1\left(\Phi_1^\dagger\Phi_1\right)^2 + \lambda_2\left(\Phi_2^\dagger\Phi_2\right)^2 + \lambda_3\left(\Phi_1^\dagger\Phi_1\right)\left(\Phi_2^\dagger\Phi_2\right) + \lambda_4\left|\Phi_1^\dagger\Phi_2\right|^2 \\ &+ \left[\lambda_5\left(\Phi_1^\dagger\Phi_2\right)^2 + \lambda_6\left(\Phi_1^\dagger\Phi_1\right)\left(\Phi_1^\dagger\Phi_2\right) + \lambda_7\left(\Phi_2^\dagger\Phi_2\right)\left(\Phi_1^\dagger\Phi_2\right) + h.c.\right] \\ &+ \left(\Phi_x^*\Phi_x\right)\left[\lambda_{1x}\left(\Phi_1^\dagger\Phi_1\right) + \lambda_{2x}\left(\Phi_2^\dagger\Phi_2\right)\right] + \left[\lambda_{12x}\left(\Phi_1^\dagger\Phi_2\right)\left(\Phi_x^*\Phi_x\right) + h.c.\right] \\ &+ \left[\lambda'_{1x}\left(\Phi_1^\dagger\Phi_1\right)\Phi_x + \lambda'_{2x}\left(\Phi_2^\dagger\Phi_2\right)\Phi_x + \lambda'_x\left(\Phi_1^\dagger\Phi_2\right)\Phi_x + h.c.\right]. \end{aligned} \tag{11}$$

The terms proportional to λ'_{1x}, λ'_{2x}, λ'_x do not guarantee the stability of the DM candidate. Proper selection of the values of the $U(1)_X$ charges succeeds in eliminating these terms. We fix the charges as $x_1 = x_2$ and $x_1 \neq x$. Other selection has been considered previously [7, 18]. Then, the mass matrices for charged and neutral scalars are

$$M^2_{H^\pm} = \begin{pmatrix} \mu_2^2 + \frac{1}{2}(\lambda_3 v^2 + \lambda_{2x}v_x^2) & 0 \\ 0 & \mu_2^2 + \frac{1}{2}(\lambda_3 v^2 + \lambda_{2x}v_x^2) \end{pmatrix} \tag{12}$$

$$M_0^2 = \begin{pmatrix} M_{11} & M_{12} & M_{13} & 0 \\ M_{12} & M_{22} & M_{23} & 0 \\ M_{13} & M_{23} & M_{33} & M_{34} \\ 0 & 0 & M_{34} & M_{44} \end{pmatrix}, \tag{13}$$

where $M_{11} = 2\lambda_1 v^2$, $M_{12} = \lambda_{1x} v v_x$, $M_{13} = \frac{1}{2}\lambda_6 v^2$, $M_{22} = 2\lambda_x v_x^2$, $M_{23} = \frac{1}{2}\lambda_{12x} v v_x$, $M_{22} = 2\lambda_x v_x^2$, $M_{33} = \mu_2^2 + \frac{1}{2}(\lambda_3 + \lambda_4 + \mathrm{Re}[\lambda_5])v^2 + \frac{1}{2}\lambda_{2x}v_x^2$, $M_{34} = -\mathrm{Im}[\lambda_5]v^2$, $M_{44} = \mu_2^2 + \frac{1}{2}(\lambda_3 + \lambda_4 - \mathrm{Re}[\lambda_5])v^2 + \frac{1}{2}\lambda_{2x}v_x^2$. The M_{13} and M_{23} matrix elements are removed to avoid mixing between DM candidates and neutral scalars like Higgs. A discrete symmetry Z_2 is introduced to eliminate these matrix elements; if Φ_1 is even and Φ_2 is odd, then $M_{13} = M_{23} = 0$.

The M_0^2 matrix can be diagonalized by

$$\begin{pmatrix} H_1 \\ H_2 \end{pmatrix} = \begin{pmatrix} \cos\alpha_1 & -\sin\alpha_1 \\ \sin\alpha_1 & \cos\alpha_1 \end{pmatrix} \begin{pmatrix} \mathrm{Re}[\phi_1^0] \\ \mathrm{Re}[\Phi_x] \end{pmatrix} \tag{14}$$

and

$$\begin{pmatrix} H_3 \\ H_4 \end{pmatrix} = \begin{pmatrix} \cos\alpha_2 & -\sin\alpha_2 \\ \sin\alpha_2 & \cos\alpha_2 \end{pmatrix} \begin{pmatrix} \mathrm{Re}[\phi_2^0] \\ \mathrm{Im}[\phi_2^0] \end{pmatrix}, \tag{15}$$

where $\tan\alpha_{1,2} = \frac{r_{1,2}}{1+\sqrt{1+r_{1,2}^2}}$ with $r_1 = \frac{\lambda_{1x} v v_x}{\lambda_1 v^2 - \lambda_{1x} v_x^2}$ and $r_2 = \frac{-\mathrm{Im}[\lambda_5]}{\mathrm{Re}[\lambda_5]}$. Therefore the masses for the scalars are

$$m_{H_1,H_2}^2 = \lambda_1 v^2 + \lambda_{1x} v_x^2 \pm (\lambda_1 v^2 - \lambda_{1x} v_x^2)\sqrt{1 + r_1^2}, \tag{16}$$

$$m_{H_3,H_4}^2 = \mu_2^2 + \frac{1}{2}(\lambda_3 + \lambda_4) v^2 + \frac{1}{2}\lambda_{2x} v_x^2 \pm v^2 \sqrt{\mathrm{Re}[\lambda_5]^2 + \mathrm{Im}[\lambda_5]^2}, \tag{17}$$

$$m_{H^\pm}^2 = \mu_2^2 + \frac{1}{2}(\lambda_3 v^2 + \lambda_{2x} v_x^2). \tag{18}$$

4. Dark matter candidates and their portals

The scalars fields Φ_1 and Φ_x have interactions with fermions and their VEVs are different from zero, therefore none of them could provide a DM candidate. However Φ_2 has no Yukawa interactions and has VEV equal to zero, so this field can provide DM candidates. The H_3, or H_4 could play the role as dark matter candidate.

The Dark matter abundance can be obtained by solving the Boltzmann equation for the number density rate [19]. The Boltzmann equation can be written as

$$a^{-3}\frac{d}{dt}\left(n a^3\right) = \langle\sigma v\rangle \left(n_{eq}^2 - n^2\right), \tag{19}$$

where n is the DM number density and a is a scale factor. All information about the model interactions is contained in the thermally averaged cross section, defined as

$$\langle\sigma v\rangle = \frac{(2\pi)^4}{(n_{eq})^2}\prod_{i=1}^{4}\int \frac{d^3 p_i}{2E_i} e^{-(E_1+E_2)/T}\delta^4\left(p_1 + p_2 - p_3 - p_4\right)|\mathcal{M}|^2. \tag{20}$$

The contributions for scattering from this DM candidate with portals are shown in the Feynman diagrams in figures 1 and 2.

5. Conclusions

We consider an extended scalar sector with two doublet and one singlet. The DM candidates arise from the second doublet which has VEV equal to zero in order to guarantee the stability of DM. The scalar fields with non zero VEV could contribute to new portals for the DM annihilation.

Acknowledgments

This work was supported by projects PAPIIT-IN113916 in DGAPA-UNAM, PIAPI1628 in FES-Cuautitlan UNAM and *Sistema Nacional de Investigadores* (SNI) México. E.A.G. thanks the CONACYT Project No. FOINS-296-2016 (Fronteras de la Ciencia).

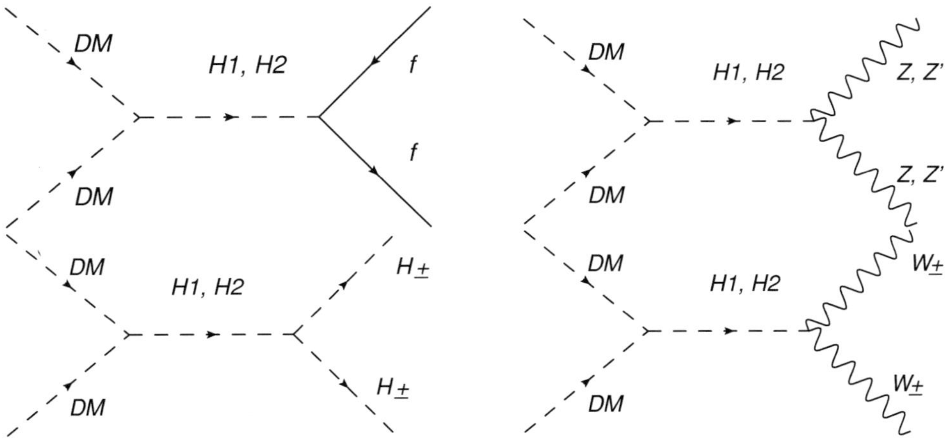

Figure 1. DM annihilation through neutral Higgs portals. The final states are for pair fermion in left-up, for Z or Z' in right-up, for charged Higgs in left-down and for $W^{\pm}$ bosons in right-down.

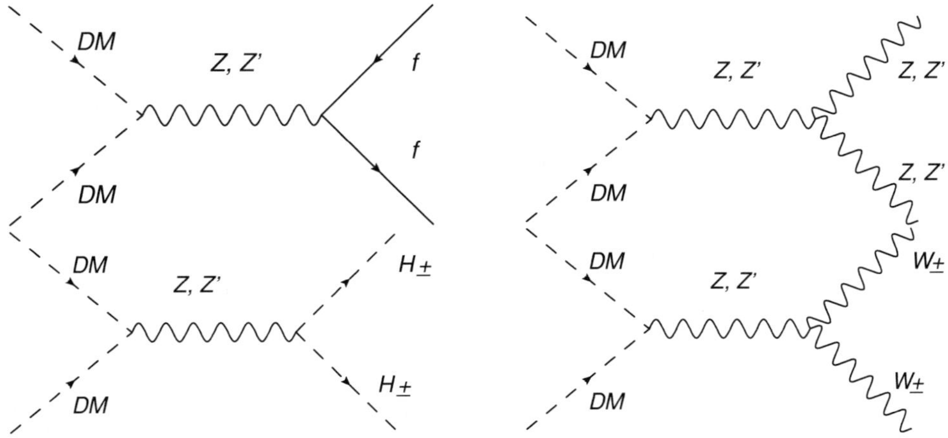

Figure 2. DM annihilation through Z or Z' portal. The final states are for pair fermion in left-up, for Z or Z' in right-up, for charged Higgs in left-down and for $W^{\pm}$ bosons in right-down.

References

[1] M. Roos, astro-ph.CO/1001.0316.
[2] G. Bertone, D. Hooper and J. Silk, *Phys. Rept.* **405** , 279 (2005), hep-ph/0404175.
[3] P. A. R. Ade *et al.* [Planck Collaboration], *Astron. Astrophys.* (2014), astro-ph.CO/1303.5076.
[4] P. A. R. Ade *et al.* [Planck Collaboration], Astron. Astrophys. (2014) [arXiv:1303.5076 [astro-ph.CO]].
[5] J. McDonald, *Phys. Rev. D* **50**, 3637 (1994), hep-ph/0702143; E. M. Dolle and S. Su, *Phys. Rev. D* **80**, 055012 (2009), hep-ph/0906.1609
[6] J. McDonald, Phys. Rev. D **50**, 3637 (1994) [hep-ph/0702143 [HEP-PH]].
[7] R. Gaitan *et al.*, J.Phys.Conf.Ser. 761 (2016) no.1, 012015.
[8] E. M. Dolle and S. Su, Phys. Rev. D **80**, 055012 (2009) [arXiv:0906.1609 [hep-ph]].
[9] L. Lopez Honorez, E. Nezri, J. F. Oliver and M. H. G. Tytgat, JCAP **0702**, 028 (2007) [hep-ph/0612275].
[10] L. Lopez Honorez and C. E. Yaguna, JCAP **1101**, 002 (2011) [arXiv:1011.1411 [hep-ph]].
[11] P. Ko, Y. Omura and C. Yu, JHEP **1401**, 016 (2014) [arXiv:1309.7156 [hep-ph]].
[12] H. Davoudiasl, H. -S. Lee and W. J. Marciano, *Phys. Rev. D* **85**, 115019 (2012), hep-ph/1203.2947; H. S. Lee and M. Sher, *Phys. Rev. D***87**, no. 11, 115009 (2013), hep-ph/1303.6653
[13] H. Davoudiasl, H. -S. Lee and W. J. Marciano, Phys. Rev. D **85**, 115019 (2012) [arXiv:1203.2947 [hep-ph]].

[14] H. S. Lee and M. Sher, Phys. Rev. D **87**, no. 11, 115009 (2013) [arXiv:1303.6653 [hep-ph]].
[15] S.A. Abel, M. D. Goodsell, J. Jaeckel, V. V. Khoze and A. Ringwald, J. High Energy Phys. 07 (2008) 124.
[16] B. Holdom, Phys. Lett. **166**, 196 (1986)
[17] C. Bouchiat and P. Fayet, Phys. Lett. B **608**, 87 (2005).
[18] R. Gaitan, E. A. Garces and J. H. Montes de Oca, arXiv:1410.5462 [hep-ph].
[19] S. Dodelson, Amsterdam, Netherlands: Academic Pr. (2003) 440 p

AMDPF2017 IOP Publishing
IOP Conf. Series: Journal of Physics: Conf. Series **912** (2017) 012048 doi:10.1088/1742-6596/912/1/012048

Conjecture about the 2-Flavour QCD Phase Diagram

M A Nava Blanco[1], W Bietenholz[2] and A Fernández Téllez[1]

[1] Facultad de Ciencias Físico Matemáticas, Benemérita Universidad Autónoma de Puebla, C.P. 72570 Puebla, Mexico

[2] Instituto de Ciencias Nucleares, Universidad Nacional Autónoma de México, A.P. 70-543, C.P. 04510 Ciudad de México, Mexico

E-mail: miguelnava2539@gmail.com, wolbi@nucleares.unam.mx, afernan@mail.cern.ch

Abstract. The QCD phase diagram, in particular its sector of high baryon density, is one of the most prominent outstanding mysteries within the Standard Model of particle physics. We sketch a project how to arrive at a conjecture for the case of two massless quark flavours. The pattern of spontaneous chiral symmetry breaking is isomorphic to the spontaneous magnetisation in an O(4) non-linear σ-model, which can be employed as a low-energy effective theory to study the critical behaviour. We focus on the 3d O(4) model, where the configurations are divided into topological sectors, as in QCD. A topological winding with minimal Euclidean action is denoted as a skyrmion, and the topological charge corresponds to the QCD baryon number. This effective model can be simulated on a lattice with a powerful cluster algorithm, which should allow us to identify the features of the critical temperature, as we proceed from low to high baryon density. In this sense, this projected numerical study has the potential to provide us with a conjecture about the phase diagram of QCD with two massless quark flavours.

1. Scales in Quantum Chromodynamics

Quantum Chromodynamics (QCD) provides a successful description of the strong interaction between quarks and gluons. If we only deal with massless quarks, then the QCD Lagrangian does not involve any dimensional parameter, which implies scale invariance at the classical level. At the *quantum level,* however, an intrinsic scale $\Lambda_{\rm QCD}$ emerges, which breaks this invariance. Recently a careful study derived, based on lattice simulations, the value $\Lambda_{\rm QCD} = 341(12)\,{\rm MeV}$ (with $N_{\rm f} = 3$ quark flavours, in the $\overline{\rm MS}$-scheme) [1].

In the real world, the quarks do have masses, but two flavours are very light compared to the intrinsic scale. At 2 GeV, again in the $\overline{\rm MS}$-scheme, their masses are [2]

$$m_u = 2.2^{+0.6}_{-0.4}\,{\rm MeV}\ , \quad m_d = 4.7^{+0.5}_{-0.4}\,{\rm MeV}\ , \quad m_u,\ m_d \ll \Lambda_{\rm QCD}\ . \tag{1.1}$$

These two flavours dominate the low-energy nuclear physics, which matters for our daily life.[1] It is therefore a sensible approximation to consider only two flavours, and — depending on the subject of interest — we may even neglect their masses.

[1] Sometimes also the s-quark is considered as light, $m_s = 96^{+8}_{-4}\,{\rm MeV}$ [2]; in fact, it contributes to some low-energy quantities at the percent level. The remaining three quark flavours c, b, t have masses well above $\Lambda_{\rm QCD}$; they are usually negligible at low energy.

2. Chiral flavour symmetry breaking

If fermions are massless, their left- and right-handed components decouple. In the case of $N_{\rm f}$ massless quark flavours, each chirality can undergo an independent unitary transform, which corresponds to a global symmetry

$$\mathrm{U}(N_{\rm f})_L \otimes \mathrm{U}(N_{\rm f})_R = \mathrm{SU}(N_{\rm f})_L \otimes \mathrm{SU}(N_{\rm f})_R \otimes \mathrm{U}(1)_B \otimes \mathrm{U}(1)_A \qquad (2.1)$$

of the QCD Lagrangian. However, also part of this symmetry breaks explicitly due to quantum effects: in particular, the invariance under the group $\mathrm{U}(1)_A = \mathrm{U}(1)_{L\neq R}$ — which corresponds to opposite phase shifts for the left- and right-handed quarks — is broken by the axial anomaly [3]. The rotations with equal phases are captured by the symmetry group $\mathrm{U}(1)_B = \mathrm{U}(1)_{L=R}$, which corresponds to baryon number conservation.

The remaining *chiral flavour symmetry* breaks spontaneously,

$$\mathrm{SU}(N_{\rm f})_L \otimes \mathrm{SU}(N_{\rm f})_R \longrightarrow \mathrm{SU}(N_{\rm f})_{L=R} \ , \qquad (2.2)$$

which gives rise to $N_{\rm f}^2 - 1$ Nambu-Goldstone bosons (NGBs). For massive but light quarks they turn into light quasi-Nambu-Goldstone bosons, which are identified with the pion triplet (for $N_{\rm f} = 2$), or the octet of pions, kaons and the η-meson (for $N_{\rm f} = 3$).

Generally, NGB fields take their values in the coset space of the spontaneous symmetry breaking. Therefore, in the case of transition (2.2), we obtain NGB fields $U(x) \in \mathrm{SU}(N_{\rm f})$. Chiral perturbation theory formulates a low-energy effective theory in terms of NGB fields [4]; in this case, its leading order Lagrangian reads (in Euclidean space)

$$\mathcal{L}(\partial_\mu U) = \frac{F_\pi^2}{4} \,\mathrm{Tr}\left(\partial_\mu U \,\partial_\mu U^\dagger\right) \ , \qquad (2.3)$$

where $F_\pi \simeq 92\,\mathrm{MeV}$ is the pion decay constant. Higher order terms in this low-energy expansion involve at least four derivatives (in addition the quark masses can be incorporated).

At fixed Euclidean time x_4, the field U maps the 3d coordinate space to the group $\mathrm{SU}(N_{\rm f})$. The identity $\Pi_3[\mathrm{SU}(N_{\rm f})] = \mathbb{Z}$ implies that these maps are divided into topological sectors, labelled by a topological charge $Q \in \mathbb{Z}$. In this case, it is given by the term

$$Q = \frac{1}{24\pi^2} \int d^3x \ \epsilon_{ijk} \,\mathrm{Tr}\Big((U\partial_i U)(U\partial_j U)(U\partial_k U)\Big) \ , \quad i,j,k \in \{1,2,3\} \ , \qquad (2.4)$$

which is conserved in time (so we are not really restricted to some instant x_4). Amazingly, it can be interpreted as the *baryon number*, although the NGB fields apparently just represent mesons [5]; for a comprehensive review we refer to Ref. [6].

Actually the leading order Lagrangian (2.3) does not allow for a stable semi-classical solution with $Q = 1$. In order to stabilise such a *skyrmion* (a type of soliton), Skyrme added a four-derivative term [5], which can be written as [7]

$$\mathcal{L}_{\rm Skyrme}(U, \partial_\mu U) = \frac{F_\pi^2}{4} \,\mathrm{Tr}\left(\partial_\mu U \,\partial_\mu U^\dagger\right) + \frac{1}{32e^2} \,\mathrm{Tr}\left[(\partial_\mu U)U^\dagger, (\partial_\nu U)U^\dagger\right]^2 \ , \qquad (2.5)$$

where e is a dimensionless parameter. The skyrmion represents a fermionic (bosonic) baryon for an odd (even) number of colors, $N_{\rm c}$. The Skyrme model works reasonably well as an effective theory for QCD, with $N_{\rm c} = 3$ and baryons included. Predictions for hadronic quantities like the nucleon and Δ mass, F_π, magnetic moments and charge radii were obtained within about 30 % of the experimental values [7], although there is some tension of this higher-order term, and of the heavy skyrmion mass, with the low-energy assumption.

3. The 3d O(4) model as a low-energy effective theory

In view of the search for a simpler, effective theory, it is of interest to describe the symmetry breaking pattern (2.2) by (locally) isomorphic orthogonal groups, $O(N) \longrightarrow O(n)$, $N > n$. This is possible *solely* for $N_f = 2$ [8], where the transition (2.2) isomorphically corresponds to $O(4) \longrightarrow O(3)$.

Therefore the O(4) non-linear σ-model represents an effective theory for low-energy QCD with $N_f = 2$ massless quark flavours. Its Euclidean action

$$S[\vec{e}] = \frac{F_\pi^2}{2} \int d^d x \; \partial_\mu \vec{e}(x) \cdot \partial_\mu \vec{e}(x) \; , \qquad \vec{e}(x) \in S^3 \; , \tag{3.1}$$

corresponds to the QCD low-energy Lagrangian for $N_f = 2$ massless quark flavors. The global O(4) symmetry may break spontaneously down to O(3), which yields three NGBs, with fields in the coset space O(3), which is isomorphic to SU(2).[2]

We focus on the O(4) model in 3-dimensional Euclidean space — the O(4) Heisenberg ferromagnet — as a static Skyrme model. In fact, the critical points — *i.e.* the second order phase transitions — in both models seem compatible [9]; for a review, which addresses this topic, we refer to Ref. [10]. The critical exponents of the 3d O(4) model were estimated some time ago (referring to the anti-ferromagnetic regime for agreement with the universality class of 2-flavour QCD) [11], and recently the scaling in the critical regime has been studied numerically [12, 13]. Precise numerical values of the critical temperature and critical exponents are summarised in Ref. [13], based on previous work [14].

However, these studies did not pay particular attention to the fact that in this model, the configurations are divided into topological sectors, as in the 4d Yang-Mills SU(N) gauge theories ($N \geq 2$), and in QCD. Thus the skyrmions are also present in the 3d O(4) formulation of the low-energy effective theory, which therefore describes QCD at finite baryon density, although this link is not rigorous. No stabilising extra term is required in this static model.

4. Skyrmions on the lattice

The goal of this project is a study of the 3d O(4) model, considering in particular the rôle of the skyrmions as carriers of the baryon number. Since these are topological properties, such a study must be non-perturbative.

The only fully controlled non-perturbative approach to quantum field theory is based on Monte Carlo simulations of the lattice regularised system (for a recent review, see *e.g.* Ref. [15]). For the O(N) models, the Wolff cluster algorithm [16] turned out to be extremely efficient: it proceeds by updating entire spin clusters in a sophisticated manner. This is far superior to local update algorithms, which suffer, for instance, from an extremely long auto-correlation time with respect to the topological charge Q as one approaches the critical point ("topological freezing").

Strictly speaking, the standard lattice action

$$S[\vec{e}] = \frac{F_\pi^2}{2} \sum_{\langle xy \rangle} (1 - \vec{e}_x \cdot \vec{e}_y) \; , \qquad \vec{e}_x \in S^3 \; , \tag{4.1}$$

(where the sum runs over all nearest neighbor lattice sites x, y) does not have topological sectors by default; all configurations can be continuously deformed into one another. Nevertheless, a variety of ways have been suggested to define a topological charge on the lattice; in particular the *geometric definition* has the virtue of guaranteeing integer charges, $Q \in \mathbb{Z}$, for all configurations (except for a subset of measure zero). It was suggested and successfully applied in the 2d O(3) model[3] [17], where the lattice plaquettes are split into triangles, and the topological charge

[2] The inclusion of quark masses corresponds here to the addition of an external "magnetic field" $\vec{H}$, *i.e.* a term $-\vec{H} \cdot \vec{e}(x)$ in the Lagrangian, which explicitly breaks the O(4) symmetry.

[3] Generally, the $O(N+1)$ model has topological sectors in N dimensions, due to $\Pi_N[S^N] = \mathbb{Z}$.

density is given by the (minimal) oriented area of the spherical triangle spanned by the spins at its vertices.

In addition to its efficiency, the cluster algorithm enables a dynamical definition of *merons* [18] (objects of half-integer topological charge). Their properties were explored in the 2d O(3) model [18] and in the 1d O(2) model (the quantum rotor) [19]. It is straightforward to apply the same concept to the 3d O(4) model.[4]

In order to extend the (sound) geometric definition of Q to the 3d O(4) model, we decompose all lattice unit cubes into five tetrahedra, as illustrated in Figure 1, as described for instance in Ref. [20] (four of them are congruent, while only the fifth one is regular).[5] For a given

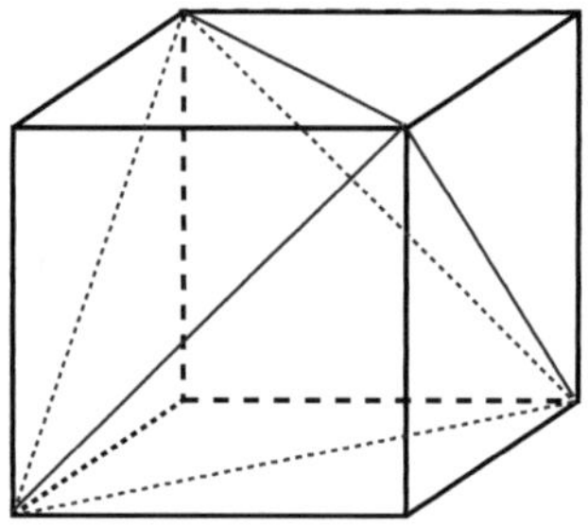

Figure 1. Illustration of the division of a lattice unit cube into five tetrahedra. This leads to space-filling set of tetrahedra. The spins at its vertices span spherical tetrahedra on S^3, and their oriented volumes represent the lattice topological charge density.

tetrahedron, the spin variables $\vec{e}_x$, $\vec{e}_y$, $\vec{e}_z$, $\vec{e}_w$ at the four vertices span a *spherical tetrahedron* on the 3-dimensional sphere S^3. Its volume can be computed according to the formulae elaborated only recently in Refs. [21]. We can assign a sign to each spherical tetrahedron according to its orientation, by invoking the sign of $\det(\vec{e}_x, \vec{e}_y, \vec{e}_z, \vec{e}_w)$.

For a given lattice configuration in some volume V with periodic boundary conditions, lattice unit cubes are split into tetrahedra, and all volumes of the corresponding oriented spherical tetrahedra are summed up to obtain the topological charge $Q \in \mathbb{Z}$ — it counts how many times the sum of the spherical tetrahedra covers the sphere S^3.

With the basic action (4.1) its expectation value vanishes, $\langle Q \rangle = 0$, due to parity symmetry. This symmetry can be broken, however, by adding a chemical potential. In this formulation it couples directly to the topological charge, without causing a sign problem. Alternatively, an isospin chemical potential has been included in the Skyrme model with the usual prescription of a covariant derivative in the Euclidean time direction [22].

Acknowledgments

WB thanks Uwe-Jens Wiese for instructive discussions. We also thank the LOC of the XXXI Reunión Anual de la División de Partículas y Campos de la Sociedad Mexicana de Física. This work was supported by DGAPA-UNAM, grant IN107915, and by CONACyT, project CB-241408.

[4] Moreover, in Refs. [18, 19] the meron cluster algorithm gave rise to an *improved estimator* as a very powerful numerical tool. However, its application to the 3d O(4) model requires rather restrictive constraints on the relative angles between neighbouring spins. This enforces a limitation to smooth configurations, which may obstruct the observation of the phase transition, hence we dispense with this technique in the ongoing study.

[5] In the 2d O(3) model, an alternative is the use of a lattice consisting of equilateral triangles [18]. The analogue recipe fails, however, in $d = 3$, since regular tetrahedra are not space-filling (contrary to a claim by Aristotle, which caused long-term confusion [20]).

IOP Conf. Series: Journal of Physics: Conf. Series **912** (2017) 012048 doi:10.1088/1742-6596/912/1/012048

References

[1] Bruno M, Dalla Brida M, Fritzsch P, Korzec T, Ramos A, Schaefer S, Simma H, Sint S and Sommer R 2017 arXiv:1706.03821 [hep-lat]

[2] Patrignani C *et al.* (Particle Data Group) 2016 *Chin. Phys.* C **40** 100001, and 2017 update

[3] Adler S 1969 *Phys. Rev.* **177** 2426
Bell J S and Jackiw R 1969 *Nuovo Cimento* A **60** 47

[4] Weinberg S 1979 *Physica* A **96** 327
Gasser J and Leutwyler H 1983 *Phys. Lett.* B **125** 321
Gasser J and Leutwyler H 1984 *Ann. Phys. (NY)* **158** 142

[5] Skyrme T H R 1961 *Proc. Roy. Soc. Lond.* A **260** 127
Skyrme T H R 1962 *Nucl. Phys.* **31** 556

[6] Zahed I and Brown G E 1986 *Phys. Rept.* **142** 1

[7] Adkins G S, Nappi C R and Witten E 1983 *Nucl. Phys.* B **228** 552

[8] Bietenholz W 2010 *Int. J. Mod. Phys.* A **25** 1699

[9] Rajagopal K and Wilczek F 1993 *Nucl. Phys.* B **399** 395

[10] Pelissetto A and Vicari E 2002 *Phys. Rept.* **368** 549

[11] Wilczek F 1992 *Int. J. Mod. Phys.* A **7** 3911 [Erratum: 1992 *Int. J. Mod. Phys.* A **7** 6951]

[12] Engels J and Karsch F 2012 *Phys. Rev.* D **85** 094506

[13] Engels J and Karsch F 2014 *Phys. Rev.* D **90** 014501

[14] Engels J and Vogt O 2010 *Nucl. Phys.* B **832** 538

[15] Bietenholz W 2016 *Int. J. Mod. Phys.* E **25** 1642008

[16] Wolff U 1989 *Phys. Rev. Lett.* **62** 361

[17] Berg B and Lüscher M 1981 *Nucl. Phys.* B **190** 412

[18] Bietenholz W, Pochinsky A and Wiese U-J 1995 *Phys. Rev. Lett.* **75** 4524

[19] Boyer T, Bietenholz W and Wuilloud J 2007 *Int. J. Mod. Phys.* C **18** 1497

[20] Senechal M 1981 *Mathematics Magazine* **54 (5)** 227

[21] Murakami J and Yano M 2005 *Communications in Analysis and Geometry* **13 (2)** 379
Murakami J 2012 *Proc. Amer. Math. Soc.* **140** 3289

[22] Loewe M, Mendizabal S and Rojas J C 2006 *Phys. Lett.* B **632** 512
Loewe M, Mendizabal S and Rojas J C 2006 *Phys. Lett.* B **638** 464

IOP Conf. Series: Journal of Physics: Conf. Series **912** (2017) 012049 doi:10.1088/1742-6596/912/1/012049

Study of atmospheric Showers Simulations, for Low Energies using Trasgos

Alejandra Parra[1], Oscar Martínez[1], Humberto Salazar[1], Juan A. Garzón[2], Alma Morales[1]

1. Facultad de Ciencias Físico - Matemáticas, (BUAP), Av. San Claudio y 18 Sur, Col. San Manuel, Edif. FM1-101B, C.P. 72570, Puebla, México,
2. LabCAF, IGFAE - Universidad Santiago de Compostela 15782, Santiago de Compostela, España.

E-mail: `alejandra.parra@alumno.buap.mx`

Abstract. In this work we present a new proposal to study the primary radiation mass composition at low energies for atmospheric extended showers (EAS), using detectors called Trasgos (TRack and time reconStructinG bOx). There consist on a stack of timing Resistive Plate Chambers (tRPC) with the advantages of high time and space resolution, as well as tracking capabilities. These characteristics allow us to analyze the arrival of the secondary particles and the fine structure of the EAS front. Our simulations consider the lateral distribution of particles near to the core of the EAS, where the electromagnetic to muonic ratio depends more strongly of the effective cross section, i.e. the mass of the primary particle.

1. Introduction

Victor Hess discovered the Cosmic Rays in 1912. He installed three electroscopes in a balloon in order to measure the rate of ionization in the atmosphere. Hess found that such rate increased to about three times with respect to the measured at sea level [1]. These results suggested that the radiation was coming from outside of the Earth. Later studies showed that there is a dependence between the energy of the primary cosmic rays and its possible sources: the Sun, at lower energies, X-ray binaries, pulsars, supernova remnants in the Galaxy and other astronomical objects at higher energies.

The Fermi Gamma-ray Space Telescope [2] measured the diffuse gamma radiation from 1 to 100 GeV approximately. For energies from 1 TeV to 10 TeV, ground based gamma ray telescopes have observed emission from supernova remnants, as expected. However the relation between these observations and the fluxes of high energy protons and nuclei, measured at Earth is not completely clear.

Direct measurements are made at high altitude with the aid of balloons, see Fig.1, and other experiments like PAMELA [4] and CREAM [5, 6], measured the spectrum of protons and helium. PAMELA reported that the spectra becomes harder above a rigidity of 200 GeV. CREAM observed such behavior only for heavier primary cosmic rays [6].

Indirect measurements on the primary cosmic ray energy spectrum for energies above $\sim 10^6$ GeV are made by large experimental arrays located on the ground. This type of experiments allow the study of the secondary particles produced along the extensive air showers. In these cases, the information about primary cosmic ray composition is limited and the relation between

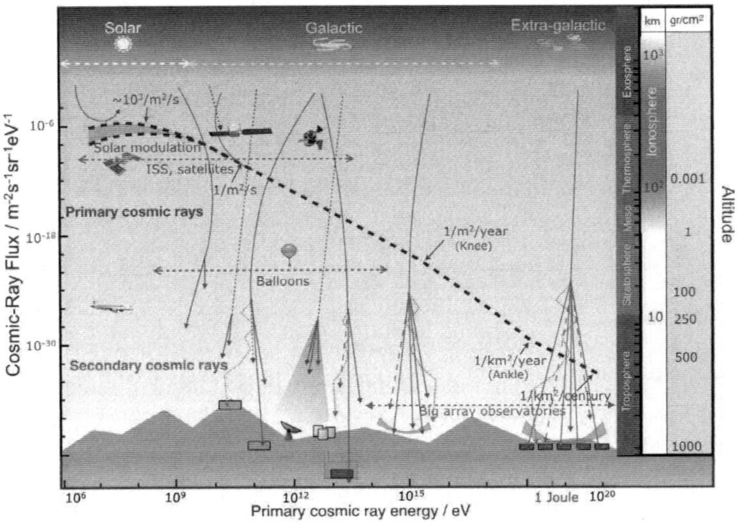

Figure 1: Cosmic rays energy spectrum and the different techniques to detect them at different energies.

the observed signal at the ground and the primary energy depends on the mass of the primary nucleus. It is important to note the relation between the measured signal and the estimated primary energy depends on the hadronic interaction models used.

There are different particle detection techniques. Scintillating material based detectors are suitable to detect mostly electrons, positrons and muons. Some other experiments, as the surface array detectors of the Pierre Auger Observatory [7] and IceTop [8] use tanks filled with water or ice to detect the Cherenkov radiation produced by charged particles inside the tanks. This method is sensitive to electrons, positrons and muons in the shower.

Experiments like HEGRA [9] and TUNKA [10] used unshielded photo-multipliers looking up at the sky to detect the atmospheric Cherenkov light generated by the extensive air shower. In these cases, the chemical composition of the primary cosmic rays is studied through the lateral distribution of the Cherenkov light at the ground. The atmospheric fluorescence technique used in Pierre Auger Observatory [7] traces the isotropically emitted nitrogen fluorescence lines excited by the passage of charged particles through the atmosphere. Thus, this method is a calorimetric measurement of the energy of each event. By tracing each shower profile, a depth of maximum is assigned to each event. This parameter is sensitive to the primary cosmic ray composition.

There are two possible scenarios two explain the flux suppression at the highest end of the energy spectrum. In one scenario, the energy spectrum around $10^{19.5}$ eV is dominated by heavy nuclei. Here, the upper limit of the particle acceleration is very close to the one from the Greisen-Zatsepin-Kuzmin (GZK) suppression. A second scenario suggest a mixed or heavy composition at the source, such that the flux suppression comes from the photo-desintegration of heavy nuclei and other GZK energy losses. Independently of the acceleration mechanism, the arrival directions of protons at sufficiently high energy are expected to be correlated with their source directions [11].

Some of the most important points of this new proposal is that Trasgos have a good space, angular and time resolution. Thus, Trasgos will allow us to study in detail the shower front

IOP Conf. Series: Journal of Physics: Conf. Series **912** (2017) 012049 doi:10.1088/1742-6596/912/1/012049

structure near the core of primary cosmic ray interaction. We will be able to perform an event by event analysis, of the electromagnetic and muonic components of the shower. We expect to improve the estimation of the primary mass composition.

2. Trasgos (TRAck reconStructigG bOxes)

The primary cosmic ray composition give us information about the origin of the cosmic rays. Depending of the cosmic rays composition, it will be possible to know which astrophysics scenario is more viable.

For these reasons, some of the most important dedicated cosmic ray experiments have plans to improve or upgrade their original detector configuration in order to improve their measurements of the mass composition. One example is the Trasgos detectors [12]. A first Trasgo detector station is now operative at the University of Santiago de Compostela [13].

Hybrid arrays complemented with a relatively small tRPC detectors will improve the determination of the muon - electromagnetic (μ/EM) ratio event by event, and may allow us to improve the measurement mass of the primary cosmic radiation, that may help us to unveil the origin of this energetic particles.

The main objective of the Trasgo detectors is to perform a detail study of the shower front, i.e. to determine the shape of the shower components with enough precision to estimate the composition of the primary particle of the shower. These detectors are based on tRPC Fig.2, which have the advantage of covering large areas in an affordable way. Some of the most important characteristics of Trasgos are:

- High temporal resolution, $\sim$ 100 ps, which allow us a good measurement of the temporal profile of the EAS.
- High granularity, to study with good accuracy the arriving particles of the shower, individually.
- Tracking, which allow us to estimate the arriving direction of single charged particles of the EAS front and particle identification by software.

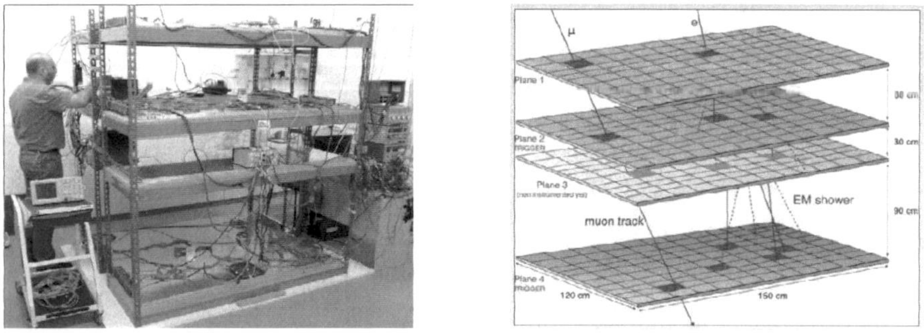

Figure 2: Left: A first prototype of Trasgo, already installed at Santiago Compostela University. Right: Particle identification capability of a Trasgo.

3. Work Proposal

With Trasgos, we will collect valuable data to study the characteristics of the EAS with a core located very close to the detector, where we expect a high density of particles. For this reason, we propose a systematic study of the main variables of the showers and the correlation between them. This study can be done for different kind of primary particles: gammas, protons, carbon

IOP Conf. Series: Journal of Physics: Conf. Series **912** (2017) 012049 doi:10.1088/1742-6596/912/1/012049

and iron nuclei at different energies and zenith angles. We will also consider in our simulations different hadronic interaction models.

To achieve our goal, we propose a global networking, called CORSAIR (Cosmic Ray SimulAtIon Research Network). To generate the simulate events we will use CORSIKA[14]. CORSIKA is specially design to simulate extensive air showers. The program allow us to fix the energy, arriving angle and primary particle of the event. The interval of energy for this study will be from 10^2 GeV to 10^5 GeV with zenith angles between 0° and 60°. Each data set will have around 10 000 simulated events, at this moment the network includes: Puebla (Mexico), Tunja (Colombia) and Santiago de Compostela (Spain).

Figure 3: Laboratorio Nacional de Supercómputo del Sureste de México (LNS) facilities located at Puebla University.

4. Work in Progress and Future Work

Up to now, all the necessary software to simulate the EAS is installed at the high performing computing Laboratory(LNS, Laboratorio Nacional de Supercómputo del Sureste de México), Fig.3. The simulated events have the following characteristics:

- Proton as primary particle.
- Primary Energies: 10 GeV, 13.3 GeV, 17.8 GeV,up to 100 GeV in logarithmic scale.
- Zenith angles: 0°-12.8°, 12.8°-22.3°, 22.3°-30.0°.

Between 10^5 and 10^4 events for each set will be simulated as a function of the energy. These libraries will help us to understand the behavior of the showers to establish some correlation between their observables to find significant trends both in energy and radial distribution.

This analysis will allow us to propose a new parameters or signature to estimate the mass of the primary cosmic ray.

We will also make a comparison between the data collected by Trasgos detectors and the Monte Carlo generated by the proposed network. A Trasgo network around the globe is also planned, (see Fig.4).

IOP Conf. Series: Journal of Physics: Conf. Series **912** (2017) 012049 doi:10.1088/1742-6596/912/1/012049

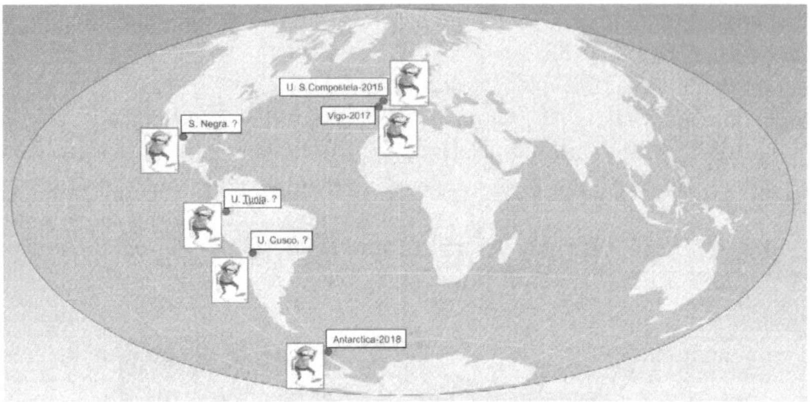

Figure 4: We show the different places where it is planned to install a Trasgo.

References

[1] A.M. Hillas, "Cosmic Rays", Serie "Selected readings in physics", Pergamon Press, (1972).

[2] M. Ackermann et al., (Fermi Collaboration), "Fermi-LAT observations of the diffuse γ-ray emission: implications for cosmic rays and the interstellar medium", Ap.J. 750, 3 (2012).

[3] E.S. Seo, "Direct measurements of cosmic rays using balloon borne experiments", Astropart. Phys. 39-40, 76(2012).

[4] O. Adriani et al. (Pamela Collaboration),"PAMELA Measurements of Cosmic-Ray Proton and Helium Spectra", Science 332, 69 (2011).

[5] H.S. Ahn et al.(CREAM Collaboration), "Energy spectra of cosmic-ray nuclei at high energies", Ap.J. 707, 593-603 (2009).

[6] H.S.Ahn et al., (CREAM Collaboration), "Discrepant hardening observed in cosmic-ray elemental spectra", Ap.J. 714, L89-L93 (2010).

[7] J.Abraham et al.(Auger Collaboration), "Observation of the Suppression of the Flux of Cosmic Rays above 4×10^{19} eV", Phys. Rev. Lett. 101, 061101 (2008).

[8] R. Abbasi et al.,(IceCube Collaboration), "IceTop: the surface component of IceCube", N.I.M. A 700, 188-220 (2013).

[9] F. Arqueros et al., "Energy Spectrum and Chemical Composition of Cosmic Rays between 0.3 and 10 PeV determined from the Cherenkov-Light and Charged-Particle distributions in Air Showers", Astron. & Astrophys. 359,682 (2000).

[10] S.F. Berezhnev et al.(Tunka Collaboration), "The Tunka-133 EAS Cherenkov light array: status of 2011", NIM A692, 98 (2012).

[11] The Pierre Auger Collaboration, "The Next Frontier in UHECR Research with an Upgraded Pierre Auger Observatory", Sept. (2013),arXiv:1307.0226v3 [astro-ph.IM]

[12] D. Belver et al., "TRASGO: A proposal for a timing RPC based detector for analyzing cosmic ray air showers", Nuc. Inst. & Meth. A661 (2012), 5163-5167.

[13] J.A. Garzón et al., "First results on cosmic ray studies and their relation with the solar activity, the Earth magnetic field and the atmospheric properties", XXV European Cosmic Ray Symposium, Turin, Sept 4 -8, 2016, arXiv: 1701.07277v1.

[14] D.Heck,G.Schatz,T.Thouw,J.Knapp, andJ.Capdevielle, "CORSIKA:A Monte Carlo code to simulate extensive air showers", Forschungszentrum Karlsruhe Report FZKA (1998) 6019.

IOP Conf. Series: Journal of Physics: Conf. Series **912** (2017) 012050 doi:10.1088/1742-6596/912/1/012050

Design, Construction, Test, Operation and Simulation of a Four Channel Cosmic Ray Detector

A. Martínez, J. Félix

Laboratorio de Partículas Elementales, Departamento de Física, División de Ciencias e Ingenierías - Campus León, Universidad de Guanajuato, León, GT 37150, MX

Email: almveg@fisica.ugto.mx, felix@fisica.ugto.mx

Abstract. Cosmic ray detectors are constructed to fit many purposes, different materials and geometries. To test materials and to measure the flux of cosmic rays, we planned, designed, constructed, tested and operated a 4 channel cosmic ray detector based on 2.54 cm X 10.32 cm X 20.64 cm Aluminum block and two 0.6 cm X 10.32 cm X 20.64 cm plastic scintillators completely covered with 0.2 cm thick Aluminum foil. The signal, produced by the passage of cosmic rays, was read out using a Hamamatsu photomultiplier in both the Aluminum block and plastic scintillator. The performance of this detector was simulated using GEANT 4. The efficiency of the cosmic ray detector was measured to give 85% approximately. Details of construction, operation, simulation, and preliminary results are presented.

1. Introduction

Today, research in cosmic rays represents an opportunity to study the physical interactions between matter and particles such as electrons, muons and neutrinos, which deposit energy within sensitive materials with which they interact. In order to quantify the deposited energy it is necessary to use particle detectors. Besides being used to measure the energy of the detected radiation, they can be also used to measure other important physical attributes such as momentum, spin and electric charge [1].

2. Planning

It is intended to build, operate and simulate a four channel cosmic ray detector based on plastic scintillator and Aluminum. Also, it is necessary to make a comparison between the experimental data with Monte Carlo simulations performed with Geant 4 [2].

3. Design

The design of the experimental prototype was made using the software Sketchup [3]. Below (figure 1) a side and a front view of detector are shown.

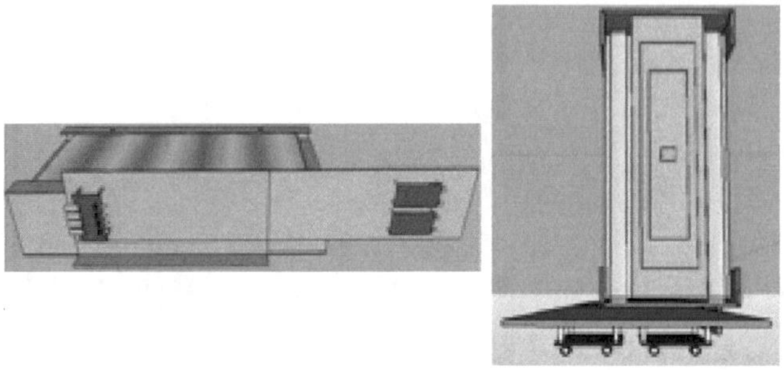

Figure 1. Side and front view of the detector. The discriminator and electronics channel boards are shown (left). The three boards are located at the bottom part of the detector (right).

The cosmic ray detector is based on three blocks, one block of Aluminum with dimensions *25.32 cm X 10.16 cm X 2.54 cm* is symmetrically placed inside two plastic scintillator blocks with dimensions *22 cm X 10.16 cm X 0.6 cm*, covering almost all the Aluminum block surface. To detect the light produced inside of the detector, a Hamamatsu avalanche silicon PMT [4] was attached at both ends of the Aluminum block and at one end of each of the plastic scintillator blocks. The geometry of the detector was simulated with Geant 4.

4. Construction

The first step consisted in sanding and polishing twelve aluminum plates, of different dimensions, used to cover to the plastic scintillator. Then, a hole was drilled to place the photodiodes in both the aluminum block and scintillator detectors. Each photodiode per channel was welded to an electronic card to be connected to the main electronics boards array, composed by a discriminator card and two electronics cards which supply voltage to the photodiodes (figure 2). In these electronic cards, an RC circuit has been established for each detection channel, in order to obtain an analog signal with exponential decay and a very small decay time (figure 3).

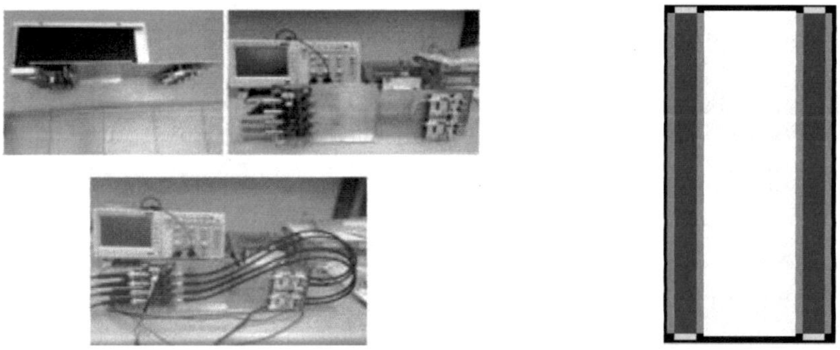

Figure 2. Particle detector experimental setup (left). Geant 4 simulation (right). A side view of the geometry of the simulated detector is shown. The block of aluminum was painted yellow, while the Aluminum plates covering the plastic scintillators were painted with navy blue, cyan and pink respectively.

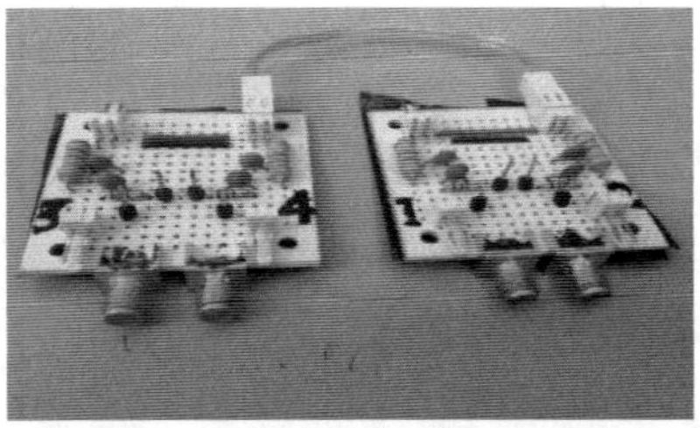

Figure 3. RC circuits on the electronic cards that supply voltage to the photodiodes.

5. Development

The operation of the four detection channels was tested, that is, using an oscilloscope was checked that an analogic signal was obtained. Also, the operation of the discriminator card was tested and subsequently it was added to the experimental setup. Doing this, both the analog signal and the digital signal were observed (figure 4). The digital signal obtained depends on the analog signal that entered the discriminator card and the threshold voltage set in that electronic board (figure 5). This threshold voltage was set so that be three times the noise of the analog signal [5].

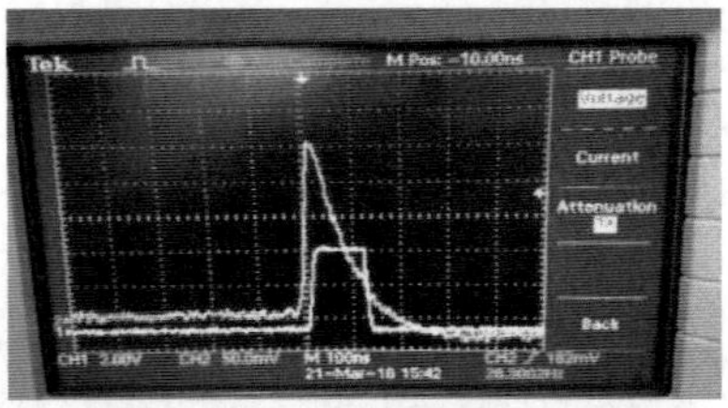

Figure 4. Oscilloscope screen showing analog (plot that has an exponential decay) and digital (square signal) signal for channel 1. Similar signals were obtained on the other channels. The fact that it has been observed both analog and digital signal pointed out that the detector works properly.

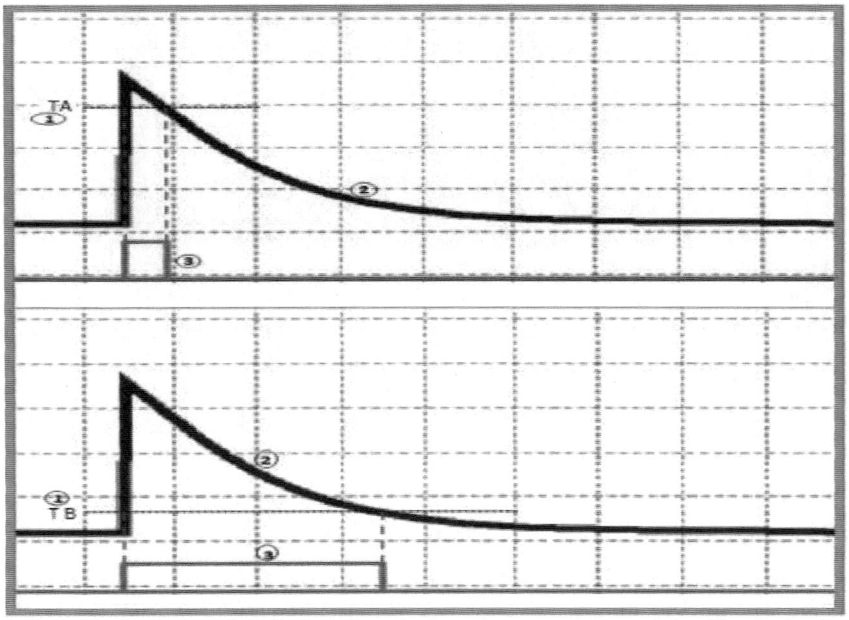

Figure 5. Relationship between the digital signal, the trigger voltage of the discriminator card and the analog signal. The number 1 enclosed in a circle represents different trigger voltages set on the discriminator card, T A and T B respectively; the number 2 represents the analog signal read on the oscilloscope; the number 3 represents the resulting digital signal.

As can be seen in figure 5, as a result of having set different trigger voltages and of the form of the input analog signal, different digital signals are obtained. In other words, if the trigger voltage is set near the peak of the analog signal (T A), the time window of the digital signal is small, in contrast, if the trigger voltage is set in T B, the time window of the digital signal is large.

On the other hand, the discriminator card was connected to a 32 channels data acquisition system (figure 6). Currently, the calibration of the detector is not within our aims, we plan to include it until we have reached a post-thesis research stage; in addition, by now, we do not have an artificial source of radiation, which could help us to calibrate the detector.

Once the proper operation of the detector and the electronics system was verified, we collected several cosmic ray data events with horizontal and vertical atmospheric muons within a time window of 1 ms.

Now, regarding the simulation, the geometry and materials that make up the detector, the primary particle gun and the relevant physical processes have been established. There were 50 events generated, generating in each of them 1000 primary particles, assigning to each of the primary particles an initial energy that is multiple of 100 and that is in the range of 100 to 5000, in units of MeV. Likewise, it was considered that $E_i < E_{i+1}$, where E_i is the initial energy of a primary particle that was generated at an event i, and E_{i+1} is the initial energy of a primary particle that was generated in a later event i+1, where i goes from 0 to 49. By the other hand, all primary particles were negative muons generated in two different event runs; in an event run, the primary particles had an initial direction of momentum perpendicular to the detector, while in the other event run the particles were shot parallel to the detector. The particles were fired from a distance of 1 mm from the detector with a uniform random spatial distribution.

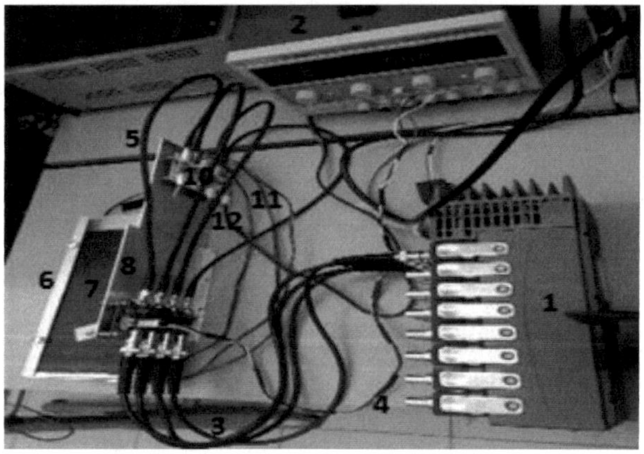

Figure 6. Experimental setup. 1: Data acquisition system, 2: Low voltage power supply, 3: BNC to BNC cables, 4: Power supply cable of the discriminator card, 5: SMA to SMA cables, 6: Mechanical press, 7: Stacking of the detection materials (Top and bottom: plastic scintillators. Middle: aluminum block), 8: Aluminum plate used to place the main electronics boards array, 9: Discriminator card, 10: Electronics cards which supply voltage to the photodiodes, 11: Bicolor cables used to connect the photodiodes with the electronics cards which supply voltage, 12: Bicolor cable used to connect the electronics cards which supply voltage with a high voltage power supply.

6. Results

In the simulation, the detector efficiency, defined as the number of detected muons divided by the number of incident muons, as function of the incident muon energy, was computed. We considered a muon as detected to the one that deposited energy in the 3 detection materials. If the muon only deposited energy in at most two of the detection materials, the muon was not considered as detected.

Therefore, only if all the primary particles (those coming out of the simulated particle gun) deposited some of its energy in the 3 detection materials, we could say that we have a 100% acceptance or efficiency in the detector. In reviewing the verbose information output of Geant 4, it was noted that the ionization process was not always produced in the 3 detection materials, that is, there were events in which a primary muon produced ionization only in 1 detection material. In other events, the primary muon produced ionization in only 2 of the 3 detection materials; there were also events in which a primary muon produced ionization in the 3 detector materials. It should also be added that in all the considered energy range (100-5000 MeV), there was always scintillation radiation and Cerenkov effect in the plastic scintillators, that is, there was always some loss of energy in the scintillating plastics as a result of the scintillation process or Cerenkov effect. By the other hand, based on the verbose information output of Geant 4, the only processes observed in the Aluminum block were the transport of particles and ionization, being the ionization a process of energy loss. The transportation process move the particles in space and time, either with or without physical fields; there is no loss of energy in the transportation process. Therefore, in order to obtain a 100% acceptance, this would imply that, of all the primary particles that are generated per event, all of them must deposit or lose some of their energy in the Aluminum block, but, in view of the before said, an acceptance of 100% could not be obtained. It is considered that the randomness of occurrence of the ionization process that was observed in the detector materials could contribute to the efficiency of the detector being maintained above 85% (see figure 7), regardless of the initial energy of the primary particles. Such randomness could be caused by the way physical processes are simulated in Geant 4. Likewise, one has the assumption of that in most events ionization was produced. That is believed to be the cause for the efficiency of the detector to remain practically constant. In order to give a more precise answer, more research is needed. However, in this work, the only energy losses that matter to us are those due to scintillation radiation and Cerenkov effect, this is because the detection channels we are using are optical channels, so the energy losses due to ionization are irrelevant. Although ionization energy losses are currently considered in the simulation, and more specifically in the efficiency of the detector, in the near future, it is planned to conserve energy losses due only to scintillation radiation and Cerenkov effect, taking into account also possible losses of energy by Cerenkov radiation in the block of Aluminum. Currently, it has only been possible to simulate the Cerenkov effect in the plastic scintillators.

As a result of the interaction of the muons with the detection materials, electrons, scintillation photons and Cerenkov photons were produced in the simulation. Qualitatively, we observed some things already known, that is, the isotropic spatial distribution of the scintillation radiation and the appearance of the Cerenkov radiation cone. It was also observed that some of the ionized electrons radiated Cerenkov light.

Now, returning to the efficiency of which we have talked, we cannot say that it is a result with which we agree completely because that result is preliminary, reason why it is necessary to make improvements in the simulation program, in addition, not having an artificial source of radiation in the laboratory, we cannot determine the efficiency of the real-life detector (not simulated) in the way it was obtained in the simulation. However, we could say that the simulated efficiency we have obtained is a good result if we take into account what was stated in the book *Measurement and Detection of Radiation* [6]. In that text, it is said that charged particle detectors have an efficiency of practically 100%, regardless of the size or density of the detector materials. Likewise, it is said that for charged particles, the detection efficiency is practically independent of particle energy except for very low energies, which is when the particles could be stopped by the detector. On the other hand, it must be said that in the simulation of our detector it was never observed that the primary muons were stopped by the detector materials.

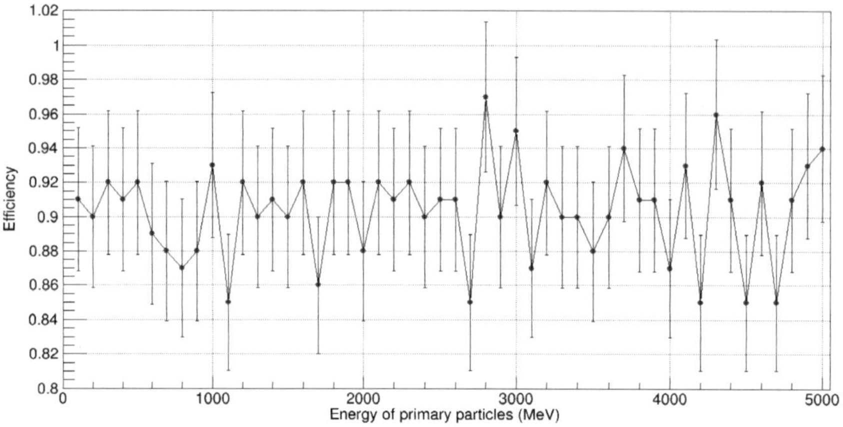

Figure 7. Efficiency of the simulated detector. Uncertainties were estimated using the error propagation approach.

On the other hand, using the data acquisition system of the laboratory, it was possible to obtain the number of counts per detection channel, that is, the number of particles detected has been obtained. This has been done per unit of time and for an extended interval of time. In our case, the number of counts for all detection channels has been recorded for 7200000 ms. Then, through the Root software, the number of counts versus statistical frequency (without units) was plotted in order to be able to observe the frequency distribution of the number of counts of each channel. In other words, we have obtained the histogram of the number of counts of each channel, obtaining at the same time the average number of counts for each channel. In order to compare the counts of all channels, the corresponding histograms have been normalized and overlapped in a single plot (see figure 8).

Because the histogram was normalized to 1, this implies that the sum of the statistical frequencies associated with each bin is equal to 1, and therefore, that the statistical frequency of each of the bins is less than 1.

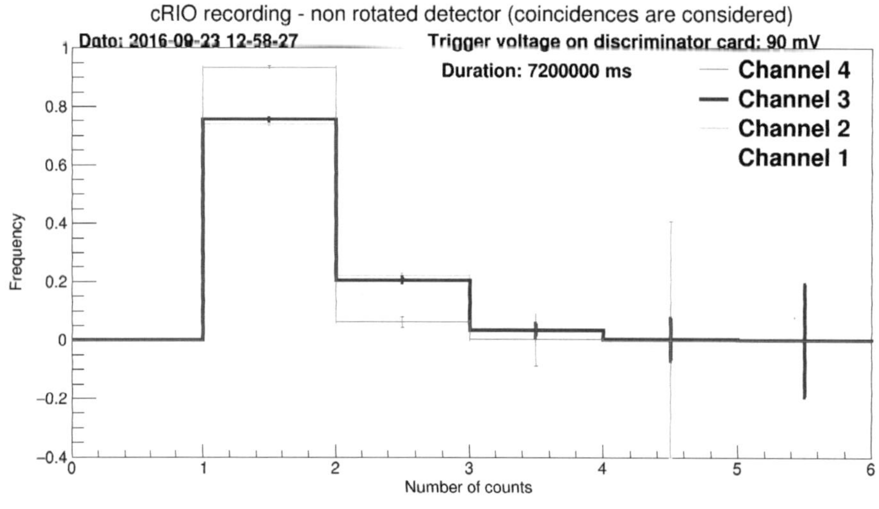

Figure 8. Histogram of the number of counts.

The detector has not yet been characterized, so we would not expect to observe a similar behavior in the channels, as can be seen in figure 7. In that figure, channels 1 and 2 represent channels located on the aluminum block, while channels 3 and 4 represent channels of the plastic scintillators. The arrangement of the detector channels is shown in figure 9.

Figure 9. Arrangement of the detector channels. Channels 1 and 2 are located on the aluminum block, while channels 3 and 4 are located on the lower and upper plastic scintillator, respectively. Channel 1 on the Aluminum block is located on the back of the detector.

7. Conclusions

We have planned, designed, constructed, tested and operated a four channel cosmic ray detector; it works properly. We have simulated the constructed detector and measured the detector efficiency.

This cosmic ray detector can be used to detect photons having a wavelength that is in the spectral range of the photodiode, from 320 to 900 nm. The physical processes that can generate such photons could be Cerenkov effect, scintillation radiation, transition radiation, etc.

Because it is a small and low-weight detector, it can be easily moved from one location to another, in case it is required to make data collection with another data acquisition system, calibrate the detector with different sources of radiation, etc. The design of the detector makes possible to obtain auto-coincidences (no other device required) both horizontal and vertical ones. Likewise, such a detector has the advantage that requires a low maintenance, is of a safe operation, does not require of special components and is of easy construction. On the other hand, despite our detector is a cosmic ray detector, this detector does not allow us to detect neutrinos, which are also cosmic rays. However, it is important to say that neutrino detection is not one of the aims of this research.

Finally, the improvements that could be made to the detector are several, for example, ionization channels could be added to measure the ionization produced in the detector, it could be used photodiodes that extend the spectral range of the photodiodes being used, it could be created and used a single electronic card that includes both the discriminator card and the electronic cards that provide voltage to the photodiodes, etc.

8. References

[1] Bichsel H, Groom D E and Klein S R 2014 *Passage of particles through matter* (International Collaboration: Particle Data Group)
[2] http://geant4.web.cern.ch/geant4/
[3] http://www.sketchup.com/
[4] http://www.hamamatsu.com/jp/en/S12572-100P.html
[5] Marshall A 2013 *Fourier, Hadamard, and Hilbert Transforms in Chemistry* (Springer Science & Business Media) p. 71
Laszlo P 2012 *Chemical and Biochemical Applications* (Elsevier) p. 65
[6] Tsoulfanidis N and Landsberger S 2015 *Measurement and Detection of Radiation* (Taylor & Francis) Second edition p 284

ACKNOWLEDGEMENTS

CONACYT for the support, project 223179, 10017, CB-2013-01 "Fondo Sectorial de Investigación para educación SEP- CONACYT".